Prefix or Suffix	Meaning	Example
glyc-	sweet	glycemia (the presence of sugar in the blood)
hetero-	other	heterotroph (organism that obtains energy from organic compounds)
homo-	common, same	homologous (similar in structure or origin)
hydr-	water	dehydrate
hyper-	excessive, above	hypersensitive
hypo-	under	hypotonic (having low osmotic pressure)
iso-	same, equal	isotonic (having the same osmotic pressure)
-itis	inflammation	appendicitis, meningitis
leuko-	white	leukocyte (white blood cell)
ly-, -lys-, -lyt-,	loosen, dissolve	bacteriolysis (dissolution of bacteria)
meso-	middle	mesophilic (preferring moderate temperatures)
meta-	changed	metachromatic (showing a change of color)
micro-	small; one-millionth part	microscopic
milli-	one-thousandth part	millimeter (10^{-3} meter)
mito-	thread	mitochondrion (small, rod-shaped or granular body in cytoplasm)
mono-	single	monotrichous (having a single flagellum)
multi-	many	multinuclear (having many nuclei)
myc-	fungus	mycotic (caused by a fungus)
myx-	mucus	myxomycete (slime mold)
-oid	resembling	lymphoid (resembling lymphocytes)
-ose	a sugar	lactose (milk sugar)
-osis	disease of	coccidioidomycosis (disease caused by *Coccidioides*)

The Microbial Perspective

Charlene Harris

Eugene W. Nester

University of Washington

Nancy N. Pearsall

University of Zambia

Jean B. Roberts

Evergreen General Hospital

C. Evans Roberts

University of Washington

Assisted by Martha T. Nester
Illustrations by Iris J. Nichols

SAUNDERS COLLEGE PUBLISHING

Philadelphia   New York   Chicago
San Francisco   Montreal   Toronto
London   Sydney   Tokyo   Mexico City
Rio de Janeiro   Madrid

Address orders to:
383 Madison Avenue
New York, NY 10017

Address editorial correspondence to:
West Washington Square
Philadelphia, PA 19105

This book was set in Aster by Waldman Graphics, Inc.
The editors were Michael Brown, Amy Satran, Janis Moore, Lynne Gery, and Michael Fare.
The art & design director was Richard L. Moore.
The text design was done by Barbara Bert.
The cover design was done by Richard L. Moore.
The artwork was drawn by Iris J. Nichols.
The production manager was Tom O'Connor.
This book was printed by Von Hoffman.

Part Opening Photographs

Part I	Scanning electron photomicrograph of streptococci. (Courtesy of D. Birdsell.)
Part II	Electron micrograph of a lymphocyte. (Courtesy of V. Chambers.)
Part III	Electron micrograph of a *Clostridium botulinum* phage. (Courtesy of E. Boatman.)
Part IV	Scanning electron micrograph of dental plaque. (Courtesy of D. Birdsell.)
Part V	Crown gall tumor on a sunflower plant induced by *Agrobacterium tumefaciens*.

THE MICROBIAL PERSPECTIVE   ISBN 0-03-047041-2

4 032 987654

CBS COLLEGE PUBLISHING
Saunders College Publishing
Holt, Rinehart and Winston
The Dryden Press

PREFACE

The subject of microbiology and the ways in which it is taught are changing at a rapid pace. This text has been written to satisfy a need for a book that can be used to teach introductory microbiology to students with little or no chemistry and biology background, especially those whose interests are oriented to the allied health professions. The book presents basic microbiological concepts with emphasis on the role of microorganisms in areas related to human health and disease. Many texts that are available for this purpose and audience are too lengthy and detailed. This book has been developed to meet the specific needs of the students who will be using it.

The Microbial Perspective draws some of its content and pedagogy from our earlier text, *Microbiology* (Nester, E. W., Roberts, C. E., Pearsall, N. N., and McCarthy, B. S., 2nd edition, 1978). The third edition of *Microbiology*, now in preparation, will be updated and expanded in scope from the second edition. It will be appropriate for introductory courses serving students with a basic biology and chemistry background and more diversified interests in the applications of microbiology.

In this text as in our others, our major concern has been to present the fundamentals of microbiology so that their relevance to the applied material is clearly evident. The sections on bacterial genetics and immunology reflect the most recent advances, and the applications of recent discoveries have been stressed. Genetic transposable elements, for example, are discussed in the context of antibiotic resistance. In the chapters on microbial diseases we have retained the "organ-system" approach of the second edition. The usefulness of this presentation is borne out by our own classroom experience and by its adaptation in other textbooks.

The present text employs many of the pedagogical aids that were so well received in *Microbiology*, including cross-references in the page margins, clear illustrations, chapter summaries, review and discussion questions, suggestions for further reading, and a glossary. Other features the two books share are an emphasis on basic principles and a reduction in minutiae, the reinforcement of important material in different contexts throughout the book (aided by cross-referencing), and a human health orientation that helps make the material interesting and meaningful to students.

In addition to these aids we have included several new features that will benefit students with little chemistry and biology preparation, who may have difficulty in appreciating the relevance of such material to their interests in microbiology. These features include one or more illustrative anecdotes for each chapter, a listing of chapter objectives, boldface type for important words, definitions of difficult terms at the bottom of the page, and a multiple-choice "Self-Quiz" for each chapter. The appendices have been expanded to include more information of reference value to students and teachers.

A package of ancillaries has been designed specifically for this text. The laboratory manual, by Marie Gilstrap and John Kleyn, emphasizes independent and accurate laboratory performance through experiments that have direct allied health applications. The study guide and the instructor's manual, both prepared by Frank Whitehouse, are keyed closely to the text and facilitate its use.

We are very much indebted to all who have given valuable criticisms and suggestions. We are especially grateful for the help of many of our colleagues at the University of Washington, including Dale Birdsell (Dentistry), Sam Eng (Clinical Microbiology), Karen Holbrook (Dermatology), Dale Parkhurst (Microbiology), Gertrude Schmidt (Laboratory Medicine), Ted Wetzler (Environmental Health), Fritz Schoenknecht, James Staley, John Sherris, Charles Evans, Thomas Stanton, and the Department of Microbiology and Immunology.

Numerous other people across the country gave invaluable advice: Norma Barrett, Galveston College; Benjamin Becker, Purdue University; Lois M. Berquist, Los Angeles Community College; David Campbell, Saint Louis Community College; J. B. Clark and John Lancaster, University of Oklahoma; Eugene Flaumenhaft, University of Akron; Marie Gilstrap, Highline Community College; Richard V. Goering, Creighton University; Marilyn Hanson and Glenn Powell, Bellevue Community College; Walter D. Hoeksema, Ferris State College; Gary E. Kaiser, Catonsville Community College; Margaret Mansel, Hinds Junior College; Eleanor K. Marr, Dutchess Community College; Joseph McGrellis, Atlantic Community College; Barbara Pogosian, Golden West College; Galen A. Renwick, Indiana University Southeast; Arlyn E. Ristau, Wartburg College; Fred A. Rosenberg, Northeastern University; and Hideo Yonenaka, San Francisco State University. Thanks are also due to JoAnn P. Fenn, University of Utah Medical Center, for the use of her thesis, "A Continuing Education Course in Microbiology for Registered Nurses."

The art work was designed and executed by Iris Nichols. We are also indebted to Kendall Getman of Holt, Rinehart and Winston, and to Michael Brown, Amy Satran, Janis Moore, and Lynne Gery of Saunders College Publishing for coaching this project through the long process leading to publication.

We very much hope that this book will prove to be interesting and educational for students and helpful to teachers of introductory microbiology. Your suggestions and comments are welcome and earnestly solicited.

E.W.N.
N.N.P.
J.B.R.
C.E.R.

CONTENTS

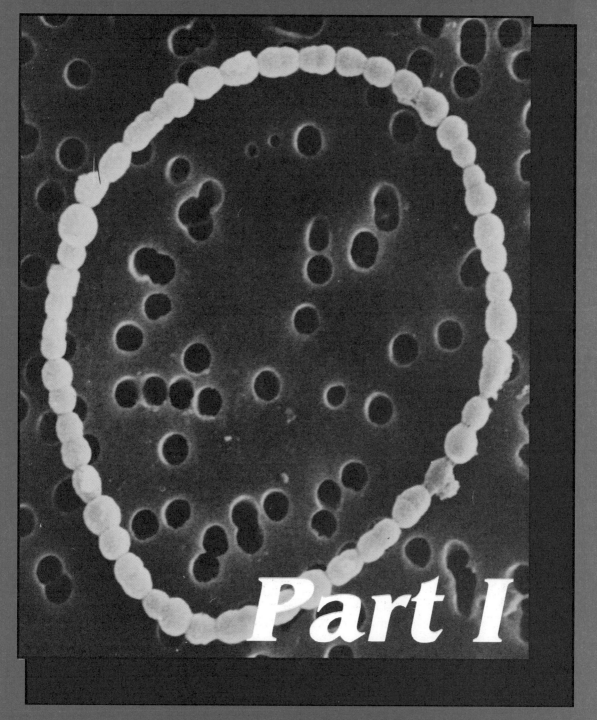

Part I

LIFE AND DEATH OF BACTERIA

Chapter 1

MICROBIOLOGY— PAST IMPACT AND FUTURE POTENTIAL

In the hot, muggy summer of 1976 in the city of Philadelphia, 221 persons came down with a mysterious ailment. Thirty-four died. Since most of the victims were attending an American Legion convention, the malady was termed "Legionnaires' disease." The symptoms of this disease—high fever, muscle aches, and coughing—strongly suggested pneumonia. However, medical technologists were unable to detect any organisms in the diseased tissues. For six months after the deaths in Philadelphia, scientists did not know whether the malady was a disease caused by an unusual biological agent or whether it was caused by a toxic chemical, perhaps originating in the air conditioning units of the hotel. The big break came when tissue specimens isolated from patients who had died of the disease were inoculated into guinea pigs. Fever developed in each animal, and organisms could be seen under the microscope when the spleen of the animal was stained. Laboratory technologists tried to grow the organisms on a variety of different nutrients under different conditions. At last a medium was found on which they would grow. Although this previously unknown organism has now been isolated and named, many questions remain unanswered. Where does the bacterium live and how is it spread? Why do only 1 percent of the people who come in contact with the organism contract the disease?

Once the organism that causes Legionnaires' disease was identified, a surprising observation was made. At least four other attacks by this organism occurred before the one in Philadelphia. The most dramatic took place in Pontiac, Michigan, in 1968. In early July of that year, 95 of 100 people working in a single building developed fever, muscle aches, and headaches, but there were no respiratory symptoms. After three to four days, all patients recovered. However,

2

no organism could be isolated, and the cause of the disease remained a mystery. Fortunately, frozen blood specimens of the patients were saved. Legionnaires' bacillus was shown to be the causative agent, since 80 percent of the patients suffering from "Pontiac fever" had antibodies* against the Legionnaires' bacillus in their blood. But why did the affected people in Michigan suffer no respiratory symptoms, and why were there no deaths? The answers are not known. Since the attack in Philadelphia, more than a thousand people are known to have contracted Legionnaires' disease, and more than 100 have died. Since one third of all respiratory diseases are caused by agents that have not been identified, probably more people are dying of Legionnaires' disease than we previously suspected.

OBJECTIVES

To know
1. The contributions made by van Leeuwenhoek, Koch, and Pasteur to the development of microbiology.
2. The major differences between eukaryotic and prokaryotic cells.
3. What functions all cells must perform.
4. Why viruses are not considered to be living cells.
5. What properties eubacteria generally possess.
6. Several ways in which microorganisms will probably contribute to solving some important human problems in the future.

ONGOING PROBLEMS IN PUBLIC HEALTH

Previously Unrecognized Infectious Diseases

The story of Legionnaires' disease illustrates that infectious diseases of unknown causes still exist even in countries with modern health facilities. Other interesting cases can be cited to emphasize this point. In 1979, a cause of bacterial pneumonia was identified. The agent, which caused deaths in 7 of 13 cases in Pittsburgh, Pennsylvania, is definitely a bacterium from its appearance, but only recently has it been grown in the laboratory. Like the Legionnaires' bacillus, the Pittsburgh agent has probably been around in the environment long before it was recognized as causing disease. Indeed, other mysterious diseases exist in which an infectious agent is suspected but has never been identified. For instance, in 1978–1979, more than 50 children in Italy died of a disease termed "dark disease" because doctors were unable

---

*antibodies—proteins produced by the body in response to the presence of a foreign substance, in this case the Legionnaires' bacillus

to identify the agent that caused it. One thought was that it was caused by a virus* that attacks the respiratory system.

Several deadly viruses have definitely been isolated and studied in recent years. These include the Marburg, Ebola, and Lassa viruses. The Marburg virus was named after the town in Germany where the disease was first recognized in 1967. It originated from the tissues of African green monkeys that had been imported from Uganda, Africa, for biomedical research and vaccine production. Every year the United States imports about 12,000 such monkeys. The monkeys show no symptoms themselves, but the disease is often fatal in humans—one in three die.

Lassa virus was named after the village in Nigeria, Africa, where the first case was reported in 1970. Four Americans who were studying the disease contracted it; two died. A headline in the *New York Times* in 1970 read "New Fever Virus So Deadly the Research Halts." "New" diseases seem to be appearing with regularity these days.

Old Diseases Making New Appearances in the United States

In addition to the appearance of "new" diseases, many infectious diseases that were on the wane in this country have started to increase again. Every day, thousands of American citizens and foreign visitors enter this country. About one in five comes from a country where malaria, cholera, plague, or yellow fever still exists. In developed countries, these diseases have been largely eliminated through sanitation, vaccination, and quarantine. An international traveler incubating a disease in his or her body could circle the globe, touch down in many areas, and expose many people before developing recognizable symptoms.

Hundreds of thousands of refugees and immigrants from such areas as Cuba, Southeast Asia, and Mexico also bring with them communicable diseases that until recently were rare in the United States. For example, although still uncommon, malaria cases in California have been doubling annually for the years 1976 through 1979. In Los Angeles County alone in 1978, 255 cases of leprosy were reported to county health officials, according to one reporter a tenfold increase over the number reported 15 years earlier. During this same period, a ninefold increase in typhoid fever was reported in Los Angeles County. In the past several years, Los Angeles County health records have also shown sharp increases in tuberculosis and other diseases, such as gastrointestinal disorders associated with poverty.

In the United States, many diseases that were previously epidemic* have become unimportant, and many kinds of infections have become rare. Nevertheless, the possibility always exists that some known or obscure infectious agent, such as the Lassa virus, may unleash a serious epidemic in part or all of the world, especially if public health measures become impaired by political or economic problems.

*virus—a nonliving, submicroscopic infectious agent

*epidemic—affecting many individuals within a region or population in a short period of time

INFECTIOUS DISEASE—PAST IMPACT

The progress that has been made in the area of human health is largely the result of the developing science of microbiology. Throughout recorded history, the number of deaths due to infectious diseases has been staggering. Over the past centuries, people constantly feared death from mysterious ailments. Between 1346 and 1350, one third of the entire population of Europe died as a result of bubonic plague. In many cases, the course of history has been determined by disease. Smallpox brought to Mexico by the invading Spaniards was one factor that made it possible for Cortez to conquer the Aztecs in 1520. Over 3.5 million native people died shortly after the disease was introduced by the invaders. The same was true of Pizzaro's conquest of the Incas of Peru in 1525. On the northwest coast of America, entire villages of Haida Indians were eliminated by epidemics of smallpox brought by European sailors trading for sea otter pelts. In the Crimean War (1854–1856), ten times more British soldiers died of dysentery than from all the Russian weapons combined. Even in the twentieth century, worldwide epidemics have occurred. Influenza spread throughout the earth in 1918–1919, infecting almost the entire population of the globe. More than 20 million people died.

IMPORTANT DEVELOPMENTS IN THE HISTORY OF MICROBIOLOGY

What were some of the key discoveries that provide the basis for the isolation and identification of the organisms that cause disease?

Microscopy

The first major discovery was made in 1674 by Antony van Leeuwenhoek, an inquisitive Dutch drapery merchant, when he peered at a drop of lake water through a glass magnifying lens that he had carefully ground. He glimpsed for the first time the world of microorganisms. Even with his lenses, which magnified only 300-fold, he was able to describe the three major shapes of bacteria (singular: bacterium). There are spherical organisms, commonly referred to as **cocci** (singular: coccus), cylindrical organisms, commonly referred to as rods or **bacilli** (singular: bacillus), and helical organisms, commonly referred to as **spirilla** (singular: spirillum). Organisms with these shapes viewed through a modern scanning electron microscope, which magnifies organisms many thousandfold, are shown in Figure 1–1. Organisms viewed at lower magnifications are also shown.

Many bacteria do not appear as isolated cells but remain attached to one another to form characteristic arrangements of cells (Fig. 1–2). The reason for such arrangements is that most bacteria divide by binary fission* to form two separate cells. However, sometimes these cells do not separate from each other. Depending on the planes in which the bacteria divide, different characteristic arrangements result. Because such arrangements characterize

---

*binary fission—a process of reproduction in which one cell splits into two independent daughter cells

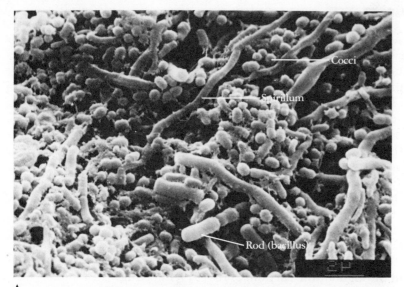

A

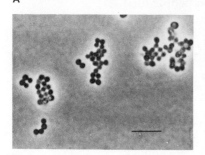

B

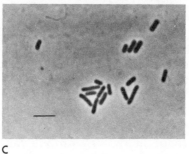

C

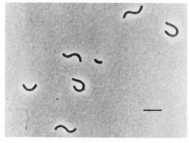

D

Figure 1–1
(a) The major shapes of bacteria. Bacteria residing in the plaque on teeth. The bar in the lower right corner represents 2 micrometers (μm). (Courtesy of D. Birdsell.) (b) Coccus. *Staphylococcus aureus.* Bar represents 5 μm. (c) Rod (bacillus). *Escherichia coli.* Bar represents 5 μm. (d) Spirillum. Bar represents 10 μm. (Courtesy of J. Staley.)

certain bacteria, they are useful for purposes of identification. For example, certain cocci tend to form long chains, a characteristic of the genus *Streptococcus* (Fig. 1–2a). Others, especially members of the genus *Staphylococcus*, tend to occur in clusters resembling bunches of grapes (Fig. 1–2b). The members of still other cocci typically form cuboidal packets of four or eight cells (Fig. 1–2c). Bacilli also often form long chains (Fig. 1–2d).

Pure Culture Methods

Before it became possible to isolate and identify the organism causing any disease, it was essential to develop techniques for growing the suspected causative agents. In the 1860's, many bacteriologists believed that bacteria could change their shape and size and perhaps even their function. We now know that this is usually not the case. For the most part, bacteria having different shapes and sizes are different bacteria, and the variety of organisms of different sizes and shapes found in most environments of nature represents

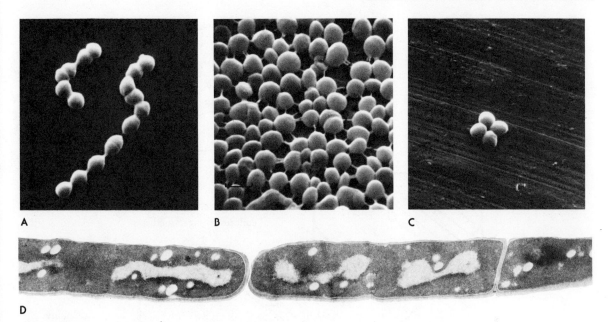

Figure 1–2
Arrangements of cocci into characteristic patterns. (a) Chains (streptococci). (b) Clusters (staphylococci). (c) Cuboidal. (a, b, and c courtesy of A. Klainer; A. Klainer and I. Geis, *Agents of Bacterial Disease*. Hagerstown, Md.: Harper & Row, 1973.) (d) Bacilli in chains. ($\times$ 22,100) (Courtesy of E. Boatman.)

a **mixed population** or mixed culture. The demonstration that each organism multiplies and gives rise to organisms of the same size, shape, and function could only be achieved if organisms could be grown independently of one another in **pure cultures.** The man who developed the pure culture techniques we use today was Robert Koch, a German physician who succeeded in combining a medical practice with a remarkably successful and productive research career. His research efforts culminated in a Nobel Prize in 1905.

Early in his career in the late 1870's, Koch realized that the isolation of pure cultures would be simplified on a solid medium on which a single isolated cell could multiply in a limited area. The population of cells that arise from such a single bacterial cell in one spot is called a **colony.** About one million cells are required for a colony to be easily visible to the naked eye.

Koch initially experimented with growing bacteria on the cut surfaces of potatoes, but he found that a lack of proper nutrients in the potato prevented the growth of some bacteria. To overcome this difficulty, Koch realized that it would be advantageous to be able to solidify any liquid nutrient medium, and he conceived the idea of adding gelatin (the same material used in gelatin desserts) as a hardening agent. Since the gelatin-containing nutrient medium is liquid at temperatures above 28°C, it was necessary only to heat

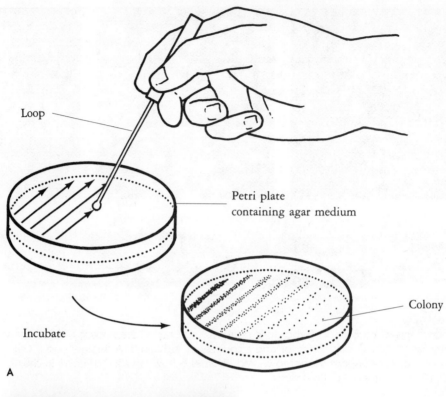

Loop

Petri plate
containing agar medium

Incubate

Colony

A

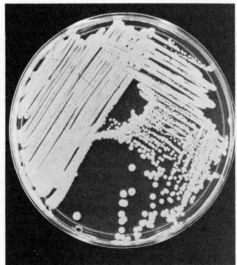

B

Figure 1–3
(a) Streak plate method. Fewer and
fewer bacteria are deposited on the
surface of the agar as the loop is
stroked across it. (b) Well-isolated colonies that result from the streak plate
method.

the medium above this temperature, pour it into a sterile* container, and allow it to cool and harden. Once hardened, a loopful of bacteria could be drawn lightly over the surface to deposit single bacterial cells at intervals (Fig. 1–3). Each of these single cells divides repeatedly, and, after a day or two, enough cells are present to form a visible colony. This technique, referred to as the "streak plate" method, is the simplest and most commonly used technique for isolating the offspring of single bacteria (Fig. 1–3). A gelatinized nutrient poses certain problems, however. The medium must be incubated at temperatures below 28°C if it is to remain solid. Furthermore, gelatin itself can be degraded by many microorganisms, resulting in the medium becoming a liquid. Thus, its use is severely restricted.

The perfect hardening agent came out of a household kitchen. Frau Hesse, the New Jersey–born wife of one of Koch's associates, suggested **agar,** a solidifying agent that her mother had used to harden jelly. Agar is a polysaccharide* extracted from certain marine algae. In contrast to gelatin, a 1.5 percent agar gel* must be heated to about 95°C before it liquifies, and it is therefore solid over the entire range of temperatures at which bacteria grow. Once melted, it can be cooled to about 40°C before it solidifies. Nutrients that might be destroyed at higher temperatures can be added aseptically* before the agar hardens. Another advantage of agar is that very few bacteria can degrade it. In all these intervening years, no better hardening agent has ever been found.

Polysaccharide, Appendix I

Another major technical advance made in Koch's laboratory was the development of a two-part glass dish that can be sterilized readily and maintained in a sterile condition. This dish serves as a convenient container during the incubation of the bacteria streaked* on the surface of the medium. The dish remains unmodified to the present day, except that plastic has replaced the glass. It retains the name of its inventor, **Petri** (Fig. 1–3).

The significance of the pure culture technique to the development of microbiology cannot be overestimated. This method proved crucial for the rapid isolation of pathogenic* as well as other bacteria.

Germ Theory of Disease

With his ability to grow bacteria in pure culture, Robert Koch was able to demonstrate convincingly that anthrax and tuberculosis were caused by different bacteria. Amazingly, however, in 1546—300 years before Koch's studies and more than 100 years before van Leeuwenhoek peered through his

---

*sterile—no living cells present

*polysaccharide—long chain of sugar molecules

*gel—material having the consistency of jelly

*aseptically—by the use of sterile techniques, to avoid introducing any organisms

*streaked—spread with a wire loop

*pathogenic—disease-causing

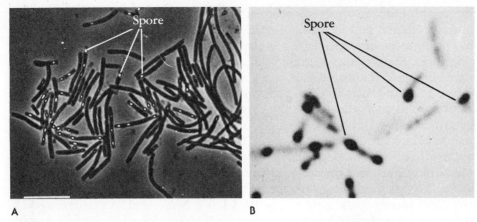

Figure 1—4
(a and b) Endospores inside *Bacillus megaterium.* Bar in (a) represents 10 μm.
(Courtesy of J. Staley.)

microscopes—the germ theory of infectious disease was proposed by Fracastoro, a classmate of Copernicus[1] at the University of Padua, Italy.

Fracastoro wrote, "Contagion is an infection that passes from one thing to another." His views were taken seriously, and in the Mediterranean region from the sixteenth century on, quarantine regulations against plague were standard. However, the infectious nature of disease was apparently discredited in 1822. After a careful study of an outbreak of yellow fever in Spain, researchers concluded that there was no possibility of contact among the different individuals who came down with the disease. No one as yet imagined that insects might be disease carriers! As a result, the popular belief arose that disease was spread by miasmata—poisonous vapors arising from decomposing bodies and swamps.

In the 1880's the germ theory of disease once again prevailed, largely because of the work of Louis Pasteur and Robert Koch. The work of Koch was especially important. His work on anthrax provided the first proof that a specific organism could cause a specific disease. In these studies he removed the spleen from infected animals and observed the growth of the anthrax bacillus under the microscope. By periodically examining the microscope slide on which he had placed the infected tissue, he discovered that the bacteria underwent developmental changes. Round bodies called **endospores** formed within rod-shaped bacteria. The endospores then developed into the rod-shaped bacteria (Fig. 1–4). Koch noted that the spores were resistant to drying, so that dried substances containing spores could cause disease when injected into mice. Koch was very fortunate to have chosen anthrax for these studies. First, the organism is relatively large and can be readily observed under the microscope. Second, the organism multiplies

Endospore, page 49

[1]Copernicus was a Polish astronomer who proved that the earth rotated around the sun. He also studied law and medicine.

inside the infected animal and reaches very high numbers in the bloodstream, where it can be seen by direct microscopic examination of the blood. Third, the organism causes disease when inoculated into mice raised in the laboratory. Thus, it was possible to study the disease in the laboratory under controlled conditions and to prove that the organism caused the disease.

The crowning achievement of Koch's scientific career was his report on the cause of tuberculosis. In 1868 Jean Antoine Villemin, a French Army surgeon, had shown that tuberculosis was an infectious disease. Therefore, it seemed quite reasonable to expect that a bacterium might be the causative agent. However, it soon became clear to Koch that unlike his straightforward studies with anthrax, the proof of this hypothesis was not going to be easy. When he looked microscopically at infected lung tissue, he was unable to demonstrate any bacteria. Only by using a special kind of staining procedure was he able to demonstrate that tiny rod-shaped bacteria (tubercle bacilli) were present in all of the infected lung tissue that he examined (see color plate 1).

INFECTIOUS AGENTS OF DISEASE

The agents that cause the diseases we have mentioned all have one property in common: They are all microscopic and can only be seen clearly if magnified at least several hundred times. All of the agents are single-celled microorganisms* except the **viruses.** The latter are **acellular** (not cellular) agents and therefore are not organisms.

General Features of Microorganisms

The microorganisms that will be emphasized in this book are the **bacteria,** which are unicellular organisms usually having a rigid cell wall that gives them their characteristic shape. Most bacteria will grow on a nutrient medium in the absence of living cells. Thus, they must have all the machinery necessary for converting the nutrients of the medium into more cells (much as the human body converts the food it consumes into cells of the body).

The second group of microorganisms that causes disease is perhaps 10 to 50 times larger than the bacteria. Included in this group are members of the **fungi** and **protozoa. Algae,** some of whose members produce deadly toxins,* are also members of the group. Photographs of some typical algae, fungi, and protozoa are shown in color plates 2 to 7. The relative sizes of bacteria, algae, fungi, protozoa, and viruses are shown in Figure 1–5. Some fungi and algae are multicellular and can be seen without a microscope.

Prokaryotic versus Eukaryotic Cells

Several features distinguish all bacteria from algae, fungi, and protozoa. These differences relate to the **size** and **complexity** of the cells. The bacterial

---

*microorganisms—living organisms that are too small to be seen without a magnifying instrument

*toxins—poisonous chemical substances

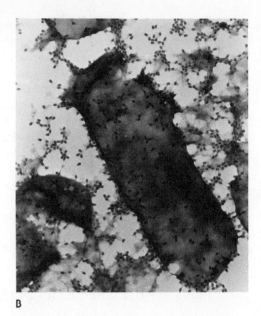

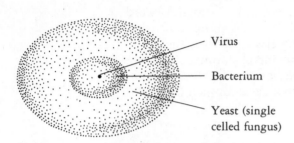

Virus

Bacterium

Yeast (single
celled fungus)

A

B

Figure 1–5

(a) The relative sizes of a bacterium, a fungus (yeast cell), and a virus. Single-celled algae
and protozoa are about the same size as the yeast cell. (b) Bacterial viruses attached to
Bacillus. Note the relative sizes of the bacteria and the viruses. (Courtesy of D. Birdsell.)

cells, which are smaller and less complex, are termed **prokaryotic.** The cells
that comprise the algae, fungi, and protozoa are called **eukaryotic.** Eukaryotic cells are far more complex than prokaryotic cells, as is obvious from
a comparison of a bacterial cell (Fig. 1–6) and a eukaryotic cell (Fig. 1–7).
Thus, once the Legionnaires' bacillus was isolated and observed with an
electron microscope, it was immediately apparent that it was a prokaryotic
microorganism (Fig. 1–8). The microscopic examination of prokaryotic and
eukaryotic cells reveals several major differences (Table 1–1). First, the chromosome in prokaryotic cells is not surrounded by a membrane, so there is
no distinct nucleus as there is in eukaryotic cells. Second, many other mem-
brane-bounded structures that occur in eukaryotic cells, such as mitochondria* and chloroplasts,* do *not* occur in prokaryotic cells. Interestingly, the
chloroplasts and mitochondria do contain some genes similar to those found
in bacteria.

Mitochondria, page 43

These two cell types have other differences that cannot readily be seen
through a microscope. Only chemical analyses or careful measurements of
the size of the same components in both cell types will reveal them. But

*mitochondria—structures concerned with the production of energy

*chloroplasts—structures concerned with the production of energy; they contain the green pigments found
in plants

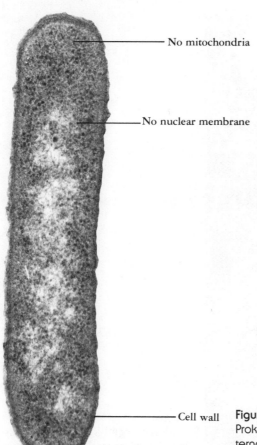

No mitochondria

No nuclear membrane

Cell wall

Figure 1–6

Prokaryotic cell (*Eikenella corrodens*). Note that there are no internal structures that are enclosed by a membrane. (×70,000) (Courtesy of A. Progulske and S. C. Holt.)

TABLE 1–1   COMPARISON OF EUKARYOTIC AND PROKARYOTIC CELLS

Features of cells	*Prokaryotic*	*Eukaryotic*
Size (diameter)	0.3 – 2 μm[a]	2 – 20 μm
Genetic structures		
Number of chromosomes	1–4 (all identical)	Many (all different)
Nuclear membrane	−	+
Cytoplasmic structures		
Cell wall	Unique chemical components	If present, composition varies between organisms
Mitochondria	−	+
Chloroplasts	−	+
Ribosomes	70S[b]	80S

[a]A μm (micrometer) is one millionth of a meter. A meter is equal to 39.37 inches, or slightly more than one yard.
[b]S (Svedberg unit) indicates the speed at which the ribosome sediments when it is centrifuged. The larger the S value, the heavier and larger the ribosome.

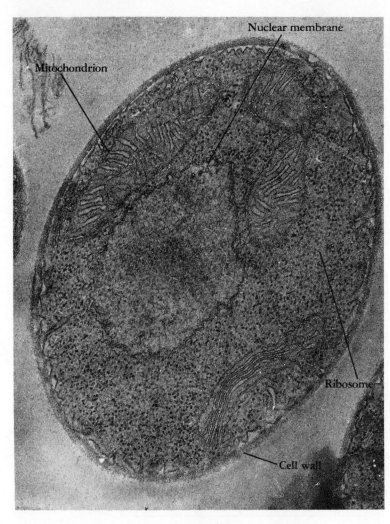

Figure 1–7
Eukaryotic cell (a yeast cell). Note that this cell contains several internal structures, the mitochondria and nuclear body, that are surrounded by a membrane. (Courtesy of H. Wolfe.)

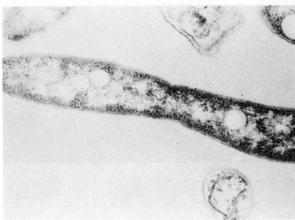

Figure 1–8
Legionnaires' bacillus. Note that the organism resembles a typical prokaryotic cell.

these subtle differences are especially important to the health of the world's population. Antimicrobial medications are useful because they selectively attack prokaryotic cells at points of difference between these two cell types. For example, penicillin selectively kills bacteria because it inhibits the synthesis of a unique structure in their cell wall that is not present in eukaryotic cells. Streptomycin selectively inhibits protein synthesis in bacteria because of differences in the structure of ribosomes* in prokaryotic and eukaryotic cells.

Penicillin, pages 31–32, 192–195

Ribosomes, pages 40–41

Functions Common to All Cells

Although two different cell types exist in terms of their structure, both must carry out the same basic function—to reproduce (replicate) exact copies of themselves. To carry out this function, a cell must contain genetic instructions for exact self-duplication. In addition, replication requires that cells be able to synthesize the constituents of living matter from the much simpler nutrients available to them (biosynthetic reactions). Because these biosynthetic reactions require energy, all cells must have mechanisms for gaining and using the energy available in nutrients they break down. All cells, whether they are prokaryotic or eukaryotic, must be able to carry out all of these functions.

Bacteria

All bacteria are prokaryotic microorganisms, but there are wide differences among bacteria. There appear to be two major groups of bacteria, based on the many subtle differences in their structure. However, one group—the **archaebacteria**—has no importance in disease and will not be discussed at any length. The other group—the **eubacteria** or "true bacteria"—represents an extremely heterogeneous population that includes all of the disease-causing bacteria. Most (but not all) eubacteria are unicellular, possess a rigid cell wall, can grow in the absence of living cells, multiply by binary fission, and (if they are motile) are propelled by means of flagella. Flagella are hairlike structures that function much like a ship's propeller.

Viruses

Viruses are the smallest of the infectious agents (Fig. 1–9). Like cells, they also must reproduce themselves. But unlike cells, viruses do not have the machinery or the chemicals necessary to carry out biosynthetic reactions (see Appendix I). Therefore they are inert, nonliving agents—a bit of DNA* or RNA* surrounded by a protein coat. However, if they enter living cells,

---

*ribosomes—cellular "workbenches" on which proteins are synthesized

*DNA—deoxyribonucleic acid; the carrier of genetic information for all organisms

*RNA—ribonucleic acid; the carrier of genetic information for some viruses

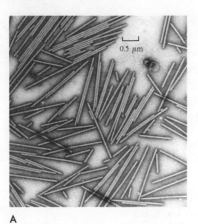
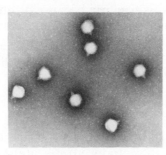
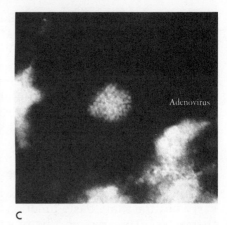

A   B   C

Figure 1–9
Viruses. (a) Tobacco mosaic virus that infects tobacco plants. A hollow protein coat surrounds a molecule of RNA. (Courtesy of E. Boatman.) (b) A bacterial virus (T7) that infects *Escherichia coli.* (×150,000) (Courtesy of R. C. Williams.) (c) Adenovirus, a virus that causes respiratory infections in humans. (Courtesy of E. Boatman.)

they are able to utilize the machinery and chemicals of the invaded cell and multiply.

NOMENCLATURE

Thus far we have not called microorganisms by their proper names, but every microorganism has one. **Nomenclature** in biology refers to the system by which organisms are named. Virtually all organisms are named according to the binomial system, in which an organism is given both a **genus** and a **species** name. The genus name is the first word in the name of the organism, and the first letter is always capitalized. The species name is the second word and is not capitalized. Both words are always italicized. For example, *Escherichia coli,* commonly called the colon bacillus, is a member of the genus *Escherichia.* The genus name is commonly abbreviated with the first letter capitalized, i.e., *E. coli.* Members of the same species differ from one another in minor ways. For example, one member may not be able to use a foodstuff that another member can. This slight difference generally does not justify giving the organisms different species names, and so they are designated as **strains** or varieties of the same species. For example, two common strains of the colon bacillus are *E. coli* K-12, isolated from a patient at Stanford University Hospital, and *E. coli* ML, which is reported to have been isolated from the bowel of a famous French scientist whose initials are ML. Both strains are identical in their major properties, but they differ in many details. The name and description of all of the bacteria that are currently recognized are given in *Bergey's Manual of Determinative Bacteriology,* a reference text.

The precise identification of an organism requires both a genus and species name. However, many organisms are commonly referred to by less precise terms that convey some information about the properties of the organisms. For example, the general term "bacillus" refers to rod- or cylinder-shaped bacteria; thus, the organism that causes Legionnaires' disease is often referred to as Legionnaires' bacillus. Rod-shaped bacteria include members of the genus *Bacillus* as well as the genera *Clostridium*, *Lactobacillus*, *Escherichia*, and *Legionella*, the genus in which the Legionnaires' bacillus has been placed. Note that the general term "bacillus" is neither capitalized nor italicized and thus can be differentiated from the genus *Bacillus*.

MICROBIOLOGY—FUTURE POTENTIAL

The "bad press" that microbiology has received in the past has been associated with disease and death. However, without the activities of microorganisms in the soil, life as we know it would not be possible on the earth. This fact is seldom publicized. We do not generally consider that our diet consists of many foods that result from bacterial activities. However, in the last several years, the importance of microbiology has become established on the front page of newspapers throughout the country. Almost every week there are newspaper stories about some aspect of microbiology and the important role that microorganisms play in our modern society. Microorganisms are being looked upon as useful tools for solving many of our severest problems.

Genetic Engineering

Genetic engineering, pages 149–151

Many human disorders can be treated by administering medicines that are isolated from their natural sources. These medicines include insulin, blood clotting factors, growth hormones, and interferon, which may prove to be an aid in the treatment of cancer. The cost of these materials often is very high because they can be isolated only in small amounts. Rather than isolate these factors from their natural sources, it has now become possible to introduce the genes that code for each of these substances into *E. coli*. As the cells of *E. coli* multiply, the genes also multiply. The bacteria synthesize the insulin and the other compounds as they multiply. It is far simpler and less expensive to isolate insulin from a culture of *E. coli* than from the pancreas of pigs and cattle.

"Man-made Life Can Be Patented." This headline appeared in a newspaper announcing the Supreme Court decision to allow a bacterium capable of "eating oil" to be patented. The investigator in this case introduced the appropriate genes into a bacterium, thus giving it the unique ability to degrade oil. By such technology, it might be possible to endow bacteria with many other useful properties that they do not now have. Perhaps organisms will be developed to degrade waste materials into alcohol, a source of energy, or into useful foodstuffs for animals.

Ames test, pages
148–149

Testing of Suspected Cancer-Causing Chemicals

A recent report estimates that 50,000 man-made chemicals exist. Every year 500 to 1000 new chemicals are introduced for industrial purposes. Do any of these cause cancer? Unfortunately, performing such determinations in animals takes several years and costs many thousands of dollars for each chemical tested. Fortunately, a bacterial test has been devised that takes a few days and costs only a few dollars to perform. This test has proved to be an excellent first screen in assessing the safety of any product. Because of this testing, a popular hair dye in the United States was ruled to be potentially cancer-causing. Its composition was altered to remove this danger. Children's pajamas were being treated with a flame-retardant chemical, which also was shown to be a potential cause of cancer by this test. The practice was stopped.

Single cell protein,
pages 102–103

Single Cell Protein

The earth's population will increase by at least 50 percent to a total of six billion people by the end of the century. As the population increases, the supply of meat and animal products available for each person to eat is likely to decrease. During the past decade, numerous processes have been developed for producing cells of microorganisms, **single cell protein,** for use as a source of protein for human and animal consumption. Since microorganisms can double their cell mass in as short a time as 20 minutes, the rate of protein production is enormous. The protein content of bacteria may be as high as 50 percent, compared with 40 percent in soybeans. Much research is now being conducted to determine the best microorganisms to use and the materials on which they should be grown.

Microorganisms are likely to play an increasingly important role in the daily lives of people around the world. In the past, microorganisms were associated primarily with disease; in the future, their vital importance in maintaining life on this planet should receive more attention.

SUMMARY

In centuries past, microorganisms played a crucial role in human history. The plagues in the Middle Ages, the deaths caused by microorganisms during wars, and the near-total destruction of civilizations by infectious disease contributed more than any other single factor to the course of events. Even today, people are afflicted with diseases of unknown causes, but their impact on civilization is only minor. Koch and Pasteur made some of the key observations in the study of the bacteria that cause disease. In addition to the prokaryotic bacteria, disease-causing agents also include fungi and protozoa, which have a eukaryotic cell structure, and viruses, which are acellular.

TABLE 1–2 PROPERTIES OF MICROORGANISMS

Feature	Bacteria	Fungi, protozoa, algae	Viruses
Cell structure	Prokaryotic (simple)	Eukaryotic (complex)	Acellular
Reproduction	Mostly free-living*	Mostly free-living*	Obligate intracellular parasites*
Cell arrangement	Single cells	Protozoa—all single cells Fungi—some single cells; mostly multicellular Algae—single cells and multicellular	

* free-living—in the absence of living cells
* obligate intracellular parasites—grow only inside living cells

Table 1–2 summarizes the properties of these various agents. There are many differences in cellular organization of prokaryotic and eukaryotic cells, but the differences are largely in size and complexity. Despite their differences in cell structure, both cell types must carry out the same functions. There is an increasing realization that microorganisms may be able to help solve many of the important problems that humans face in the near future.

SELF-QUIZ

1. The pure culture technique is associated mainly with
   a. van Leeuwenhoek
   b. Koch
   c. Pasteur
   d. Villemin
   e. Petri

2. Viruses are not living because
   a. they only grow in living cells
   b. they contain neither the machinery nor most of the chemicals required for life
   c. they contain no genetic information
   d. they contain no nucleic acid
   e. they are shapeless

3. Prokaryotic cells differ from eukaryotic cells in the following way:
   a. Prokaryotic cells contain no chromosomes; eukaryotic cells do.
   b. Prokaryotic cells cannot generate energy; eukaryotic cells can.
   c. Prokaryotic cells contain no ribosomes; eukaryotic cells do.
   d. Prokaryotic cells do not possess a nuclear membrane; eukaryotic cells do.
   e. Prokaryotic cells possess a rigid cell wall; eukaryotic cells do not.

4. In *increasing* order of complexity, the following arrangement is correct.
   a. prokaryotic > eukaryotic > viruses
   b. eukaryotic > prokaryotic > viruses
   c. eukaryotic > viruses > prokaryotic
   d. viruses > prokaryotic > eukaryotic
   e. prokaryotic > viruses > eukaryotic

QUESTIONS FOR DISCUSSION

1. Discuss reasons why new diseases seem to be more prevalent today than they were in the past.
2. Discuss the many roles that bacteria play in the various foods that we eat.
3. Why was the development of the pure culture technique so important to the development of the science of microbiology?
4. What potential dangers, if any, do you see in the genetic engineering of bacteria?

FURTHER READING

Bioscience, 30:375–407 (1980). This is a special issue devoted to Food from Microbes. All articles are written for the lay reader.

Brock, Thomas (ed.): *Milestones in Microbiology.* Washington, DC: American Society for Microbiology, reprinted 1975. A collection of historically significant papers in microbiology that cover the past 400 years. The editor briefly discusses the significance of each paper to the development of microbiology.

Cohen, S. N.: "The Manipulation of Genes." *Scientific American* (July 1975).

Devoret, R.: "Bacterial Tests for Potential Carcinogens." *Scientific American* (August 1979).

Dobell, C.: *Antony van Leeuwenhoek and His "Little Animals."* New York: Dover Press, 1960. A very readable and fascinating biography by an author who obviously greatly admired his subject. One of the best biographies ever written.

Fraser, D. W., and McDade, J. E.: "Legionellosis." *Scientific American* (October 1979).

Fuller, J. G.: *Fever! The Hunt for a New Killer Virus.* New York: Readers Digest Press, 1974. A popular story relating the detective work involved in tracking down the causative agent of a deadly virus.

McNeill, W. H.: *Plagues and People.* New York: Anchor Books, 1976. A very scholarly work describing the role that microorganisms have played in determining the course of history and the spread of the various diseases. Highly recommended for serious students.

Wetzel, R.: "Applications of Recombinant DNA Technology." *American Scientist* (November-December 1980).

Woese, C.: "Archaebacteria." *Scientific American* (June 1981).

Zinsser, H.: *Rats, Lice and History.* Boston: Little, Brown, 1934. A highly readable and entertaining account of the role of infectious disease, and typhus fever in particular, in world history.

Chapter 2

FUNCTIONAL ANATOMY OF PROKARYOTIC AND EUKARYOTIC CELLS: BASIS OF ANTIBIOTIC ACTION

"Chance favors the prepared mind." Sir Alexander Fleming's discovery of penicillin in the fall of 1928 is an excellent example of this statement's truth. Previously, Fleming had studied mercuric chloride, which killed bacteria. Unfortunately, this compound was also poisonous to the host. With his mind constantly considering ways of destroying bacteria without killing the host, it is not surprising that Fleming became intrigued when he stared at a Petri dish containing nutrient agar, colonies of staphylococci and a mold contaminant. Since the top of the Petri dish had been removed numerous times in the course of the experiment, a mold spore apparently had fallen accidentally on top of the agar and had germinated.* What caught Fleming's attention was the dissolution of the colonies of staphylococci all around this particular mold. Instead of forming opaque yellow masses, the colonies looked like drops of dew. Apparently, the mold was forming something that was released and was dissolving the bacteria. Fleming had discovered penicillin!

However, it was not until 12 years later that the true value of penicillin was

---

*germinated—began to grow

proved by two other investigators, Howard Florey and Ernst Chain. Apparently, Fleming's intuition led him astray. He and a colleague had performed experiments in which they injected penicillin into the vein of a rabbit and observed that only a trace amount was present 30 minutes later. Since they knew that penicillin took from four to eight hours to kill all of the staphylococci in a broth medium, it seemed unlikely that its short life in blood would be enough to treat deep-seated infections. Thus, Fleming believed that penicillin would be useful only for treating surface infections, and he paid little attention to it after the discovery.

OBJECTIVES

To know
1. The four functions all bacteria must be able to carry out and the additional capabilities some bacteria have.
2. The names of the structures, chemical composition, and functions of the three layers that surround and enclose the cytoplasm.
3. The mode of action of penicillin and why it is so selective against bacteria.
4. The several functions of the cytoplasmic membrane.
5. The differences between prokaryotic and eukaryotic ribosomes and the several antibiotics that affect prokaryotic ribosomes.
6. The various steps in the movement of bacteria in response to nutrients.
7. The conditions under which certain organisms form endospores, the distinctive properties of endospores, and their importance to medicine.

FUNCTIONAL ANATOMY—AN OVERVIEW OF PROKARYOTIC AND EUKARYOTIC CELLS

Fleming observed that penicillin, an antibiotic* kills some but not all bacteria. Most important, it has no effect on the cells of animals. In this chapter, we will focus on the structures responsible for carrying out the numerous functions required for the life of bacteria. Furthermore, we shall consider their counterparts—the eukaryotic cells—in humans.

Recognizing the similarities between these cells will provide an understanding of the unity of life. Appreciating the differences in their chemical structure will give an understanding of how antibiotics can affect some cells but not others, and why some antibiotics may have serious consequences in humans. In addition, we shall discuss the role that each of these structures may play in causing disease.

*antibiotic—a chemical substance produced by certain molds and bacteria that inhibits the growth of or kills other microorganisms.

The emphasis in this chapter will be on bacteria, since they are responsible for causing many diseases of microbial origin. Some of the techniques of viewing and staining bacteria are discussed in Appendix II. An electron micrograph and a diagram of a cylindrical bacterium are given in Figure 2–1.

The functional anatomy of this bacterium must provide the means to (1) enclose the internal contents of the cell and separate it from the medium in

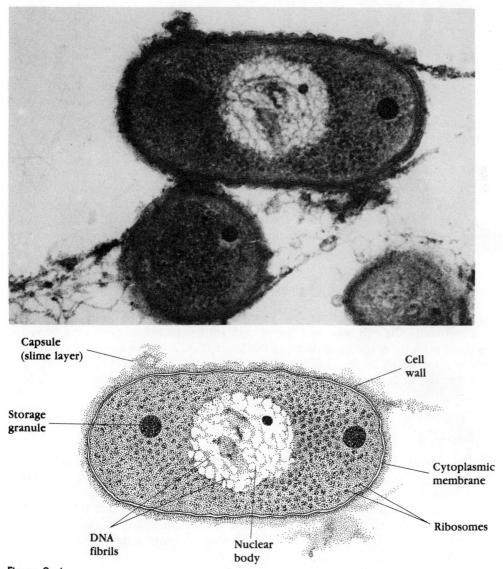

Figure 2–1

A rod-shaped bacterium, *Eikenella corrodens,* which is commonly found in the mouth. (×130,000) (Courtesy of A. Progulske and S. C. Holt.)

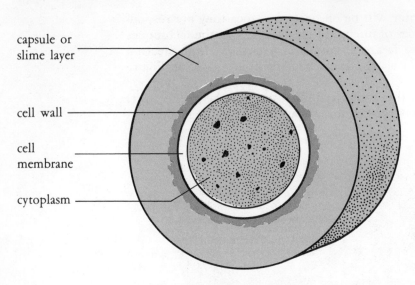

capsule or
slime layer

cell wall

cell
membrane

cytoplasm

Figure 2–2
Diagrammatic representation of
the surface layers surrounding the
cytoplasm—the cell envelope.

which the cells are growing; (2) store and reproduce genetic information; (3) synthesize cellular components; and (4) generate, store, and utilize energy-rich compounds. In addition, *some* bacteria have the means to (1) move actively; (2) transfer genetic information to other bacteria; and (3) store reserves of building blocks and energy. We will now consider in more detail the structures that are concerned with each of these functions.

ENCLOSURE OF CYTOPLASM: CELL ENVELOPE

Most typical bacteria have two structures that surround and enclose their cytoplasm: the **cell wall** and the **cytoplasmic membrane.** Some may have a third structure, the **capsule** or slime layer (Fig. 2–1). As a group, the layers are referred to as the cell envelope (Fig. 2–2). The discussion will proceed from the outermost (capsule) to the innermost layer (cytoplasmic membrane).

Capsule or Slime Layer

The capsule is a loose-fitting, gelatinous structure that surrounds some bacteria (Fig. 2–3). Colonies of bacteria that are synthesizing capsules usually appear moist, glistening, and slimy. Capsular material may adhere to a transfer loop when the loop touches a bacterial colony and is pulled away. Capsules are produced only in certain species of bacteria and often only when they are grown under certain nutritional conditions (Fig. 2–4).

Composition. Capsules vary in their chemical composition, depending on the particular organism. Most are composed of polysaccharide chains, but

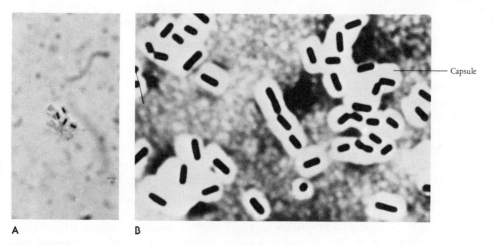

A   B

Figure 2–3

(a) Capsules surrounding *Streptococcus pneumoniae*. (Courtesy of Goodhart.) (b) Capsules surrounding bacilli. (Courtesy of V. Chambers.)

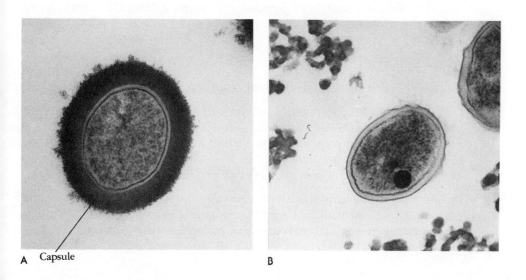

A   Capsule   B

Figure 2–4

(a) *Leuconostoc mesenteroides* grown on medium containing sucrose. The capsular material is an insoluble polysaccharide. (b) The same organism grown on medium lacking sucrose. Note the absence of the capsule. (Courtesy of B. E. Brooker. B. E. Brooker, *J. Bacteriol.*, *131*:288, 1977.)

Amino acids, Appendix I
others consist of one or two kinds of amino acids* polymerized* into long polypeptide* chains.

Function. A capsule confers certain capabilities under certain situations for some species. One of the best-studied capsules belongs to an organism that can cause bacterial pneumonia, *Streptococcus pneumoniae.* If the organism does not produce a capsule, it is quickly destroyed by the defenses of the infected host. If the bacterium synthesizes a capsule, the phagocytic cells* of the host cannot engulf it, and pneumonia may result. In this case, the capsule protects the cell.

Another function of capsular material is attachment. In nature, many bacteria form a mass of tangled fibers of polysaccharide that originate from the surface of the cell. These fibers serve to attach bacteria to a variety of surfaces, such as other bacteria or animal and plant cells, often with great specificity (Fig. 2–5). The formation by these fibers of a capsule, termed the **glycocalyx,** is well illustrated by *Streptococcus mutans,* the principal cause of dental caries. This organism must accumulate in large masses on the surface of the teeth in order to cause dental decay. This attachment is made possible by a glycocalyx that is synthesized from glucose molecules, which are in turn produced from sucrose.* These fibers of glucans* continue to accumulate on the surface of the teeth, trapping more bacteria and building up the yellowish film called plaque (Fig. 2–6). Although *Streptococcus mutans* can multiply in the absence of sucrose, this sugar allows the organism to synthesize the glycocalyx, attach to the teeth, and cause decay.

Polysaccharide, Appendix I

Bacterial Cell Wall

The bacterial cell wall deserves special attention for several reasons:

1. Differences in the composition of the cell wall determine the staining characteristics of the cell using certain dyes. (See Appendix II.)
2. The cell wall is the part of the bacterium attacked by some of the most effective antibiotic medicines.
3. The cell walls of certain organisms can produce symptoms of disease.

The cell wall also maintains the shape of the organism: cylindrical cells have a cylindrical cell wall; spherical organisms have a spherical cell wall. If the cell is broken, the cell wall does not collapse totally but maintains its rigid shape (Fig. 2–7).

---

*amino acids—the subunits of a protein molecule

*polymerized—joined together into a larger molecule by chemical bonds

*polypeptide—referring to the bonds joining the amino acids

*phagocytic cells—cells that protect the host by ingesting foreign particles such as bacteria

*sucrose—a sugar consisting of a molecule of glucose bonded to fructose, i.e., common table sugar

*glucans—a polysaccharide built up of glucose molecules

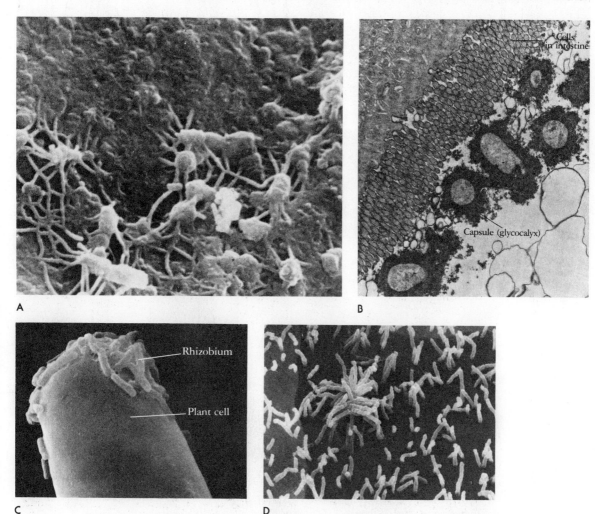

Figure 2–5
Attachment of bacteria to various surfaces. (a) Masses of cells of *Eikenella corrodens* covered with slime, which joins the cells together. (Courtesy of A. Progulske and S. C. Holt. A. Progulske and S. C. Holt, *J. Bacteriol., 143*:1003, 1980.) (b) Cells of *Escherichia coli* attached to cells in the small intestine (ileum). (Courtesy of K. J. Cheng and J. W. Costerton.) (c) Cells of *Rhizobium* attached to plant roots. (Courtesy of F. Dazzo. F. Dazzo and W. Brill, *J. Bacteriol., 137*:1362, 1979.) (d) Cells of *Seliberia stellata* attached to a surface by means of their ends. (×5000) (Courtesy of J. R. Swafford.)

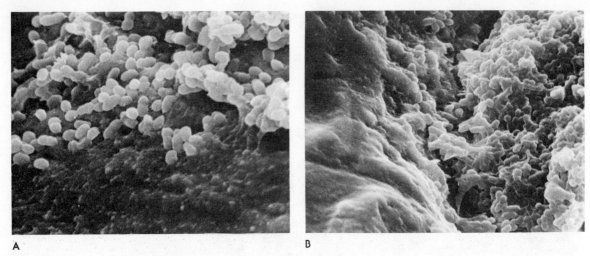

A   B

Figure 2—6
Streptococcus mutans in a lesion of an enamel surface. (a ×6000, b ×3600) (Courtesy of Z. Skobe.)

Structure. The cell wall can be considered as a strong, rigid "corset" that surrounds the cell. The strength of this wall is based on the physical and chemical properties of a component that is found only in bacteria, the **peptidoglycan layer.**

The peptidoglycan layer is a very large molecule composed of two repeating molecules, *N*-acetylmuramic acid and *N*-acetylglucosamine.* These two molecules alternate with one another to form one large molecule that surrounds the cytoplasm (Fig. 2–8). This molecule is very strong because many cell walls are composed of layer upon layer of peptidoglycan which are joined together by bridges composed of amino acids. Thus, the multilayers are really one very big molecule (Fig. 2–9).

Cell Wall of Gram-positive Bacteria. The Gram staining procedure separates bacteria into two groups, Gram-positive and Gram-negative, based upon their ability to take up and retain certain dyes. (See Appendix II.) Bacteria retaining the purple dye that is first applied (gentian or crystal violet) are termed "Gram-positive" (color plates 8 and 9). Those bacteria that lose this dye are termed "Gram-negative" (color plate 10). The reason why some bacteria retain the purple dye and others do not is related to the chemical structure of the cell wall. It is very important to realize that many other properties of bacteria correlate with whether an organism is Gram-positive or negative; therefore, knowing the staining characteristics of an

---

N-acetylmuramic acid and *N*-acetylglucosamine—molecules related to the sugar glucose but containing several additional chemical groups

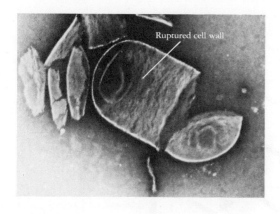

Ruptured cell wall

Figure 2—7
Rigidity of the bacterial cell wall.
(Courtesy of D. Birdsell.)

organism will provide considerable information about the organism other than the nature of its cell wall.

The cell wall of most Gram-positive bacteria contains as its major component layer upon layer of peptidoglycan; each layer is connected to the one above and the one below by amino acid bridges (Fig. 2–9).

Cell Wall of Gram-negative Bacteria. The wall of Gram-negative bacteria also contains a peptidoglycan layer, but it is much thinner and therefore not as strong as the multilayered structure found in Gram-positive bacteria.

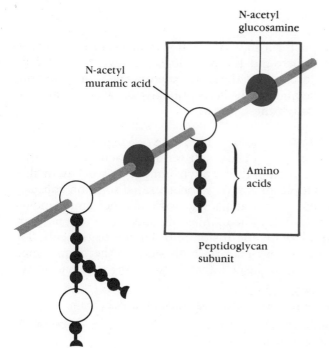

N-acetyl
glucosamine

N-acetyl
muramic acid

Amino
acids

Peptidoglycan
subunit

Figure 2—8
Peptidoglycan layer.

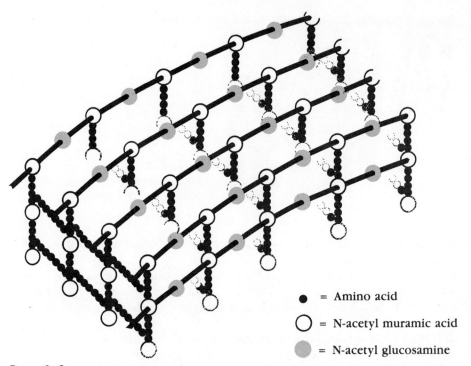

= Amino acid

◯ = N-acetyl muramic acid

⬤ = N-acetyl glucosamine

Figure 2–9
Representation of the three-dimensional nature of the peptidoglycan layer of the cell wall.

However, Gram-negative bacteria have thick layers of lipoprotein* and lipopolysaccharide,* which surround the peptidoglycan layer (Fig. 2–10). These two layers together are called the **outer membrane.** Pores extend through the outer membrane and reach the peptidoglycan layer. Small but not large molecules pass through these pores.

Functions of the Cell Wall. The cell wall holds the cell together and protects it from bursting in a solution of low salt (dilute). Bacteria normally grow in a dilute solution in their natural environment as well as in the laboratory. The bacterial cytoplasm is a very concentrated solution consisting of inorganic salts, sugars, amino acids, and various other small molecules. The concentration of particles (the number of molecules and ions* per unit volume) tends to equalize on the inside and outside of the cell. To achieve equal concentrations of particles on both sides of the membrane, water flows from the medium into the cell, thereby reducing the concentra-

---

*lipoprotein—the macromolecule formed by complexing lipid and protein (see Appendix I)

*lipopolysaccharide—the macromolecule formed by the bonding of lipid to polysaccharide (see Appendix I)

*ions—atoms or groups of atoms that carry a positive or negative charge

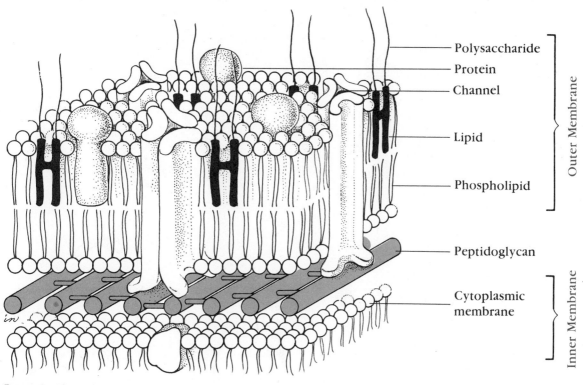

Figure 2–10
Representation of the cell wall of a Gram-negative organism.

tion of particles on the inside of the cell (Fig. 2–11). The inflow of water exerts tremendous pressure, called osmotic pressure,* on the envelope that encloses the cytoplasm. The innermost layer, the **cytoplasmic membrane,** is generally not strong. Therefore, bacteria growing in a dilute solution will burst unless the cell has a rigid, very strong cell wall that cannot expand. In the absence of a rigid cell wall, most cells simply balloon in size until they burst (Fig. 2–11). The peptidoglycan layer is largely responsible for providing this strength in both Gram-positive and Gram-negative organisms.

Penicillin. Anything that weakens the peptidoglycan layer will kill the bacteria. Penicillin exerts its effect by affecting the *synthesis* of the peptidoglycan. It does this by interfering with the activity of an enzyme* concerned with joining layers of peptidoglycan together. Therefore, the wall is very weak, and the bacteria balloon up and burst (Fig. 2–12). Its effect on the

Penicillin, pages 192–195

*osmotic pressure—the pressure exerted by water on the cytoplasmic membrane as a result of the difference in the concentration of dissolved substances on each side of the membrane

*enzyme—a protein that speeds up a chemical reaction in a biological system

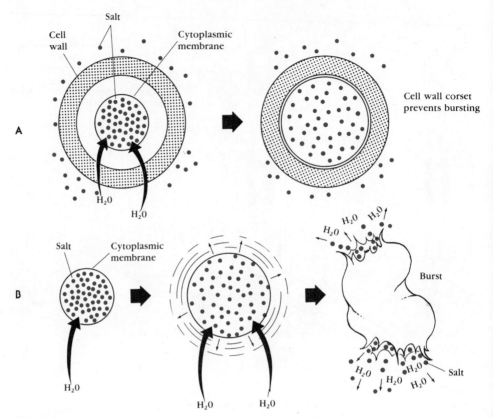

Figure 2–11
Functioning of bacterial cell wall. (a) Cell wall functions to prevent cell from bursting. (b) In absence of cell wall, cell bursts.

peptidoglycan layer explains many of the older observations about penicillin. For example, penicillin affects only growing cells. The cells must be synthesizing their cell walls in order to be affected. Furthermore, Gram-positive organisms tend to be more susceptible to penicillin than many Gram-negative organisms, since many Gram-positive organisms have their peptidoglycan layer very close to the outer edge of the cell where it can readily be reached by penicillin. On the other hand, many Gram-negative organisms have their peptidoglycan layer surrounded by thick layers of lipopolysaccharide and lipoprotein. These external layers hinder the penicillin from reaching its target. Unless the drug can do this, it is not effective.

Why is penicillin such a good antibiotic? It attacks the bacterial cell at a site that has no counterpart in the cells of the host. The peptidoglycan layer is unique to bacteria. Since this layer is not present in the host, there is no danger that both the host and the bacteria will be killed. Penicillin is discussed further in Chapter 8.

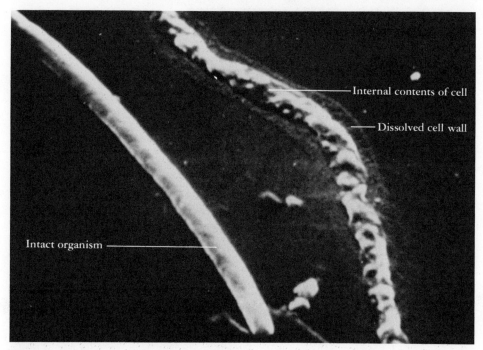

Internal contents of cell

Dissolved cell wall

Intact organism

Figure 2–12
Effect of penicillin on a bacterium. (Courtesy of S. Falkow.)

Lysozyme. Early in his career, Alexander Fleming discovered another substance affecting the peptidoglycan layer. This substance is the enzyme **lysozyme.** Unlike penicillin, which prevents the synthesis of strong peptidoglycan layers, lysozyme *degrades** this layer when it is already present in the cell wall. Therefore, cells exposed to lysozyme will experience the same effect as those exposed to penicillin. They will balloon up and burst. Unlike penicillin, however, lysozyme is effective on both growing and nongrowing cells. Lysozyme is found in a variety of body tissues and fluids and plays a role in defending the host against bacterial infection. What happens to cells if they are treated with lysozyme in a medium containing a high concentration of salt or sugar so that the pressure on the inside and outside of the cell is equal? Under these conditions, water has no tendency to enter the cell. However, the lack of the peptidoglycan layer results in such a weak cell wall that it is no longer able to maintain the shape of the cell. Therefore, cells of all shapes become spherical, their most stable shape (Fig. 2–13). This wall-less cell, called a **protoplast,** can carry out metabolic processes.*

---

*degrades—breaks up or destroys

*metabolic processes—biochemical reactions catalyzed by the enzymes of the cell, involving both the synthesis and breakdown of organic materials

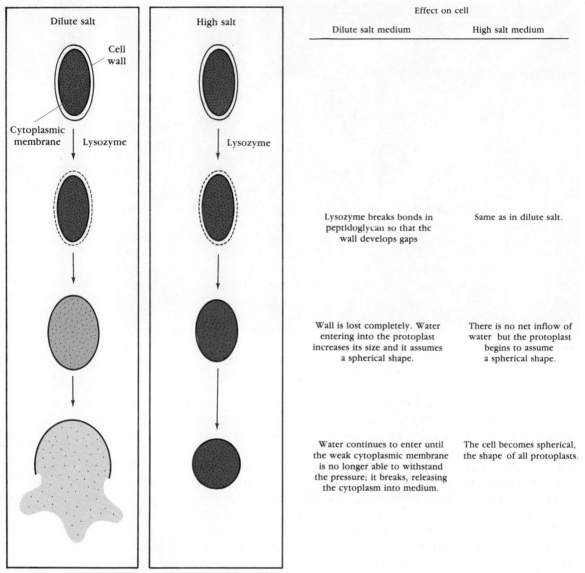

	Effect on cell	
	Dilute salt medium	High salt medium
	Lysozyme breaks bonds in peptidoglycan so that the wall develops gaps	Same as in dilute salt.
	Wall is lost completely. Water entering into the protoplast increases its size and it assumes a spherical shape.	There is no net inflow of water but the protoplast begins to assume a spherical shape.
	Water continues to enter until the weak cytoplasmic membrane is no longer able to withstand the pressure; it breaks, releasing the cytoplasm into medium.	The cell becomes spherical, the shape of all protoplasts.

Figure 2–13
Effect of disruption of the cell wall by lysozyme in a dilute salt and a high salt medium.

Wall-deficient Organisms

Lysozyme and penicillin accomplish in the laboratory what has already occurred in nature. It is possible to isolate from the host bacteria that have lost the ability to synthesize the peptidoglycan portion of their cell wall. These bacteria, termed **L-forms,** are derived from either Gram-positive or Gram-negative bacteria. As predicted, they are resistant to penicillin, grow

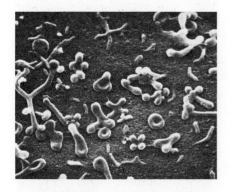

Figure 2–14
Mycoplasma pneumoniae has a plastic shape
because it lacks a rigid cell wall. (Courtesy of E.
Boatman.)

only in a high salt medium, and have no fixed shape. Although most bacteria
have a rigid cell wall, one important group lacks it. Included are members
of the genus *Mycoplasma*. Apparently, these organisms have an unusually
strong cytoplasmic membrane, which explains why some of them can grow
in dilute media. These organisms tend to grow slowly, and generally require
an enriched* medium containing 20 percent serum on which to grow. Some
species, however, can grow rapidly in dilute media. The expected properties
are present: resistance to penicillin and lack of a definite shape (Fig. 2–14).
One species of *Mycoplasma* causes primary atypical pneumonia in humans.

Cytoplasmic or Plasma Membrane (Inner Membrane)

Identification. This membrane is the outer envelope in cells such as members of the genus *Mycoplasma*, which lack a cell wall. In all other bacteria,
this thin, expandable membrane lies close to the inner surface of the cell
wall (Fig. 2–1), although there is generally a very narrow space, the **periplasmic space**, between these layers (Fig. 2–15).

Chemical Composition. The cytoplasmic membrane contains approximately 60 percent protein and 40 percent lipid. The membranes from bacteria and eukaryotic cells have the same microscopic appearance and a similar chemical composition. For this reason, as a general rule, this structure
is not a selective target for antibiotic action. Some antibiotics damage the
cytoplasmic membrane of eukaryotic pathogens, such as fungi, since they
bind to sterols* that most bacteria lack, but they are also toxic to mammalian hosts.

Function. The cytoplasmic membrane of bacteria performs several functions absolutely essential to the life of the cell. For example, many enzymes
concerned with the degradation of foodstuffs and the production of energy

---

*enriched—having supplementary nutrients such as those found in nutrient broth

*sterols—special components of some lipids; cholesterol is a common sterol

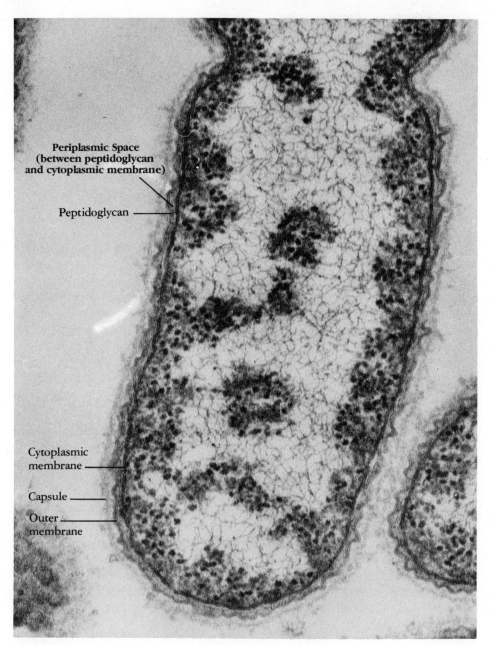

Periplasmic Space
(between peptidoglycan
and cytoplasmic membrane)

Peptidoglycan —

Cytoplasmic
membrane —

Capsule —

Outer —
membrane

Figure 2–15
Surface structures. The periplasmic space is evident between the peptidoglycan layer and the cytoplasmic membrane of this gram-negative organism (Aeromonas). (Courtesy of Dr. E. Boatman.)

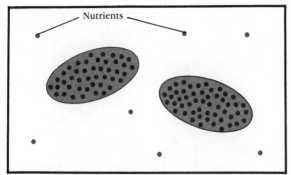

Active transport: because of the process, the internal environment of a cell is generally much different from the external environment.

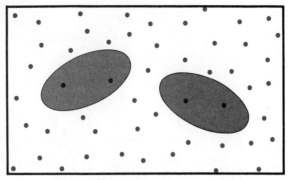

Passive diffusion: the internal and external concentrations are identical.

Figure 2–16
Active transport and passive diffusion.

are attached to the membrane. Thus, it is an indispensable structure for all bacteria and, indeed, all cells.

The membrane is semipermeable, meaning that only certain kinds of molecules can pass through it. Generally, only low molecular weight* materials (those with molecular weights no greater than several hundred) can penetrate to the inside of the cell. These molecules must also be soluble in water.

Most materials enter the cell by one of two distinct processes: **passive diffusion** or **active transport.** In passive diffusion, the molecules flow freely into and out of the cell, so that the concentration of any particular molecule is the same on the inside as it is on the outside of the cell. In active transport, the cell uses energy to transport molecules into and out of the cell. In general, the cell transports more molecules in than out, so that molecules accumulate inside the cell (Fig. 2–16). This ability to concentrate nutrients allows the bacterial cell to maintain a relatively constant intracellular composition, even though the external environment changes drastically. Since bacteria live in environments in which many of the nutrients that they require for growth are present in extremely low concentrations, passive diffusion would not be sufficient to provide the high intracellular concentration of nutrients required for rapid growth. Thus, mechanisms have evolved by which bacteria can actively transport and accumulate the majority of nutrients that they need.

In active transport, the concentration of a particular nutrient may be more than a thousand times greater inside the cell than it is in the external environment. Such active transport requires the participation of a complex transport system. The transport systems, called **permeases,** consist of several enzymes that are located in the membrane. Generally, a separate permease

*molecular weight—the relative weight of an atom or molecule based on a scale in which the H atom is assigned a weight of 1.0

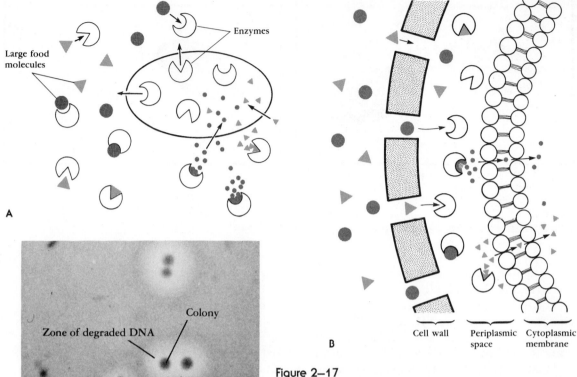

Figure 2–17

(a) Degradation of large molecules by enzymes excreted by bacteria. (b) Degradation of large molecules by enzymes in the periplasmic space. (c) Degradation of DNA by *Serratia liquefaciens*. The colonies excrete the enzyme deoxyribonuclease, which degrades the DNA in the medium. This leads to a zone of clearing around the colony. (Courtesy of S. Kominos and Wright.)

exists for each nutrient, although the same permease may transport several compounds that have a similar chemical structure.

Bacteria also have a means for using the high molecular weight compounds that are often present in the environment but that are too large to penetrate the cell wall or cytoplasmic membrane. The components of these molecules are potential nutrients. Most bacteria secrete enzymes into the medium to degrade the large food molecules into their subunits, which are then small enough to be transported into the cell (Fig. 2–17a). Some bacteria contain degradative enzymes in their periplasmic space. Thus, any large molecules that pass through channels in the cell wall will come in contact with these enzymes, be broken down into smaller molecules, and be transported through the cytoplasmic membrane via the permease (Fig. 2–17b and c).

STORAGE OF GENETIC INFORMATION

Chromosomes

Chromosome, page 110

The major structure in which the genetic information of bacteria is stored is the **chromosome.** In bacteria, the chromosome is *not* surrounded by a nuclear membrane and so does not occupy a clearly defined area in all cells (Fig. 2–1). Each chromosome is composed of a single long molecule of DNA, which exists as a circular molecule. When extended to its full length, it is about 1 millimeter long, approximately 1000 times longer than the entire bacterium (Fig. 2–18). When tightly packed into the cell, it takes up only about 10 percent of the volume of the cell.

Plasmids

Plasmid, pages 135–136

Most bacteria contain additional genetic information that is carried on small circular DNA molecules called **plasmids.** These are only about 0.1 to 1.0 percent the size of the chromosome, and some cells may carry as many as nine different ones. The significance of many plasmids is not understood, but it is known that the information that determines cellular resistance to

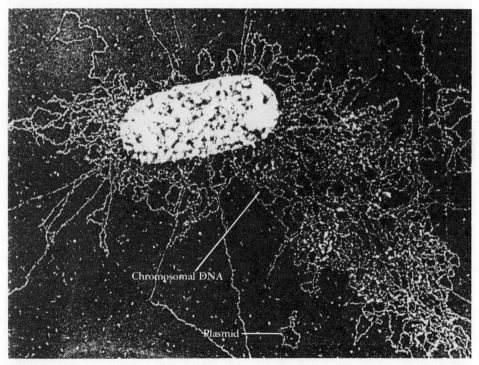

Figure 2–18
Chromosomal DNA and plasmid leaking out of a cell. Note the relative sizes of the two kinds of DNA. (Courtesy of Griffith.)

Figure 2-19
A plasmid. Note that this molecule is in the form of a circle. (Courtesy of T. Currier.)

certain antibiotics is encoded by genes* on these small circular DNA molecules (Figs. 2-18 and 2-19).

Vital genetic information without which the cell cannot live, is coded by the *chromosome;* plasmids code for useful supplementary information.

SYNTHESIS OF PROTEIN

Ribosomes

The most conspicuous structures in bacteria are the **ribosomes** (Fig. 2-1). The ribosomes are the "workbenches" on which amino acids are joined together to form proteins. Protein synthesis is a complex assembly line process involving the chromosome, various molecules of RNA, and the ribosome. (The details of this process will be discussed in Chapter 5.) As many as 15,000 of these ribosomes are found in the cytoplasm of a single cell. The faster the bacteria are growing, the greater the rate of protein synthesis and the larger the number of ribosomes in each cell. Each ribosome is composed of two parts or subunits, one large and one small. Each subunit is composed of both RNA and protein (Fig. 2-20).

Although both bacteria and eukaryotic cells have ribosomes, there are distinct differences between ribosomes of these two types of cells. Prokaryotic ribosomes are smaller, and the proteins that make up their structure are distinctly different from the proteins making up the ribosomal structure

---

*genes—the subunits of chromosomes, each of which codes for one protein

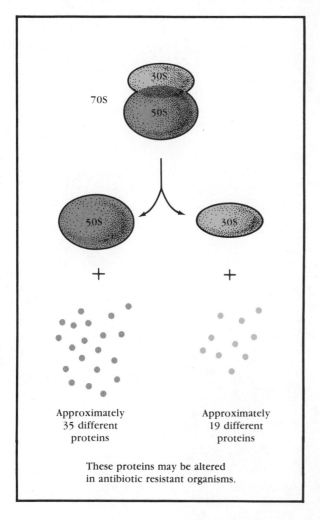

70S

30S

50S

50S

30S

+

+

Approximately
35 different
proteins

Approximately
19 different
proteins

These proteins may be altered
in antibiotic resistant organisms.

Figure 2–20
The prokaryotic ribosome and
its constituent parts.

in eukaryotic cells. These differences represent points of attack by several common antibiotics, all of which inhibit protein synthesis. Streptomycin, kanamycin, and tetracycline, to name a few, attach to the protein in the small subunit of the bacterial ribosome and inhibit the synthesis of protein. Chloramphenicol, erythromycin, and lincomycin attach to the large subunit of the bacterial ribosome. Other antibiotics interfere with other parts of this very complex process of protein synthesis. Although the overall aspects of protein synthesis are the same in all cells, small differences in the ribosomes provide selective sites for antibiotic action. Some antibiotics inhibit protein synthesis only in bacteria and other prokaryotic cells; others inhibit protein synthesis only in eukaryotic cells; and some antibiotics inhibit protein synthesis in both types of cells.

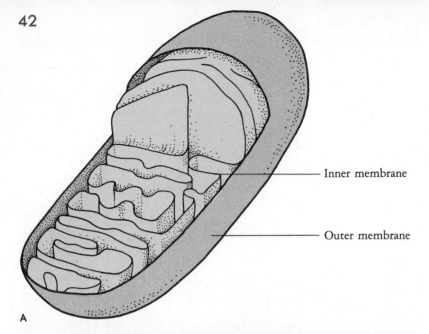

Inner membrane

Outer membrane

A

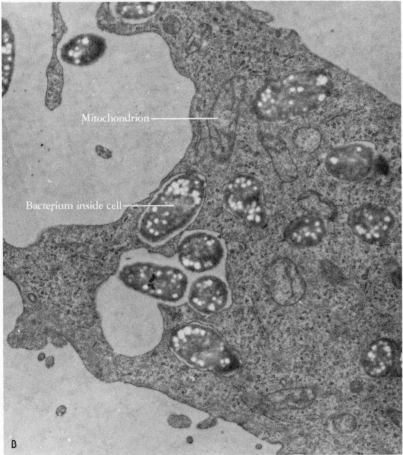

Mitochondrion

Bacterium inside cell

B

Figure 2–21
(a) A mitochondrion. The enzymes of energy metabolism are either located in the fluid inside or attached to the inner membrane. (b) A mitochondrion inside a mammalian cell that also contains bacteria. Note that they are about the same size. (Courtesy of D. Portnoy.)

GENERATION OF ENERGY

Cytoplasmic Membrane

Most cells obtain their energy by degrading foodstuffs through a series of metabolic steps. Each step requires a specific enzyme. In bacteria many of the enzymes involved in the generation of energy are attached to the cytoplasmic membrane.

In eukaryotic cells, energy is generated in **mitochondria,** highly complex structures about the size of bacteria (Fig. 2–21). Enzymes concerned with energy metabolism are attached to the folded membranes. Mitochondria possess DNA, ribosomes, and other components required for protein synthesis, which they do carry out. Of considerable interest is the fact that the ribosomes are of the prokaryotic type. This is an important consideration in the use of certain antibiotics. (See Chapter 8.)

CELL MOVEMENT

Flagella

Structure. The **flagella** (singular: flagellum) are long, helical protein filaments that are responsible for the motility of most bacteria. The flagellum is a very complex structure composed of a number of parts: the filament, the hook, and the basal body. This latter structure anchors the flagellum to the cell wall and cytoplasmic membrane (Fig. 2–22).

Bacteria can have different arrangements of flagella characteristic of the particular species of bacteria. Some have a single flagellum at one end of the

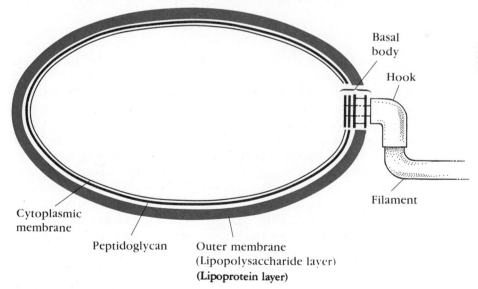

Figure 2–22
The attachment of the flagellum to the bacterial cell.

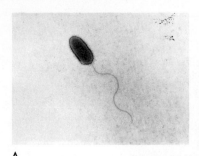 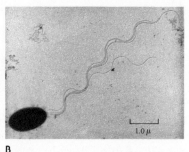 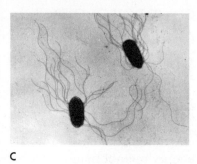

A   B   C

Figure 2–23
Arrangements of flagella. (a) Single polar flagellum. *Pseudomonas aeruginosa* (polar flagellation). (×5250) (Courtesy of V. Chambers.) (b) Tuft at one end (polar flagellation—lophotrichous). (Courtesy of E. Boatman.) (c) Flagella inserted throughout *Proteus mirabilis* (peritrichous flagellation). (×3500) (Courtesy of V. Chambers.)

cell; some have a tuft. Other bacteria have a flagellum at either end, and still others have their flagella inserted at one or many points along the side of the cell (Fig. 2–23). Since the arrangements of flagella are characteristic of the particular species, they are often useful in helping to identify an organism.

Function. The flagella propel the bacterium by pushing the cell, much as a propeller pushes a ship through water (Fig. 2–24). In bacteria that have many, the flagella bind together to form a single bundle to push the cell in one direction.

Chemotaxis*

Clearly it is beneficial for a cell to be able to move in a directed fashion—in the direction of nutrients and away from toxic materials. Bacteria have a primitive sensory system that is able to detect the presence of nutrients. This information is transmitted to the flagella. The flagellum or bundle of flagella rotates in a counterclockwise direction and moves the bacteria in the direction of the nutrients (Fig. 2–25). If they wander off course so that they can no longer detect the nutrients, the flagella lose their tendency to stay together. Rather, the flagellar bundle flies apart, and the cell wanders aimlessly until it once again detects the nutrients. This information is then sent from the sensors of the bacteria to the flagella, which once again stick together to form a coordinated unit. Toxic compounds can also be detected. In general, these chemicals have the opposite effect. Their presence causes the flagellar bundle to fly apart, so bacteria move away from the danger (Fig. 2–25). How bacteria sense chemicals and how this information is transmitted to flagella are not well understood.

*chemotaxis—movement of an organism in response to chemicals in the environment

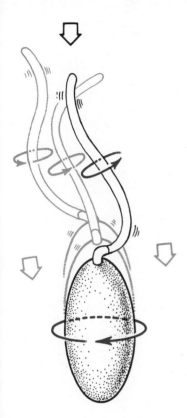

Figure 2–24
Flagellum pushing a bacterium. The flagellum and the cell rotate in opposite directions.

Other Methods of Locomotion

Other bacteria use different methods for locomotion. For example, gliding bacteria (Fig. 2–26a) have flexible cell walls that somehow propel the cells. Spirochetes (Fig. 2–26b) move by means of an axial filament attached to both ends of the organism. In each of these cases, the mechanisms that move the cell are not well understood. However, for pathogenic members of this group, there are suggestions that motility may sometimes be necessary for the pathogenic organisms to reach the cells they invade.

ORGANELLES OF ATTACHMENT

Pili

Many Gram-negative bacteria (but few Gram-positive organisms) possess hundreds of hairlike appendages called **pili,** which are composed of protein and are considerably shorter and thinner than flagella (Fig. 2–27). Pili are involved in the transfer of genetic material (DNA) from one bacterium to another when two bacterial cells come into contact. This contact most often occurs through pili holding the cells together (Fig. 2–28); thus, pili serve the same function as the glycocalyx. How the DNA passes from one cell to an-

DNA transfer, pages 136–144

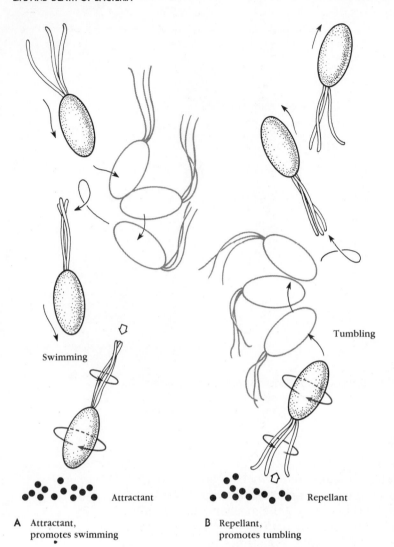

Swimming

Tumbling

A Attractant,
promotes swimming

Attractant

B Repellant,
promotes tumbling

Repellant

Figure 2–25
(a) Cell movement. The cell moves in the direction of the attractant since it promotes swimming. (b) Cells move away from a repellant because it promotes tumbling.

other is still a mystery, even though the system has been studied for many years. Like the glycocalyx, pili also attach some bacteria to other objects such as animal cells or may also serve to keep bacteria near the surface of a liquid if the bacteria use oxygen in their nutrition. Attachment may be very specific, in that certain bacteria will attach only to certain surfaces. Some bacteria, such as *Neisseria gonorrhoeae*, which cause gonorrhea, will only cause disease if they have pili on their surface. *Bordetella pertussis*, the causative agent of whooping cough, adheres selectively to the lining of the respiratory tract. Evidence suggests that in both cases the pili allow the bacteria to attach to the cells they will infect.

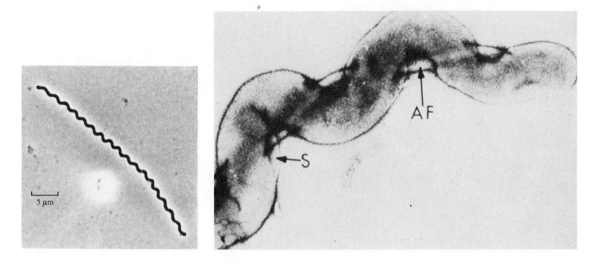

Figure 2–26
(a) A gliding bacterium. (b) Spirochete. The axial filament (AF) is covered by a layer of cell wall material termed the sheath (S).

STORAGE MATERIALS

Many bacteria store high molecular weight compounds in their cytoplasm in the form of **granules.** These granules, which serve as reserve food supplies, can be stained with special stains and are generally large enough to be readily detected by light microscopy after they are stained (Fig. 2–1 and color plates 11 and 12). Some common storage products are glycogen and hydroxybutyric acid, which can be degraded to provide carbon and energy, and volutin (meta-chromatic granules), which can be degraded to provide phosphate. The synthesis of storage materials depends on the environment and the organism. If the environment contains an excess of a nutrient so that not all of it is needed at the moment, the cell will often convert the surplus into large polymers and store it until it is needed.

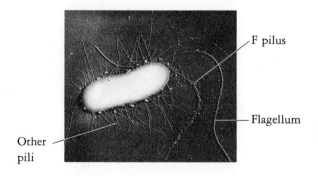

F pilus

Other pili

Flagellum

Figure 2–27
Pili of *Escherichia coli.* Note that there are many different kinds. The ones that are concerned with the transfer of DNA are the long F pili. (× 11,980) (Courtesy of C. Brinton, Jr.)

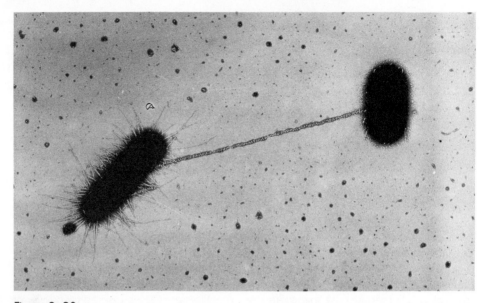

Figure 2–28
An F pilus holding two cells of *Escherichia coli* together. (Courtesy of C. Brinton, Jr., and J. Carnahan.)

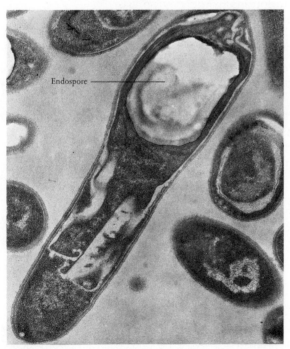

Endospore

A

Figure 2–29
(a) Endospore inside a vegetative cell of *Clostridium*. (Courtesy of L. Santo. L. Santo, H. Hohl, and H. Frank, *J. Bacteriol.*, *99*:824, 1969.) (b) Endospores inside cells of *Clostridium tetani*. Note the bulging of the cells at the end as a result of the endospore.

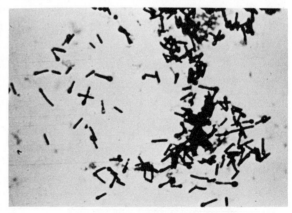

B

ENDOSPORES

A very distinctive structure, the **endospore** (sometimes simply called a spore) is often visible inside certain species of Gram-positive bacilli, members of the genera *Bacillus* and *Clostridium* (Fig. 2–29), or in the medium in which such bacteria are growing. The structure is a unique type of cell that is formed inside the normal bacterial cell termed the **vegetative cell.** The process involves a complex, highly ordered sequence of morphological changes called **sporogenesis** (Fig. 2–30). This process represents an example of cellular differentiation in bacteria.

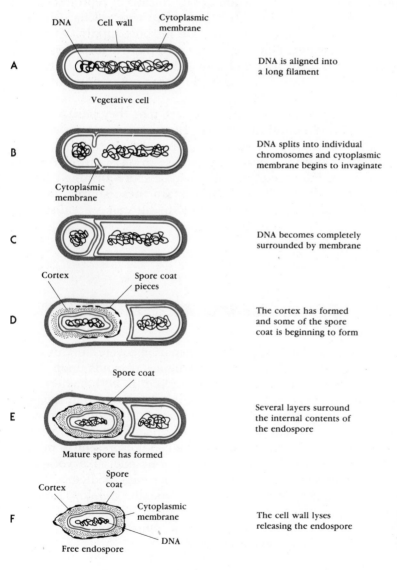

A — DNA / Cell wall / Cytoplasmic membrane / Vegetative cell
DNA is aligned into a long filament

B — Cytoplasmic membrane
DNA splits into individual chromosomes and cytoplasmic membrane begins to invaginate

C — DNA becomes completely surrounded by membrane

D — Cortex / Spore coat pieces
The cortex has formed and some of the spore coat is beginning to form

E — Spore coat / Mature spore has formed
Several layers surround the internal contents of the endospore

F — Cortex / Spore coat / Cytoplasmic membrane / DNA / Free endospore
The cell wall lyses releasing the endospore

Figure 2–30
Schematic representation of sporogenesis.

Spores can remain as endospores for long periods of time (reportedly as long as 150,000 years), but under certain conditions a spore can go through a sequence of events in which it is transformed back into a vegetative cell identical to the one from which it developed.

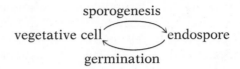

This is the process of **germination.** Since one vegetative cell gives rise to one endospore, sporogenesis and germination are *not* means of reproduction in bacteria. The process of spore formation is a protective mechanism that bacteria use during periods when nutrients are in short supply. When nutrients once again become available, the spores germinate to form vegetative cells that multiply. Thus, cells will multiply as long as conditions are suitable for growth. However, if the bacteria begin to run out of nutrients, they undergo sporogenesis.

The most distinctive features of endospores are that they have virtually no metabolic activity and contain no water. They are usually resistant to treatments that kill vegetative cells, such as heating, drying, freezing, chemical treatment, and irradiation. Because spores are so resistant to common killing agents, and since the vegetative cells of some spore-forming bacteria produce deadly toxins, foods that might contain these spores must be processed with special care. The techniques used to kill spores are discussed in Chapter 7.

SUMMARY

Table 2–1 summarizes the most important information covered in this chapter. The selective activity of antimicrobial medicines depends upon differences in the chemical structure of components present in both bacteria and the host or on the presence of structures found in bacteria but not the host.

The host cells are eukaryotic in structure and are considerably more complex. In particular, they have chromosomes surrounded by a nuclear membrane and possess organelles (such as mitochondria) not present in prokaryotes. The structural differences between prokaryotic and eukaryotic cells provide the basis for the selective action of antimicrobial medicines.

A few bacteria form endospores, a unique cell type that develops from the vegetative bacterial cell by a sequence of morphological changes. This cell type is dormant and extremely resistant to high temperatures and drying conditions, which kill ordinary vegetative cells.

TABLE 2–1   MAJOR STRUCTURES OF PROKARYOTIC CELLS

	Structure	Function	Chemical composition	Distinguishing features	Basis for antimicrobial action	Major antibiotics
Cell envelope	Capsule	Protective	Varies; polysaccharides or polypeptides	Synthesized only under certain conditions by some bacteria	—	—
	Cell wall	Encloses cytoplasm; shapes cell	Mainly peptidoglycan	Structure varies in Gram-positive and Gram-negative cells	Inhibition of synthesis of peptidoglycan	Penicillin
	Cyto-plasmic membrane	Metabolic activity; active transport	Lipoprotein	Unit membrane structure common to all membranes	—	—
	Ribosome	Workbench for protein synthesis	Protein and ribonucleic acid (RNA)	Size varies in eukaryotic and prokaryotic cells	Inhibition of protein synthesis because of differences in structure of subunits	Streptomycin Kanamycin Tetracycline Lincomycin
	Chromo-some	Carrier of genetic information	Deoxyribonucleic acid (DNA)	Circular, double-stranded DNA molecule	—	—
	Flagellum	Motility	Protein	Complex structure, consisting of several parts; propels cell by pushing	—	—

SELF-QUIZ

1. Peptidoglycan
   a. is present in all prokaryotic cells
   b. is present in all bacteria
   c. is broken down by lysozyme
   d. is broken down by penicillin
   e. is a component of the cytoplasmic membrane in some bacteria
2. The cytoplasmic membrane
   a. is very different in eukaryotic and prokaryotic cells
   b. is semipermeable
   c. is always surrounded by a cell wall

     d. is a selective site for attack of bacteria by most antimicrobial medicines

     e. may be lacking in some bacteria

3. Endospores

     a. are formed by prokaryotic and eukaryotic cells

     b. are a means of reproduction in certain bacteria

     c. are formed by most genera of bacteria

     d. are usually Gram-positive

     e. are unusually resistant to heating and drying

4. Capsules

     a. are present in all bacteria

     b. are synthesized only under certain conditions

     c. differ in their composition depending on the medium on which the cells are growing

     d. are important in determining which molecules enter the cell

     e. confer heat resistance on bacteria

5. Ribosomes

     a. are involved in protein synthesis

     b. are identical in eukaryotic and prokaryotic cells

     c. carry most of the genetic information of the cell

     d. occur in the same number in bacteria under all conditions of growth

     e. represent a poor target for antibiotic activity

QUESTIONS FOR DISCUSSION

1. What features might be present in a cell that is intermediate between a eukaryote and a prokaryote?
2. What are the properties of cells that are Gram-positive? Gram-negative?

FURTHER READING

Adler, J.: "The Sensing of Chemicals by Bacteria." *Scientific American* (April 1976).

Berg, H.: "How Bacteria Swim." *Scientific American* (August 1975).

Capaldi, R.: "A Dynamic Model of Cell Membranes." *Scientific American* (March 1974).

Costerton, J., Geesey, G., and Cheng, K.-J.: "How Bacteria Stick." *Scientific American* (January 1978).

Everhart, T., and Hayes, T.: "The Scanning Electron Microscope." *Scientific American* (January 1972).

Fox, C. F.: "The Structure of Cell Membranes." *Scientific American* (February 1972).

Goodenough, U., and Levine, R.: "The Genetic Activity of Mitochondria and Chloroplasts." *Scientific American* (November 1970).

Hare, R.: "Penicillin—Setting the Record Straight." *New Scientist* (February 15, 1979).

Margulis, L.: "Symbiosis and Evolution." *Scientific American* (August 1971).

Rowland, J.: *The Penicillin Man.* New York: Roy Press, 1957. A biography of Sir Alexander Fleming written for students at the high school level.

Satir, P.: "How Cilia Move." *Scientific American* (October 1974).

Sharon, N.: "The Bacterial Cell Wall." *Scientific American* (May 1969).

Chapter 3

MICROBIAL GROWTH IN THE LABORATORY

Finding the cause of Legionnaires' disease is a medical detective story. Shortly after the convention closed in Philadelphia, reports to the Pennsylvania Department of Health revealed that an epidemic had developed among those who had attended the convention. It became apparent very quickly that the disease was not confined only to conventioneers, since 72 cases were reported in people who were not attending the convention. But all people who contracted the disease had one feature in common. They had been either in or near the convention center, the Bellevue Stratford Hotel. The people who got the disease spent more time on the sidewalk in front of the hotel than those who had not experienced any symptoms. Furthermore, Legionnaires who experienced symptoms after returning home did not transmit the disease to members of their families. These observations suggested that the causative agent was airborne but that the disease was not contagious.

The clinical symptoms could have been caused by a variety of agents, both living and nonliving. The nonliving agents included such materials as heavy metals* and toxic organic substances. A microorganism or virus was also possible, but all initial attempts to demonstrate such agents gave negative results. Sections of lung taken from people who had died of the disease were treated with dyes by the usual techniques for revealing bacteria. No bacteria were evident. Furthermore, when lung tissue from infected patients was inoculated onto a variety of media, no bacteria grew. However, the search for a bacterial agent continued on the assumption that perhaps the organism was so unusual that it had special requirements for growth not met by the usual media used to grow other pathogenic bacteria. This proved to be the case. The bacteria causing this disease will

---

*heavy metals—metals such as lead, mercury, and zinc

grow only if the medium contains high concentrations of the amino acid cysteine and iron. Being able to grow the organism was the first step in its study. The numerous questions about this organism will probably be answered in the near future now that the key problem of how to grow it has been solved.

OBJECTIVES

To know
1. The three major physical factors that influence growth of microorganisms and what structures are affected by each of them.
2. How bacteria are classified according to their optimum temperature for growth.
3. How the osmotic pressure of the medium may inhibit growth of bacteria.
4. The classification of bacteria according to their major source of carbon.
5. The classification of bacteria based on their requirement for oxygen.
6. Differences between defined and enriched media.
7. How anaerobic bacteria are grown.
8. The steps in the isolation of bacteria by enrichment techniques.
9. The four stages of bacterial growth.
10. Why bacteria stop increasing in numbers.
11. The difference between open and closed systems of growth.

DEFINITION OF GROWTH

The growth of bacteria is measured by an increase in the number of cells, in contrast with a multicellular organism whose growth is reflected primarily by an increase in the size of a single organism. After a bacterial cell has almost doubled in size and in amount of each of its components, it divides into two daughter cells by the process of **binary fission** (Fig. 3–1). The time required for one cell to divide into two is the **doubling** or **generation time.** In the case of bacteria, there is a tremendous variation in the doubling time, depending on the species and growth conditions. A few organisms have doubling times as short as 10 minutes; the organism causing tuberculosis doubles every 800 minutes (Table 3–1).

FACTORS INFLUENCING MICROBIAL GROWTH

General Considerations

A variety of environmental factors influence the growth of microorganisms. These can be divided into two broad categories: **physical environment** and **nutritional factors.** The physical environment includes temperature, pH, and osmotic pressure. The nutritional environment comprises the material from

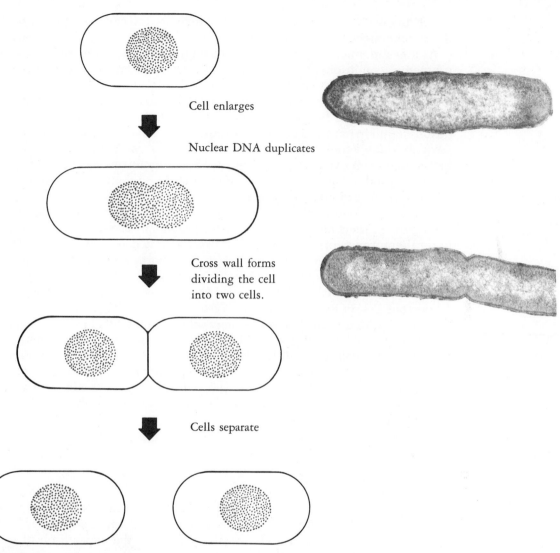

Cell enlarges

Nuclear DNA duplicates

Cross wall forms
dividing the cell
into two cells.

Cells separate

Figure 3–1
Cell division. (Photographs courtesy of A. Progulske and S. C. Holt.)

TABLE 3–1   DOUBLING TIMES OF VARIOUS BACTERIA

Species	Doubling time (min)	Time required for one cell to grow to a visible colony (hr)
Clostridium perfringens	10	8
Beneckea natriegens	10	8
Escherichia coli	20	16
Mycobacterium tuberculosis	800 (>13 hours)	336 (2 weeks)

which the cells derive energy and the nutrients the cells use to synthesize their components.

As a group, microorganisms will grow under an enormous range of environmental conditions. Some microorganisms can adapt to a wide range of conditions, but most species can multiply only under a limited range of environmental conditions and therefore are found only in restricted natural habitats. No single species is able to multiply over the entire range of conditions found in nature. As expected, most species live under those conditions most commonly found on the earth's surface, but microorganisms can often be isolated from extreme environments. Such organisms have generally become adapted to their particular environment and do not grow well if cultivated under ordinary conditions.

Laboratory conditions are usually quite different from those under which cells grow in their natural habitat. Once organisms are removed from their natural environment, they must adapt to the new conditions of the laboratory. Once organisms adapt to growth in the laboratory, they may be unable to survive if they are returned to their original habitat. For example, *Brucella abortus*, the bacterium responsible for brucellosis in humans, requires 5 to 10 percent CO_2 in its environment to grow when it is initially isolated from a diseased patient. After a few transfers on medium while in the presence of CO_2, however, this requirement is lost. Another example is the famous strain of *Escherichia coli*, K-12. This strain, originally isolated from a human intestinal tract and now grown for years in the laboratory, can no longer grow in the intestinal tract. Apparently, the strain lost the genetic information necessary to synthesize materials needed for its growth in what was once its natural habitat.

Physical Factors

Temperature and enzyme activity, page 90

Temperature. Most bacteria grow over a temperature range of approximately 30°C, with each species having a well-defined upper and lower limit. As a general rule, bacteria grow most rapidly near the upper limit of their range. Above the upper limit there is a very sharp drop in the speed with which bacteria grow (Fig. 3–2). Growth slows as the temperature approaches the lower limit. These differences in growth rate reflect the effects of temperature on the enzymes of the cell. The rise in temperature increases enzyme activity, making the cells grow faster. If the temperature rises too high, enzymes critical to the life of the cell are denatured,* and the cells grow more slowly or die.

Protein denaturation, page 90

Bacteria are customarily divided into three groups according to their optimal growth temperature—the temperature at which they have the *shortest* generation time (Fig. 3–3). **Psychrophiles** grow best between −5°C and 20°C.

---

*denatured—destroyed by disrupting the three-dimensional structure of the protein molecule

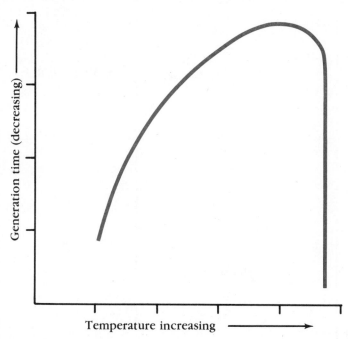

Figure 3–2
Effect of temperature on the generation time. A decreasing generation time means that
the bacteria are multiplying more rapidly.

Some actually die when placed at room temperature. **Mesophiles,** which
constitute the majority of bacteria, grow best within the range of 20°C to
about 50°C. Most disease-causing bacteria adapted to growth in the human
body have their optimal temperatures between 35°C and 40°C. Therefore,
when specimens are taken from the human body to demonstrate the presence
of bacteria, the inoculated nutrients are generally incubated* at 37°C. However, samples taken from the surface of the skin, which has a temperature
close to 30°C, are incubated at 30°C. Bacteria found in the soil also generally
have a temperature optimum close to 30°C. The **thermophiles,** the third
group, have their optimum temperature between 50°C and 60°C. Thermophilic bacteria isolated from some hot springs can actually grow at temperatures above 90°C, only 10° below boiling! As a general rule, protein molecules from thermophiles differ from those from mesophiles in not being
denatured at high temperatures.

Although it is convenient to classify organisms into these three groups
according to their growth temperatures, in actual practice there is no sharp
dividing line that separates the temperatures at which organisms in each

---

*incubated—allowed to grow

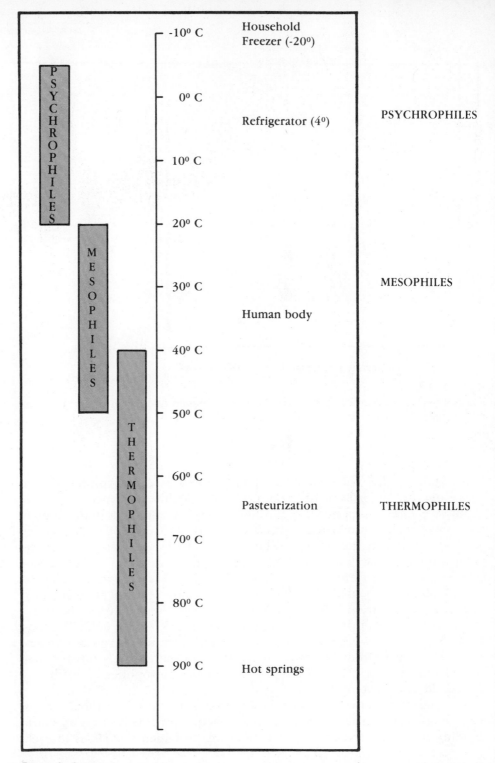

Figure 3–3
Temperature growth range of microorganisms. Organisms in each category cannot grow over the entire range for the group.

group will grow (Fig. 3–3). Furthermore, not every organism can grow over the entire range indicated for its group. The upper limit of growth is partly related to the acidity of the environment. Thus, in hot springs the upper temperature at which bacteria are found decreases as the springs become more acid. In highly acid water, the upper limit of growth is 80°C; in neutral* or alkaline springs, bacteria are found at temperatures above 90°C.

Since the microorganisms that cause food spoilage are in most cases mesophiles, storage of food at low temperatures, approximately 10 to 15°C, is a commonly used method of preserving food. Fruits, vegetables, and cheeses are stored at these low temperatures to retard their spoilage. Refrigerators, which maintain temperatures around 4°C, have become a major tool in preserving foods. However, it is important to recognize that organisms can multiply at refrigerator temperatures. This can cause problems in storage of food and other materials, such as blood and medicines, which are nutritious enough to support microbial growth. Freezing has often been employed as a means for long-term storage of those items that can withstand freezing. Growth of all organisms stops a few degrees below freezing, although the organisms are not necessarily killed. However, the freezing temperature of a liquid is determined by the concentration of dissolved materials in the liquid, so in fact freezing temperatures (like boiling temperatures) can vary.

pH. The pH is a measure of the degree of acidity or alkalinity of a solution (Fig. 3–4). Most bacteria grow best at the neutral pH of 7, although they can tolerate ranges from pH 5 (acidic) to pH 8 (basic). Most general bacteriological media have a pH of 6.8 to 7.4. Yeasts and molds grow best in an acid medium, and most media for the growth of fungi have a pH below 6.0. Some bacteria can tolerate very high acid concentrations, such as those that form vinegar, which is about 5 percent acetic acid. Members of the genus *Thiobacillus*, which produce sulfuric acid as a result of metabolic processes, can grow at pH values of 1 and even lower. The pH inside the cell must be considerably higher than this, and the ability of this organism to grow depends on its ability to keep the acid (H^+) ions out of the cell.

Many bacteria produce enough acid as a byproduct of the breakdown of foodstuffs to inhibit their own growth. To prevent the accumulation of acid (H^+) or alkali (OH^-) ions, chemicals called **buffers** are added to "soak up" these ions. Buffers are often supplied to the cell as a mixture of two salts of phosphoric acid, the sodium phosphates Na_2HPO_4 and NaH_2PO_4. These salts resist pH changes because they have the ability to combine chemically with the H^+ ions of acids and the OH^- ions of alkaline solutions according to the following equations:

$$H^+ + HPO_4^{2-} \longrightarrow H_2PO_4^- \tag{1}$$
$$OH^- + H_2PO_4^- \longrightarrow HPO_4^{2-} + HOH \tag{2}$$

---

*neutral—neither acid nor alkaline

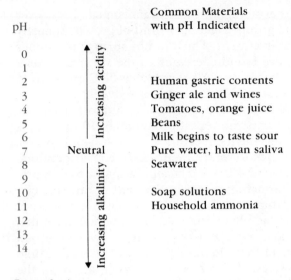

pH

Common Materials
with pH Indicated

pH		
0		
1		
2		Human gastric contents
3		Ginger ale and wines
4		Tomatoes, orange juice
5		Beans
6		Milk begins to taste sour
7	Neutral	Pure water, human saliva
8		Seawater
9		
10		Soap solutions
11		Household ammonia
12		
13		
14		

Figure 3–4
The pH scale is defined as the negative logarithm of the H^+ ion concentration. Note that there is a tenfold variation in the H^+ ion concentration between any two successive figures in the pH scale.

Many other compounds can serve as buffers. They all have the ability to combine with H^+ (acid) or OH^- (alkali) ions and thus change the pH to a neutral value. Many media have natural buffers already in them, and it is not necessary to add additional buffers. For example, the amino acids added to certain media as nutrients can also serve as buffers.

Osmotic Pressure. The **osmotic pressure** of a medium depends on the concentration of dissolved substances in solution. Such a substance commonly encountered in nature is sodium chloride (common table salt). Most bacteria can grow over a broad range of salinity* because the cell can maintain a relatively constant internal salt concentration through the activity of specific

Permeases, pages 37–38

permeases. If the salt concentration on the outside of a cell becomes too high, however, water is lost from the cell, the cytoplasmic membrane with the cytoplasm shrinks away from the cell wall, and cell growth is inhibited (Fig. 3–5). This mechanism is the basis of food preservation by salting, used in such foods as beef jerky, bacon, salt pork, and anchovies.

Besides the preservative effect of salts, many foods such as jams, jellies, honey, preserves, and sweetened condensed milk have a high sugar content,

*salinity—amount of salt

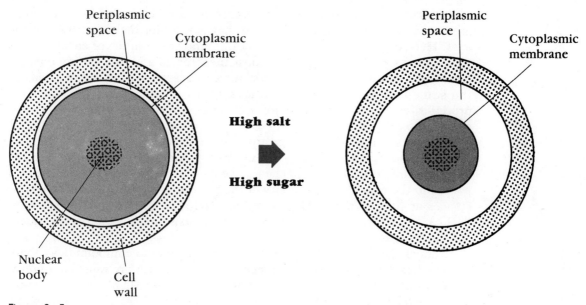

Periplasmic space

Cytoplasmic membrane

High salt

High sugar

Periplasmic space

Cytoplasmic membrane

Nuclear body

Cell wall

Figure 3–5
Effect of increasing the salt concentration on the outside of a cell. Water leaves the cell, and the elastic cytoplasmic membrane shrinks.

which leads to the same result. The process of pickling combines salt with a very acid environment and is quite effective in controlling the growth of most microorganisms.

However, some organisms, called **halophiles**, actually require a high salt concentration for growth. These organisms, which are members of the Archaebacteria, lack peptidoglycan in their cell walls. The high salt concentration prevents the weak cell walls from bursting. Some of these organisms may grow only in environments in which the salt concentration approaches saturation (above 30 percent). Halophiles isolated from the Dead Sea (salt concentration of 29 percent) have a very high internal salt concentration, which is required for the functioning of many of their enzymes as well as their ribosomes. Many bacteria isolated from the ocean require the salt concentration found in their natural environment (about 3.5 percent) in order to multiply. If the concentration varies much from this level, they may die.

Archaebacteria, page 15

Nutritional Factors

For an organism to grow, it must have a suitable physical environment, and foodstuffs must be available that can serve as raw materials for the synthesis of cell components and as a source of energy. Water is a necessity for all cells.

Spectrum of Nutritional Types. Organisms can be divided into two large groups based on the source of carbon that is used for making cell constituents. **Heterotrophs** utilize organic compounds as sources of carbon from which they synthesize their own constituent parts and also obtain energy. A variety of organic compounds* can serve as energy sources. In many cases, the same compounds that serve as sources of carbon for biosynthetic processes can also serve as sources of energy. **Autotrophic** organisms use carbon dioxide as the major source of carbon for the synthesis of their cell components. Generally, this group gains its energy either from the sun (through photosynthesis) or by metabolizing inorganic compounds. Several different inorganic compounds can serve as sources of energy, and this fact points up the nutritional diversity of the microbial world. Such simple substances as hydrogen gas (H_2), ammonia (NH_3), reduced iron (Fe^{2+}), nitrite (NO_2^-), and hydrogen sulfide (H_2S) can serve as sources of energy for different microorganisms.

Although heterotrophs can use a variety of organic compounds as their source of carbon for biosynthetic needs, these bacteria are most commonly grown in the laboratory on media containing glucose. Members of the genus *Pseudomonas* are amazingly versatile. Certain members of this genus are able to degrade more than 90 different organic compounds. Most organic molecules, regardless of complexity, can be degraded by at least one species of microorganism. This versatility does not hold for any single species of microorganism but rather for the microbial world as a whole. Any single species can degrade only a limited number of organic compounds.

However, recent exciting developments are beginning to change this story. Scientists are learning to create organisms with increased degradation capabilities that do not exist in nature. For example, in nature there are organisms that can degrade one component but not other components of crude oil. However, the genes that produce proteins responsible for degrading each of these components can be introduced into a single bacterium. This bacterium now can break down all the components of oil (Fig. 3–6). The single organism represents the sum of the selected abilities of the various individual organisms. Although most biologically generated compounds in the environment can be broken down by groups of organisms working together, a large number of man-made products cannot be degraded. Perhaps the genetic engineering techniques used in creating organisms with new degradative capabilities for oil will help in developing organisms with the ability to degrade compounds that are nondegradable at the present time.

Mineral Requirements. In addition to a source of carbon, all organisms require sources of each of the other elements found in cell constituents—primarily nitrogen, sulfur, and phosphorus. Carbon, hydrogen, oxygen, nitrogen, phosphorus, and sulfur make up 99 percent of the dry weight of the cell and therefore must be available in foodstuffs in the largest amounts.

---

*organic compounds—any carbon-containing compounds except CO_2 and CO

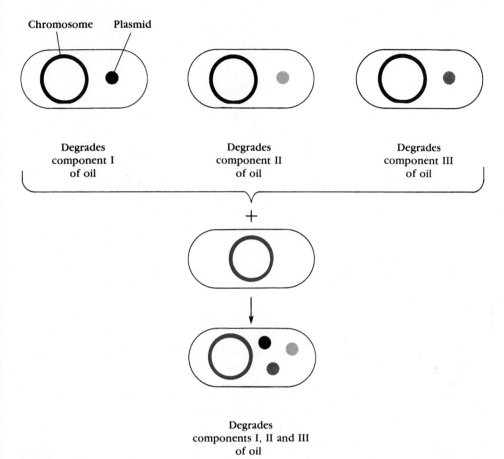

Figure 3–6
The development of an organism that has increased abilities to degrade many components of oil.

Except for carbon, most microorganisms—both heterotrophs and autotrophs—can utilize inorganic salts as a source of each of these elements.

Microorganisms obtain nitrogen in a variety of forms, which they can incorporate into amino acids, purines, and pyrimidines.* Most organisms cannot utilize the nitrogen gas in the atmosphere and require other nitrogen-containing compounds, such as ammonium chloride (NH_4Cl) and sodium nitrate ($NaNO_3$). Bacteria can degrade nitrogen-containing compounds such as proteins and nucleic acids and release ammonia (NH_3). The ammonia can then be recycled and used for the synthesis of the same molecules in the bacteria.

In the laboratory, the element sulfur is often provided in the medium as a sulfate salt, such as ammonium sulfate $(NH_4)_2SO_4$, which is then used to

Amino acids, purines, and pyrimidines, Appendix I

*purines and pyrimidines—organic bases that serve as the building blocks of DNA and RNA molecules

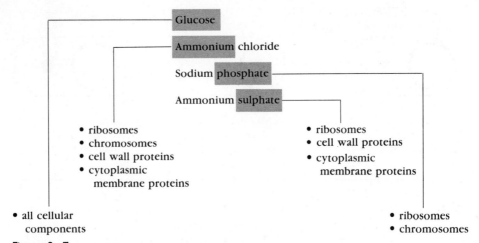

Figure 3-7
The fate of the small molecules on which many bacteria can grow.

Coenzymes, pages 86–88

make sulfur-containing amino acids, coenzymes, and other cellular constituents. Bacteria can also degrade sulfur-containing organic matter with the release of hydrogen sulfide (H_2S). This compound is then used for the synthesis of sulfur-containing compounds in the bacteria.

ATP, page 91

Phosphorus is present in nucleic acids and phospholipids, as well as in the key molecule of energy metabolism, adenosine triphosphate (ATP). To meet the phosphorus requirements, salts of phosphoric acid, such as the sodium phosphates Na_2HPO_4 and NaH_2PO_4, are generally included in laboratory media. These compounds thus can serve two functions, as a buffer and as a source of phosphate. The cellular components into which these inorganic compounds and the carbon source are converted are shown in Figure 3–7.

Oxygen and metabolism, pages 96–97

Oxygen Requirements. Oxygen gas is an essential nutrient for many bacteria, but for others it is lethal. Between these two extremes are bacteria that require oxygen, but at concentrations of about 5 percent, which is lower than that found in the atmosphere (20 percent). Thus, there exists an entire range of oxygen requirements, and bacteria are often classified according to these requirements.

1. **Obligate** *(strict)* **aerobes.** These organisms have an absolute requirement for oxygen gas. They grow best if they are grown in flasks that are continuously shaken to mix oxygen gas with the medium and thus make the oxygen continuously available.
2. **Obligate** *(strict)* **anaerobes.** These organisms cannot utilize oxygen. Some members of this group are actually killed by traces of oxygen, whereas others are inhibited in their growth.
3. **Facultative anaerobes.** These organisms can utilize oxygen gas if it is

present but can grow in its absence. Growth is generally but not always more rapid with oxygen.

4. **Microaerophilic organisms.** These organisms require oxygen gas in concentrations of 2 to 10 percent. Some of these organisms also require an increased carbon dioxide concentration, which is conveniently provided by growing the organisms in a closed jar that contains a burning candle. The burning candle uses up oxygen and releases CO_2.

5. **Indifferent** or **aerotolerant microorganisms.** The growth of these organisms is neither aided nor inhibited by oxygen. A major group of pathogenic organisms falls into this classification.

The response of organisms to oxygen is illustrated by the growth positions of different organisms in "shake tubes" (Fig. 3–8). The bacteria, which are shaken and dispersed throughout the entire tube before the nutrient agar solidifies, are trapped in place by the solidification of the agar. The top of the tube is highly aerobic and the bottom is anaerobic, with gradations of oxygen between these extremes.

The different responses of bacteria to oxygen result from two considerations: whether the metabolic pathways that the cell possesses for degrading foodstuffs require oxygen gas and whether the bacteria have the enzymes to dispose of the highly toxic compounds produced from oxygen by cell metabolism. These considerations will be discussed further in Chapter 4.

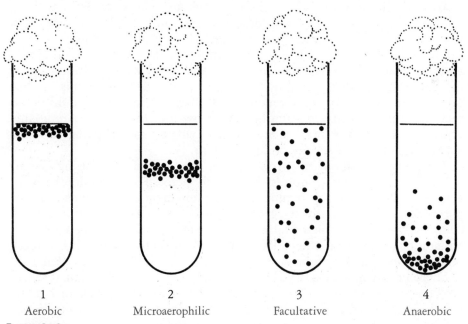

1	2	3	4
Aerobic	Microaerophilic	Facultative	Anaerobic

Figure 3–8
Shake tubes. The bacteria are trapped throughout the agar but grow only where the oxygen (air) concentration is suitable for the particular organism.

Growth Factors. Despite their similarity in chemical composition, microorganisms display a wide spectrum of growth requirements. The differing growth requirements reflect the different enzymatic capabilities of various microorganisms. The more enzymes the cell has, the fewer nutrients must be added. At one extreme are obligate intracellular bacteria, such as the members of the genera *Rickettsia* and *Chlamydia*, which lack some of the enzymes or structures necessary for survival in the absence of living cells. At the other extreme are the autotrophs, such as the photosynthetic bacteria, which have enzymes that allow them to multiply using inorganic salts and sunlight. Many heterotrophs, although able to live in the absence of living cells, do not have all the enzymes present in autotrophs and therefore require certain organic compounds that they cannot synthesize themselves. However, all three groups have many enzymes in common. Studies on a wide variety of organisms have shown that most organisms can carry out a core of enzymatic reactions. Many heterotrophs (and a few autotrophs) do not have all the enzymes necessary for synthesizing some of the small molecules required for growth. These small organic molecules, called **growth factors** (such as amino acids, vitamins, purines, and pyrimidines), must therefore be provided in the medium. The more enzymes an organism lacks for the biosynthesis of small molecules, the more factors it requires for growth. The lactic acid bacteria must be supplied with about 95 percent of the subunits that comprise their cell material. Growth requirements have proved useful in identifying some bacterial pathogens in hospital laboratories.

Microbiological Assay. The technique whereby microorganisms are used to detect the presence of any material is termed **microbiological assay.** The requirements of bacteria for growth factors were used to isolate several vitamins long before anything was known about these vitamins' chemical structure and metabolic function in humans. Several such vitamins were first recognized as unknown growth factors for certain microorganisms; their presence in extremely small amounts (10^{-12} gram) in media could be detected by their ability to stimulate the growth of these microorganisms. Thus, vitamin B_{12} was first isolated by purifying a substance that served as a growth factor for a species of *Lactobacillus*.

The exact quantity of any growth factor can be assayed by using a strain of microorganism that requires the growth factor. For example, the lactic acid bacteria as a group require a large number of amino acids, purines, pyrimidines, and vitamins for growth. If any one of these growth factors is completely left out of the medium, the bacteria will not grow. If an unknown amount of the factor is then added, the final level of cell growth is directly related to the amount of the growth factor present in the medium (Fig. 3–9).

Microbiological assays have been used in the diagnosis of certain diseases, such as phenylketonuria. This is a serious congenital* disease of humans who lack an enzyme that converts phenylalanine to tyrosine. Thus, in persons with phenylketonuria the concentration of phenylalanine increases dra-

*congenital—existing at birth

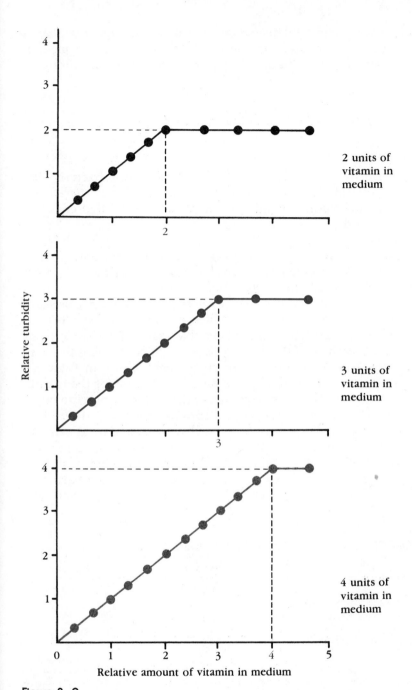

2 units of
vitamin in
medium

Relative turbidity

3 units of
vitamin in
medium

4 units of
vitamin in
medium

Relative amount of vitamin in medium

Figure 3–9
Microbiological assay. The turbidity of the medium is a measure of cell growth. The growth
of the cells, which require the vitamin, is proportional to the amount of vitamin in the
medium. Thus, it is possible to determine how much of the vitamin is present in any
solution by how much the cells grow in the medium.

matically. Its presence can be detected easily if bacteria requiring phenylalanine grow when the blood of such an individual is added to a medium that is lacking phenylalanine. If untreated, phenylketonuria can cause mental retardation.

CULTIVATION OF BACTERIA IN THE LABORATORY

Basic Media

Microbiologists use two basic types of media for cultivating bacteria. One contains known amounts of pure chemical compounds; it is therefore a chemically **defined** or **synthetic medium.** The other type, called **enriched,** contains complex and poorly defined materials, such as digests of milk, ground meat, or plants.

The composition of a synthetic medium commonly used to grow the bacterium *Escherichia coli* is given in Table 3–2. This species can obviously synthesize all its cellular components from glucose and a few inorganic* salts. Glucose also serves as the species' energy source. Such a glucose-salts medium can serve as a basic medium to which various growth factors can be added to support the growth of organisms that cannot synthesize them. For example, some members of the genus *Neisseria* (which includes the causative agents of gonorrhea) will grow if the supplements listed in Table 3–3 are added to a glucose-salts medium. The function of many of the organic supplements is unknown. Other bacteria may have different nutritional requirements and will not grow even in this enriched medium. It often becomes necessary to add undefined nutrients, such as yeast and meat extracts, to permit growth of some organisms. As a group, the pathogenic organisms are particularly exacting in their nutritional requirements. Their fastidious nature probably is a result of their having become adapted to a particular environment—the animal body—which contains a variety of "rich" nu-

TABLE 3–2 COMPOSITION OF SYNTHETIC MEDIUM FOR GROWTH OF *ESCHERICHIA COLI*

Ingredient	Source of	Amount
Glucose[a]	carbon	5 gm
Dipotassium phosphate (K_2HPO_4)	phosphate and	7 gm
Monopotassium phosphate (KH_2PO_4)	buffer	2 gm
Magnesium sulfate ($MgSO_4$)[b]	magnesium	0.08 gm
Ammonium sulfate ($(NH_4)_2SO_4$)	nitrogen and sulfur	1.0 gm
Water	—	1000 ml[c]

[a]The glucose is generally sterilized separately from the salts because heating the glucose with the salts converts glucose to materials that may be toxic to cell growth.
[b]Magnesium is required for the action of many enzymes.
[c]For the preparation of solid medium, 1.5 percent agar (final concentration) is generally added.

*inorganic—not containing carbon

TABLE 3–3 COMPOSITION OF SYNTHETIC MEDIUM FOR GROWTH OF *NEISSERIA*

Inorganic components	Organic components
Sodium chloride (NaCl)	Glucose
Potassium sulfate (K_2SO_4)	Uracil
Magnesium chloride ($MgCl_2$)	Hypoxanthine
Ammonium chloride (NH_4Cl)	Spermine
Potassium monohydrogen phosphate (K_2HPO_4)	Hemin
Potassium dihydrogen phosphate (KH_2PO_4)	Nitrilotriethanol
Calcium chloride ($CaCl_2$)	Polyvinyl alcohol
Ferric nitrate [$Fe(NO_3)_3$]	Sodium lactate
	Glycerine
	Oxaloacetate
	Sodium acetate
	Amino acids:
	20 different
	Vitamins:
	7 different

Adapted from Catlin, B. W.: *Journal of Infectious Disease, 128*:178, 1973.

trients. It is more economical for them to use the available nutrients than to synthesize their own.

It is generally much more convenient to use a single enriched medium that meets the growth requirements of many species than to use many different defined media to grow the same species. This is especially true in laboratories in which a variety of different organisms are grown. For this reason, hospital diagnostic microbiology laboratories rely almost solely on undefined media.

The composition of a typical enriched broth (nutrient broth) is shown in Table 3–4. The peptones in the medium are breakdown products of plant, fish, and meat protein and consist primarily of peptides* of various sizes. Beef extract is a water extract of lean meat and provides vitamins, minerals, and other nutrients. To make the medium richer, yeast extract may be added. Yeast extract, the broth of autolyzed* yeast cells, serves as an excellent source of vitamins.

TABLE 3–4 COMPOSITION OF ENRICHED UNDEFINED MEDIUM: NUTRIENT BROTH

Ingredient	Amount
Peptone (0.5%)	5.0 gm
Meat extract	3.0 gm
Distilled water	1000 ml

*peptides—short chains of amino acids

*autolyzed—broken up or destroyed by means of a cell's own enzymes

It is not possible to prepare a single enriched medium that would support the growth of all microorganisms. Indeed, materials in enriched media are toxic to various bacteria. Certain pathogenic microorganisms seem to be especially sensitive, and in such cases other material must be added to neutralize the toxins. For example, some organisms, notably some mycoplasmas, grow only in media containing a high concentration of serum (20 percent). One of the major functions of serum is to supply serum albumin, which binds small molecules such as fatty acids, metal ions, and detergents, all of which are toxic to many cells.

Members of the genus *Haemophilus* and certain other bacteria that are difficult to grow are generally grown on chocolate agar, a medium enriched with blood and then heated sufficiently to release the hemin* from the denatured hemoglobin of the blood (giving a chocolate-brown color to the medium). The heating also serves to denature an enzyme, present in blood, that destroys a growth factor required by many bacteria. The manufacture of hundreds of different types of media is the primary concern of several companies. Each of these media has been specially formulated to permit luxuriant growth of one or several groups of organisms.

Even with the availability of these enriched media, however, some microorganisms still have never been cultivated in the laboratory in the absence of living cells. These include the rickettsiae and the spirochete causing syphilis, *Treponema pallidum.*

Cultivation Methods for Anaerobes

One large group of organisms that present special problems for cultivation is the anaerobes. Special techniques are required for cultivating anaerobes, since their growth is inhibited to varying degrees by free oxygen gas. Two general methods are available. One involves incubating cultures of these bacteria in jars from which oxygen has been removed. One common type of anaerobic jar contains a disposable hydrogen generator and a catalyst,* which combines the atmospheric oxygen in the jar with the hydrogen to form water. Thus, no oxygen gas is available to inhibit growth (Fig. 3–10).

Another technique is to add to the media various chemicals that reduce the level of oxygen by combining with it chemically. These compounds include sodium thioglycollate, cysteine, and ascorbic acid. The medium is often boiled first to drive out air before inoculation. The surface can then be covered with sterile petroleum jelly (Vaseline) and paraffin to limit the reentry of oxygen.

The cultivation of anaerobic organisms presents a great challenge to the microbiologist. It is often difficult to provide strict anaerobes with suitable conditions quickly enough to keep them alive once they have been removed

---

*hemin—the crystalline salt derived from blood by heating

*catalyst—a substance that speeds up the rate of a chemical reaction without being altered or depleted in the process

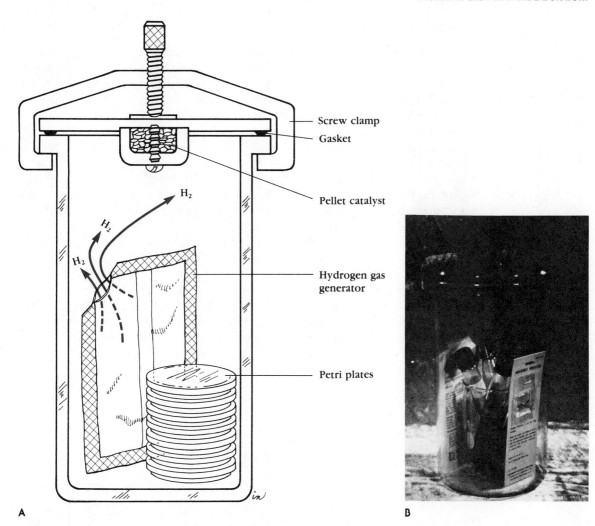

Screw clamp

Gasket

Pellet catalyst

H_2

H_2

H_2

Hydrogen gas generator

Petri plates

A   B

Figure 3–10
Anaerobic jar. The hydrogen released combines with any oxygen to form water. (Photograph courtesy of Leon J. Le Beau, Ph.D., University of Illinois Medical Center, Chicago.)

from their natural source. Unquestionably, as techniques for culturing anaerobes are improved, new and previously unsuspected species will be isolated from humans and other natural habitats.

From this discussion of the growth requirements of both aerobes and anaerobes, it should be clear that the inability to demonstrate the growth of an organism does not guarantee that the organism is not there. It may merely mean that the proper nutritional and environmental conditions required for its laboratory cultivation have not been met. The length of time required to demonstrate the causative agent of Legionnaires' disease, *Legionella pneumophila*, provides a good example.

Enrichment Cultures

An **enrichment culture** enhances the growth of one particular organism in a population of different bacteria. This method can be used to isolate in pure culture an organism from natural sources. The technique is based on the principle that the best-adapted organism in any environment will multiply most rapidly and become the dominant organism. The first consideration in setting up an enrichment culture is selecting the sample from which the organism is to be isolated. It should be selected from an area in nature in which the organism is most likely to occur. The sample is next inoculated into a medium containing nutrients that favor the growth of the desired organism. By such culture methods, a skilled microbiologist can sometimes manipulate the growth of certain species so dramatically that a single species will predominate. It is then relatively simple to inoculate a portion of the culture onto a solid medium and to select a single colony of the predominant species, which can be transferred to another medium to obtain a pure culture (Fig. 3–11).

The isolation of an organism capable of degrading crude oil in sea water illustrates the use of enrichment culture techniques. Crude oil was added to unsterilized sea water, to which phosphate and ammonium sulfate were then added to provide phosphate, nitrogen, and sulfur. After incubation for a week, the oil became evenly dispersed. Bacteria were using the oil as their

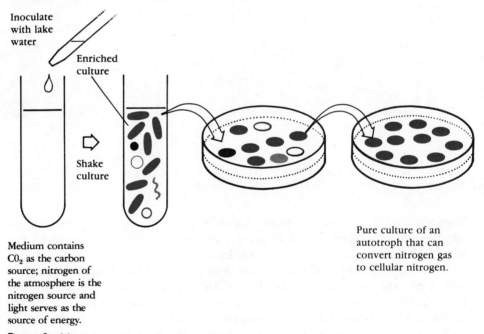

Inoculate with lake water

Enriched culture

Shake culture

Medium contains CO_2 as the carbon source; nitrogen of the atmosphere is the nitrogen source and light serves as the source of energy.

Pure culture of an autotroph that can convert nitrogen gas to cellular nitrogen.

Figure 3–11

Schematic diagram for promoting the growth of a particular organism by manipulating the environmental conditions under which the bacteria are grown.

sole source of added carbon and energy. When this culture was then inoculated onto agar containing oil, different types of colonies arose. Some of these bacteria may be useful in preventing the major cause of oil pollution—the discharge of sea water that serves as ballast in oil tankers. This sea water is contaminated with about 1 percent of the oil left over from the ship's last oil cargo. Oil tankers dump about 1 million tons of oil into the oceans annually from this source alone.

There is a problem in selecting a very fastidious organism when most organisms in the environment are less demanding in their growth requirements. One solution to the problem involves using **selective inhibitors** in media to suppress the growth of some organisms in the population if the organism being sought is resistant to the inhibitor. For example, one diagnostic procedure for identifying typhoid fever involves isolating the causative organism, *Salmonella typhi*, from feces, a habitat containing a large number of organisms related to *Salmonella*. The fecal specimens are first incubated in selenite or tetrathionate broth. These compounds inhibit the growth of the flora* normally found in the feces far more than they inhibit multiplication of *Salmonella*. Another example is the use of medium containing sodium azide. This compound selectively inhibits the growth of Gram-negative bacteria while allowing Gram-positive organisms to grow. Many other examples involving other pathogens will be considered later.

Once an organism becomes the dominant species, it may be tentatively identified by its pattern of growth on **indicator** or **differential agar medium.** These media contain a component that is changed in a recognizable and unique way. Certain dyes added to media will change color if a medium becomes acidic. For example, on MacConkey's agar, lactose-fermenting* colonies turn pink. In cases of suspected typhoid fever, clinical microbiologists often plate the organisms on bismuth sulfite agar. *Salmonella typhi* form hydrogen sulfide, which turns the colonies black. Blood agar* is another common indicator medium used in clinical microbiology laboratories. Some bacteria produce enzymes called hemolysins, which break down blood. Colonies of the organism causing "strep" throat are readily detected on blood agar medium by their ability to destroy red blood cells, thereby producing a clear zone around the colony (beta hemolysis). A wide variety of other materials can serve as diagnostic aids when incorporated into media.

Beta hemolysis, color plate 20

DYNAMICS OF POPULATION GROWTH IN THE LABORATORY

The increase in cell number in a growing bacterial population is **logarithmic** or **exponential,** because each cell gives rise to two cells, each of which in turn gives rise to two cells—a total of four—and so on (Fig. 3–12). When the

---

*flora—the living organisms found in a particular environment

*lactose-fermenting—capable of degrading the sugar lactose with the resulting production of acid

*blood agar—agar medium containing blood

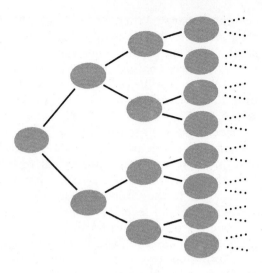

Arithmetic number	1	2	4	8 - - -	16 - - -	32 - - -

Logarithm to base 2	2^0	2^1	2^2	2^3 - - -	2^4 - - -	2^5 - - -

Figure 3–12
Logarithmic growth of bacteria.

number of cells in a population is graphed against the time of incubation, a straight line is generated if the ordinate (y-, or vertical, axis) representing number of cells is a logarithmic scale and the abscissa (x-, or horizontal, axis) representing time of incubation is an arithmetic scale. The straight line indicates that there is an identical percentage of increase in the number of viable* cells during any constant time interval (Fig. 3–13). Thus, if the number of cells increases from one to four in one hour (a generation time of 30 minutes), there will be another fourfold increase in the population after an additional hour and yet another fourfold increase after each succeeding hour. Note, however, that the increase in the number of cells varies within the three intervals. Thus, in the early stages of growth, 100 cells will increase to 400 in an hour, but in a late stage 100 million (10^8) cells will increase to 400 million (4×10^8)! Since the number of bacteria are usually important, the growth of bacteria in the later stages of exponential growth is usually the most important. The total number of cells present at any time depends on

---

*viable—capable of living

Cell number

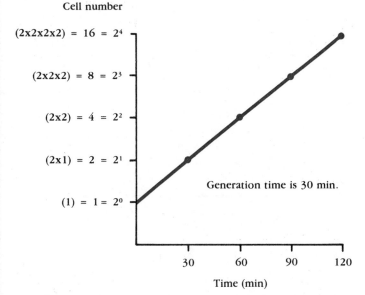

Figure 3–13
The plotting of logarithmic growth on a semi-log graph.

the number of cells present initially and the number of cell generations. Since many bacteria can divide every 30 minutes (and a few species even divide every 10 minutes), the number of these cells in a population reaches high levels in a relatively short time. This is why it often takes only a few days for a person to become ill after infection with a small number of pathogenic bacteria.

Phases of Bacterial Growth—A Closed System

In the laboratory, bacteria are grown in a closed system. The amount of nutrients available to the organism is fixed in the medium in which the bacteria are growing. The products of metabolism accumulate in the medium and are not removed. In such a closed system, there are four well-defined phases of growth (Fig. 3–14): (1) the **lag phase,** in which there is no increase in the number of viable cells; (2) the **log phase,** in which the cell population increases logarithmically or exponentially with time; (3) the **stationary phase,** in which the total number of living cells remains constant; and (4) the **death phase,** which is characterized by an exponential decrease in the number of viable cells.

Lag Phase. Usually, when bacteria are placed into fresh medium, the number of viable cells does not increase immediately. The cells must first

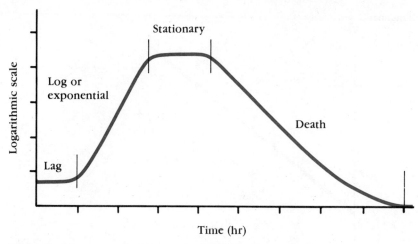

Figure 3–14
The phases of bacterial growth plotted on semi-log graph paper.

prepare for growth by "tooling up." During this "tooling up" stage, the cell is synthesizing and accumulating molecules it will require for cell division. Especially important are molecules of ATP, which represent the supply of energy; ribosomes, required for protein synthesis; and any enzymes that may be required for growth.

Log Phase. During the log phase, the cells divide at their fastest rate. The increase in cell number does not occur in discrete jumps but rather gradually with time, thus indicating that individual cells are at different stages in their division cycle. Some cells have just divided, others have enlarged to the size at which they are about ready to divide, and still others are at intermediate stages in the division cycle. Thus, cells in the population do not all divide at the same time.

For cells in the logarithmic phase of growth, the following simple growth equation expresses the relationship between the original number of cells (zero time), the number of cells in a population at a later time, and the number of divisions (or generations) the cells have undergone. If any two values are known, the third can easily be calculated from the equation

$$b = B2^n$$

where B = number of cells at zero time, b = number of cells at any given later time, and n = number of cell generations.

Two features characterize the exponential growth phase of any culture: the growth rate and the duration of exponential growth. The faster cells multiply, the faster the growth rate, and, in terms of the graph, the steeper the line (greater the slope of the line) representing exponential growth.

The growth rate is influenced by all the environmental conditions previously mentioned in this chapter. Obviously, the optimal conditions for growth will vary for each particular organism. One important generalization that can be made is that the growth rate of a bacterium is proportional to its rate of energy metabolism. All the environmental and nutritional factors that influence the growth rate do so primarily by influencing the rate of ATP production by the cell. During the log phase, bacteria are most susceptible to antimicrobial medication and other agents that inhibit their growth or kill them.

The size that a cell population reaches before it stops growing is controlled primarily by the environment but also depends on the genetic makeup of the organism. The generation time of many bacteria is about 30 minutes and is as short as 10 minutes in a few cases. With such organisms, starting with a single cell, the population will reach 10^9 cells in five to ten hours. Since the volume of an average bacterial cell is 1 μm^3, the total cell volume (size of each cell times the number of cells) of a culture that continued growing exponentially for 45 hours would be greater than the volume of the earth! Fortunately, bacterial cultures stop growing once they have reached a population of approximately 10^9 cells per milliliter.

Generally, cells stop growing for one of two reasons: either they exhaust nutrients or they accumulate toxic products. Bacteria can exhaust a required nutrient even when growing in an enriched medium. The material most commonly exhausted by aerobic organisms is oxygen. This gas is not very soluble in water, and once the population reaches a high cell number, there are enough cells at the air-liquid interface* in a liquid culture to use all of it immediately. As a result, even the cells just beneath the surface of the liquid cannot grow.

Cells growing under anaerobic conditions generally exhaust a usable energy supply when the population reaches high cell numbers. They also produce a variety of waste products in their metabolism, many of which are toxic.

As a general rule, toxic materials produced by cells growing under anaerobic conditions limit the final level that the population can attain under otherwise optimal growth conditions. This explains why some facultative anaerobes will generally grow to a higher cell density under aerobic conditions. For reasons that will be discussed in Chapter 4, not as much toxic material is likely to accumulate under aerobic conditions.

Products of metabolism, Figure 4–9, page 96

Stationary Phase. The stationary phase is reached when the total number of living cells in the population no longer increases. Thus, in this stage some cells may be multiplying and an equal number may be dying. However, most of the cells in the population probably stop multiplying without dying. In

---

*interface—the surface forming the shared boundary between the air and liquid

the first situation, the total number of cells, both living and dead, generally increases unless autolysis occurs.

The duration of the stationary phase varies, depending on the species of organism and environmental conditions. Some organisms remain in this phase for only a few hours, others for several days. Generally if toxic products have accumulated in the exponential phase, the cells begin to die rapidly. Some cells lyse* very quickly once they stop multiplying because of the action of a lytic* enzyme present in the cell wall. This same enzyme may function in extending cell walls during cell division. If the enzyme functions even though the cells are not dividing and extending their cell walls, then the walls will be destroyed.

Death Phase. A decrease in the total number of viable cells in the population signals the onset of the death phase. A cell is dead if it is no longer capable of multiplying. In this case, just as in the growth phase, a graph of viable cells on a logarithmic plot versus time on an arithmetic plot normally yields a straight line (Fig. 3–14), but the slope of the line is in the other direction. The same mathematical treatment given to growth in the logarithmic phase holds true here, but now there is an exponential decrease in the number of viable cells rather than an exponential increase. Thus, if there is a tenfold decrease in the number of viable cells during the first hour of the death phase, an additional tenfold decrease is observed by the end of the next hour. There is no relationship between how rapidly cells grow in the exponential phase and how rapidly they die in the death phase.

The majority of organisms in a culture are similar enough that each organism is equally likely to die. Which cells actually do die in any time period depends entirely on chance. The likelihood of a cell dying remains constant during most of the death phase, and so the cells continue to die at a constant rate. However, as soon as the majority of the cells have died, the death curve flattens out (Fig. 3–14). This flattening indicates that a few cells in the population are more resistant than others to dying. Why a few cells are more resistant than all other cells is not understood. This resistance is not transferred to the cells' progeny,* however, since survivors that are diluted into fresh medium under the same stages of growth die at the same rate as the original population.

If a population of cells in the death phase is diluted and inoculated into fresh medium, the cells must go through a lag phase before their log phase occurs. Since cells in the death phase are not "tooled up" for rapid multiplication, a period of time will elapse before they can multiply. In contrast, cells transferred from a log phase stage into fresh medium of the same composition do not exhibit a lag phase, since their metabolic machinery is already geared to multiplication. If the media are markedly different in composition, then a "tooling up" period will be required.

---

*lyse—dissolve; break up

*lytic—degradative, leading to destruction

*progeny—descendants

GROWTH IN AN OPEN SYSTEM

In an open system of growth, relatively constant environmental conditions are maintained, nutrients are continuously being added, and the products of metabolism are constantly being removed. This kind of system is often approached in nature. However, although there may be a continuing food supply in nature, it may not be constant. Rather, there are often periods of excess followed by periods of famine. Under these conditions bacteria multiply, but at different rates. Bacteria have evolved mechanisms for dealing with changing conditions. Open systems can be simulated in the laboratory by continually adding nutrients and removing cells and waste materials. Under these conditions the bacteria remain in the log phase of growth.

SUMMARY

This chapter has focused on the principles of microbial growth in the laboratory. Bacteria can be grown in a variety of media, each of which must provide sources of energy as well as the elements carbon, nitrogen, phosphorus, and sulfur in forms that can be used by the organism for the synthesis of cell material. An unusually broad spectrum of nutrients and energy sources can serve these purposes for different microorganisms. At one end of the spectrum are the autotrophs—organisms that derive their energy from sunlight or from the metabolism of inorganic compounds and their carbon from carbon dioxide. At the other end are obligate intracellular parasites—bacteria that cannot multiply unless they are inside living cells. The majority of bacteria gain carbon for the biosynthesis of their cellular components from organic sources and are termed heterotrophs. Heterotrophs gain their energy by metabolizing organic materials. Many organisms can use inorganic sources of nutrients such as nitrates, phosphates, and sulfates for the synthesis of proteins and nucleic acids, although some cells require certain preformed subunits called growth factors. Oxygen is a required nutrient for some cells and a toxic element for others, and still others can grow in either its presence or its absence. The complexity of nutrient requirements is inversely related to an organism's biosynthetic capabilities.

Several environmental factors are also important to the growth of bacteria, including the temperature, pH of the medium, and the osmotic pressure. Any single species can grow only within a limited range of conditions. However, as a group, microorganisms can grow over a very broad range.

The growth of bacteria in the laboratory follows a predictable pattern, with four readily identifiable phases: lag phase, log phase, stationary phase, and death phase. During these phases the culture undergoes a period of active metabolism without cell multiplication (lag), logarithmic increase in the number of viable cells (log), no further increase in the number of viable cells (stationary), and exponential decrease in the number of viable cells (death).

SELF-QUIZ

1. Organisms causing disease in humans are most likely to be
   a. psychrophiles
   b. mesophiles
   c. thermophiles
   d. halophiles
   e. autotrophs
2. The group of organisms that characteristically prefers to grow at acid pH is
   a. algae
   b. fungi
   c. protozoa
   d. archaebacteria
   e. eubacteria
3. The bacterial genus famous for its ability to degrade a wide variety of compounds is
   a. *Escherichia*
   b. *Salmonella*
   c. *Thiobacillus*
   d. *Pseudomonas*
   e. *Neisseria*
4. Shaking a sample of soil in a glucose-salts medium favors the growth of
   a. aerobic heterotrophs
   b. anaerobic heterotrophs
   c. aerobic autotrophs
   d. anaerobic autotrophs
   e. facultative anaerobic autotrophs
5. The phase of the growth curve in which bacteria are most susceptible to penicillin is
   a. lag
   b. log
   c. stationary
   d. death
   e. equally susceptible in all stages

QUESTIONS FOR DISCUSSION

1. Discuss how you would attempt to isolate a spore-forming thermophilic heterotroph able to grow on nitrogen gas as its source of nitrogen.
2. The concept that bacteria, like all organisms, need "living room" could explain why bacteria stop growing when they reach a high cell density. Do you think this is a reasonable explanation? Why or why not? Cite evidence to support your view.

FURTHER READING

Difco Manual of Dehydrated Culture Media and Reagents for Microbiological and Clinical Laboratory Procedures, 9th ed. Detroit: Difco Laboratories, 1971. Difco, one of the leading manufacturers of bacteriological media, has published a book that describes the composition of all the media that it sells as well as some of the salient features of these media. Many references are given on how the media were formulated and what they can accomplish. To keep the book up to date, the company publishes supplementary literature on new products that it introduces.

Finegold, S., Martin, W., and Scott, E.: *Bailey and Scott's Diagnostic Microbiology,* 5th ed. St. Louis: C. V. Mosby, 1978. A classic text useful as a reference in medical bacteriology laboratories and as a textbook for courses in diagnostic bacteriology.

Jannasch, H., and Wirsen, C.: "Microbial Life in the Deep Sea." *Scientific American* (June 1977).

Chapter 4

METABOLIC PROCESSES—ENERGY METABOLISM AND BIOSYNTHESIS

In the diagnostic microbiology laboratory of a large city hospital, a medical technologist frowns as she inspects a number of test tubes containing turbid liquids of different colors. Each tube contains bacteriological growth medium containing a different potential nutrient and a colored indicator* dye. The tubes had been inoculated the previous night with a pure culture of a bacterium recovered from a patient with a puzzling case of fever. The technologist has become accustomed to looking for color changes in the media, which indicate that bacteria have broken down nutrients and have produced new substances from them. These tests have become an important aid in helping to identify the causative agents of infectious disease. Fortunately, the pattern of utilization of different nutrients is so constant that the microorganisms can be identified whether the disease occurs in Europe, Southeast Asia, or the United States. For the second time, the technologist mentally checks off the results so there can be no mistake: glucose, acid production without gas; mannitol, not broken down; sucrose, not broken down; lactose, not broken down; hydrogen sulfide, produced in a small amount; tryptophan, converted to indole; urea, no ammonia produced. Now definitely concerned, the technologist turns to her colleagues. "Looks like we may have another case of typhoid fever," she says.

---

*indicator—a substance that changes from one color to another as a result of a specific chemical reaction

OBJECTIVES

To know
1. The major properties of enzymes.
2. The way enzymes work.
3. The molecule that stores energy in the cell and how this energy is stored.
4. The general features of the glycolytic pathway, especially the steps concerned with energy metabolism.
5. The general features of the tricarboxylic acid (TCA) cycle.
6. The features of cell growth under aerobic and anaerobic conditions.
7. A reaction that links energy metabolism to biosynthesis.
8. The relationship between a cell's response to oxygen and its metabolic pathway of energy formation.

GENERAL ASPECTS OF METABOLISM

Many bacteria can grow on very simple media containing glucose and a few simple inorganic salts. Therefore, these bacteria must be able to transform these few simple molecules into bacterial cells. The conversion of these small molecules into more cells requires energy. The food provides a source of energy as well as the starting material for the synthesis of more cells. There are approximately 5000 different kinds of molecules inside a bacterial cell, and to attempt to learn how the cell transforms its food into each of these molecules is an impossibly difficult task. Fortunately, a knowledge of a number of rules by which cells operate provides a general understanding of these transformations. This chapter will explain the ground rules all cells— whether they are from plants, animals, or bacteria—employ for converting foodstuffs into their own components. The pathways by which these conversions occur can be divided into two major categories: those yielding energy—the degradative (breakdown) reactions—and those requiring energy—the biosynthetic reactions.

FEATURES COMMON TO ENERGY GENERATION AND BIOSYNTHESIS

When a few yeast cells are placed in a glucose-salts medium in a flask and incubated for a few days, a cursory glance at the flask and a smell of its contents are enough to convince anyone that profound changes have taken place. Whereas the contents of the flask were perfectly clear at the beginning of the incubation, they are now turbid, with bubbles of gas at the surface. In addition, the odor of ethyl alcohol emanates from the previously odorless liquid. The increase in turbidity and the formation of ethyl alcohol (ethanol) and gas reflect the entire spectrum of metabolic activity of the yeast cells.

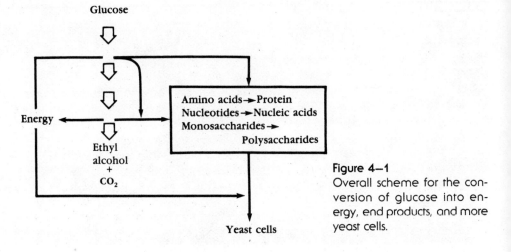

Glucose

Energy

Ethyl
alcohol
+
CO_2

Amino acids → Protein
Nucleotides → Nucleic acids
Monosaccharides →
Polysaccharides

Yeast cells

Figure 4–1
Overall scheme for the conversion of glucose into energy, end products, and more yeast cells.

$$\text{glucose} + \text{yeast cells} \rightarrow \text{more yeast cells} + \underset{\text{dioxide}}{\text{carbon}} + \underset{\text{alcohol}}{\text{ethyl}}$$

$C_6H_{12}O_6$ CO_2 C_2H_5OH

The cells have transformed some of the glucose and inorganic salts into more yeast cells through biosynthetic reactions. This transformation explains the turbidity of the medium. In addition, the cells have degraded some of the glucose to ethyl alcohol and CO_2 by a pathway that provides energy to the cells.

There are hundreds of steps in the conversion of the food (glucose) provided for the yeast to the products of metabolism (ethanol, CO_2, and more yeast cells) (Fig. 4–1). Each step requires an **enzyme** that converts one substance into another. Before we begin to discuss the pathways cells use to obtain energy and synthesize cellular components, it is necessary to know about enzymes—the protein molecules that make metabolic reactions possible. Their properties, in large part, determine the conditions under which cells live and die.

ENZYMES

Description

The growth of any organism depends on the functioning of enzymes. Enzymes are protein catalysts that convert one molecule (the substrate) into another molecule (the product). This can be diagramed as follows:

$$\text{substrate} \underset{\longleftarrow}{\overset{enzyme}{\longrightarrow}} \text{product}$$

Because of the presence of enzymes, the reaction takes place at the temperature of the living organism. Without the enzyme, the substrate would be converted to the product, but at an extremely slow rate, so slow that cells could not live without enzymes.

As the two arrows indicate, some enzyme reactions are reversible, and the same enzyme functions in converting the product (which now becomes the substrate of the reaction) into the substrate (which now becomes the product). Like all catalysts, the enzyme is not used up in the reaction; therefore, one enzyme molecule can convert millions of substrate molecules to molecules of product. Accordingly, cells need very few molecules of any particular enzyme.

Each enzyme consists of from 100 to about 600 amino acids joined together to form a protein molecule with a unique shape (Fig. 4–2). Each one of the hundreds of reactions in the cell is catalyzed by a different enzyme, which differs from all other enzymes in the order of amino acids that make up the molecule. The amino acid sequence in turn gives the enzyme a shape that differs from that of all other enzymes. If this shape is destroyed by high temperatures or acid, the enzyme cannot function but is irreversibly denatured. The most important part of the molecule's shape is a cavity that is almost, but not quite, the exact size and shape of the substrate molecule (Fig. 4–2). This cavity is called the **active site** or **catalytic site.** The substrate molecule, which can be considered analogous to a key, adjusts its shape enough to be able to fit into the cavity of the enzyme, which is analogous to a lock. In fitting into the cavity, stress is placed on the bonds that hold the substrate together. This stress helps to break the bonds of the substrate. In other kinds of reactions, the active site may hold two substrate molecules very close together so that a bond may readily form between them. The fit between the enzyme and its substrate must be very precise; this explains why each reaction with its different substrate molecule (or molecules) requires a different enzyme.

All cells must have the ability to carry out the reactions necessary for life, such as the synthesis of DNA and RNA. However, many reactions are not essential to life. Some cells are able to carry out some reactions; other cells can carry out other reactions. Since the ability to synthesize an enzyme is determined by the genetic composition of the organism and is therefore a stable characteristic of the cell, it is possible to identify an organism by its ability to carry out certain reactions, which result in the formation of certain identifiable products. Thus, the ability of some but not other species to degrade lactose with the production of acid is used to help distinguish closely related Gram-negative organisms. The production of acid is conveniently determined by incorporating acid-base indicators that change color when acid is produced in the culture medium. (See color plates 13 and 14.)

Amino acids, Appendix I

Protein synthesis, page 114

Coenzymes

Some enzymes require a small nonprotein portion that is essential for their functioning. One very important group of such molecules are the **coenzymes,** small organic molecules that readily separate from the enzyme. Coenzymes

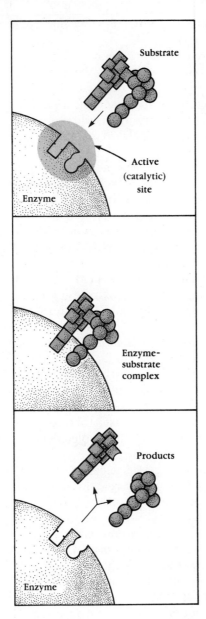

Figure 4–2
Schematic diagram of the way an enzyme functions.

generally function in the transfer of small molecules from one molecule to another so that they interact with many different enzymes. The coenzyme associates with the active site of one enzyme, attaches to a small unit such as a hydrogen atom, which is part of the substrate, and then leaves the enzyme (Fig. 4–3). The coenzyme carries the hydrogen atom to another mol-

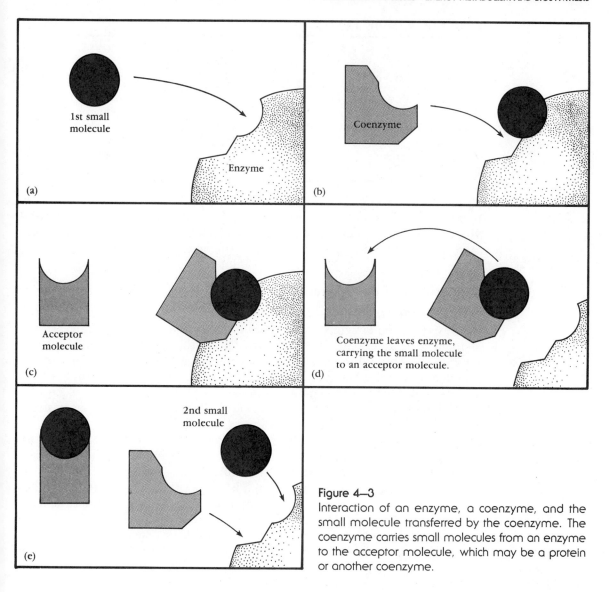

(a) 1st small molecule / Enzyme

(b) Coenzyme

(c) Acceptor molecule

(d) Coenzyme leaves enzyme, carrying the small molecule to an acceptor molecule.

(e) 2nd small molecule

Figure 4–3

Interaction of an enzyme, a coenzyme, and the small molecule transferred by the coenzyme. The coenzyme carries small molecules from an enzyme to the acceptor molecule, which may be a protein or another coenzyme.

ecule, which accepts the hydrogen atom. The coenzyme is then free to associate with the first enzyme and attach to and carry another hydrogen atom. However, each coenzyme is able to transfer only one kind of small molecule or atom. Coenzymes are synthesized from vitamins, and this is why vitamins are required in the diet of human beings and other organisms (Table 4–1). Since coenzymes, like enzymes, are recycled, they are needed only in extremely minute amounts. Therefore, vitamins, whether in the diet of micro-

Vitamins in media, Table 3–3, page 69

TABLE 4–1 VITAMINS AND COENZYMES

Name of coenzyme	Vitamin from which it is synthesized	Entity transferred
Nicotinamide adenine dinucleotide (NAD)	Niacin	Hydrogen atoms
Flavin adenine dinucleotide (FAD)	Riboflavin	Hydrogen atoms
Coenzyme A	Pantothenic acid	2 carbon atoms
Thiamine pyrophosphate	Thiamine	Aldehydes
Pyridoxal phosphate	Pyridoxine	Amino group
Tetrahydrofolic acid	Folic acid	1-carbon group

organisms or humans, are required in very small amounts. If an essential vitamin is not provided, then *all* the different enzymes that require the coenzyme cannot function, and serious consequences result.

Inhibition of Enzyme Activity

Competitive Inhibition—Reversible. Although the substrate must fit into the cavity of the enzyme, *any* molecule that has a closely similar size and shape will be able to fit. This fact provides another potential vulnerable site for a chemical to inhibit the growth of bacteria. The sulfa drugs* provide an excellent example. Many bacteria must synthesize their own supply of the vitamin folic acid. The synthesis involves the conversion of the substrate *para*-aminobenzoic acid (PABA). The compound sulfanilamide has a structure very similar to that of PABA, the natural substrate (Fig. 4–4). The enzyme cannot differentiate one molecule from the other. Therefore, if both PABA and sulfa drug are present, the enzyme often binds to the sulfa drug rather than to PABA. The result is that folic acid is not synthesized, and the bacteria are inhibited in their growth. The higher the ratio of sulfa molecules to PABA molecules, the greater the inhibition (Fig. 4–5). Since the sulfa drug competes with the PABA for binding to the active site of the enzyme, this mode of inhibition is termed **competitive inhibition.** This type of inhibition is *reversible*, since removal of the drug results in the complete restoration of

Sulfa drugs, pages 191–192

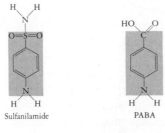

Figure 4–4
Structure of sulfanilamide (sulfa) and of *para*-aminobenzoic acid (PABA). The portions of the molecules that are almost identical are shaded.

---

*sulfa drugs—drugs that have a close chemical relationship to sulfanilamide

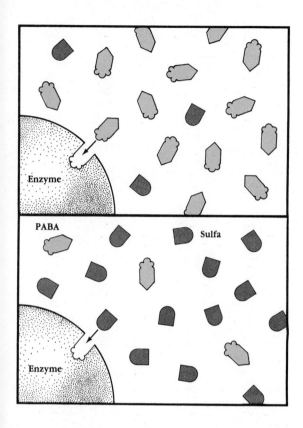

Figure 4–5
Competitive inhibition of folic acid synthesis by sulfa. The higher the concentration of sulfa molecules relative to PABA, the more likely that the enzyme will bind the sulfa, and the greater the inhibition of folic acid synthesis.

enzyme activity. The selective action of sulfa on certain microorganisms is due to the fact that only certain bacteria, fungi, and protozoa have the enzymes for converting PABA to folic acid; humans do not. Humans require folic acid in their diet because they cannot synthesize it from PABA.

    Irreversible Inhibition. The shape of an enzyme molecule depends on many bonds between amino acids in the protein molecule. Many heavy metals, such as silver and mercury (in Mercurochrome), can combine with the atoms involved in these bonds (such as the —SH group of cysteine), thereby breaking the bond and changing the shape of the enzyme (Fig. 4–6). The enzyme becomes denatured and cannot function. The action of these heavy metals is *irreversible* since removal of the heavy metals does not restore the activity of the enzyme. The heavy metals are inhibitory to *all* cells to which they are applied because all enzymes are affected. Therefore these denaturing agents must be used cautiously.

Mercury as a disinfectant, page 170

Factors Influencing Enzyme Activity

The speed at which cells grow and the conditions under which they multiply depend on the proper functioning of their enzymes. Thus, the optimum temperature, pH, and salt concentration for growth of any cell represent the best

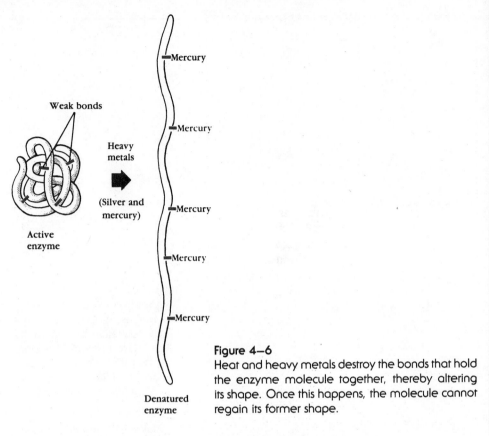

Figure 4-6

Heat and heavy metals destroy the bonds that hold the enzyme molecule together, thereby altering its shape. Once this happens, the molecule cannot regain its former shape.

average for the functioning of all its enzymes. In general, increasing the temperature promotes more rapid bacterial growth until a temperature is reached at which indispensable enzymes lose their shape (become denatured) and cease functioning. Then the organism dies.

Naming Enzymes

Unfortunately, enzymes have not always been named according to a systematic rule. However, a few general rules may help in understanding the function of a particular enzyme from its name. In general, enzymes are named by taking the name of the substrate and adding the suffix *-ase* to it. For example, galactosidase is an enzyme that degrades galactosides.* Sometimes the *-ase* is added to only a part of the name of the substrate. For example, an enzyme that degrades DNA is deoxyribonuclease. Enzymes that build up large molecules (polymers) are termed "polymerases." Those that synthesize DNA are termed "DNA polymerases"; those for RNA are termed "RNA polymerases." The presence of bacterial enzymes such as dehydrogenases,

---

*galactosides—molecules that contain the sugar galactose (the enzyme breaks the bond joining this sugar to the other portion of the molecule)

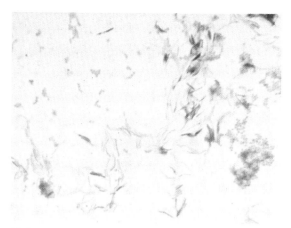

COLOR PLATE 1 Acid-fast stain of *Mycobacterium*. (×1200) (Courtesy of Leon J. Le Beau, Ph.D., University of Illinois Medical Center, Chicago.)

COLOR PLATE 2 A dish of plum pudding was left open to the air. A *Penicillium*-type mold grew over the entire surface (note greenish color). (Courtesy of M. Nester.)

COLOR PLATE 3 *Rhizopus* mold growing on the surface of bread. Note the black sporangia at the end of the stalks. These contain asexual spores. (Courtesy of M. Nester.)

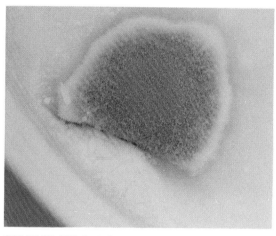

COLOR PLATE 4 A *Penicillium* mold found growing on the surface of some refrigerated sour cream. *Penicillium* molds are quite adaptable and can grow at a variety of temperatures and on a variety of organic materials. (Courtesy of M. Nester.)

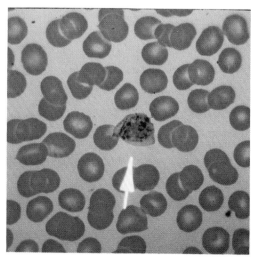

COLOR PLATE 5 Blood smear from a patient with malaria. One cell contains a trophozoite of *Plasmodium vivax*. (Courtesy of C. E. Roberts.)

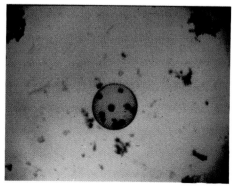

COLOR PLATE 6 *Volvox*, a microscopic alga.

COLOR PLATE 7 Four species of algae found on the shores of Puget Sound, Washington. Clockwise starting at the bottom: *Gigartina exasperata*, *Fucus distichus*, *Porphyra perforate*, and *Ulva lactuca* (sea lettuce). (Barnacles are the white dots to the right.) (Courtesy of M. Nester.)

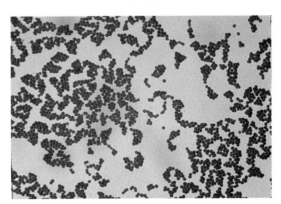

COLOR PLATE 9 *Staphylococcus aureus* from blood agar culture. Gram stain. (×1000) (Courtesy of Leon J. Le Beau, Ph.D., University of Illinois Medical Center, Chicago.)

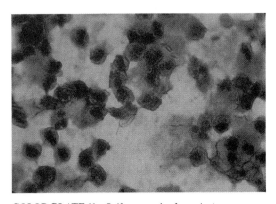

COLOR PLATE 11 Sulfur granules from *Actinomyces*.

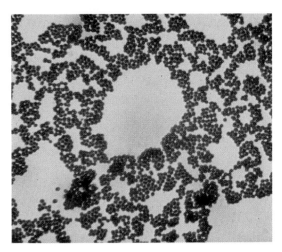

COLOR PLATE 8 Smear of *Gaffkya tetragena* from nutrient agar. Gram stain. (×1000) (Courtesy of Leon J. Le Beau, Ph.D., University of Illinois Medical Center, Chicago.)

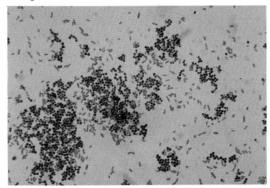

COLOR PLATE 10 Mixed *Escherichia coli* and *Staphylococcus aureus* smear showing Gram-positive (dark purple) and Gram-negative (red) stains. (×1000) (Courtesy of Leon J. Le Beau, Ph.D., University of Illinois Medical Center, Chicago.)

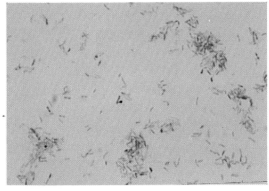

COLOR PLATE 12 Fat droplets in the Legionnaires' disease bacillus, stained with Sudan black.

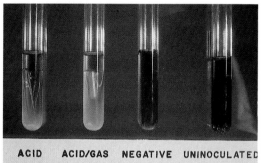

ACID ACID/GAS NEGATIVE UNINOCULATED

SUGAR FERMENTATION

COLOR PLATE 13 Sugar fermentation test with brown cresol purple indicator (1–8). (Courtesy of Leon J. Le Beau, Ph.D., University of Illinois Medical Center, Chicago.)

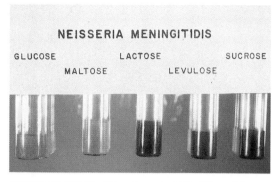

NEISSERIA MENINGITIDIS

GLUCOSE   LACTOSE   SUCROSE
MALTOSE   LEVULOSE

COLOR PLATE 14 *Neisseria meningitidis* differential sugar reactions. The bacteria are able to degrade glucose and maltose with the production of acid (yellow color). (Courtesy of Leon J. Le Beau, Ph.D., University of Illinois Medical Center, Chicago.)

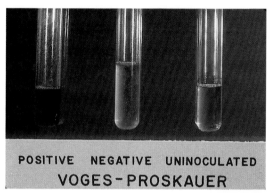

POSITIVE NEGATIVE UNINOCULATED

VOGES-PROSKAUER

COLOR PLATE 15 The Voges-Proskauer test distinguishes *Escherichia coli* from *Enterobacter aerogenes*. *Enterobacter aerogenes* turns the solution red (1–6). (Courtesy of Leon J. Le Beau, Ph.D., University of Illinois Medical Center, Chicago.)

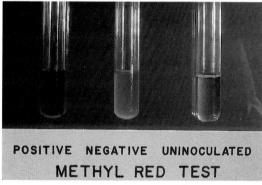

POSITIVE NEGATIVE UNINOCULATED

METHYL RED TEST

COLOR PLATE 16 The methyl red test also distinguishes *Escherichia coli* from *Enterobacter aerogenes*. The *Escherichia coli* turns the solution red (1–6). (Courtesy of Leon J. Le Beau, Ph.D., University of Illinois Medical Center, Chicago.)

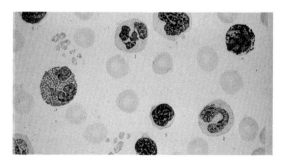

COLOR PLATE 17 Normal cellular constituents of adult human blood: 1. Polymorphonuclear neutrophil. 2. Monocyte. 3. Eosinophil. 4. Basophil. 5. Lymphocyte. 6. Lymphocyte. 7. Mononuclear phagocyte. 8. Platelets. 9. Erythrocyte.

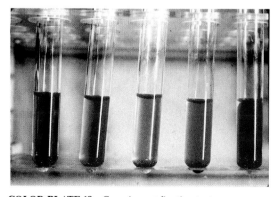

COLOR PLATE 18 Complement fixation test.

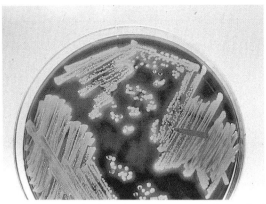

COLOR PLATE 19 *Staphylococcus aureus* colonies streaked onto blood agar plate showing zones of beta hemolysis (clear areas around the colonies). (Courtesy of M. Lampe.)

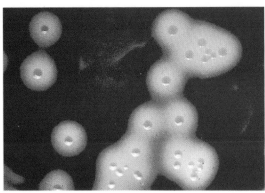

COLOR PLATE 20 Colonies of group A *Streptococcus pyogenes* showing zones of beta hemolysis. (Courtesy of J. Portman.)

COLOR PLATE 21 Alpha hemolysis of *Streptococcus* on blood agar. Note the greenish color of the alpha hemolysis. Optochin disc (white disc) shows zone of no growth of bacteria.

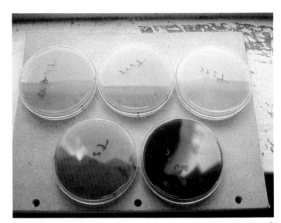

COLOR PLATE 22 Cultures from different strains of *Pseudomonas aeruginosa* from infected patients. Note the differing colors produced by water-soluble pigments. (Courtesy of C. E. Roberts.)

COLOR PLATE 23 *Escherichia coli* colonies on blood agar (1:1). (Courtesy of Leon J. Le Beau, Ph.D., University of Illinois Medical Center, Chicago.)

COLOR PLATE 24 *Escherichia coli* showing lactose fermentation on EMB agar. Note the green sheen on the colonies. (Courtesy of Leon J. Le Beau, Ph.D., University of Illinois Medical Center, Chicago.)

deaminases, and decarboxylases helps to identify specific bacteria in diagnostic clinical laboratories. What these enzymes do can also be surmised from their names. The prefix *de-* means "from" and suggests that something is removed from the substrate. The middle portion of the word (-hydrogen-) indicates what is being removed. Deaminases remove amino groups (—NH$_2$), and decarboxylases remove carboxyl groups (—COOH). Often the specific substrate is named. Thus, tryptophan decarboxylase is an enzyme that removes the carboxyl group (—COOH) from tryptophan. Lysine deaminase is an enzyme that removes an amino group (—NH$_2$) from lysine.

ENERGY METABOLISM

Energy metabolism can be summarized as follows. The original nutrient (glucose in the case of the yeast cells mentioned above) contains a great deal of energy in the bonds between its atoms. However, this energy is spread throughout the molecule, a little bit in each bond. The energy is not present in a readily usable form. The pathway of degradation of the nutrient proceeds through a series of steps in which the small amounts of energy scattered throughout the molecules of food are first concentrated in a few high-energy bonds. The energy in these bonds is then converted into a form that is readily used by the cells: the phosphate bond energy of **adenosine triphosphate** (ATP) (Fig. 4–7). An analogy for these conversions might be a large

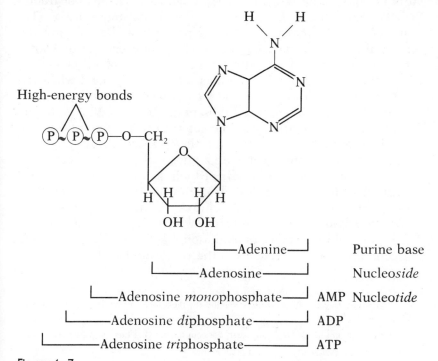

Figure 4–7
The formula of adenosine triphosphate (ATP). The two high-energy bonds are denoted by a wavy line.

number of used postage stamps, each of which is worth less than a penny. Taken together, however, they can be traded to a stamp dealer for a few valuable stamps. These valuable stamps can then be sold to a serious collector who is willing to pay cash, the form of currency used to buy clothes, food, and services. The cellular currency, which corresponds to our money, is ATP. The energy in this molecule is present in the form of two high-energy bonds indicated by the symbol ~ (Fig. 4–7). When these bonds are broken they release large amounts of energy, which then can be used to supply the energy required for biosynthesis, or the formation of bonds between atoms. This energy is also used for many other processes associated with life, such as transporting nutrients from the outside to the inside of the cell through the action of permeases.

Glycolysis*

Bacteria can degrade a tremendous variety of compounds to obtain energy. Such compounds include organic and, occasionally, inorganic materials. In the laboratory the most common food is glucose. This sugar can be degraded along a number of different pathways depending on the species of bacteria. The most common is the **glycolytic** pathway that yeasts employ to degrade glucose. This pathway is diagrammed in Figure 4–8. The complete pathway is in Appendix III. In summary, the 6-carbon glucose molecule is degraded into two 3-carbon molecules called pyruvic acid and four molecules of ATP. We shall now describe the major steps in the pathway and emphasize the steps that either alter the carbon skeleton* or result in energy gain or loss.

The first step in the pathway actually results in the *loss* of a high-energy bond. One high-energy bond of ATP becomes a low-energy bond when the phosphate of ATP becomes attached to the 6-carbon sugar to form sugar-6-phosphate (step 1). In step 2, another high-energy bond of ATP is lost as another phosphate molecule attaches. The sugar now has two phosphate molecules attached. In a succeeding step (3), the 6-carbon sugar splits into two molecules of 3 carbon atoms, each with a phosphate molecule attached.

Step 4 is a very important step. A pair of hydrogen atoms are removed from each 3-carbon compound. Each hydrogen atom consists of one electron* and one hydrogen ion (H^+).* The removal of the hydrogen atom is termed an **oxidation,** and the compound from which it is removed becomes oxidized. The biological importance of oxidation reactions is that one of the first steps in generating ATP molecules is the removal of hydrogen atoms.

---

*glycolysis—breakdown of carbohydrates; the initial sequence by which this occurs is often called the Meyerhof-Embden-Parnas pathway after the scientists who did the most to determine the order of its reactions

*carbon skeleton—the carbon atoms bonded to one another; does not include the other atoms in the molecule

*electron—a negatively charged particle that is a part of all atoms

*hydrogen ion (H^+)—a hydrogen atom that has lost its electron, that is, its negative charge, and thus has a positive charge

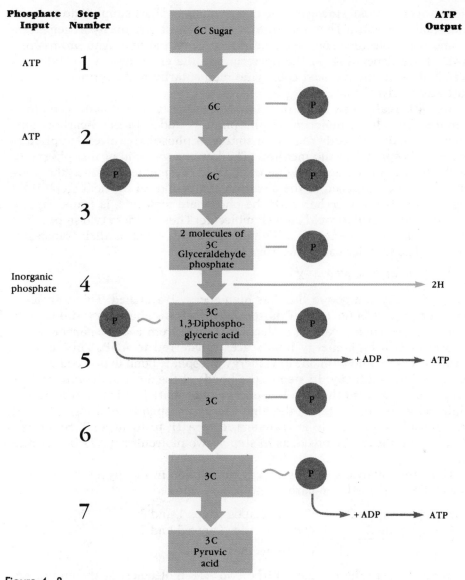

Figure 4—8

The glycolytic pathway. Note that oxygen is not required for any part of this pathway. Only two molecules of ATP are produced for every molecule of sugar degraded.

The hydrogen atoms removed from each 3-carbon compound initially combine with a coenzyme, NAD.* The NAD is said to be **reduced** when it has hydrogen atoms attached and can be designated NAD_{red}. Oxidation (the removal of hydrogen atoms) is always accompanied by reduction (the addition

---

*NAD—abbreviation for nicotinamide adenine dinucleotide

of hydrogen atoms). However, the coenzyme is in short supply and must be continually recycled. Therefore, the coenzyme must give the hydrogen atoms to another molecule. The molecule that accepts the hydrogen atoms from NAD varies depending on the bacteria and the environmental conditions. We shall leave the reduced coenzyme momentarily and continue with the pathway of glycolysis.

Step 4 is really a two-step reaction. Not only is a pair of hydrogen atoms removed but also a molecule of phosphate is added to the 3-carbon compound. Recall that cells require a source of phosphate (often supplied as sodium phosphate) in the medium. The low-energy bond of this phosphate in the medium becomes a high-energy bond (designated ~) once the phosphate becomes attached to the 3-carbon compound. In the next step (5) the high-energy bond, together with the phosphate molecule, is transferred *as a single unit* to ADP to yield an ATP molecule. Thus, in these two steps a low-energy bond is converted to a high-energy bond, which is then transferred to ADP. The transfer can be written as follows:

$$X \sim P + ADP \rightarrow ATP + X$$

where X is any compound that has high-energy P attached. This transfer of the high-energy bond to ADP is termed **substrate phosphorylation.** Since there are two such 3-carbon molecules formed from every 6-carbon sugar, a total of two high-energy bonds are transferred to ADP. This leaves a 3-carbon compound with one low-energy phosphate bond to proceed through the glycolytic pathway. In step 6, the bonds of the 3-carbon compound become rearranged, so that the low-energy phosphate bond is converted to a high-energy bond. In step 7, the high-energy phosphate bond together with the phosphate molecule is transferred to ADP to form ATP. Since two 3-carbon atoms are involved, as in step 5, two molecules of ATP are synthesized.

Thus, the balance sheet for energy production in the glycolytic pathway is as follows. For every 6-carbon molecule degraded:

Energy expended — 2 ATP molecules (steps 1 and 2)
Energy gained   + 4 ATP molecules (steps 5 and 7)

Net gain   + 2 ATP molecules

In addition to the net gain of the two ATP molecules, at the end of glycolysis the cell has generated two molecules of pyruvic acid and two pairs of hydrogen atoms, which we left attached to the coenzyme NAD (NAD_{red}).

1 glucose + 2 ADP $\rightarrow$ 2 pyruvic acid + 2 NAD_{red} + 2 ATP

Bacteria produce a tremendous variety of compounds as the products of metabolism. The nature of these products is determined by which compounds finally accept the hydrogen atoms from the coenzyme. If the compound is organic, then the *overall* process of the breakdown is termed **fermentation.** If the compound is inorganic, then the process is termed **respiration.** The degradation of a sugar to ethanol and carbon dioxide (CO_2)

is an example of fermentation. When yeast degrades sugar to yield water and CO_2, this is an example of respiration.

Fermentation

In these reactions, the NAD_{red} gives the hydrogen atoms to a variety of organic molecules, depending on the microorganism in which the reaction is occurring. The fermentation is named after the final organic compound that is formed, as illustrated by the following examples. All fermentations can take place in an anaerobic environment.

Lactic Acid Fermentation. In this fermentation, the two molecules of NAD_{red} generated in step 4 (Fig. 4–8) reduce the two molecules of pyruvic acid formed in glycolysis to two molecules of lactic acid. NAD is regenerated.

$$2\ CH_3\!-\!\overset{\overset{O}{\|}}{C}\!-\!\overset{\overset{O}{\|}}{C}\!-\!OH + 2\ NAD_{red} \rightleftarrows 2\ CH_3\!-\!\overset{\overset{O-\boxed{H}}{|}}{\underset{\boxed{H}}{C}}\!-\!\overset{\overset{O}{/\!/}}{C}\!-\!OH + 2\ NAD$$

pyruvic acid   lactic acid

This fermentation is important in both the spoilage and production of a number of food products. Depending on the microorganisms present, fermentation of the milk sugar lactose results in a variety of foodstuffs, including yogurt, buttermilk, and cheese. Sauerkraut is produced by the lactic acid fermentation of fresh cabbage, and pickles result from a similar fermentation of cucumbers. *Streptococcus mutans*, probably the chief culprit in tooth decay, does its damage by producing lactic acid, which demineralizes* tooth enamel.

Alcoholic Fermentation. Two steps are involved in the production of alcohol from pyruvic acid. The first reaction is the conversion of pyruvic acid to acetaldehyde and CO_2.

$$2\ CH_3\!-\!\overset{\overset{O}{\|}}{C}\!-\!\overset{\overset{O}{\|}}{C}\!-\!OH \rightarrow 2\ CH_3\!-\!\overset{\overset{O}{\|}}{C}\!-\!H + 2\ CO_2$$

pyruvic acid   acetaldehyde

The acetaldehyde then accepts the hydrogen atoms from NAD_{red} to form ethyl alcohol.

$$2\ CH_3\!-\!\overset{\overset{O}{\|}}{C}\!-\!H + 2\ NAD_{red} \rightarrow 2\ CH_3\!-\!\overset{\overset{O-\boxed{H}}{|}}{\underset{\boxed{H}}{C}}\!-\!H + 2\ NAD$$

acetaldehyde   ethyl alcohol

These two reactions are carried out by yeast *(Saccharomyces)* and form the basis for wine and beer production.

---

*demineralizes—removes the mineral that forms part of the tooth enamel

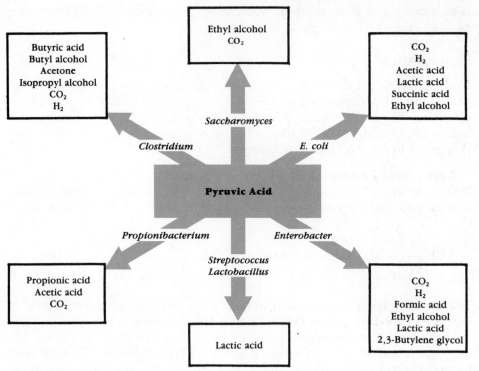

Figure 4–9
Possible conversions of pyruvic acid by different bacteria.

Other Products of Fermentation. Pyruvic acid can be converted to a wide variety of other products by reactions summarized in Figure 4–9. In all of these cases the metabolism of pyruvic acid involves the same principles. The NAD_{red}, which is in limited supply, must give up hydrogen atoms to an organic molecule (reduce that molecule) so that it can participate further in glucose metabolism by picking up more hydrogen atoms.

Tricarboxylic Acid (TCA) Cycle (Krebs or Citric Acid Cycle)

Energy Aspects. If cells are growing under aerobic conditions so that oxygen is present, the NAD_{red} will give up its hydrogen atoms to another coenzyme, FAD,* and eventually the hydrogen atoms will combine with the O_2 to form H_2O. In addition, pyruvic acid will enter into another pathway for further degradation, the **tricarboxylic acid (TCA) cycle.** In glycolysis, high-energy bonds of ATP are derived from the breakdown of glucose by first forming high-energy phosphate bonds with several compounds. These high-energy phosphate bonds are transferred intact to ADP to yield ATP. In the TCA cycle, energy is generated in a different fashion, involving the transfer

---

*FAD—flavin adenine dinucleotide, a derivative of the vitamin riboflavin

of hydrogen atoms through a series of coenzymes to the final hydrogen acceptor, oxygen, to form H_2O.

It is advantageous for a cell to degrade glucose further than pyruvic acid because most of the energy in the original sugar molecule is still retained in the pyruvic acid. The cell gains no additional energy in any of the conversions of pyruvic acid we have discussed. The cell has tapped less than 10 percent of the available energy of the glucose molecule through the processes of glycolysis and fermentation.

A great deal of energy exists in the hydrogen atoms bound to the coenzyme NAD. To convert this energy to ATP bond energy, it must be released slowly, step by step, rather than in one big jump. Thus, NAD_{red} does not give its hydrogen atoms to oxygen directly. Rather, the hydrogen atoms are given to another coenzyme, FAD. The passing of the hydrogen atoms to a lower energy level results in a release of energy that is captured in a high-energy bond of ATP (Fig. 4–10). The FAD_{red} in turn passes hydrogen atoms to other coenzymes.

Two additional molecules of ATP are synthesized for every pair of hydrogen atoms (a total of three) that are passed through this "bucket brigade." Finally, the hydrogen atoms combine with the final hydrogen acceptor, gaseous oxygen, to form H_2O. The series of enzymes and coenzymes through which hydrogen atoms are transferred is termed the **respiratory chain.** A total of three high-energy bonds of ATP and one molecule of H_2O results from each pair of hydrogen atoms that pass through it. The generation of ATP during the passage of hydrogen atoms through the respiratory chain is termed **oxidative phosphorylation** (Fig. 4–10).

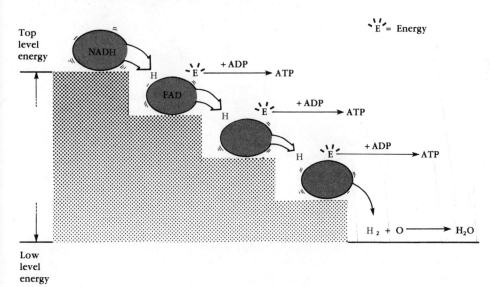

Figure 4–10
Diagrammatic representation of the formation of ATP molecules as hydrogen atoms move from a high energy level to a lower energy level. Note that there are discrete steps for the transfer of energy. The energy is not generated all at once.

Carbon Metabolism. The pyruvic acid enters the TCA cycle by its conversion to a 2-carbon compound, acetic acid, with the removal of hydrogen atoms. These hydrogen atoms also pass through the respiratory chain with the production of three high-energy bonds and one molecule of H_2O. Carbon dioxide is also released in the conversion of the 3-carbon compound (pyruvic acid) to the 2-carbon compound (acetic acid).

The TCA pathway is shown in abbreviated form in Figure 4–11. The com-

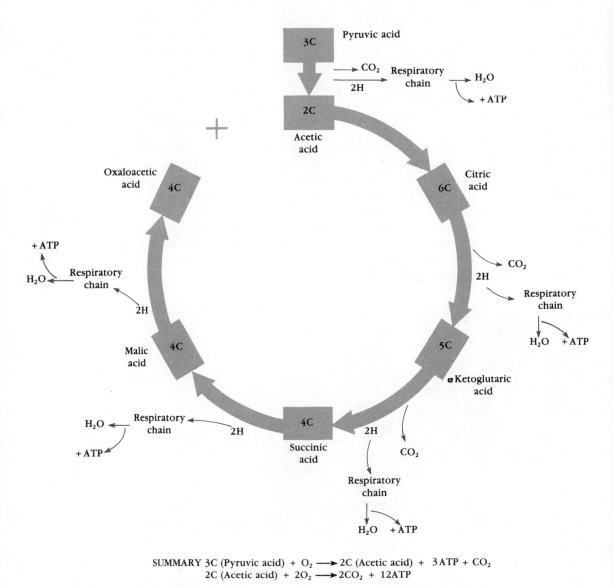

SUMMARY 3C (Pyruvic acid) + O_2 ⟶ 2C (Acetic acid) + 3 ATP + CO_2
2C (Acetic acid) + $2O_2$ ⟶ $2CO_2$ + 12 ATP

Figure 4–11
Tricarboxylic acid (TCA) cycle.

plete pathway is given in Appendix III. It is a cyclic pathway that starts with the joining of a 2-carbon compound, acetic acid, to a 4-carbon compound, oxaloacetic acid, to form a 6-carbon compound, citric acid.* In every turn of the cycle, two carbon atoms are removed as CO_2, resulting in the regeneration of the 4-carbon compound (oxaloacetic acid). This 4-carbon compound then picks up another molecule of the 2-carbon compound (acetic acid), which funnels in through the glycolytic pathway to begin the cycle over again. Thus, the acetic acid is converted to two molecules of CO_2.

The large amounts of energy gained in respiration as in the TCA cycle result from the removal of hydrogen atoms at four steps in the cycle. Each pair of hydrogen atoms removed passes through the respiratory chain to yield three molecules of ATP. Thus, 12 high-energy bonds are generated for every turn of the cycle. These high-energy bonds in ATP can be used by the cell to do work, such as synthesis of cell constituents.

Economics of Fermentation versus Respiration

The breakdown of glucose through the glycolytic pathway followed by the TCA cycle can be summarized by the following equation:

$$\underset{\text{glucose}}{C_6H_{12}O_6} + 6\ O_2 \rightarrow 6\ CO_2 + 6\ H_2O$$

The oxygen serves as the final acceptor for the hydrogen atoms. Thus, the process is respiration. In the course of this breakdown, there is a net gain of 38 molecules of ATP, only two of which are generated in glycolysis. Thus, a cell growing under anaerobic conditions must degrade about 20 times more glucose than a cell growing under aerobic conditions to obtain the same amount of energy. Therefore, cells growing under anaerobic conditions generally grow more slowly than the same cells growing under aerobic conditions. A practical example of the economics of fermentation versus respiration is illustrated by the growth of yeast under the two conditions. When yeast cells are used for the production of alcohol, the vats are kept anaerobic, and ethanol and CO_2 are the major products. However, when yeast cells are being grown to be sold as baker's yeast, then it is most important to obtain the maximum yield of yeast cells from the substrate. For this purpose, the yeast cells are grown under highly aerobic conditions. This situation illustrates that many organisms can carry out either fermentation or respiration, depending on whether or not oxygen is available. The relationship between the growth of cells and the pathway of degradation is summarized in Table 4–2.

The bacteria classified as indifferent or aerotolerant do not have the enzymes necessary to go through the TCA cycle. Therefore they can only carry out fermentation, even if they are growing under aerobic conditions.

---

*citric acid—a 6-carbon compound with three carboxyl groups (—COOH) (thus the name "tricarboxylic acid" has been given to the pathway)

TABLE 4–2   GROWTH OF CELLS UNDER AEROBIC AND ANAEROBIC CONDITIONS

Aerobic	Anaerobic
Cells grow most rapidly because most ATP is generated	Cells grow less rapidly because less ATP is generated in glycolysis
Cells reach high cell density* because toxic materials are not accumulated—only CO_2 and H_2O	Many toxic products are accumulated (i.e., acids, alcohols), so cells do not reach a high cell density

*density—number of cells per volume of medium

Breakdown of Compounds Other Than Glucose

In addition to glucose, many other compounds can be degraded to provide energy. Indeed, if enzymes are available to catalyze the degradation, anything that has chemical bond energy* can be degraded and its energy converted into the high-energy phosphate bond energy of ATP. Most naturally occurring compounds are degradable by some microorganism, although the pathways by which these compounds are degraded are in most cases unknown. It is certainly true, however, that degradation of organic compounds will often generate ATP, most likely through oxidation-reduction reactions. There is great interest in developing bacteria that are able to degrade substances that are harmful to the environment or to humans. These substances include not only oil spilled in the environment but also herbicides* and pesticides,* which have been implicated in causing cancer and birth defects.

Most organic compounds that are degraded, including some amino acids, lipids, and carbohydrates, are converted into the intermediates* of either the glycolytic pathway or the TCA cycle and thus enter the pathways that provide energy to the cell (Figs. 4–8 and 4–11).

Utilization of Inorganic Compounds as Energy Sources

Many microorganisms can utilize a variety of inorganic compounds to gain energy. Thus hydrogen gas, hydrogen sulfide gas, iron, and ammonia can be used by certain bacteria as a source of energy. Just as the heterotrophs oxidize organic compounds, autotrophs oxidize inorganic chemicals for their energy. The importance of these organisms and these metabolic processes will be considered elsewhere in the text.

Photosynthesis

The discussion thus far has focused on the mechanisms that bacteria use to gain energy by converting the chemical energy present in the bonds of either

---

*chemical bond energy—the energy present in the bonds joining atoms together

*herbicides—chemicals used to kill plants

*pesticides—general class of agents that kill either plant or animal pests

*intermediates—any compounds of a biochemical pathway

inorganic or organic molecules into ATP. The bacteria that get their energy in this way are the most common. Some organisms, however, can convert the radiant or light energy absorbed by chlorophyll into useful chemical energy in the form of ATP molecules, through the process of **photosynthesis.** This chemical energy is then available to the organism itself as well as to heterotrophic organisms, which can degrade the organic material in photosynthetic organisms as a source of energy.

The overall reaction of photosynthesis carried out by green plants, eukaryotic algae, and cyanobacteria* is summarized in the equation:

$$6 \; CO_2 \; + \; 6 \; H_2O \; \xrightarrow{\text{light}} \; C_6H_{12}O_6 \; + \; 6 \; O_2$$

These organisms utilize CO_2 and H_2O, from which they synthesize carbohydrate molecules and release oxygen in the presence of light. The carbohydrate ($C_6H_{12}O_6$) in this equation represents not only the starch* that these organisms synthesize but also all structures that the cells produce as they grow. This reaction therefore summarizes all the biosynthetic reactions of the cell. It is the *reverse* of the reactions by which glucose is oxidized completely via glycolysis and the TCA cycle:

$$C_6H_{12}O_6 \; + \; 6 \; O_2 \rightarrow 6 \; CO_2 \; + \; 6 \; H_2O \; + \; \text{energy}$$

Just as energy is gained when glucose is broken down, energy must be provided when it is synthesized from its constituents. Because hydrogen atoms are removed in the degradation of carbohydrates, hydrogen atoms must be supplied for their synthesis.

Applied Aspects of Degradative Metabolism

The metabolism of pyruvic acid has important consequences for many areas of microbiology. To some degree, the determination of whether an organism can degrade a particular compound and, if so, what metabolic products are generated may help to identify particular organisms, because the ability to carry out any one of these metabolic reactions is specific to a particular group of organisms. (See color plates 15 and 16.)

Many of the metabolic products of microorganisms have great commercial value. Needless to say, the ability of yeast to degrade a wide variety of grains with the production of ethyl alcohol has had profound effects in all parts of the world. In alcoholic fermentation of natural products, it is important that the proper environment be carefully controlled so that all the desired end products be produced. All home brewers of beer and wine recognize that it is important to maintain the proper anaerobic conditions lest the aerobic vinegar bacteria, members of the genus *Acetobacter*, oxidize the ethanol to vinegar (acetic acid). Louis Pasteur in 1866 first showed that these bacteria

*cyanobacteria—blue-green algae

*starch—a large molecule consisting of glucose subunits polymerized into long chains

were responsible for spoiling the wines that the French were hoping to export.

Many foods are prepared by the action of a wide variety of bacteria and fungi on different plant materials. In many cases, these products are used only in their native countries. Such products as soy sauce, miso, sufu, natto, and Ang-kak are products of fermentation of soybeans or rice and are primarily used in the Orient. In all these foods, the microorganisms that multiply as they degrade the substrate, the undegraded substrate, and the products of fermentation are eaten.

Although microorganisms are being grown and processed as supplements for animal feed, as yet there are no proteins developed for human consumption. However, this is likely to change in the future. For example, a mold grown on glucose under aerobic conditions contains 45 percent protein and has a low fat and cholesterol content. It may be marketed in the very near future for human consumption as a substitute for meat (Fig. 4–12).

BIOSYNTHETIC METABOLISM

The synthesis of the bacterial cell obviously is extremely complex and has never been accomplished except by a living cell as it reproduces. However, scientists have come a long way in recent years in unraveling some of the steps in the synthesis of various cellular components. Thus, genes and proteins have been synthesized inside a test tube starting from simple chemicals. Within your lifetime it is even possible that an entire cell may be synthesized in the laboratory.

It is far beyond the scope of this text to go into the details of the complex and myriad enzymatic reactions required for the synthesis of cell components. However, a few simple generalizations can be made. The synthesis of the constituents of the cell occurs in several steps. First, the subunits of the macromolecules* are synthesized; then, the subunits are joined together to form macromolecules. These large molecules in turn combine with other large molecules in specific ways to form cellular structures such as ribosomes, the cytoplasmic membrane, and so forth. In this chapter we will consider the biosynthesis of small molecules, the subunits of the macromolecules. In the following chapter we will discuss the biosynthesis of macromolecules.

In general, bacteria will synthesize the materials they need for growth only if they cannot be obtained from the environment. If the small molecules are provided in the medium in which bacteria are growing, the bacteria will take them in by means of permeases and use them to synthesize cell constituents. Bacteria consume far less energy if they can use what is available to them rather than synthesize the materials necessary for macromolecule production. Therefore, bacteria multiply faster in an enriched nutrient me-

---

*macromolecules—large molecules composed of repeating small molecules, the subunits. For example, the macromolecule RNA is composed of nucleotides; proteins are composed of amino acids

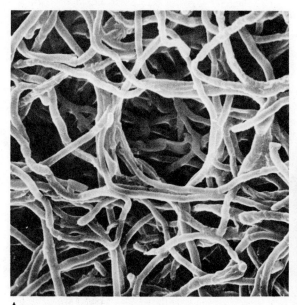

Figure 4–12
Mycoprotein. The fungus *Fusarium* grows as shown in (a) (×1000). In (b) the appearance of the processed material is shown. Perhaps it will serve as a meat substitute. Since fungi are multicellular, this is not technically single cell protein. (Courtesy of RHM Research Ltd.)

A   B

dium than in a glucose-salts medium. The energy saved by growth in the enriched medium is available for carrying out other energy-requiring processes.

Biosynthesis of Carbon Subunits

If the only source of carbon provided to the cell is sugar, then the carbon of these molecules goes into all the organic molecules of the cell. In this case, both the glycolytic pathway and the TCA cycle serve a biosynthetic as well as an energy-generating function. This is true because many of the compounds in these pathways can serve as starting materials for the synthesis of small molecules, which are the subunits of macromolecules (Fig. 4–13). Of course, if the compound is shunted out of the glycolytic pathway or TCA cycle, it cannot be used to generate the high-energy bonds of ATP. Thus, the cell loses a potential source of energy. However, it does not behoove a cell to generate a great deal of ATP unless it has a supply of small molecules from which it can synthesize its components. An oversupply of one or the other would be wasteful if the cell is to grow at the fastest speed that the environment will allow.

Overview of Biosynthesis of Carbon-, Nitrogen-, Sulfur-, and Phosphorus-Containing Compounds

As pointed out in Chapter 3 (Table 3–2), many bacteria can grow on a glucose-salts medium. All the subunits of macromolecules are synthesized from the simple compounds present in the medium. Table 4–3 lists the compo-

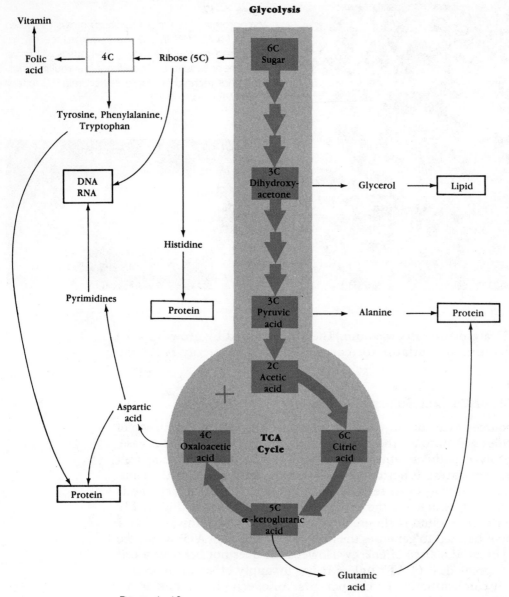

Figure 4–13
The overall view of the conversion of sugar to macromolecules.

nents of a glucose-salts medium and the cell components into which the larger molecules are transformed. This table complements the information provided in Figure 3–7.

Utilization of Ammonium Salts. No purpose is served in detailing all the enzymatic steps necessary for the components of the medium to be converted

TABLE 4—3 COMPOSITION OF GLUCOSE-SALTS MEDIUM AND CELL COMPONENTS THAT CAN BE FORMED FROM IT

Medium constituent	Important element(s) or group	Major subunits	Macromolecules
Glucose	Carbon (C)	All amino acids (20)	Protein
		Nucleotides (4)	Ribonucleic acid (RNA)
		Deoxyribonucleotides (4)	Deoxyribonucleic acid (DNA)
Dipotassium phosphate	Phosphate (PO_4^{3-})	Nucleotides	Ribonucleic acid
Monopotassium phosphate		Deoxyribonucleotides	Deoxyribonucleic acid
Magnesium sulfate	Sulfur (S)	Several amino acids (cysteine, methionine)	Protein
Ammonium sulfate	Ammonium (NH_4^+)	All amino acids	Protein
		Nucleotides	Ribonucleic acid
		Deoxyribonucleotides	Deoxyribonucleic acid

to the subunits of macromolecules. However, one enzymatic reaction is especially important because it links biosynthetic and degradative reactions. This is the reaction in which nitrogen, often provided as $(NH_4)_2SO_4$ in the medium, is incorporated into the amino group ($-NH_2$) of amino acids. In this reaction, an intermediate in the TCA cycle, α-ketoglutaric acid (Fig. 4–11), reacts with NH_4^+, and the amino acid glutamic acid is synthesized.

$$NH_4^+ + HO-\overset{O}{\underset{}{\overset{\|}{C}}}-\overset{H}{\underset{H}{\overset{|}{C}}}-\overset{H}{\underset{|}{\overset{|}{C}}}-\overset{O}{\underset{}{\overset{\|}{C}}}-\overset{O}{\underset{}{\overset{\|}{C}}}-OH + NAD_{red} \leftrightharpoons HO-\overset{O}{\underset{}{\overset{\|}{C}}}-\overset{H}{\underset{H}{\overset{|}{C}}}-\overset{H}{\underset{H}{\overset{|}{C}}}-\overset{NH_2}{\underset{}{\overset{|}{C}}}-\overset{O}{\underset{}{\overset{\|}{C}}}-OH + NAD$$

ammonium ion α-ketoglutaric acid   glutamic acid

The reaction is reversible. Therefore, glutamic acid, an amino acid, can be converted to α-ketoglutaric acid, which is then able to proceed through the TCA cycle. Thus, this reaction is a key step linking energy metabolism and biosynthetic reactions.

SUMMARY

The metabolic transformations that cells carry out increase the quantity of all cellular components. Since the synthesis of these components requires energy, the cell must also derive energy from foodstuffs. Energy conversion involves the conversion of diffuse chemical bond energy into the concentrated high-energy bonds of ATP. Heterotrophs obtain energy by oxidizing reduced organic compounds through a series of enzymatic reactions; autotrophs generally oxidize inorganic compounds to obtain energy. One of the most common pathways for the breakdown of carbohydrates is the glycolytic pathway, which functions under both aerobic and anaerobic conditions. For every molecule of glucose metabolized, two molecules of pyruvic acid, two molecules of ATP (net), and two molecules of NAD_{red} are formed. Pyruvic acid can be metabolized further in a variety of ways, which yield a variety of products. These include ethanol, CO_2, lactic acid, acetic acid, and several other acids and alcohols. Under aerobic conditions and provided the cell has the necessary enzymes, the pyruvic acid may be metabolized to acetic acid, which can be metabolized further through the TCA cycle. If glucose is oxidized only to organic compounds (fermentation), most of the energy available in the glucose molecule remains untapped. However, if it is oxidized completely to CO_2 and H_2O (respiration), then about 20 times more energy is gained. This energy is generated in the course of the transfer of hydrogen atoms along the respiratory chain. The ability of an organism to carry out fermentation or respiration depends on the enzymes the cells contain; this in turn depends on whether the organisms are aerobic, anaerobic, facultative anaerobic, or indifferent.

The degradation of glucose provides the starting compounds from which many small molecules are synthesized. Many steps in these biosynthetic reactions require energy, and the ATP generated in the breakdown of sugar "drives" these reactions. A key reaction that links energy metabolism involves the synthesis of glutamic acid from α-ketoglutaric acid. Glutamic acid can also be converted to α-ketoglutaric acid.

SELF-QUIZ

1. What is *not* true for enzymes?
   a. They always are composed of proteins.
   b. Some function best at high temperatures; others function best at low temperatures.
   c. Their function may be destroyed by heat.
   d. Many enzymes can catalyze the same reactions.
   e. Enzymes function by putting stress on the bonds of substrate molecules.
2. The cell gains the most energy from the degradation of glucose by converting glucose to
   a. ethanol and CO_2
   b. lactic acid
   c. cell components
   d. pyruvic acid
   e. CO_2 and H_2O
3. The introduction of NH_4^+ into an amino acid
   a. involves a compound in the TCA cycle

    b. is an energy-yielding reaction
    c. requires ATP
    d. results in the synthesis of pyruvic acid
    e. is an irreversible reaction
4. Spoilage of wine can be due to
    a. yeast cells
    b. presence of air
    c. conversion of pyruvic to acetic acid
    d. *Escherichia coli*
    e. enzyme denaturation
5. All enzymes have (Choose the *wrong* answer.)
    a. a coenzyme
    b. an active site
    c. an optimum temperature for activity
    d. amino acids
    e. a substrate on which they act

QUESTIONS FOR DISCUSSION

1. Anaerobic bacteria have certain enzymes of the TCA cycle. Why?
2. Why do diseases of vitamin deficiency have widespread and far-ranging effects on the body?

FURTHER READING

Arms, K., and Camp, P.: *Biology*, 2nd ed. Philadelphia: Saunders College Publishing, 1982. A well-written, beautifully illustrated text in introductory biology. Chapter 8 covers most of the material discussed in this chapter.

Hinkle, P., and McCarty, R.: "How Cells Make ATP." *Scientific American* (March 1978).

Green, D., and Goldberger, R.: *Molecular Insights into the Living Process.* New York: Academic Press, 1967. An extremely well-written and understandable presentation of metabolic principles, both degradative and biosynthetic. Chapters 4 to 8 are especially relevant to this chapter.

Lechtman, M., Rookk, B., and Egan, R.: *The Games Cells Play.* Menlo Park, Calif.: Benjamin/Cummings Publishing Co., 1979. The basic concepts of cellular metabolism presented in an interesting and unusual manner.

Lehninger, A.: *Bioenergetics*, 2nd ed. Menlo Park, Calif.: W. A. Benjamin, 1971. Probably the best single volume on the generation, storage, and utilization of energy in eukaryotic and prokaryotic cells. For students who have had some chemistry.

Chapter 5

MACROMOLECULES: SYNTHESIS AND CONTROL

Today it is a well-known fact that deoxyribonucleic acid (DNA) is the basic material of life. However, there were many bitter arguments among distinguished scientists before data were obtained that proved conclusively that DNA, and not protein, controlled the destinies of cells. What finally settled the controversy was the DNA from dead bacteria. These experiments were based on a discovery made in England by medical microbiologist Dr. Fred Griffith in 1928. Before the discovery of penicillin, bacterial pneumonia was one of the most challenging problems in infectious disease. In experiments aimed at developing a better vaccine for immunization against this disease, Griffith injected into mice pneumococci without capsules, which were not virulent, together with heat-killed pneumococci, which possessed capsules (termed Type III capsules on the basis of their composition). These latter bacteria, if alive, would kill the mice. To Griffith's great surprise, the mice died of pneumonia. Upon autopsy, their tissues were found to be teeming with live pneumococci that had the Type III capsule! The simple explanation—that the heat-killed bacteria were not all dead—was not correct, since if the same batch of heat-killed bacteria were injected into mice without the living nonvirulent bacteria, the mice lived. Therefore, the only other explanation, incredible as it seemed, was that the dead bacteria had somehow transformed the capsuleless bacteria into bacteria with capsules. Furthermore, the ability to synthesize the capsules could be passed along to their offspring.

The proof that DNA was the transforming material resulted from the studies of Dr. Oswald Avery and his colleagues at the Rockefeller Institute in the 1940's. Their approach was to dismantle systematically the dead bacteria that had a capsule and determine whether the bacteria could still transform the nonvirulent bacteria. First, the capsule was removed; the resulting bacteria could still transform. Next, the protein was removed, and the cells did not lose their transforming ability. DNA isolated* from the bacterial cells could also transform, but protein isolated from the same organisms could not. Therefore, the basic material of life must be DNA!

But how can such a simple molecule as DNA, consisting of only four subunits, carry enough information to control the destinies of cells? The answer to that question required another two decades to unravel and culminated with the cracking of the genetic code.

The information provided in this chapter and in Chapter 6 will give you some appreciation of how DNA, the master molecule of life, carries out its functions.

OBJECTIVES

To know
1. The structure of DNA.
2. The structure and types of RNA.
3. How DNA reproduces itself.
4. The steps in the synthesis of a protein molecule.
5. Several antibiotics that interfere with protein synthesis.
6. Two ways in which the end product of a biosynthetic pathway controls its own synthesis.

GENERAL ASPECTS OF MACROMOLECULE SYNTHESIS

In the previous chapter we noted the synthesis of the small molecules that join together to form the major macromolecules DNA, RNA, and protein. We mentioned that the breakdown of glucose relies on enzymes whose ability to catalyze a specific reaction is determined by the sequence of amino acids making up the protein. Furthermore, all the properties of the cell depend on its enzymes, whether it is Gram-positive or Gram-negative, what shape it is, and whether it can cause a specific disease. What determines the enzymes a cell has? What determines the sequence of amino acids in a protein molecule? The answers to these questions provide information crucial to an understanding of life itself.

CONCEPT OF INFORMATION STORAGE AND TRANSFER

All the information about how a cell is formed from the food supplied to it is stored in the sequence of purines and pyrimidines in its DNA. This infor-

---

*isolated—purified from contaminating material

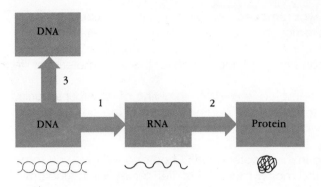

Figure 5–1
Flow of information in biological systems.

mation in DNA is transferred by several steps into the sequence of amino acids in protein molecules (Fig. 5–1, steps 1 and 2). Thus, the cell can be looked upon as a factory that synthesizes proteins. What proteins are synthesized depends on the DNA the cell contains. In addition, the sequence of purines and pyrimidines in the cells' DNA is copied precisely into an identical sequence, which is passed on to the daughter cells in the course of cell duplication (Fig. 5–1, step 3). Note that the sequence of purines and pyrimidines of DNA is transferred first into RNA (the process of **transcription**) before it is transferred into the amino acid sequence of a protein (**translation**).

BIOCHEMISTRY OF DNA AND RNA

DNA, Appendix I

DNA exists as a long double-stranded helical structure (Fig. 5–2). Each strand has a backbone of alternating sugar and phosphate molecules. Attached to each sugar molecule is a purine or pyrimidine base. This subunit of sugar, phosphate, and purine or pyrimidine base is called a nucleotide. The two strands of the helix are held together by hydrogen bonds joining each of the purine molecules (adenine or guanine) to its complementary pyrimidine molecule (thymine or cytosine). The purine molecule guanine is complementary to the pyrimidine cytosine and so will form hydrogen bonds only with this pyrimidine; likewise, adenine bonds only to thymine. Although each hydrogen bond is weak, the large number of them in a molecule of DNA (more than a million in a bacterial chromosome) holds the two strands together very firmly.

RNA, Appendix I

Like DNA, the RNA molecule consists of a sequence of nucleotides, but unlike DNA it usually exists as a single strand. There are three major types of RNA, all of which have important roles in protein synthesis: messenger RNA (mRNA), transfer RNA (tRNA), and ribosomal RNA (rRNA). More information about the chemical structure of DNA and RNA can be found in Appendix I.

The long DNA molecule forming the bacterial chromosome contains a large number of **genes.** Each gene is the *sequence of purine and pyrimidines that codes for one protein molecule.*

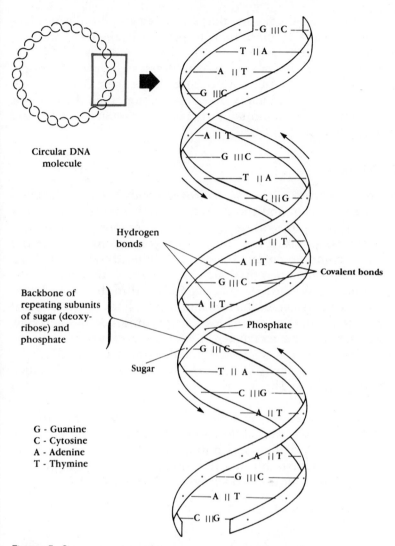

Circular DNA
molecule

Hydrogen
bonds

Covalent bonds

Backbone of
repeating subunits
of sugar (deoxy-
ribose) and
phosphate

Phosphate

Sugar

G - Guanine
C - Cytosine
A - Adenine
T - Thymine

Figure 5–2
Structure of a circular DNA molecule and its helical structure.

The percentage guanine + cytosine (GC) content of the DNA, defined as

$$\frac{\text{moles* G + moles C}}{\text{moles G + moles C + moles A + moles T}} \times 100$$

differs between different species. In bacteria, this value varies between 25 and 70 percent. However, since all the properties of the cell are determined by the proteins in the cell, which in turn are coded for by the DNA, closely

---

*mole—number of grams equal to the molecular weight of a chemical compound. For example, if guanine has a molecular weight of 54, then 54 grams of guanine is one mole of guanine.

related bacteria must have GC contents that are either very similar or identical. If two organisms vary by more than 10 percent in their GC content, they cannot be closely related.

The relationship between two organisms can be assessed most accurately by determining how closely the nucleotide *sequences* of their DNA's match each other. If the sequences have many bases in common, then the proteins that are synthesized will be very similar, so the organisms must also be very similar and therefore closely related.

REPLICATION* OF DNA

Complementarity,
Appendix I

In the replication of a chromosome, the hydrogen bonds between the complementary bases are broken, and the two strands separate to form a fork (Fig. 5–3). Two new strands are synthesized, each one complementary to one of the original strands. The fork continues to progress along the original DNA molecule, and complementary strands continue to be synthesized. Since adenine will pair only with thymine and guanine with cytosine, the two new strands that are synthesized must be identical to the two original strands. Note that the sequence of purines and pyrimidines in the original DNA determines the sequence of purines and pyrimidines in each new strand. Therefore, the two new molecules of DNA are each identical to the original molecule. Because each of the new double-stranded molecules contains one old strand and one newly synthesized strand, the process is called **semiconservative replication**. The joining together of the subunits of DNA requires the energy of ATP.

Although the overall process of DNA replication in bacteria and humans is very similar, there are several subtle differences that provide points of attack for antimicrobial* medicines. One antimicrobial medicine that inhibits DNA replication in bacteria but not in humans is nalidixic acid. This drug inhibits the unwinding of the two strands of the helical DNA.

PROTEIN SYNTHESIS

Specificity of Antimicrobial Medicines in Protein Synthesis

Some of the most useful antimicrobial medicines interfere with protein synthesis in bacteria. The points of attack are the synthesis of mRNA and various steps involving the functioning of ribosomes. This selectivity for mRNA synthesis results from the fact that RNA polymerase, the enzyme that synthesizes mRNA, is different in bacteria and humans. Also, each of the two subunits of the ribosome is different in prokaryotic and in mammalian (eukaryotic) cells.

---

*replication—production of an exact copy; duplication

*antimicrobial—destroying or inhibiting the growth of microorganisms

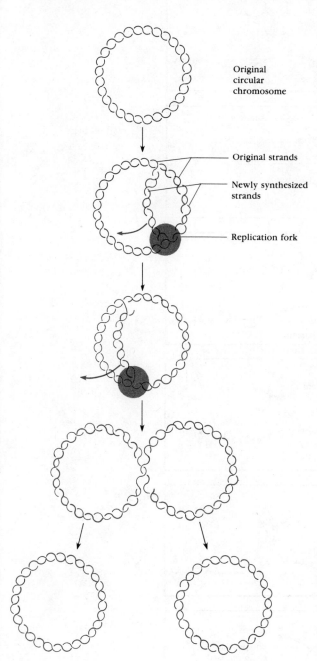

Original
circular
chromosome

Original strands

Newly synthesized
strands

Replication fork

Figure 5–3
Replication of a circular DNA
molecule. Note that each of
the two molecules that result
contains one of the original
strands and one newly synthesized strand of DNA.

114

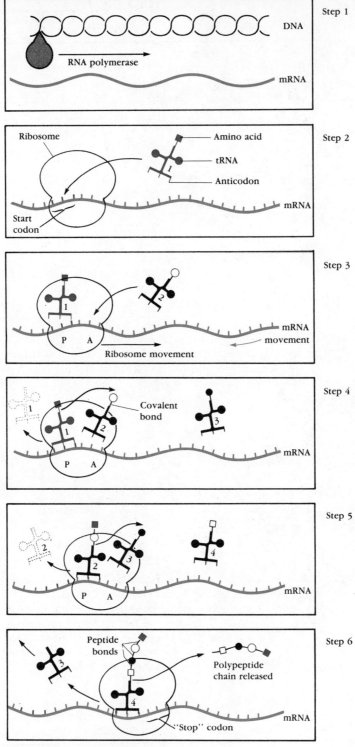

Figure 5–4
Stages of protein synthesis.

Stages of Protein Synthesis

Step 1 (Fig. 5–4). The first step in protein synthesis is the transfer of genetic information from the DNA into a molecule of messenger RNA (mRNA). This process is called **transcription.** The enzyme RNA polymerase attaches loosely to a site on one strand of the DNA just before the start of a gene. The enzyme moves down the length of the gene or sometimes several genes in a row, catalyzing the synthesis of a molecule of RNA, which is **complementary** to that strand of DNA. The joining of the subunits of RNA requires the use of ATP bond energy.

Complementarity, Appendix I

The antimicrobial medicine rifampin binds specifically to the bacterial RNA polymerase but not to the corresponding mammalian enzyme and therefore specifically inhibits transcription in bacteria.

Step 2 (Fig. 5–4). The **ribosome**—the "workbench" on which proteins are synthesized—binds to the end of the mRNA, which represents the beginning of the gene. The sequence of nucleotides in the mRNA is **translated** into a sequence of amino acids three nucleotides at a time, since three nucleotides, a **codon,** specify one amino acid. The decoder is another kind of RNA, **transfer** RNA (tRNA), a molecule which has two important ends. One end of the molecule termed the **anticodon** is a sequence of three nucleotides complementary to those of the codon. The other end of the tRNA molecule binds a specific amino acid (Fig. 5–5). Since there are 20 amino acids, there must be

Photograph of ribosomes, page 23

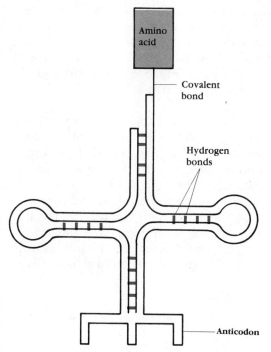

Figure 5–5
Transfer RNA (tRNA) molecule. Note the two important parts. One end binds to an amino acid; the other end, the anticoden, binds to the codon in mRNA.

at least 20 different tRNA molecules, each one having a different anticodon at one end and carrying a different amino acid at the other end. The codons for every amino acid have now been determined. This set of codons is the genetic code.

Step 3 (Fig. 5–4). The tRNA carrying an amino acid searches out its complementary codon on the mRNA at the ribosome. If the codon is properly positioned on the ribosome, then the anticodon will recognize it and bind to it using the energy of ATP. This first binding site is termed the "P site."

Step 4 (Fig. 5–4). The anticodon of a second tRNA molecule binds to the codon adjacent to the first at the A site. Thus, two sites on the mRNA can bind to the anticodon of tRNA molecules. A peptide bond* forms between the two adjacent amino acids carried by the two tRNA molecules. The first tRNA, the one at the P site, gives up its amino acid and leaves the ribosome to attach to another amino acid.

Step 5 (Fig. 5–4). A "ratchet" type mechanism moves the mRNA along the ribosome a distance of one codon so that the second tRNA molecule, now carrying two amino acids joined together, moves to the position formerly occupied by the first tRNA (the P site). The second codon site (site A) becomes vacant and therefore available to bind another tRNA molecule with its attached amino acid. The second amino acid in the growing chain binds to the end of the amino acid on the tRNA at site A, and again the tRNA at site P is free to leave the mRNA and pick up another amino acid (Fig. 5–6). Site P becomes occupied by the tRNA having the three amino acids attached, leaving site A open. As the mRNA moves along the ribosome, other ribosomes attach to the end of the mRNA and undergo steps 3 to 5. Five or six ribosomes can be attached to the same mRNA molecule (Fig. 5–6); each ribosome is concerned with the synthesis of one protein molecule.

Step 6 (Fig. 5–4). The peptide chain grows by the addition of amino acids as the mRNA moves along the ribosome. The formation of peptide bonds requires ATP. A typical protein requires steps 2 to 5 be repeated 300 times, since there are 300 amino acids in an average protein. A complete protein is synthesized on each ribosome, beginning when the ribosome attaches to one end of the mRNA, proceeding as the mRNA moves along the ribosome three nucleotides at a time, and ending when the ribosome reaches a codon that codes for no amino acid but serves as a "stop" signal. At this "stop" signal, the protein dissociates from the tRNA molecule and the mRNA from the ribosome. Thus, the mRNA is synthesized from one gene and is involved in the synthesis of one specific protein.

*peptide bond—bond formed between the —COOH group of one amino acid and the —NH$_2$ group of the adjacent amino acid during protein synthesis

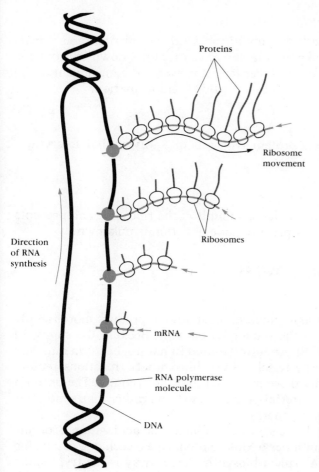

Proteins

Ribosome
movement

Ribosomes

Direction
of RNA
synthesis

mRNA

RNA polymerase
molecule

DNA

Figure 5–6
Diagram of many ribosomes
attached to the same mRNA
molecule.

Ribosome structure, Fig-
ure 2–20, page 41

SPECIFIC ANTIMICROBIAL MEDICINES AFFECTING PROTEIN SYNTHESIS

Aminoglycosides

Aminoglycosides, pages
197–198

A large group of antimicrobial medicines interfere with protein synthesis in
bacteria. The group of antibiotics called the aminoglycosides, which includes
streptomycin, kanamycin, and gentamicin, apparently affects protein synthesis at several stages. These drugs bind to the small subunit of the bacterial
ribosome but not to the host's and prevent its proper functioning.

Tetracyclines

Tetracyclines, page 196

The tetracyclines prevent the binding of tRNA at the ribosome, so that peptide bonds between amino acids cannot form.

Erythromycin, page 198

Erythromycin

Erythromycin binds to the large subunit of the bacterial ribosome, but its exact mode of action is unknown. In some way it slows down or stops the translation of mRNA into protein. Perhaps it prevents the movement of the mRNA along the ribosome or inhibits peptide bond formation.

Lincomycin, page 198

Lincomycin

Lincomycin probably affects protein synthesis in a way similar to erythromycin.

Chloramphenicol, pages
196–197

Chloramphenicol

This antibiotic also binds to the large subunit of the ribosome and prevents its proper functioning, but its precise mode of action is unknown.

REGULATION OF CELLULAR ACTIVITIES

Control of Biosynthetic Pathways

To survive in any environment, bacteria must often reproduce more rapidly than any other organism in the same environment. Because the supply of energy available for growth is generally limited in nature, bacteria must use the limited supply of energy wisely. Since biosynthetic reactions require energy, cells synthesize only those materials that they require. The bacteria will use preferentially any materials available to them rather than synthesize their own, thereby conserving energy.

As discussed in Chapter 4, the synthesis of an amino acid, vitamin, or any other material requires a number of different enzymes, each one catalyzing one step in the pathway. A typical biosynthetic pathway is the tryptophan pathway (Fig. 5–7). A compound in the medium, such as an amino acid, vitamin, or any final product of a biochemical pathway, signals the cell that it does not need to synthesize this particular compound. One way that bacteria shut off the synthesis of unneeded materials is to shut off the synthesis of all the enzymes of the particular pathway. In Figure 5–8, this is illustrated for the pathway of tryptophan biosynthesis. The end product of the pathway, in this case tryptophan, controls the binding of RNA polymerase to the DNA. If mRNA is not synthesized, the enzymes required for the synthesis of the particular end product are not synthesized, and the end product is not made. The end product of the biochemical pathway **represses** the synthesis of enzymes required for its synthesis (Fig. 5–8). The process is reversible. If the end product is in short supply and no longer available to repress RNA po-

Enzyme

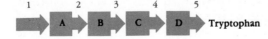

Figure 5–7
The pathway of tryptophan biosynthesis—a typical biosynthetic pathway.

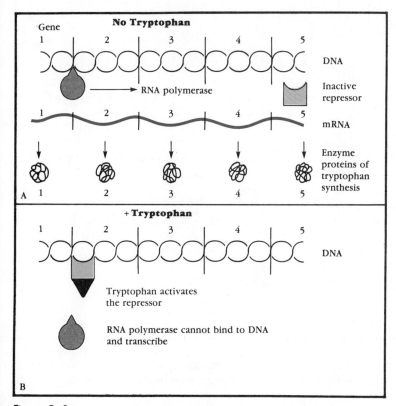

Figure 5—8
Repression of a biosynthetic pathway. The end product of the pathway (tryptophan in this case) binds to a protein molecule called a repressor. This protein now gains the ability to bind to the DNA and prevents the RNA polymerase enzyme from binding. Therefore, the genes of tryptophan biosynthesis are not transcribed. In this way, tryptophan shuts down the synthesis of the enzymes of tryptophan biosynthesis. When the tryptophan is removed, the repressor protein falls off the DNA, and RNA polymerase can bind and transcribe the genes of tryptophan synthesis. Note that each biosynthetic pathway has a different repressor protein.

lymerase binding, then RNA polymerase can bind to DNA and transcribe the DNA into mRNA. By this mechanism of control, the cell can conserve the energy required for the synthesis of enzymes as well as the materials from which the end product is synthesized. Note that tryptophan *only* represses the synthesis of enzymes involved in tryptophan production and does not affect any other pathway of amino acid synthesis.

The end product of a pathway also controls its own synthesis in a way completely unrelated to repression. This mechanism is termed **feedback inhibition.** The end product combines with a site on the first enzyme of the biosynthetic pathway termed the **allosteric site,** which is different from the **catalytic site** at which the substrate would attach. This alters the shape of the enzyme so that it cannot combine with its substrate (Fig. 5–9). As in

120

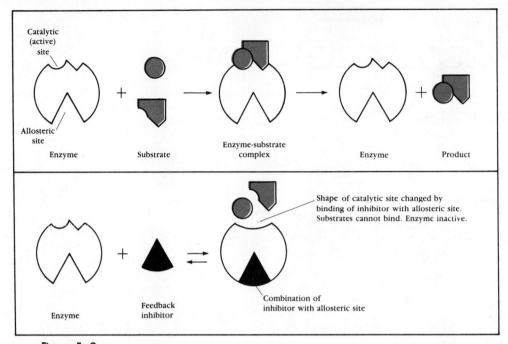

Figure 5–9
Inhibition of enzyme activity. The end product combines with the first enzyme of the pathway concerned with the synthesis of that product, thereby changing the shape of the catalytic (active) site.

repression, the effect is reversible, so that the enzyme can revert to its original shape and combine with its substrate if the end product is not available. Repression and feedback inhibition are compared in Table 5–1.

Control of Degradative Pathways

Bacteria can degrade a large number of different materials from which they derive energy and nutrients. However, if these compounds are not present in the environment, then it is wasteful in energy and nutrients if the orga-

TABLE 5–1 FEATURES OF ENZYME REPRESSION AND FEEDBACK INHIBITION

Features	Repression	Feedback inhibition
Mechanism of control	Transcription—enzyme synthesis	Enzyme activity
Enzymes involved	All enzymes	First enzyme of biosynthetic pathway
Speed of process	Slow—preformed enzymes still function	Immediate
Reversibility	Yes	Yes
Specific for particular pathway	Yes	Yes

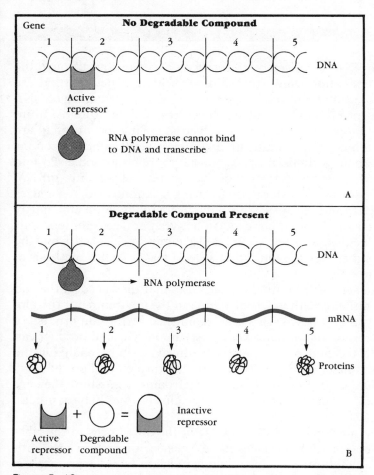

Figure 5—10
Control of a degradative pathway. The compound to be degraded combines with the repressor protein and inactivates it. Therefore, it cannot bind to the DNA, thereby allowing the RNA polymerase to bind and initiate transcription. In the absence of the degradable compound, the repressor binds to the DNA, thereby preventing transcription.

nism synthesizes enzymes involved in their degradation. Therefore, only when the degradable compound is present are the genes coding for the enzymes that degrade the compound actually transcribed by RNA polymerase (Fig. 5–10). The compound **induces** the synthesis of enzymes involved in its degradation.

Enzyme induction may also protect the cell from harmful antimicrobial medicines. Many bacteria can degrade or alter certain antibiotics, such as penicillin and chloramphenicol. However, the enzymes that act on the antibiotics are synthesized only when the substrate (the antibiotic) is present. In this case, the antibiotic induces the synthesis of enzymes that degrade or modify it.

SUMMARY

The information in DNA is passed on to succeeding generations through the process of DNA replication. New strands of DNA are synthesized that are complementary to each of the original strands. The base sequence of DNA codes for the sequence of amino acids in the corresponding proteins. In increasing order of complexity, the units involved in coding for a protein are nucleotides; three nucleotides make up a codon; several hundred codons make up a gene; about a thousand genes make up a chromosome. Protein synthesis begins with the transcription of the genetic message into mRNA, which binds to ribosomes. Transfer RNA, carrying an amino acid, recognizes a codon of the mRNA, and the anticodon of the tRNA binds complementarily with this mRNA codon. Peptide bonds form between adjacent amino acids on the tRNA molecules, and the completed protein is released from the ribosome. ATP bond energy is required at several stages in the synthesis of DNA, RNA, and protein.

Bacteria tend to synthesize only those enzymes that they need. If the end product of a biosynthetic pathway is available in the environment, the biosynthetic pathway is shut down. This is accomplished by the end product of the pathway controlling the *initiation* of transcription. The end product can also combine with the first enzyme of the pathway, thereby changing the enzyme's shape and inactivating it. Thus, cells can turn off and on the *functioning* of individual genes according to the environment in which they are growing. They also can control the *activity* of certain enzymes in biosynthetic pathways (Fig. 5–11).

Many antibiotic medicines take advantage of the differences in the ribosomes and RNA polymerases of bacteria and eukaryotic cells. These antibiotics preferentially inhibit protein synthesis in bacteria at various stages.

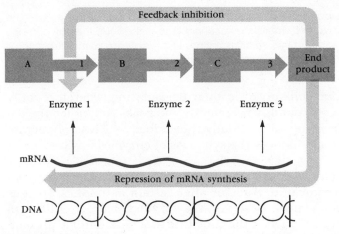

Figure 5–11
Summary of regulatory circuits by the end products of biosynthetic pathways.

SELF-QUIZ

1. DNA functions directly (Choose *two* of the following answers.)
   a. as the storage form of genetic information
   b. as the site of action of permeases
   c. in the control of enzyme activity
   d. as the material from which RNA is transcribed
   e. in the formation of peptide bonds
2. Transcription of DNA is associated with
   a. DNA replications
   b. synthesis of mRNA
   c. ribosome binding to mRNA
   d. peptide bond formation
   e. control by feedback inhibition
3. Complementarity is *not* associated with
   a. synthesis of mRNA
   b. DNA replication
   c. amino acids binding to tRNA
   d. tRNA binding to mRNA
4. The antimicrobial medicine that inhibits DNA replication is
   a. nalidixic acid
   b. streptomycin
   c. tetracycline
   d. erythromycin
   e. penicillin
5. Feedback inhibition directly involves
   a. control of mRNA synthesis
   b. DNA replication
   c. all enzymes of a biosynthetic pathway
   d. the control of activity of the first enzyme of a biosynthetic pathway
   e. transcription in degradative pathways

QUESTIONS FOR DISCUSSION

1. When the end product of a biosynthetic pathway is added to a culture of *E. coli* growing in a glucose-salts medium, is the synthesis of the end product stopped immediately? Why or why not? What is the first control mechanism that would have an effect?
2. Messenger RNA is degraded a few minutes after it is synthesized. Why is this important for bacteria?

FURTHER READING

Cairns, J.: "The Bacterial Chromosome." *Scientific American* (January 1966).

Clark, B. F. C., and Marker, K.: "How Proteins Start." *Scientific American* (January 1968).

Crick, F. H. C.: "The Genetic Code: III." *Scientific American* (October 1966).

Gorini, L.: "Antibiotics and the Genetic Code." *Scientific American* (April 1966).

Holley, R.: "The Nucleotide Sequence of a Nucleic Acid." *Scientific American* (February 1966).

Kornberg, A.: "The Synthesis of DNA." *Scientific American* (October 1968).

Miller, O.: "The Visualization of Genes in Action." *Scientific American* (March 1973).

The Molecular Basis of Life. San Francisco: W. H. Freeman, 1968. A collection of offprints of articles originally appearing in *Scientific American*, including most of those listed here.

Nirenberg, M.: "The Genetic Code: II." *Scientific American* (March 1963).

Nomura, M.: "Ribosomes." *Scientific American* (October 1969).

Watson, J. D.: *Molecular Biology of the Gene*, 3rd ed. New York: W. A. Benjamin, 1975. Presents the fundamentals of molecular biology with unusual clarity. The insight and perception that the author brings to the discussion of biological phenomena make this book a classic. It was written for an undergraduate course in molecular biology and biochemistry.

———. *The Double Helix*. New York: New American Library (Signet Books), 1969. Fast-moving story of the discovery of the structure of DNA told by one of the participants.

Yanofsky, C.: "Gene Structure and Protein Structure." *Scientific American* (May 1967).

Chapter 6

HEREDITY AND GENE TRANSFER IN BACTERIA

At the end of World War II, bacterial dysentery (caused by *Shigella*) was prevalent in Japan. With the introduction of treatment using the sulfonamides, the number of cases dropped dramatically as this antibacterial medicine proved to be highly successful. However, after 1949 the incidence of dysentery again rose. Coincident with this increase, investigators observed that most of the strains of *Shigella* were now resistant to the sulfonamides. Fortunately, the organisms were still susceptible to such antibiotics as streptomycin, chloramphenicol, and tetracycline. In 1955 a Japanese woman who had recently returned from Hong Kong developed *Shigella* dysentery that did not respond to antibiotic treatment. When the organism was isolated, it was found to be resistant not only to sulfa drugs but also to streptomycin, chloramphenicol, and tetracycline. In the next several years, several dysentery epidemics developed in Japan. In some patients the causative organisms were sensitive to the antibiotics; in others the organisms were resistant. Curiously, following treatment of certain patients with one antibiotic, bacteria were isolated from those individuals and found to be resistant not only to that antimicrobial medicine but also to several others that had not been used. In addition, *E. coli* residing in the colon* of many of those patients who carried antibiotic-resistant organisms were also resistant to the same antimicrobial medicine. Clearly, the advent of antimicrobial medicines reduced the incidence of numerous microbial infections. However, the development and spread of resistant strains of bacteria have created unique problems for the treatment of infectious disease. The understanding and explanation of these problems are the themes of this chapter.

---

*colon—large intestine

OBJECTIVES

To know
1. The difference between genotype and phenotype.
2. Two ways in which the genotype of an organism can be altered.
3. How one selects directly and indirectly for mutants.
4. Several ways by which changes can occur in the DNA of bacteria.
5. The three mechanisms by which DNA can be transferred between bacteria and the basic features of each process.
6. Three areas in which microbial genetics is important in industry and medicine today and ways in which information gained in studies of microbial genetics contributes to medicine and industry.
7. The steps for cloning foreign genes in bacteria.

GENERAL PRINCIPLES

The preceding story dramatically illustrates one of the most important features of bacteria: their ability to respond to changes in their environment by altering their properties. The fact that the properties of bacteria can change has very important consequences in the treatment of diseases with antibacterial medicines. Since organisms that were once killed by such medications are now resistant, new antibiotics must constantly be developed to keep pathogenic bacteria under control. Furthermore, changes in the properties of bacterial populations can occur with amazing speed since bacteria multiply so quickly. In this chapter we shall focus on the mechanisms by which the properties of bacteria become altered and the ways these altered properties can be transferred rapidly to other bacteria. The underlying theme will be the reasons why antibiotic resistance in a large number of species has become so widespread.

GENOTYPE VERSUS PHENOTYPE

The properties that bacteria display at any one time represent the information coded in the DNA coupled with the *expression* of that genetic information. The **genotype** of the organism is all the genetic information that an organism contains. The **phenotype** represents the characteristics that the organism displays in any particular environment. Thus phenotype depends on both the genetic information in an organism and the environment in which it lives. As pointed out in Chapter 5, bacteria can turn on and shut off gene transcription depending on the physical and chemical characteristics of the environment.

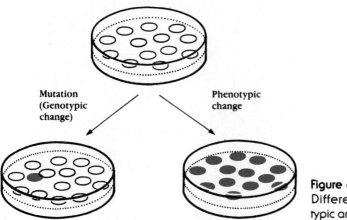

Figure 6-1
Difference between geno-
typic and phenotypic change.

ALTERATIONS IN THE GENOTYPE OF THE CELL

A bacterial cell can have its genotype altered in two different ways. First, the existing DNA of the organism can be altered. For example, the substitution of a different purine or pyrimidine for one already present in a DNA molecule can result in the synthesis of a protein that has a different amino acid. This type of modification of the DNA is called a **mutation.** A second way of altering the genotype of an organism is by adding new DNA to the organism. If this DNA contains information different from that originally present in the cell, then the cell will have an altered genotype. This mechanism was directly responsible for the rapid rise in *Shigella* strains resistant to so many different antimicrobial medicines, discussed in the opening of this chapter.

A change in the phenotype of a cell can be distinguished from a change in the genotype in at least two ways. First, if a modification of a cell's genome* occurs, then only a fraction of the cells in the population will be altered, not all of the cells. Second, unlike phenotypic changes, alterations in the genome are stable and do not change with changes in the environment. These differences are diagramed in Figure 6-1.

Gene Mutation

Since all properties of a cell depend on the proteins that a cell possesses, the substitution of even one amino acid for another among the several hundred in a protein molecule may alter the ability of a protein to carry out its function. For example, an alteration in the gene coding for the synthesis of a protein that is part of the bacterial ribosome confers streptomycin resist-

---

*genome—all the genetic material in the cell

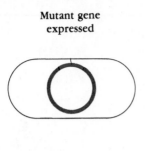

Mutant gene
expressed

Bacterial Cell
(Haploid)

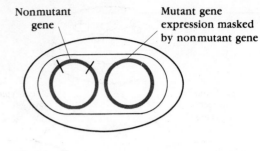

Nonmutant
gene

Mutant gene
expression masked
by nonmutant gene

Eukaryotic Cell
(Diploid)

Figure 6–2

Haploid versus diploid cell. Many mutations are recessive so that their presence is masked by the presence of the same gene which is nonmutant.

Prokaryotic cell proper-
ties, Table 1–1, page 13

ance on the cell. The alteration of a gene that codes for the enzyme used in histidine* biosynthesis may result in the synthesis of a nonfunctional enzyme. Therefore, the cell will require histidine in the medium in order to grow. A cell that has a specific nutritional requirement for an amino acid or vitamin is called an **auxotroph.** A cell that does not need supplementary nutrients is termed a **prototroph.** Since bacterial cells are haploid,* a mutation cannot be masked by an unmutated or normal gene (Fig. 6–2). In eukaryotic cells, which are usually diploid,* an unmutated gene *can* mask a mutant gene (Fig. 6–2), meaning that it can code for the protein that the mutant gene cannot specify. For example, a mutation in a gene coding for hemoglobin may result in a hemoglobin molecule that is unable to function properly in carrying oxygen. If the homologous gene* in the cell codes for a fully functional protein, then the anemia that results is not very severe. If mutations occur in both genes, then sickle cell anemia* may result.

Production of Mutations. Gene mutations can arise in a number of different ways. In all cases, the DNA of the cell is changed, and this change is passed on to all its progeny. Most mutations involve the incorporation of an incorrect purine or pyrimidine during the synthesis of new strands of DNA during DNA replication (Fig. 6–3). In this example, cytosine (the wrong base) is incorporated opposite adenine instead of thymine, the pyrimidine complementary to adenine. This results in the incorporation of the amino acid

DNA replication, pages
112–113

---

*histidine—an amino acid

*haploid—condition in which each type of chromosome is represented only once

*diploid—condition in which each type of chromosome is represented twice

*homologous gene—one of the pair of genes (or chromosomes) in diploid eukaryotic cells that carry information for the same trait(s)

*sickle cell anemia—a condition in which an individual produces abnormal hemoglobin molecules, resulting in altered red blood cell shape and decreased oxygen-carrying capacity

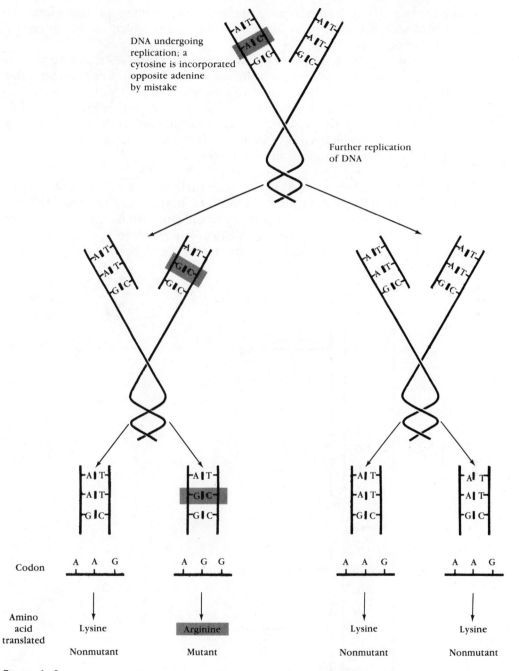

Figure 6–3
Production of a mutant by incorporation of the wrong base (cytosine instead of thymine).

arginine in place of lysine in the protein coded by this gene. Mutations are also produced if one or more nucleotide bases are added to or subtracted (deleted) from a gene (Fig. 6–4).

Replication of DNA is amazingly accurate, and mistakes that lead to mutations in any particular gene occur in only a very few of the cells undergoing replication. The frequency of mutations varies from 10^{-6} to 10^{-10} (a chance of from one in a million to one in ten billion) mutations per cell per division for any particular gene. However, since bacteria can reach concentrations of more than 10^9 cells per ml of medium, it is likely that almost every cell in the culture has a mutation in one or more genes!

Selection. To identify a cell with a particular mutation, it is necessary to be able to **select** for the mutation. Indeed, a mutation can be shown to have occurred only if the cell undergoes a recognizable change. It is relatively simple to select for certain characteristics. For example, a cell carrying a mutation that confers resistance to streptomycin can be readily detected by

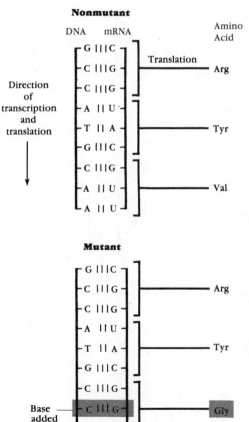

Figure 6–4

Production of a mutant by the addition of an extra base (cytosine) in the DNA.

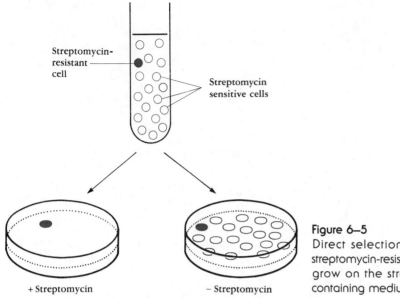

Streptomycin-resistant cell

Streptomycin sensitive cells

+ Streptomycin

– Streptomycin

Figure 6–5
Direct selection. Only the streptomycin-resistant cells will grow on the streptomycin-containing medium.

growing the culture on a solid medium that contains streptomycin. Only the streptomycin-resistant cells will grow to form colonies (Fig. 6–5). This **direct selection** is convenient for detecting any mutants that can grow under conditions in which the vast majority of cells will not grow. Note that streptomycin does not cause the mutation; it merely is the means for identifying the mutants that already exist in the population. Streptomycin *selects* for pre-existing mutants.

It is far more difficult to detect those cells in a population that *cannot* grow under conditions under which the vast majority of cells *will* grow. For example, how is it possible to detect the one mutant cell in a million cells that requires histidine for growth? An ingenious method termed **replica plating** was devised for this purpose. This is an **indirect selection** method (Fig. 6–6). Velvet cloth is touched to an enriched nutrient agar plate* containing well-separated colonies. Individual cells from each colony will stick to the velvet, which is then touched to a glucose-salts agar medium (which lacks histidine) as well as to the same medium supplemented with histidine. Any colonies that grow on the latter but not on the former medium must require histidine. Any colony not growing on either medium must be a mutant that requires a nutrient other than histidine. This method is a clever way of identifying the mutant colonies. But since a mutation is such a rare event, a great amount of replica plating is still required before a mutant colony can be found.

It is also possible to increase the percentage of auxotrophic mutants in a population by using penicillin. Since penicillin kills only cells that are grow-

Petri dish, page 9

Action of penicillin, pages 30–32

---

*plate—the common laboratory name for a Petri dish

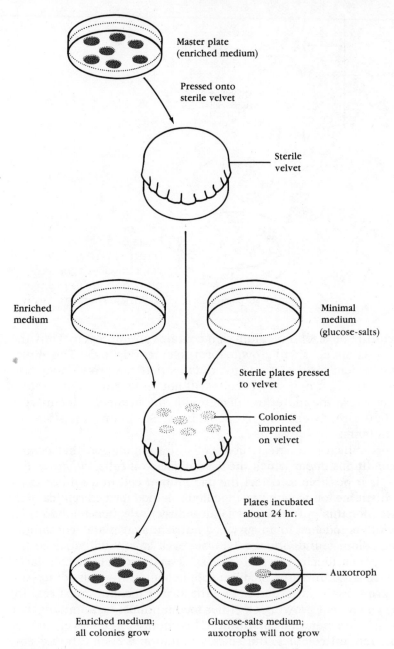

Figure 6–6
Replica plating.

Glucose-salts
minimal medium

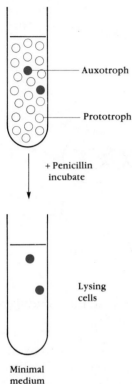

Auxotroph

Prototroph

+ Penicillin
incubate

Lysing
cells

Minimal
medium

Figure 6–7
Penicillin enrichment. Since penicillin kills only growing cells, any auxotrophic mutants will not be killed in a glucose-salts medium, but the unmutated cells will be.

ing, only prototrophs will be killed in a glucose-salts minimal medium. Since the auxotrophs are not killed, they will become a proportionately larger part of the total population. This technique of **penicillin enrichment** can increase the percentage of mutants in a population several thousandfold (Fig. 6–7). After this procedure, the cells are grown on a solid enriched medium and replica plated to identify the mutant colonies.

Mutagenic Agents. To increase the frequency of mutants in a population even more, geneticists have used agents that alter the structure or change the shape of DNA so that there is a thousandfold or so greater probability of the wrong purine or pyrimidine being incorporated in the course of replication. These are called **mutagenic agents** or **mutagens.** There are two general types: chemicals and radiation. Radiation includes ultraviolet light and x-rays. The chemicals include nitrous acid, which alters the hydrogen bonding properties of the purine and pyrimidine bases, and compounds that are similar but not identical to the purines and pyrimidines. These compounds "fool" the enzymes and are incorporated into DNA in place of the correct purine or pyrimidine; once incorporated, they have properties slightly dif-

Purines and pyrimidines, Appendix I

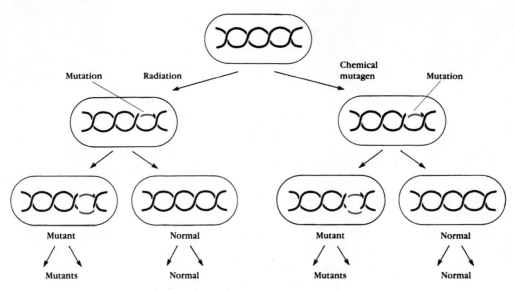

Figure 6–8
Permanent changes as a result of mutations.

ferent from those of the natural purines and pyrimidines. All these mutagenic agents produce permanent changes in the DNA (Fig. 6–8).

Frequency of Mutations. Since spontaneous mutations—those that occur without the use of known mutagenic agents—occur at a very low frequency, the chance that two mutations will occur in the same cell is extremely small. The frequency is the *product* of the frequency of each individual mutation. Thus, if the mutation rate to streptomycin resistance is 10^{-6} per cell division and to penicillin is 10^{-8}, then the probability that both mutations will occur in the *same* cell is the product of these two numbers, or 10^{-14},[1] an infinitesimally small number. For this reason **combined therapy,** the simultaneous addition of two or more antimicrobial medicines, is sometimes used in the treatment of certain diseases (tuberculosis in particular) that require treatment for long periods of time. Any of the infecting cells resistant to one antibiotic will probably still be sensitive to the second antibiotic and therefore will probably be killed.

From this discussion, it should not be surprising that bacteria resistant to one antibiotic can be isolated from natural* sources. Spontaneous mutations occur in the bacterial chromosome, which confers this resistance on the bacterial cells, and this is one important mechanism by which bacteria become resistant to an antimicrobial medicine. These mutant genes give no particular advantage to a bacterium in the *absence* of the antimicrobial med-

---

[1]Multiplying numbers with exponents is done by adding the exponents.

*natural—occurring in nature

icine to which it is resistant. Actually, the organism may be at a disadvantage, since it must use part of its energy to synthesize a material for which it has no use if the antimicrobial medicine is not present. However, the observation that organisms arise that are resistant to four or more unrelated antibiotics, such as the dysentery *Shigella*, cannot be explained by mutation. Some other mechanism must be involved to account for this phenomenon.

Plasmids

Photograph of plasmid, Figures 2–18 and 2–19, pages 39 and 40

We now know the mechanism by which cells become resistant to several quite different antimicrobial medicines simultaneously. An organism can become resistant because it receives from another bacterium a circular molecule of DNA, a miniature chromosome, which carries genes that code for resistance to several antimicrobial medicines. Another part of this DNA molecule contains the genes that allow it to replicate and also to be transferred from one bacterial cell to another (Fig. 6–9). This DNA molecule is given the general name **plasmid,** defined as a molecule of DNA that exists and reproduces independently of the cell's chromosome. It is now known that plasmids are very widespread in a large number of bacteria. Plasmids that carry resistance to antibiotics are termed **R plasmids** or **R factors**. Many R factors also carry genes that confer resistance to heavy metals such as mercury and cadmium. Other kinds of plasmids confer other abilities on bacteria.

All plasmids have one feature in common: They code for properties not required for the growth of the cell, but rather that might be extremely useful to the cell in certain environments (Table 6–1). The hospital environment is an excellent example of a situation in which R factors might prove valuable to a bacterial strain. In hospitals it is a common practice to clear air bubbles from syringe needles by expelling a small portion of the syringe contents into the air. Although the resultant aerosol* rarely exceeds 0.1 ml, it has nevertheless been estimated that this procedure could add 15 to 30 liters* of antimicrobial medicines to the environment of a hospital every year. The aerosol is breathed in by nurses and patients and thus becomes part of the

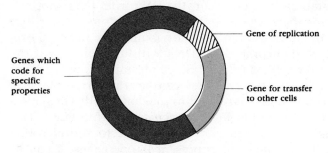

Figure 6–9
Diagram of a plasmid.

*aerosol—a suspension of liquid droplets in air

*liter—metric unit of volume slightly greater than a quart

TABLE 6–1   SOME PLASMID-CODED TRAITS

Trait	Organism
Antibiotic resistance	E. coli, Salmonella, Shigella, Neisseria, Staphylococcus
Pili synthesis	E. coli
Enterotoxin production	E. coli
Tumor induction in plants	Agrobacterium tumefaciens
Nitrogen fixation (symbiotic)	Rhizobium
Oil degradation	Pseudomonas
Gas vacuole production	Halobacterium
Insect toxin synthesis	Bacillus thuringiensis
Plant hormone synthesis	Pseudomonas
Antibiotic synthesis	Streptomyces
Increased virulence	Yersinia enterocolitica

secretion of the nose. In addition, generally about 40 percent of the patients receive antibacterial medicines. In such an environment there is a strong selection pressure for antibiotic-resistant organisms to survive and for antibiotic-sensitive organisms to be killed.

MECHANISMS OF GENE TRANSFER

Both plasmid and chromosomal genes can be transferred from one bacterial cell to another. There are three different mechanisms by which DNA transfer occurs from one cell, the **donor,** to another cell, the **recipient.** They are (1) **transformation,** (2) **conjugation,** and (3) **transduction.**

The DNA of the donor cell enters the cytoplasm of the recipient following its transfer by one of these three mechanisms. What happens to the donor DNA once it is inside the recipient cell depends on whether chromosomal or plasmid DNA is transferred. Since plasmids can exist independently of the chromosome, they can begin to multiply and code for proteins as soon as they enter (Fig. 6–10a). In contrast to plasmid DNA, chromosomal DNA cannot function independently of the recipient chromosome. This DNA must be integrated* into the recipient chromosome, a process termed **recombination** (Fig. 6–10b). Before recombination can occur, the DNA from the donor cell must pair with a complementary region on the recipient chromosome. Since complementary regions of chromosomal DNA are found only in closely related organisms, transferred chromosomal DNA will function only if the donor and recipient cells are closely related. Since plasmids usually do not integrate into the recipient chromosome, they can be transferred between and will function in genera that are unrelated to the donor cell. Thus, plasmids from *Pseudomonas* will function in *Escherichia* and *Rhizobium*. This has great significance as far as the transfer of resistance to antimicrobial medicines is concerned. If an R plasmid is present in one genus, it can be transferred to and function in about ten other genera.

*integrated—joined by strong chemical bonds

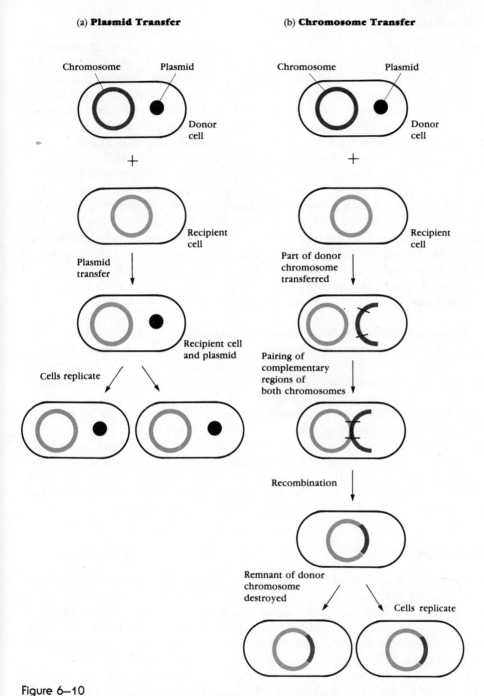

Figure 6–10
DNA transfer from one bacterium to another. (a) Plasmid transfer. (b) Chromosome transfer.

DNA-Mediated Transformation

In this mechanism of transfer, the DNA, whether plasmid or chromosomal, is transferred from donor to recipient cell as a "naked" piece of DNA. The cell walls of the donor cells are degraded, and the cells lyse. Since the DNA is tightly compacted into the cell, breakage of the cell membrane results in an explosive release of the chromosomal DNA, which fragments into about a hundred pieces. (Plasmids, on the other hand, are released intact as small circular chromosomes.) A fragment of donor DNA, which is about 20 genes in size, enters through the cell wall and cytoplasmic membrane of the recipient cell. Only some cells in the population can take up such large molecules. These cells are said to be **competent.** Chromosomal DNA, when integrated, displaces a piece of recipient cell DNA, which is then degraded by enzymes in the cell. The degradation products are recycled* (Fig. 6–11a). If the DNA that becomes integrated carries different genetic information than the DNA it replaces, the transformed cell (transformant) will have different traits if the gene is still functional. If the transforming DNA is a plasmid, it does not displace any information from the recipient cell; rather, it adds to the information already present (Fig. 6–11b).

Importance of DNA-Mediated Transformation in Nature. This mechanism of gene transfer may play a role in transferring antibiotic resistance coded by chromosomal genes from one organism to another, although this is not at all certain. Strains of *Streptococcus pneumoniae,* one causative agent of bacterial pneumonia, are appearing that are ten times more resistant to penicillin than previously isolated strains. Evidence suggests that this resistance is due to the selection of resistant mutants by the use of penicillin in treating patients, followed by the transfer of this resistance to formerly sensitive strains, perhaps by DNA-mediated transformation. It is interesting that in nature some cells in a population of pneumococci lyse naturally at the same time that other cells in the population become competent. Thus, transformation could occur in people colonized* with resistant and sensitive strains.

Conjugation

Conjugation is a process of DNA transfer that requires physical contact between donor and recipient cells. It is common in Gram-negative but rare in Gram-positive organisms. This contact is generally made by **pili,** which are coded by a plasmid carried by donor cells. After the donor and recipient cells touch each other, the entire plasmid is transferred to the recipient cell, conferring on it the properties encoded by the plasmid (Fig. 6–12).

Chromosomal DNA is not usually transferred during conjugation. However, some plasmids have the ability to integrate into the chromosome of the

Pili photograph, Figures 2–27 and 2–28, pages 47 and 48

---

*recycled—in this example, used as subunits for the synthesis of nucleic acids

*colonized—harboring a site of bacterial reproduction without necessarily exhibiting tissue damage

DNA-Mediated Transformation

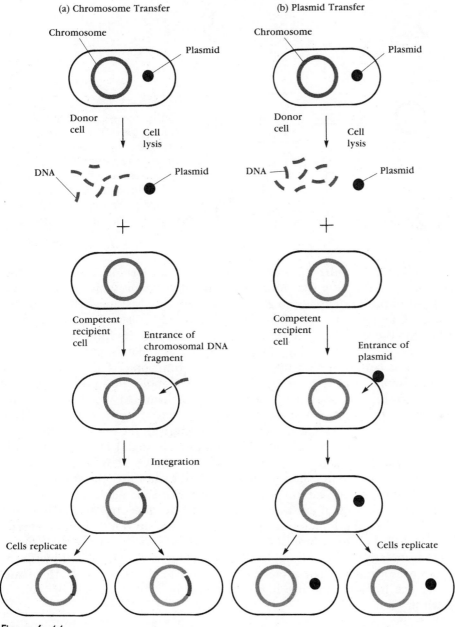

Figure 6–11
DNA-mediated transformation. (a) Chromosome transfer. (b) Plasmid transfer.

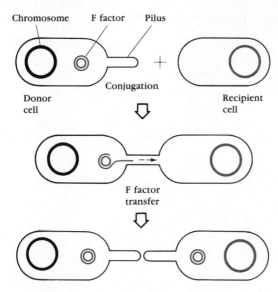

Chromosome   F factor   Pilus

Donor cell

Conjugation

Recipient cell

F factor transfer

Figure 6–12
Conjugation—plasmid transfer. Although there are many different plasmids, the one that has been studied most is the F factor, which codes for the synthesis of the sex pilus. This F factor confers male properties on the cell. Other plasmids include antibiotic resistance factors.

donor cell and "mobilize" it so that it becomes transferable. The best-studied plasmid for mobilizing the chromosome is the **F factor,** which codes for the synthesis of pili but carries no genes for antibiotic resistance. The donor cell with the F factor in its cytoplasm is termed an "F^+" or **male cell.** The recipient cell is the F^- or **female cell.** The donor cell with its integrated F factor is termed an "Hfr cell," for "high frequency of recombination," and it is these donor cells that can transfer their chromosome. The circular chromosome breaks and is transferred as a linear* molecule into the recipient cell (Fig. 6–13). The end of the donor DNA molecule at which the F factor integrates is transferred first; the integrated F factor is transferred last. Thus, there is an order in which the genes are transferred; the entire chromosome takes about 90 minutes to be transferred. Chromosomal DNA transfer continues only as long as the two cells remain in contact. Thus if the pili break, as they are prone to do, then the transfer of the chromosome stops.

Just as the F factor can integrate into the chromosome, it can also do the reverse and once again exist as a free F factor in the cytoplasm (Fig. 6–14). However, when the F factor excises* from the chromosome, it often carries with it a bit of chromosomal DNA. This DNA becomes a part of the F factor and is transferred just as though it were the plasmid. This plasmid with the bit of chromosomal DNA attached is given a new name, the **F′ (F prime) factor.** The interrelationship between F^+ factor, F′ factor, and Hfr cells is shown in Figure 6–15.

There are several differences in the transfer of plasmids and chromosomes. Plasmids are transferred within minutes to all the recipient cells, so that

---

*linear—having two ends

*excises—becomes free of; the opposite of integration

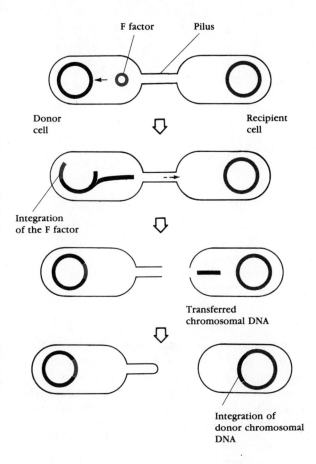

F factor   Pilus

Donor
cell

Recipient
cell

Integration
of the F factor

Transferred
chromosomal DNA

Integration of
donor chromosomal
DNA

Figure 6—13
Conjugation—transfer of chromosomal DNA. The DNA is transferred as a single-stranded DNA molecule, with the other strand remaining in the donor cell, and a complementary strand is synthesized. Thus the donor cell does not lose any genetic information.

mixing F^+ with F^- cells quickly results in a population of all F^+ cells. Any chromosomal genes attached to the plasmid will be transferred to all the recipient cells in the population. On the other hand, the chromosome is transferred to only a small percentage of the recipient population. So, mixing Hfr donor cells with F^- cells results in recombination in only a small percentage of the recipient cells. Furthermore, since the F^+ factor, which is

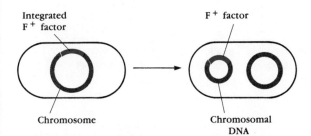

Formation of F'

Integrated
F^+ factor

F^+ factor

Chromosome

Chromosomal
DNA

Figure 6—14
Formation of F' factor.

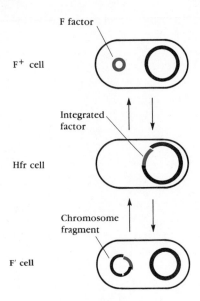

F factor

F⁺ cell

Integrated
factor

Hfr cell

Chromosome
fragment

F' cell

Figure 6–15
Interconvertibility between F⁺ factor, F' factor,
and Hfr cell.

integrated into the donor chromosome, enters last as the chromosome is
transferred, most of the cells that receive some of the donor chromosome
still remain F⁻. Plasmids, which include F' factors, can be transferred to and
function in a wide variety of unrelated bacterial genera. Chromosomal DNA
can function only in bacteria closely related to the species in which it originated. The properties of chromosome and plasmid transfer are summarized
in Table 6–2.

TABLE 6–2 COMPARISON OF CHROMOSOME AND PLASMID TRANSFER BY
CONJUGATION

Feature	Chromosome	Plasmid
Time required for transfer of entire molecule	90 minutes	Few minutes
Percentage of recipient cells that receive donor DNA	Low (0.1%)	High (1–100%)
Transfer between unrelated organisms	No	Yes
Transfer of entire molecule	Rarely occurs; specific order of gene transfer	Always occurs; no order of gene transfer
Number of different organisms in which it occurs	Few	Many
Directionality of transfer	Unidirectional; Hfr → F⁻	Unidirectional; F⁺ → F⁻

Transduction

In transduction, bacterial DNA is transferred from the donor to the recipient cell enclosed in the protein coat of a **bacterial virus (bacteriophage** or **phage).** In one common type of transduction (Fig. 6–16), the bacteriophage infects the bacterium and causes the bacterial chromosome to be broken into about a hundred pieces (almost the same number that the donor chromosome breaks into during DNA transformation). The bacteriophages multiply, and in the course of this multiplication one of the fragments of bacterial chromosomal DNA or a bacterial plasmid may become packaged accidentally into the bacteriophage particle. After the bacteriophages have finished multiplying, they often lyse the bacterial cells that they have infected (donor cells) and are released into the surrounding medium. The bacteriophages

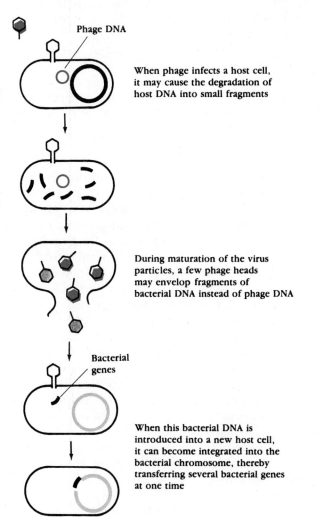

Phage DNA

When phage infects a host cell, it may cause the degradation of host DNA into small fragments

During maturation of the virus particles, a few phage heads may envelop fragments of bacterial DNA instead of phage DNA

Bacterial genes

When this bacterial DNA is introduced into a new host cell, it can become integrated into the bacterial chromosome, thereby transferring several bacterial genes at one time

Figure 6–16

Transduction. In this type of transduction, called "generalized transduction," any piece of donor DNA can be transferred.

infect other cells, and the bacterial DNA in the bacteriophage is transferred to these newly infected cells (recipient cells). If the transferred DNA is chromosomal, then it must undergo recombination with the recipient cells before the donor DNA can be expressed. If the donor DNA is a plasmid, then it can be expressed without delay. Of course, it is important that the recipient cells not be lysed by the phage, or the bacterial DNA to be transferred will be useless to the recipient cells.

Chromosomal DNA as well as plasmids can be transferred by transduction. Perhaps the best-known example of genes transferred by this mechanism involves the plasmids of *Staphylococcus aureus*. These plasmids code for a variety of antibiotic resistances, primarily to penicillin and erythromycin. Staphylococci are now frequently resistant to antibiotics because resistance is transferred to sensitive strains in the natural environment by transduction. These transductants* are able to multiply and become an important part of the total population of bacteria because of the widespread use of antimicrobial medicines.

Comparison of the Mechanisms of Gene Transfer

The major features of these gene transfer mechanisms are compared in Table 6–3.

GENE MOVEMENT

From the preceding discussion it should be apparent that mutations occurring in one organism can be transferred in many cases to a wide variety of other organisms. In addition to gene transfer *between* organisms, another type of gene movement was discovered recently. Individual genes can move from one DNA molecule (such as the chromosome) to another DNA molecule (such as a plasmid) in the *same* organism. The genes that "jump" from one DNA molecule to another are called **transposons** (Fig. 6–17). If another plasmid enters this organism, the gene can "jump" onto the new plasmid. If this plasmid is then transferred to other bacteria of a different genus, as many plasmids readily are, the new characteristic can become widely disseminated (Fig. 6–18). Exactly why a few genes are able to move in this fashion is not known. However, some of the kinds of genes that are able to "jump" are known. These include many genes for antibiotic resistance as well as a gene that codes for a heat-stable* toxin in *E. coli*. This toxin is responsible for diarrhea in humans and certain animals. Other transposons include genes concerned with degradation of some organic compounds.

The actual movement of one particular transposon, the gene coding for ampicillin resistance, has been followed. The same gene has been found in a number of different R factors present in a variety of unrelated bacteria isolated from around the world. Apparently, this piece of DNA jumped from

*transductants—recipient cells that have received and express donor DNA gained by transduction

*heat-stable—resistant to destruction by heating; heat-stable toxin of *E. coli* resists boiling for 30 minutes

TABLE 6–3   COMPARISON OF THE MECHANISMS OF GENE TRANSFER IN BACTERIA

Feature	*Transformation*	*Conjugation*	*Transduction*
Mode of DNA transfer	Across the cell membrane of the recipient	Through pili after cell-to-cell contact	Within the protein coat of a bacteriophage
Bacteria involved	Both Gram + and Gram −	Almost only Gram −	Both Gram + and Gram −
Amount of donor DNA transferred	About 20 genes	From 20 genes to entire chromosome	About 20 genes
Plasmid transfer	Yes	Yes	Yes
Resistance to deoxy-ribonuclease (degrades DNA)	No	Yes	Yes
Unidirectional transfer	No	Yes	No
Important as exchange mechanism in nature	Probably	Yes	Yes

Figure 6–17
The transposon moves from one DNA molecule to another in the same cell.

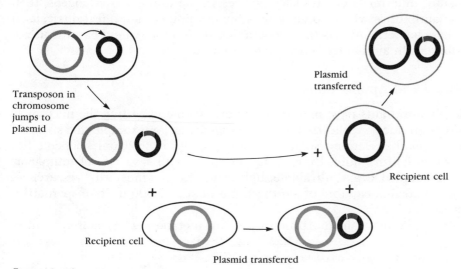

Figure 6–18
The movement of a transposon throughout a bacterial population.

TABLE 6–4 IDENTITY OF THE SAME AMPICILLIN RESISTANCE GENE IN A VARIETY OF ORGANISMS

Organism in which isolated	Country in which isolated
E. coli	Japan
Pseudomonas rettgeri	Greece
Pseudomonas aeruginosa	
Salmonella paratyphi	United Kingdom
Klebsiella pneumoniae	France
Providencia species	Canada
E. coli	Greece

one R factor to another, and the R factors were then transferred from one organism to another (Table 6–4). This transposon has also been shown to "jump" from a plasmid to a chromosome.

PRACTICAL ASPECTS OF MICROBIAL GENETICS

Development of Antibiotic Resistance

Undoubtedly, the use of antibiotics has greatly increased the number of antibiotic-resistant organisms in the environment (Fig. 6–18). Not only are antibiotics used to treat infections in humans, but antibiotics such as tetracycline are given to chickens and other food animals. Approximately 40 percent of the antibiotics used in the United States are added directly to livestock feed, and the biggest source of antibiotics for most people is not medication but the meat they eat. There is great concern that the use of these antibiotics in such a manner is increasing the incidence of organisms resistant to these antibiotics in the environment. If the bacteria that become resistant in animals can transfer this resistance to human pathogens, then these antibiotics will become useless for treating human infections caused by these bacteria. At this time, federal legislation that would ban the use of antibiotics in animal feeds has been proposed.

Commercial Applications

Microbial mutants have played an important role in getting the maximum yield of products synthesized by microorganisms. For example, the strain of *Penicillium* that Fleming isolated initially produced 2 to 20 units of penicillin per ml of culture fluid. By treating the mold with a variety of mutagens, which included x-rays, ultraviolet light, and nitrogen mustard,* researchers isolated strains capable of synthesizing about 10,000 units of penicillin per ml.

Many mutants that overproduce certain metabolites* because of muta-

*nitrogen mustard—a toxic chemical used as a chemical warfare agent in World War I

*metabolites—any products of metabolism

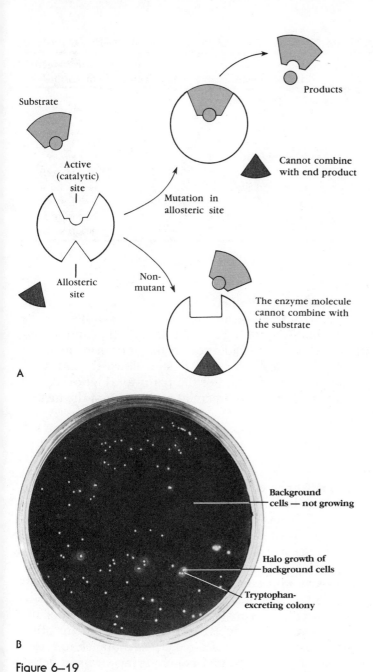

Substrate

Active
(catalytic)
site

Allosteric
site

Mutation in
allosteric site

Non-
mutant

Products

Cannot combine
with end product

The enzyme molecule
cannot combine with
the substrate

A

Background
cells — not growing

Halo growth of
background cells

Tryptophan-
excreting colony

B

Figure 6–19

Altered regulation resulting in the overproduction of an end product. (a) The theoretical basis for overproduction of an end product as a result of a mutation in an allosteric site. (b) The background cells require tryptophan and are not growing except around those colonies which are excreting tryptophan. The cells in these colonies overproduce and excrete tryptophan because they have a mutation in the allosteric site in the first enzyme of tryptophan synthesis.

TABLE 6–5 SOME MATERIALS TESTED FOR MUTAGENIC AND CARCINOGENIC ACTIVITY

Material	Area of use	Mutagenic	Carcinogenic in animals
Vinyl chloride	Chemical industry	+	+
Captan	Fungicide	+	+
Cigarette smoke condensate	Recreation	+	+
Nicotine	Tobacco	–	–
Marijuana	Recreation	–	Uncertain
Moon Haze No. 32[a]	Hair dye	+	Not tested
AF-2[a]	Food additive	+	+
Charred fish and beef	Food	+	Not tested
Coffee	Food	+	Not tested
Tea	Food	+	Not tested

[a]These compounds have been banned or modified.

tions involving the regulation of biosynthetic pathways have been isolated. As pointed out in Chapter 5, bacteria are able to control the synthesis of the end products of metabolic pathway through feedback inhibition and repression of enzyme synthesis. However, a mutation in the allosteric site of an enzyme results in a cell that is no longer able to shut down the pathway. Such strains overproduce the end product of the pathway (Fig. 6–19) and prove to be extremely valuable in the commercial production of the metabolite.

Testing for Potential Carcinogens*

A relatively inexpensive, rapid, and accurate test has been devised based on the observation that about 90 percent of the agents that induce cancer in mice also induce mutations in bacteria. Therefore, Dr. Bruce Ames at Berkeley, California, devised a simple test to determine whether a compound can cause mutations in bacteria. This test, called the Ames test, is based on the fact that a mutagen can change a cell that has a certain nutritional requirement to one that has no such requirement. Just as a mutation can alter the sequence of purines and pyrimidines in DNA, thereby resulting in the synthesis of a nonfunctional protein, another mutation can modify the sequence of nucleotides back to its original form. This results in a **reversion** of the mutation so that the cell then has a properly functioning enzyme or other protein. Thus, in the Ames test the ability of a suspected mutagen to revert a histidine-requiring strain of *Salmonella* to a nonrequiring strain is determined. Since the nonrequiring strain can grow on a medium lacking histidine, whereas the histidine-requiring strain cannot, it is possible to select *directly* for the reversion by plating on a medium that lacks histidine. The

*carcinogens—cancer-causing agents

simple technique of the Ames test is diagramed in Figure 6–20. The sensitivity, inexpensiveness, and speed of this test result from the characteristics of bacteria. They can be grown in very large numbers in a small volume, so that rare revertants* can be detected. Furthermore, bacteria grow so rapidly that revertants can be detected within a few days. In this way, numerous compounds have been tested for their mutagenic activity. A positive test suggests, but does not prove, that the compound may be a potential cancer-causing agent in humans. Some of the compounds that have been tested are listed in Table 6–5. Unfortunately, a negative test does not completely exclude the possibility that a substance will be mutagenic even though it does not give a positive reaction in the Ames test. It is also conceivable that compounds that are mutagenic in the Ames test will not be carcinogenic. Therefore, additional tests must be performed to substantiate the results of this test.

Genetic Engineering

As pointed out in Chapter 5, the DNA of bacteria directs the synthesis of all the protein made by the cell. It is now possible to take advantage of the bacterial cell as a factory that synthesizes protein. A wide variety of foreign

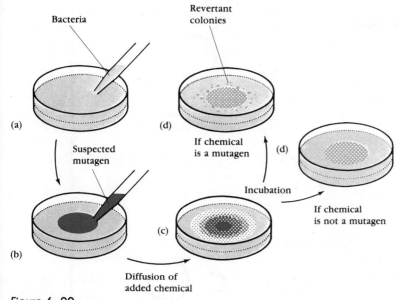

Figure 6–20
The Ames test.

*revertants—organisms in which a second mutation has restored one of their original properties, in this case the ability to grow in the absence of histidine

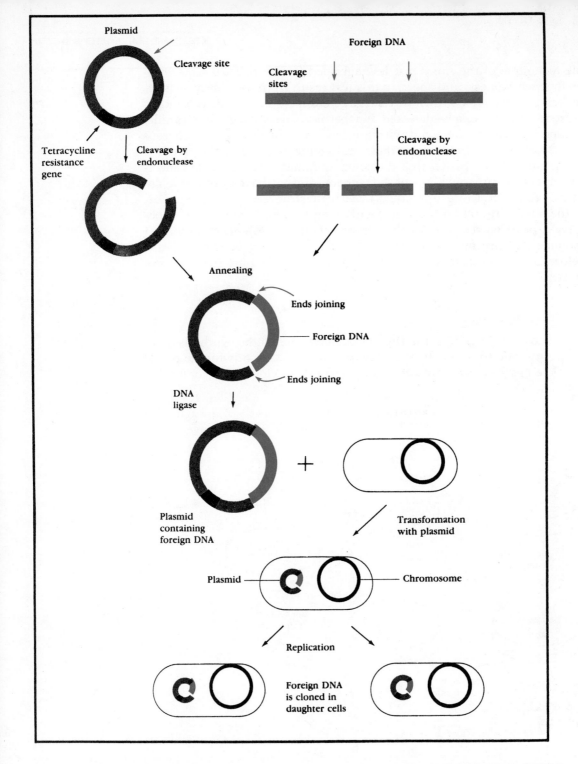

Figure 6–21
Cloning of foreign genes in bacteria.

genes from both animals and humans have been introduced into bacteria. The bacteria do not distinguish between these foreign genes and their own genes but synthesize the proteins coded by these genes. Active research is being conducted to have bacteria produce in large amounts such materials as insulin, interferon, somatostatin,* and blood clotting factors.

Genetic engineering involves first breaking open a circular plasmid molecule and then inserting the foreign genes. The plasmid is resealed by connecting its broken ends to the foreign genes. The plasmid containing the foreign DNA is transformed into *E. coli*. As the bacteria multiply, the plasmid with its foreign DNA also multiplies, producing exact copies of itself. A foreign gene, thus established in a strain of bacteria, is said to be **cloned,** since exact copies are synthesized every time the bacteria reproduce (Fig. 6–21). As the genes of the plasmid are expressed, the materials coded by the foreign genes are synthesized.

SUMMARY

Variations can arise in a microorganism as a result of two phenomena: changes in the nucleotide sequence of its DNA (mutations) and changes in the environment that determine which genes will function. Changes in the nucleotide sequence of a DNA molecule occur as a result of a nucleotide substitution or the addition or deletion of nucleotides. These changes in the DNA are rare, but their frequency can be increased many thousandfold by mutagenic agents. Most cancer-inducing agents are also mutagenic.

Three mechanisms of genetic exchange have been demonstrated in bacteria: transformation, in which "naked" donor DNA is taken up by the recipient cell; transduction, in which the donor DNA is carried to the recipient DNA inside a bacteriophage; and conjugation, in which the donor DNA passes from the donor to the recipient during cell-to-cell contact. Plasmid DNA can be transferred to more distantly related bacteria than can chromosomal DNA because plasmids function without becoming integrated into the DNA of the recipient cell. Some genes have the ability to "jump" from one DNA molecule to another DNA molecule within the same cell. Such gene movement coupled with the ability of DNA to be transferred to other bacteria explains how antibiotic resistance spreads among distantly related bacteria. Microbial genetics plays an important role in various facets of industry and medicine.

SELF-QUIZ

1. Streptomycin added to a culture of *E. coli*
   a. provides a selective mechanism to find bacteria that are resistant to streptomycin

---

*somatostatin—a growth hormone

    b. serves as a mutagen and causes an increase in the frequency of strep-
tomycin-resistant mutants in the medium

    c. inhibits growth of cells resistant to streptomycin

    d. selects for cells defective in cell wall formation

    e. increases the number of streptomycin-resistant cells in the population

2. An Hfr cell is one in which

    a. the F factor is transferred at a high frequency

    b. the genetic material replicates at a high frequency

    c. the F factor is incorporated into the bacterial chromosome

    d. the F factor exists as an independent unit in the cytoplasm

    e. the F factor is transferred immediately to an F^- cell

3. R factors are *not*

    a. transmissible by conjugation

    b. plasmid DNA molecules possessing antibiotic resistance genes

    c. chromosomal resistance genes

    d. often able to transfer resistance to four or more antibiotics simulta-
neously

    e. any of these

4. The longest piece of donor DNA can be transferred by the process of

    a. conjugation

    b. transduction

    c. transformation

    d. depends on the species of organism

5. A geneticist mixes together 10^8 mutant cells that require methionine and
are streptomycin-sensitive and 10^8 cells that require phenylalanine and
thiamine and are streptomycin-resistant. To detect whether conjugation
has occurred, it would be best to plate the mixture of cells on

    a. enriched medium

    b. methionine + phenylalanine

    c. phenylalanine + streptomycin

    d. minimal medium + streptomycin

    e. minimal medium + methionine + phenylalanine + thiamine

QUESTIONS FOR DISCUSSION

1. Why is it reasonable to find that plasmids code only for functions that
are not essential to the life of the cell?

2. Which kind of mutation—base substitution or addition—would have the
most severe consequences for the cell? Why?

FURTHER READING

Broda, P.: *Plasmids.* San Francisco: W. H. Freeman and Co., 1979. A well-written
book presenting very broad coverage of the field.

Brown, D.: "The Isolation of Genes." *Scientific American* (August 1973).

Clowes, R.: "The Molecule of Infectious Drug Resistance." *Scientific American* (April 1973).

Cohen, S.: "The Manipulation of Genes." *Scientific American* (July 1975).

Cohen, S., and Shapiro, J.: "Transposable Genetic Elements." *Scientific American* (February 1980).

Devoret, R.: "Bacterial Tests for Potential Carcinogens." *Scientific American* (August 1979).

Falkow, S.: *Infectious Multiple Drug Resistance.* London: Pion Limited, 1975. An advanced text that covers in considerable detail most aspects of plasmids. Written by one of the major research investigators in this field.

Gilbert, W., and Villa-Komaroff, L.: "Useful Proteins from Recombinant Bacteria." *Scientific American* (April 1980).

Grobstein, C.: "The Recombinant DNA Debate." *Scientific American* (July 1977).

Novick, R. P.: "Plasmids." *Scientific American* (December 1980).

Tomasz, A.: "Cellular Factors in Genetic Transformation." *Scientific American* (June 1958).

Wollman, E., and Jacob, F.: "Sexuality in Bacteria." *Scientific American* (July 1956).

Chapter 7

STERILIZATION AND DISINFECTION

Pseudomonas aeruginosa, page 311

Not long ago, patients receiving treatment on kidney machines in a large city hospital abruptly experienced high fevers and teeth-rattling chills. Because septicemia* was suspected, their blood was cultured in the microbiology laboratory and was shown to contain the Gram-negative rod-shaped bacterium *Pseudomonas aeruginosa*. The same microorganism was also found to be present in large numbers in the kidney machines. The presence of the bacteria was due to a failure of a pump designed to force a cleansing and bacteria-killing fluid through the machines after each use.

In an unrelated incident miles away, an elderly man experienced an upset stomach followed by symmetrical paralysis of his face and double vision. He was taken to the local hospital, where he suddenly became unable to breathe and died despite all efforts to save him. Subsequently, his wife and two friends, who along with the patient had eaten a salad containing home-canned string beans, also developed botulism.* Fortunately, their doctors, alerted by the diagnosis of the previous case, were able to save the man's wife and friends.

Both these events represent failures of microbial control measures; the first, because of inadequate delivery of a bacteria-killing chemical and the second, because of improper use of heat in canning string beans. Understanding some basics of microbial control helps ensure that such failures do not occur.

---

*septicemia—illness caused by microorganisms circulating in the bloodstream; "blood poisoning"

*botulism—the disease caused by the toxin of *Clostridium botulinum*

OBJECTIVES

To know
1. The distinctions between sterilization and disinfection.
2. The differences in time needed to kill different types of organisms.
3. The differences among the various control measures that employ heat—boiling, autoclaving, pasteurization, and tyndallization.
4. The limitations of filters.
5. The uses and limitations of chemical agents and radiation.
6. How to decide which control agent is the best one to use for a given purpose.

APPROACHES TO CONTROL

Sterilization

Removing or killing *all* microorganisms and viruses on an object or in any material is called **sterilization.** "Killing" means making them unable to reproduce *under any conditions* and thus is an irreversible process. Microorganisms can appear to be dead but may grow if the environment is changed. An example would be organisms in an acid medium that appear to be dead after heating but are able to grow when the medium is neutralized by the addition of an alkali.

It is common practice to speak of "killing" viruses, even though they are not "living" in the sense that microorganisms are. The word "killing" is applicable to viruses because it specifies only that they are no longer able to reproduce.

Disinfection

Disinfection means reducing the number of pathogens on a material until they are no longer a hazard. Unlike sterilization, disinfection implies that some living microbes may persist. A **disinfectant** is a chemical used for disinfection of inanimate objects, whereas a disinfectant nontoxic enough to be used on human tissues is called an **antiseptic. Decontamination** is often used interchangeably with disinfection, but it implies a broader role, including inactivation or removal of both microbial toxins and living microbial pathogens themselves. The term "sanitize" is often used in food preparation and housekeeping activities that involve disinfection and cleaning to make something aesthetically pleasing. Technically, **sanitization** is the process of substantially reducing microbial populations on objects to acceptably safe public health levels.

A disinfectant capable of killing microbes rapidly is a **germicide.** Special terms, such as bactericide, fungicide, and viricide, are used to indicate killing action against specific microbial groups. Related terms, such as bacteriostatic and fungistatic, indicate that an antimicrobial agent is primarily inhibitory in its action, preventing growth without substantial killing. Growth can resume if the inhibitory agent is neutralized or diluted. An agent that is germicidal against one species of microorganism may be only inhibitory against another, and the nature of its action may change depending on its concentration, pH, temperature, and other factors.

SOME BASIC PRINCIPLES

Only a fraction of the microbes present die during any given time interval. For example, if 60 percent of the organisms are killed in the first two minutes by a given treatment, then 60 percent of those surviving the first two minutes will be killed during the next two minutes. It follows from the first principle that the time it takes to achieve sterility depends on the number of organisms that existed on a material at the beginning. This is one reason why prevention of contamination and mechanical methods of removing microorganisms (such as scrubbing, mopping, and washing) are important to the successful use of disinfecting or sterilizing agents.

Various microorganisms and viruses differ in susceptibility to sterilizing agents (Table 7–1). There is variation among and within species. There is also variation with growth phase, most organisms being at peak susceptibility when actively growing. Spores of microorganisms are generally more resistant than the vegetative forms. Fungal spores are not as resistant as bacterial endospores, although many of those responsible for food spoilage easily resist pasteurization.* Also, some viruses (such as the hepatitis virus) are killed relatively slowly by heat and disinfectants. Many other factors, including temperature, influence the rate of killing of microorganisms. The higher the temperature, the more rapidly the microorganisms are killed. For example, a population of tubercle bacilli killed in 30 minutes at 58°C could

Endospores, page 49

Picornaviruses, Appendix VIII

TABLE 7–1 RANKING OF DIFFERENT TYPES OF MICROORGANISMS AND VIRUSES ACCORDING TO RESISTANCE TO STERILIZING AGENTS[a]

1. Bacterial endospores
2. Mycobacteria such as *Mycobacterium tuberculosis*
3. Certain fungal spores
4. Non-lipid-containing viruses (such as picornaviruses)
5. Vegetative fungi
6. Lipid-containing viruses (such as influenza viruses)
7. Vegetative bacteria

[a]Ranging from most resistant (bacterial endospores) to least resistant (vegetative bacteria). The table depicts the general pattern, but there are a number of exceptions.

---

*pasteurization—the process of heating food or other substances under controlled conditions of time and temperature (for example, 63°C for 30 min) to kill pathogens and reduce the total number of microorganisms without damaging the substance pasteurized

be killed in only 20 minutes at 59°C, while only 2 minutes would be needed to kill them at 65°C.

The rate of killing is also influenced by the characteristics of the suspending fluid, such as pH and viscosity.*

HEAT AS AN ANTIMICROBIAL AGENT

The use of heat for microbial control long antedates the discovery of the microbial world and remains to this day the most generally satisfactory control method. Heat is fast, reliable, and cheap, and it does not introduce toxic substances into the material being treated. It kills by coagulating the proteins of cells. The use of heat is limited, however, to substances that can withstand the temperatures required for microbial killing and to objects small enough to fit into a suitable container. Cooking of food is a familiar example of the use of heat for microbial killing. However, a "well done" appearance does not always guarantee that a food is safe to eat. Temperatures sufficient to kill the microbes in question must be maintained for an adequate time *throughout* the material being heated and not just on its surface.

Dry Heat

Heating is frequently used to kill microorganisms in materials other than food. Sterilization of laboratory equipment, for example, the direct flaming of wire transfer loops, is a common application. To sterilize glass Petri dishes and pipettes, the glassware is put into ovens with static* air and left at temperatures of 160° to 170°C for two to three hours. Circulating hot air sterilizes in half the time of static air because of the more efficient transfer of heat to laboratory glassware. Dry heat requires much more time than wet heat to kill microorganisms (200°C for 1½ hours of dry heat to give the killing equivalent of 121°C for 15 minutes of moist heat). Dry heat technology has been used for many years in the sterilization of spacecraft.

Boiling

The use of boiling to prevent disease extends back at least to the time of Aristotle, who advised Alexander the Great to have his armies boil their drinking water. Under ordinary circumstances, with concentrations of microorganisms of less than one million per milliliter, suspensions of vegetative cells and eukaryotic spores can be sterilized in boiling water at 100°C in about ten minutes. Most viruses also die, although if large concentrations of hepatitis virus are present, some may possibly survive boiling for longer times. Some bacterial endospores, including certain strains of the food poi-

*viscosity—state of being sticky or gummy

*static—not moving

A

C

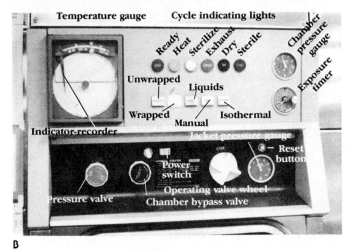

B

Temperature gauge   Cycle indicating lights

Ready   Heat   Sterilize   Exhaust   Dry   Sterile   Chamber pressure gauge

Unwrapped   Liquids

Wrapped   Manual   Isothermal   Exposure timer

Indicator-recorder

Jacket pressure gauge

Power switch   Reset button

Pressure valve   Operating valve wheel
Chamber bypass valve

Figure 7–1
(a) Modern autoclave. On left are records of autoclave function. (b) Close-up view of controls. (c) Temperature recording gauge. (Photographs courtesy of C. Iden Roberts.)

Clostridium species, page 303

soning bacteria *Clostridium perfringens* and *Clostridium botulinum,* survive boiling for hours.

The problem with boiling is that water temperatures never get hot enough to kill many endospores. When heat is applied to water at sea level, the temperature rises until it reaches 100°C (212°F). An additional amount of heat energy (heat of vaporization) is then taken up by the water without a change in temperature, whereupon boiling begins to occur. The temperature does not go above 100°C, and at high altitudes, water boils several degrees lower than this because of the reduced atmospheric pressure. Alkali is sometimes added to boiling water to increase its killing power against bacterial endospores. Even with such methods, boiling cannot be relied upon to kill the most resistant strains of spores. It is therefore *not* a reliable sterilizing technique.

Pressure Cooking and Autoclaves

Pressure cookers were probably first used for sterilization in about 1860. These devices simply enclosed the vessel containing heated water so that pressures above those of the atmosphere might be attained. A suitable safety valve was included to prevent an explosion. Water and steam at temperatures above 100°C could thus be used, killing even the most heat-resistant microorganisms in a few minutes.

The modern autoclave (Fig. 7–1) is, in principle, a sophisticated pressure cooker with mechanisms for regulating the steam pressure and for ensuring complete evacuation of air from the chamber (Fig. 7–2). The presence of air

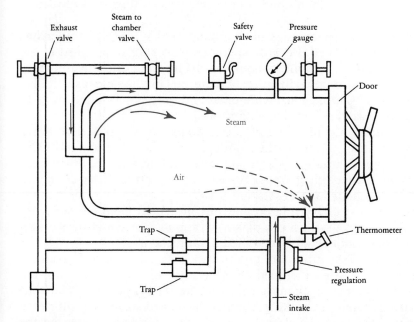

Figure 7–2
Steam-jacketed autoclave. Entering steam displaces air downward and out through a port in the bottom of the chamber. Objects should be placed in the autoclave so as to avoid trapping air.

would allow an increase in pressure within the chamber without a corresponding increase in temperature. Thus, it is important always to check the temperature as well as the pressure, because the latter is of little or no significance in killing microorganisms. Superheated steam (dry steam) is likewise unsatisfactory and kills at slow rates, similar to those of hot air. The conditions usually employed for sterilization are 15 pounds pressure above atmospheric pressure and 121°C (250°F) for 15 minutes. In operating rooms, where more rapid sterilization of instruments is sometimes important, a higher temperature and pressure are commonly employed. This is called flash autoclaving.

The main reason for the great effectiveness of the autoclave compared with the hot air oven relates to the fact that heat coagulates proteins, including essential enzymes, more readily under moist than dry conditions. Release of the heat of vaporization by steam condensing on organisms may play a role in the coagulation of proteins and rapid killing. On the other hand, dry proteins resist denaturation, and the killing of organisms with dry heat occurs by unknown mechanisms. Killing appears to take place almost equally well whether or not oxygen is present, indicating that the chemical reaction of microbial substances with atmospheric oxygen is not the primary cause of death.

The autoclave is the simplest and most consistently effective means of sterilizing most objects. The following are some practical aspects of its use.

1. Air trapping. The importance of removing air from the chamber has already been mentioned. Because steam displaces air downward in the chamber, long, thin containers should not be inserted into the autoclave in an upright position. Sterilization of materials in plastic bags will likewise fail, unless the bags are wide open and only partly filled so as to allow easy access of steam and escape of air.

2. Heat penetration. A cold object placed in an autoclave takes time—over and above the time at which the chamber temperature reaches 121°C— to warm up. For example, a 1-liter container of fluid must be autoclaved for 25 minutes after the chamber reaches 121°C, but a 4-liter container may require an hour for sterilization.

3. Contact with steam. Dry microorganisms protected from contact with steam (for example, within oil, plastic wrappings, or containers of talcum powder) cannot reliably be killed by autoclaving.

4. Heat indicators. Test tubes and tapes containing a heat-sensitive chemical indicator are often included along with objects when they are autoclaved. The indicator changes color during the autoclaving and thus gives a visual means of checking whether the objects have been subjected to adequate heat. A changed indicator does not always indicate that the object is sterile, because heating may not have been uniform, and even stored wrapped objects may have organisms reintroduced (see No. 7).

Tubes or envelopes containing large numbers of heat-resistant spores of the nonpathogenic bacterium *Bacillus stearothermophilus*, which acts as a biological indicator, are also frequently included with packs of materials being autoclaved. Death of these spores indicates adequate killing at the point where the tube or envelope was placed, which should be near the center of the pack.

5. Elevated boiling points under pressure. Fluids in autoclaves are prevented from boiling, even though well above their normal boiling point, by the elevated pressure. If at the end of the period of autoclaving, valves are opened to release this pressure, these fluids will immediately boil and may even explode their containers. The pressure must be maintained to prevent them from boiling until temperatures have dropped below the boiling point at atmospheric pressure.

6. Deleterious effects of 121°C heat on some materials. Most autoclaves can operate at temperatures lower than 121°C. The use of lower temperatures is satisfactory for materials that can be tested for successful sterilization before being used. For example, certain heat-sensitive bacteriological media are sterilized at 115°C. This temperature is usually effective because heat-resistant microbial contaminants are rarely present in the media. Moreover, samples of such autoclaved media are tested for sterility before being used.

7. Prevention of recontamination. Objects to be autoclaved are usually wrapped in paper or cloth to allow penetration of steam during sterilization and to prevent recontamination thereafter. When wet, these coverings are readily permeable to bacteria and should therefore be allowed to dry before removal from the autoclave chamber. After that they should be stored in a closed cupboard or drawer to prevent the reintroduction of contaminants.

Pasteurization

In the 1800's, the French wine industry was simultaneously threatened by foreign competition from without and by a microbial invasion from within. Good wine is the result of properly controlled conversion of fruit sugars to ethyl alcohol by yeast. Bacteria and molds from the environment may become established in the wine, degrading the alcohol and producing various ill-flavored metabolites. Pasteur found that with moderate heating of wine, at just the right conditions of time and temperature, these spoilage microbes could be killed without significantly changing the taste of the wine. He verified his findings by supplying heated and unheated wines to sailors of the French Navy, who consistently found the heated wines to be superior.

This process of controlled heating at temperatures below boiling (now called pasteurization) helped save the French wine industry and is widely used today for ridding milk and other foods of bacterial pathogens, such as those causing tuberculosis and typhoid fever. The high-temperature, short-time process is called flash pasteurization (72°C for 15 seconds). The longer method takes 30 minutes at 63°C. Although one does not ordinarily think of pasteurizing such things as linen, this can easily be done by regulating the temperature of the water in a washing machine. The times and temperatures required may be different from those required for pasteurizing milk. For example, in some hospitals anesthesia masks are pasteurized at 80°C for 15 minutes. The temperatures and times used vary according to the organisms present and the heat-stability of the material. Pasteurization is *not* equivalent to sterilization. However, pasteurization causes a substantial reduction in the numbers of microbes present and thereby often retards spoilage in foods.

Tyndallization

Repeated controlled heating at relatively low temperatures can sterilize, because it will kill even spore-forming bacteria, provided the suspending medium enables their spores to germinate. Recommended procedures for tyndallization specify heating at 100°C for 30 minutes on three successive days or at 60°C for one hour on five successive days. The material is usually incubated at 37°C following each period of heating. Tyndallization is reliable only with liquids that are suitable growth media.

FILTRATION

Fluids that cannot tolerate heat can be sterilized by filtration. This procedure has many uses such as in space technology and in the production of medications such as eye drops. Filters capable of removing bacteria from fluids were devised during the last decade of the nineteenth century (Fig. 7–3). They were made from metallic compounds of silica,* including porcelain, diatomaceous earth, and asbestos, partially fused by intense heat. Filters of this type are still used today. Their action is often thought to be that of a microscopic sieve, holding back microbes while letting the suspending fluid pour through the small holes in the sieve. In fact, the passages that run through such filters are very tortuous; some have large open areas, and many others are blind. The diameter of the passage is often considerably larger than that of the microbes they retain, and trapping of microbes by electrical forces may be involved because the walls of the filter passages are electrically charged.

From these considerations, it is easy to see that the effectiveness of the microbial filters just discussed varies not only with their pore size but also with the chemical nature of the suspending fluid and the amount of pressure used to transfer it across the filter. These filters are often used satisfactorily to remove infectious agents larger than viruses from aqueous fluids. They are relatively inexpensive and unlikely to clog. They are not suitable for some other purposes, however, because they absorb relatively large amounts

*metallic compounds of silica—substances formed by the reaction of the element silicon with metals

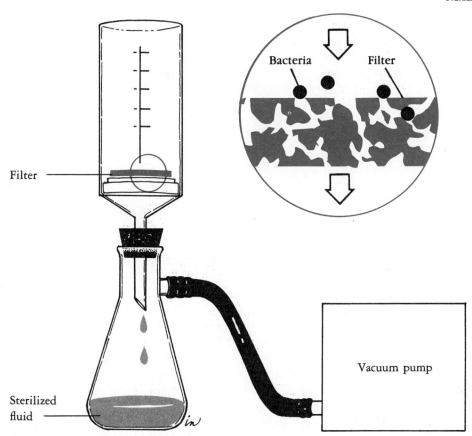

Figure 7–3
Filtration apparatus. Liquid to be sterilized flows through the filter on top of the flask in response to a vacuum produced in the flask by means of a pump.

of filtrate; they may introduce metallic ions into the filtrate, or they may adsorb biological substances such as enzymes from the filtrate.

Enzymes, page 84

In recent years, membrane filters composed of compounds such as cellulose acetate have been widely used in the laboratory and industry. Beer may now be filtered by this means rather than being pasteurized. The filters are paper-thin and are produced with graded pore sizes extending below the dimensions of the smallest known viruses. They are relatively inert chemically and adsorb very little of the suspending fluid or its biologically important constituents. Furthermore, they become semitransparent when immersion oil is applied, thus allowing microscopic inspection for any microbes that adhere to their surface. Their chief disadvantages are that they are expensive and that they clog easily.

Filters also figure in plans for space exploration, being used to sterilize heat-susceptible fluids in spacecraft. Because scientists want to avoid introducing microorganisms from earth onto other planets, the International

Committee on Space Research has specified that when a vehicle lands on another planet, the probability of its carrying a single microbe must be less than 1 in 10,000. As indicated previously, heat is a highly reliable method of ridding the metallic surfaces of space vehicles of living microbes. But what about the reliability of the filters needed to free heat-labile fluids of microbes? Studies comparing different kinds of filters have indicated that the older types—diatomaceous earth, porcelain, and asbestos—were not sufficiently reliable. Only membrane filters gave satisfactory results for use in sterilizing heat-susceptible fluids in space vehicles. Even so, it is very difficult to guarantee the complete absence of every microorganism, although we can usually state the probability of a microorganism escaping the procedure designed to remove it.

Except for the smallest pore size membranes, filters will not sterilize fluids containing viruses. Bacteriophages and other viruses readily pass through most bacteriological filters and, if present in the original suspension, will be present in the filtrate.

CHEMICALS

Chemicals are generally unreliable sterilizing agents, the only exception being ethylene oxide, a toxic gas used for materials that cannot be autoclaved. Chemicals do, however, have value as disinfectants and preservatives. The major groups of disinfectants are described in subsequent paragraphs and consist of the alcohols, the halogens (chlorine and iodine), the formaldehyde group, phenolics, quaternary ammonium compounds, and metallics (Table 7–2). Most of these are inactivated by organic materials; this emphasizes the need for mechanical cleaning before disinfectants are used. Even ethylene oxide may fail to kill vegetative bacteria and other types of infectious agents if they are enclosed in pus or blood.

Heat generally enhances the action of these disinfectants. Chemical disinfectants work very slowly against bacterial endospores, against the bacteria that cause tuberculosis, and against some viruses. The hepatitis viruses represent a special problem, because they often reach high concentrations in blood, and their response to disinfectants is difficult to study. Extended treatment with selected viricidal agents is presumed to be effective (Table 7–3). Carefully cleaned instruments that allow free access of chemicals to all their parts can probably be disinfected by less stringent techniques (Table 7–4). The choice of the best disinfectant for a given purpose should be based on knowledge of its chemical nature and on testing under the conditions in which it will be used. Appendix IV gives the recommended chemical and times for various instruments.

The ingredients of soaps, medicines, deodorants, contact lens solutions, foods, and many other everyday items often include an antimicrobial agent that acts as a preservative. Heparin (a substance used to prevent blood clots) may contain a phenol derivative, contact lens solutions may contain a quaternary ammonium compound, and a leather belt may be treated with one

TABLE 7–2 MAJOR GROUPS OF CHEMICAL AGENTS USEFUL IN DISINFECTION

Type of disinfectant	Probable mode of action	Level of activity[a]
Alcohols	Coagulate proteins	
Ethyl		Intermediate
Isopropyl		Low
Halogens	Oxidize proteins	
Iodine in alcohol or water		Intermediate
Iodophores		Low to intermediate[b]
Chlorine, chlorine compounds		Intermediate
Formaldehyde	Precipitate proteins	
Formaldehyde in alcohol		High
Glutaraldehyde (alkaline)		High
Phenolics	Destroy cell membrane	Intermediate
Quaternary ammonium compounds	Destroy cell membrane	Low
Mercurials (a group of metallics)	Precipitate proteins; react with enzymes or other essential cellular components	Low
Gaseous agents	Destroy cell (by reacting with	
Ethylene oxide	the structural and enzymatic components)	High

Action of mercury, page 89

[a]Activity at the concentrations generally employed:

High: -cidal activity against vegetative bacteria, bacterial endospores, fungi, and human viruses except hepatitis.

Intermediate: -cidal activity against vegetative bacteria, fungi, and most human viruses except hepatitis; activity inadequate against endospores.

Low: -cidal activity against most vegetative bacteria and fungi; activity inadequate against tubercle bacilli, endospores, and many viruses.

[b]Activity of iodophores at a concentration of 75–150 ppm of available iodine. Concentrations of 750–5000 ppm available iodine give intermediate to high activity.

or more phenol derivatives. The same preservatives and disinfectants may be sold under a variety of trade names. The purpose of these agents is to prevent or retard spoilage by microbes that are inevitably introduced from the environment. Unfortunately, some people experience rashes and other forms of hypersensitivity as a result of repeated exposures to such preservatives and disinfectants.

TABLE 7–3 CHEMICAL AGENTS USEFUL AGAINST HEPATITIS VIRUSES[a]

Agent	Time
Sodium hypochlorite (5000–10,000 ppm available chlorine)	30 minutes
Formaldehyde 16% in water	12 hours
Glutaraldehyde 2% (alkaline)	10 hours

[a]For use only when heat or ethylene oxide gas sterilization is impossible.

From: "Perspectives on the Control of Viral Hepatitis, Type B," *Morbidity and Mortality Weekly Report.* Vol. 25 (supplement), 7 May 1976.

TABLE 7—4   DISINFECTION OF FIBEROPTIC BRONCHOSCOPES
USED TO EXAMINE PATIENTS WITH HEPATITIS B[a]

Agent	Time
Iodophore (5000 ppm available iodine)	10–30 minutes
Formaldehyde 8% in water	10–30 minutes
Glutaraldehyde 2% (alkaline)	10–30 minutes

[a]These instruments, used for looking into people's lungs, are damaged by heat and by ethylene oxide. They must be thoroughly cleaned before using the above disinfectants.

From: *Hepatitis Surveillance Report Number 41.* Communicable Disease Center, Atlanta, Georgia, September 1977.

Alcohols

Ethyl and isopropyl alcohols rapidly kill vegetative bacteria and fungi at 50 to 80 percent concentration but are of little value against spores and some medically important viruses such as hepatitis viruses. In the past, some doctors, dentists, and other medical personnel prepared their instruments by soaking them in alcohol. This procedure sometimes resulted in serious illness and death because of the failure of alcohol to kill hepatitis viruses on the instruments. Alcohols act by coagulating essential proteins, and since proteins can only be coagulated when they are in solution, water must be present for alcohol to work. The final concentration of alcohol in the material being disinfected should be about 75 percent by volume for optimal results. Because of the difficulty in hydrating spores, however, they are frequently resistant to alcohols. Valuable as disinfecting agents by themselves, alcohols can also enhance the activity of other chemical agents, such as iodine, chlorhexidine, and quaternary ammonium compounds. The use of alcohols is, of course, limited to materials resistant to their solvent action. Alcohol swabs are often used as an antiseptic on the skin before an injection is given.

Halogens

Chlorine. The use of chlorine as a disinfectant in municipal drinking water and in the water of swimming pools is familiar to most people. The amount that must be added depends on the amount of organic matter in the water, since organic matter binds the chlorine and makes it unavailable for action against microorganisms and viruses. Properly chlorinated drinking water requires about 0.5 part per million (ppm)[1] of free chlorine. Chlorine is widely used in disinfection in the form of free gas and in solutions containing hypochlorite ion. Chlorine gas reacts with water to produce hypo-

[1]The concentrations of chlorine used for killing microorganisms and viruses are usually measured in parts per million (ppm). A 0.5 ppm solution would be equivalent to 0.5 gram of chlorine in a million grams (1000 liters) of water.

chlorite, which is thought to act on microorganisms by oxidizing essential proteins.

Household bleaches (such as Clorox) consist of about 5 percent sodium hypochlorite solution, and three quarters of a cup of such preparations in a gallon of water results in a solution of approximately 1000 ppm of chlorine. This concentration is several hundred times the amount lethal to most pathogenic microorganisms, but these high concentrations are usually necessary for faster killing and also because organic material is often present. Where hepatitis B virus is a constant danger, such as in hemodialysis areas or autopsy rooms, a sodium hypochlorite solution of 5000 ppm of chlorine should be used for cleaning. Use of chlorine disinfectants is limited to materials resistant to the corrosive action of these compounds. Certain kinds of rubber and certain metals, for example, are broken down by chlorine. Chlorine is also irritating to the skin and mucous membranes.

Iodine. Iodine, like chlorine, is very active against most microbial species, irreversibly oxidizing essential molecules in the cell. Like several other disinfectants, iodine has enhanced activity when dissolved in alcohol to produce a tincture. However, experiments show that simple swabbing of the skin with tincture of iodine does not reliably kill bacterial endospores, such as the spores of *Clostridium perfringens*. Fortunately, high concentrations of the endospores of bacterial pathogens are not generally found on clean skin; other antiseptics are no more sporocidal than tincture of iodine. Compounds of iodine with surface-active agents (iodophores) have been very popular because they do not irritate the skin or sting as much as tincture of iodine when applied to wounds, and they are not as likely to stain. However, the alkaline water characteristic of some parts of the country can result in inadequate antimicrobial activity of iodophores when concentrated stock solutions of these disinfectants are diluted for use. Tincture of iodine and the iodophores are widely used as antiseptics but must be avoided in those individuals (estimated as 1 in 10,000) known to have an allergy to iodine.

Formaldehyde

Eight percent formaldehyde in water is an extremely active disinfectant, killing most forms of microbial life in minutes. Lower concentrations, however, are only slowly active against some bacterial endospores and some viruses. Failure to recognize this fact has caused several near-disasters from the use of formaldehyde to kill living agents in vaccines. Solutions of formaldehyde also have limited use because of their irritating vapors, and in recent years other aldehydes have been used instead. For example, glutaraldehyde can be used as a disinfectant and a sterilizing agent if a treatment time of 10 to 12 hours is employed. After treatment, glutaraldehyde is easily removed by rinsing in water.

Phenolics

Phenolics form a very large group of compounds chemically related to phenol (carbolic acid). The group includes cresols and xylenols.[2] Most commercial preparations are used at a concentration of 2 percent or less. They kill most vegetative bacteria, and in high concentrations (from 5 to 10 percent) they even kill the tubercle bacillus (*Mycobacterium tuberculosis*). They act by destroying the cell membrane. Five percent phenol is also effective against all the major groups of viruses, but some of the more commonly used derivatives of phenol lack sufficient antiviral activity, and phenolics are not generally employed for viral disinfection. Lower concentrations of phenol and its derivatives may be ineffective. In fact, some Gram-positive cocci and Gram-negative rod-shaped bacteria grow in the presence of 0.1 percent phenol and actually use it as a source of carbon. The major advantages of phenolics include their wide spectrum of activity, reasonable cost, and ability to remain active in the presence of soaps and other detergents.

M. tuberculosis, page 325

Phenolic disinfectants are now commonly used to wipe all surfaces of operating rooms after each use. This practice effectively kills and removes *Pseudomonas aeruginosa*, a Gram-negative rod-shaped bacterium to which burned patients are particularly susceptible.

A patient with burns infected with *P. aeruginosa* was taken to the operating room for cleaning of the wounds and removal of dead tissue. After the procedure was completed, tests were performed to see to what extent the organism had spread. *P. aeruginosa* was recovered in all parts of the room (Fig. 7–4). This is an example of how readily and extensively an operating room can become contaminated by an infected patient. This is why operating rooms and other patient care rooms are cleaned thoroughly after use. The process is known as **terminal cleaning** (Table 7–5).

Hexachlorophene and Chlorhexidine. One phenol derivative, hexachlorophene, deserves special mention. In many hospitals, the surgeons and nursery personnel are required to scrub their hands for five minutes with a hexachlorophene solution before treating patients. This chemical has substantial bacteriostatic and bactericidal activity against *Staphylococcus aureus* and tends to be retained by the skin, so that its antimicrobial activity increases with repeated use. However, it is not effective against many Gram-negative bacteria. The use of hexachlorophene during the two decades before the late 1960's was of major importance in controlling infections caused by *Staphylococcus aureus*. Since then, the use of hexachlorophene has been restricted, since infants repeatedly immersed in soaps containing this disinfectant were shown to absorb quantities of the disinfectant. This absorption produced brain damage in test animals. Hexachlorophene, however, is of proven value in preventing nursery staphylococcal infections, and with its judicious use the risk of brain damage in infants would appear to be too

---

[2]See Appendix V for chemical formulas.

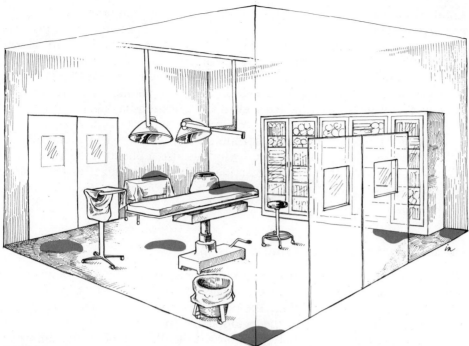

Figure 7—4
Diagram of an operating room in which a patient's burns infected with *Pseudomonas aeruginosa* were debrided of dead tissue. The colored areas indicate places where *P. aeruginosa* was recovered following the surgical procedure.

TABLE 7—5 PROCEDURE FOR WEEKLY TERMINAL CLEANING OF AN OPERATING ROOM

Equipment
 Germicidal detergent cleaner (disinfectant)
 Floor machine
 Scrubbing and stripping pads
 Wax for conductive floors
 Cloths
 Pails
 Wet mop
 Bucket with wringer
 Ladder
 Vacuum, wet pick-up

Procedure
 Mix disinfectant with water in amount indicated on container. Using a wall mop, clean the ceiling with this solution. Then clean all lights, lighting equipment, and vents. Next wash the walls, windows, and doors.
 Change the water and proceed to wash all tables, counter tops, stools, and examining tables.
 Flood the floor and let stand 10 minutes, then mop up. If floor is badly marked or scuffed, spray with wax and buff.

small to justify its abandonment. Immersion in hexachlorophene baths is now avoided, as is the repeated use of hexachlorophene in conditions that facilitate its absorption, such as on extensive areas of broken skin. Hexachlorophene absorbed through the skin of pregnant women has been suspected of causing birth defects. For this reason, the Food and Drug Administration suggests avoidance of the use of these compounds by those who are or may become pregnant, at least until the question has been resolved by scientific studies.

There is a tendency for hexachlorophene to be replaced by a chemically different antiseptic—chlorhexidine—which is also a phenol derivative. This agent has the advantage of being active against Gram-negative organisms while possessing the prolonged action of hexachlorophene against staphylococci.

Quaternary Ammonium Compounds

Quaternary ammonium compounds, commonly called "quats," represent a large group of compounds called surface-active agents, which reduce the surface tension* of liquids. Like the phenolics, they act against many vegetative bacteria by attacking the cytoplasmic membrane. Quaternary ammonium compounds are widely used for the disinfection of clean inanimate objects. They are also used as preservatives in nonfood materials, an example being eye drops, where their use prevents bacterial growth. These uses have sometimes led to infections in human beings because some bacteria, notably species of *Pseudomonas*, are resistant to the quaternary ammonium compounds and often grow in their presence. In other instances, there has been a failure to recognize that these substances are easily inactivated by soaps, detergents, and organic materials such as gauze. Nevertheless, quaternary ammonium compounds are economical and effective sanitizing agents. Their effectiveness and safety increase when they are combined with other disinfecting agents.

Metallic Compounds

Mercury compounds are largely bacteriostatic. They were widely used as disinfectants and antiseptics for many years. Formerly they were thought to have strong antimicrobial properties, including the ability to kill spores. Because of this mistaken idea, many people were infected by medical instruments that had been soaked in mercury compounds and that were erroneously thought to be sterile. Trade names of mercurial antiseptics include Mercurochrome and Merthiolate.

---

*surface tension—a property of liquids whereby their surfaces resemble a thin elastic membrane under tension

Compounds of mercury, tin, arsenic, copper, and other metals were once widely used as preservatives and to prevent microbial growth in recirculating cooling water and industrial processes. Their extensive use has resulted in serious pollution of natural waters and has led to strict controls on their use. For example, mercurials may no longer be employed in pulp mills for controlling slime-producing microorganisms.

Gaseous Agents

Ethylene oxide is the most useful gaseous antimicrobial agent in commercial and hospital practice. It penetrates well into fabrics, anesthesia equipment, and intravenous catheters* (in contrast to formaldehyde) and is one of the very few chemical agents that can be relied on for sterilization. Because it is explosive, it must be mixed with some inert gas such as carbon dioxide or Freon. The effectiveness of ethylene oxide also depends importantly on temperature and relative humidity, which must be carefully controlled. Its use is therefore restricted to a special chamber where temperature, pressure, and relative humidity can be regulated. In practice, 3 to 12 hours are generally required to sterilize.[3] With the use of heated, forced air cabinets, the time for airing of the sterilized material is 8 to 12 hours, depending on the temperature of the cabinet. If the materials are left at room temperature, it might take a week or more for the absorbed gas to be released. One must allow absorbed ethylene oxide to dissipate because of its irritating effect on tissues or because of the undesired persistence of its antimicrobial effect. Disposable plastic dishes, syringes, rubber, and other heat-sensitive items used in laboratories or for medical purposes are commonly sterilized with ethylene oxide. It should not be used for items employed for testing carcinogenicity, however, because traces of residual ethylene oxide can produce false positive results.

Fumigation has been used as a weapon against infection at least as far back as in ancient Greece, where burning sulfur was apparently used for purification. Formaldehyde gas has likewise had many years of use and is still recommended for decontaminating air filters on the cabinets used for culturing *Mycobacterium tuberculosis*.[4] Heating the powdery polymer of formaldehyde (paraformaldehyde) is a safe and effective way of generating the gas if the concentration and relative humidity are controlled to prevent explosions and repolymerization. Formaldehyde generated in this manner can be used in laboratories in which botulism is being studied, not only to kill the endospores of *Clostridium botulinum* but also to neutralize effectively the powerful nervous system toxin produced by this organism.

---

[3]See Appendix IV for recommended treatment and times for disinfection and sterilization of instruments.
[4]In the handling of *M. tuberculosis* and other especially dangerous infectious agents, special cabinets are often used in the laboratory to prevent accidental dispersal of organisms into the air.

*intravenous catheters—plastic tubes inserted into a vein to provide medications, fluid, or nutrients

RADIATION

Electromagnetic radiations can be thought of as waves having energy but no mass. Examples include x-rays, gamma rays, and ultraviolet and visible light rays. The energy that electromagnetic rays possess is proportional to the frequency of the radiation (the number of waves per second). Thus, electromagnetic radiation of short wavelength, such as gamma rays, has much more killing power than that of long wavelength, such as visible light. Gamma rays and ultraviolet radiation have proved to be valuable tools for microbial control.

Gamma Rays

Salmonella species, page 315

Gamma rays are an example of ionizing radiation that causes biological damage by producing hyperreactive ions and other molecular forms when they give up their energy to a microorganism. Gamma radiation from the radioisotope cobalt-60 has proved to be of practical value in the control of medically important microorganisms. Gamma irradiation, for example, is used for killing pathogens such as *Salmonella* in food products. Such applications are analogous to the use of heat in the pasteurization process, with complete sterilization being impractical because of undesirable changes in color, flavor, or consistency with higher doses of radiation. A number of biological materials (such as penicillin) and numerous disposable plastic items (such as hypodermic syringes) can be sterilized effectively with high-dose gamma irradiation without altering the material. Sterilizing plastic with gamma rays is an alternative method to sterilizing with ethylene oxide, and it has the advantage of allowing immediate use of the sterilized objects without the extended wait required after gas sterilization. Large commercial firms sterilize many types of medical equipment and supplies by irradiation. These items can be stored for a long time and are generally used once and then discarded.

Various substances and conditions alter the susceptibility of microbes to ionizing radiation. Oxygen and certain chemicals may decrease microbial resistance, whereas vitamin C, ethyl alcohol, and glycerol may increase resistance. Protective chemicals are, of course, of great interest in view of modern-day radiation hazards.

Bacterial endospores are the most radiation-resistant microbial forms, whereas the Gram-negative rod-shaped bacteria, such as species of *Salmonella* and *Pseudomonas*, are among the most sensitive. Some bacteria that do not form spores are peculiarly radiation-resistant; for example, *Micrococcus radiodurans* and some laboratory-derived mutants of other bacteria have enzymes that can repair moderate amounts of radiation damage and are therefore unusually resistant.

Ultraviolet Radiation

Certain wavelengths of radiation are much more effective antimicrobial agents than those immediately shorter or longer. The most important example is the band of wavelengths from 200 to 310 nanometers in the ultraviolet zone. This electromagnetic zone of enhanced killing includes the wavelengths optimally absorbed by nucleic acids with resulting damage to their structure and function. The absorbed energy causes this damage by producing changes in the purines and pyrimidines so that their normal function is impaired. However, in microbial populations that have apparently been killed by ultraviolet radiation, some organisms will recover if irradiated with longer wavelengths of light. Furthermore, recovery of damaged viruses may also occur if they infect cells containing repair enzymes.*

Ultraviolet light, page 133

In practice, ultraviolet light with satisfactory germicidal properties can be produced by passing an electric current through vaporized mercury in a special glass tube similar to a fluorescent light bulb. Ultraviolet rays penetrate very poorly, so that a film of grease on the bulb may markedly reduce effective microbial killing, as will extraneous materials covering the microorganisms. Therefore, ultraviolet lamps are of greatest value against exposed microorganisms in air or on clean surfaces and at close range. One should also be aware that ultraviolet rays can cause damage to the skin and eyes and promote the development of skin cancers.

The growth phase of the microorganism influences the effectiveness of ultraviolet irradiation. Actively multiplying organisms are the most easily killed, while bacterial endospores are the most resistant. Most types of glass and plastic effectively screen out ultraviolet radiation. It has been assumed for years that the microbial killing effect of sunlight is due entirely to ultraviolet rays. However, much of the sun's ultraviolet radiation is absorbed by the atmosphere, and the bactericidal effect of sunlight on earth is primarily due to another mechanism, **photo-oxidation.** In photo-oxidation, the light energy of wavelengths longer than ultraviolet is absorbed by microorganisms, producing lethal oxidation in the presence of atmospheric oxygen. In contrast, microbial killing by ultraviolet irradiation does not require the presence of oxygen.

COMPARISON OF AGENTS USED AGAINST MICROORGANISMS

Different means of killing microorganisms are often compared in the laboratory by counting the number of surviving organisms after varying time intervals or by counting the survivors at a fixed time. The count must be achieved after neutralizing or removing the killing agent. Different dilutions are inoculated on appropriate media, and the number of survivors able to grow into a visible colony is determined. Curves can then be plotted, and

---

*repair enzymes—enzymes that can correct damage done to DNA by radiation

"D" or "Z" values can be determined. The D value, or "decimal reduction time," is the time it takes for 90 percent of the organisms to be killed at a given temperature, while the Z value represents the temperature required to obtain 90 percent killing in a given time period. These values, rather than the time required for complete killing of the population, are used to compare different methods of microbial killing. The complete-killing time varies widely because of the shape of the killing curves and is therefore of little use in scientific studies.

Disinfectants may be evaluated by comparing the ratio of their antimicrobial activity with that of pure phenol. The larger the resulting value—known as the **phenol coefficient**[5]—the more active the agent under those conditions. The phenol coefficient has applicability in government control and labeling of phenolic disinfectants but may be highly misleading in evaluating these substances for actual use, because differences in killing rates depend on the species of microorganism, stage of growth, nature of the killing agent, suspending medium, pH, salt concentration, and many other factors. For these reasons any agents being compared for possible use in a given situation (such as disinfection of a hospital floor or preparing the skin for an injection) must be evaluated under the same conditions in which they will actually be used. Widely conflicting information on the value of disinfecting agents has often arisen because various and often inappropriate conditions of testing were used.

For example, in counting surviving microbes, evaluators may fail to employ a medium that neutralizes the disinfectant being tested. If this happens, traces of disinfectant carried over to the counting medium may prevent growth of the microbes and falsely indicate that they have been killed. Disinfectants shipped in interstate commerce must have federal approval, and their labeling is subject to evaluation in laboratories of the Environmental Protection Agency. By law, the evaluation procedures are those defined in a manual of the Association of Official Analytical Chemists (AOAC tests). Similarly, the Food and Drug Administration has responsibility for antiseptics shipped in interstate commerce.

SUMMARY

"Sterilization" means killing or removing all viable organisms and viruses, whereas "disinfection" means reducing or eliminating agents that might represent a health hazard. Heating is one of the most practical methods of controlling unwanted microbes and is also the most reliable. Heat is most effective when applied under moist conditions. Autoclaving and tyndallization can bring about sterilization, while pasteurization and boiling can reduce or eliminate unwanted microorganisms. Filtration can be used to ster-

[5]See Appendix V.

ilize fluids, although with less reliability than heating. In using filters it is important to know their approximate pore size, response to solvents and filtration pressure, and their affinity for biologically active substances such as enzymes. Most chemical agents are unreliable for sterilizing but have wide use as disinfectants. The major classes of chemical disinfectants include alcohols, halogens, formaldehyde, phenolic substances, and quaternary ammonium compounds. Gaseous agents used in sterilization include ethylene oxide and formaldehyde vapor. Electromagnetic radiation is useful both for sterilizing and for reducing the number of potentially harmful microbes.

Germicidal agents generally cause a rapid decrease in number of susceptible microorganisms, but all require time to sterilize or satisfactorily reduce the number of viable organisms. Microorganisms vary in their susceptibility depending on pH, temperature, growth phase, strain, and species differences.

Laboratory comparisons of germicidal agents may be made by analyzing their plotted curves of microbial killing. However, the most meaningful comparisons are made *under conditions of their actual use.*

SELF-QUIZ

1. Pasteurizing foods is a good idea because it
   a. kills all microorganisms and viruses present
   b. kills pathogenic microorganisms
   c. helps delay spoilage
   d. b and c
   e. all of the above
2. The most reliable indicator to use for testing the adequacy of a sterilization procedure would be
   a. bacterial endospores
   b. non-lipid-containing viruses
   c. fungal spores
   d. lipid-containing viruses
   e. vegetative bacteria
3. A veterinarian wishes to disinfect some instruments used to drain an abscess on a cow. Which of the following measures would aid the action of the disinfectant?
   a. careful washing and rinsing of the instruments before disinfection
   b. lowering the temperature of the disinfectant
   c. adding soap to the disinfectant
   d. all of the above
   e. none of the above
4. The reason autoclaving is more effective than boiling in killing microorganisms is that
   a. air is pumped out of the autoclave
   b. the pressure produced in the autoclave enhances killing
   c. steam can reach a higher temperature in the autoclave

   d. the autoclave is not influenced by atmospheric pressure
   e. all of the above
5. The best available alternative for sterilizing an object that cannot withstand autoclaving temperatures is
   a. a quaternary ammonium compound
   b. ethylene oxide
   c. isopropyl alcohol
   d. a phenolic
   e. a mercurial
6. Ultraviolet light would not be useful for sterilizing a fluid in a glass jar
   because ultraviolet light
   a. causes photo-oxidation
   b. has very poor penetrating power
   c. reacts only with nucleic acids
   d. is ineffective against vegetative bacteria
   e. requires the presence of oxygen

QUESTIONS FOR DISCUSSION

1. A fluid has been passed into a sterile container through a filter with pores
   small enough to retain the smallest bacterium. Is this fluid sterile?
2. A gallon jar of liquid is autoclaved at 121°C for 15 minutes. Can we be
   sure that it is sterile?

FURTHER READING

Cundy, K. R., and Ball, W.: *Infection Control in Health Care Facilities*. Baltimore:
   University Park Press, 1976.
Favero, M. S.: "Sterilization, Disinfection, and Antisepsis in the Hospital." *In* Len-
   nette, E. H., Balows, A., Hausler, W. J., Jr., and Truant, J. P. (eds.): *Manual of
   Clinical Microbiology*, 3rd ed. Washington, D.C., American Society for Micro-
   biology, 1980.
Infection Control in the Hospital, 4th ed. Chicago: American Hospital Association,
   1979.
Lawrence, C. A., and Block, S. S.: *Disinfection, Sterilization and Preservation*. Phila-
   delphia: Lea and Febiger, 1968. A large and comprehensive reference book.
National Nosocomial Infections Study Report. Center for Disease Control, Annual Sum-
   mary 1975 (Issued October 1977.)
Sykes, G.: *Disinfection and Sterilization*. Princeton, New Jersey: D. Van Nostrand
   Co., Inc., 1958.

Chapter 8

ANTIMICROBIAL AND ANTIVIRAL MEDICINES

On the 17th of May 1933, at the meeting of the Dusseldorf (Germany) Dermatological Society, Drs. Schreus and Foerste gave an astonishing report. They had had under their care a ten-month-old boy with a skin infection caused by staphylococci. Despite repeated efforts to cure him using all known methods, the bacteria invaded his bloodstream, his condition deteriorated, and he was near death. By chance at that time they received a small supply of a mysterious red medicine from the giant I.G. Farben Chemical Company with the suggestion that it had been shown to be effective for treating infections in mice. The new medicine was given to the child in a last-ditch effort to save his life. Miraculously, his temperature promptly subsided to normal and he returned rapidly to good health. There were no ill effects whatsoever from the medicine.

This child ushered in the modern age of infectious disease treatment. He was cured by a sulfa drug, the first of a continuing series of highly effective and safe antimicrobial medicines.

OBJECTIVES

To know
1. The basis for selective toxicity and the meaning of the therapeutic ratio.
2. The sources of antimicrobial medicines.
3. Some mechanisms by which antimicrobial medicines work.
4. How bacteria are tested for susceptibility to different antibacterial medicines.
5. The limitations of antimicrobial medicines.
6. Examples of medicines useful against prokaryotic and eukaryotic cells and against viruses.

ANTIMICROBIAL CHEMOTHERAPY

From discussions in earlier chapters, it is clear that all living cells show a remarkable similarity in their biochemical makeup. The same types of chemical substances—nucleic acids, proteins, lipids, and carbohydrates—are found in all, genetic information is stored and interpreted in almost identical fashion, and energy generation and transfer employ similar chemical pathways. It is not surprising, then, that substances that poison one kind of cell are generally poisonous for others, too—the disinfectants being prime examples. This "unity of biochemical principle" poses the central dilemma in the treatment of infectious diseases.

On the other hand, although similar in many regards, important differences exist between our own cells and the cells of a large number of pathogenic microorganisms. During the "Golden Age" of bacteriology in the late 1800's, Paul Ehrlich, a German physician, became convinced that differences in the way cells stained with dyes would lead to the discovery of "magic bullets," medications that would react with and destroy a vital function of pathogenic microorganisms but would not react against human cells. Ehrlich synthesized many chemicals in an attempt to find "magic bullets" but was only marginally successful. His greatest achievement was the development of an arsenic compound that, when administered to patients, killed the spirochete of syphilis at concentrations that did not kill the patients (although it often made them very sick). Many patients whose infections were previously considered hopeless were cured by the new medicine, and the idea of **selective toxicity** became firmly established.

The first real breakthrough in the developing science of antimicrobial chemotherapy* came almost 20 years after the death of Ehrlich. Stimulated by his work, the German chemical industry began screening the effectiveness of thousands of dyes for the treatment of streptococcal infections in animals. One of them was a red dye called Prontosil; it was found to be dramatically effective without producing toxic effects in the infected animals. Surprisingly, Prontosil had no effect on streptococci in test tubes. The reason soon became apparent when it was discovered that once inside the animal body the Prontosil molecule was split apart by enzymes in the blood and a smaller molecule called sulfanilamide, a sulfa drug, was produced. Sulfanilamide acted against the infecting streptococci. Thus, the discovery of sulfanilamide was based on luck as well as persistence. The idea that dyes would be the "magic bullets" was wrong from the start, and if Prontosil had been screened against bacteria in test tubes instead of in animals, its effectiveness would never have been discovered.

In 1929, Alexander Fleming reported the discovery of penicillin. Like Pasteur more than 50 years earlier, he was alert to the possibility that one microorganism could be used to fight another. Therefore, when, through an extraordinary sequence of accidents, a penicillin-producing mold spore from

---

*chemotherapy—treatment of disease by use of chemicals

the air landed on one of his cultures of staphylococci, grew, and caused death of the staphylococci, he recognized the tremendous potential of his observation. Within a decade, penicillin was purified and used in treating infections and produced dramatic cures without harm to the host animal.

Sulfanilamide and penicillin differed greatly from Ehrlich's arsenic compound. With the latter medication, the dose required to kill the syphilis spirochetes was only slightly lower than the dose that made the patient sick. Syphilis, page 501 With sulfanilamide and penicillin, however, doses many times greater than those required to kill the pathogen have little or no effect on the patient. Indeed, each antimicrobial medicine can be characterized by its **therapeutic ratio,** defined as the highest dose a patient can tolerate without toxic effects divided by the dose required to control a microbial infection (Fig. 8–1). Thus, Ehrlich's arsenic compound has a low therapeutic ratio, whereas the ratios

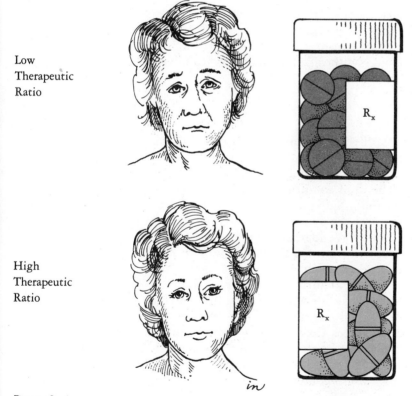

Low
Therapeutic
Ratio

High
Therapeutic
Ratio

Figure 8–1

The therapeutic ratio. Patients receiving antimicrobial medicines having a low therapeutic ratio are likely to experience toxic side effects, whereas those who receive a medicine with a high therapeutic ratio are unlikely to do so. The toxic side effects become important because, in practice, more medicine is generally given to a patient than is needed to treat the invading microorganisms. This is to allow for individual variations in absorption and other factors.

for sulfanilamide and penicillin are high. Very high therapeutic ratios imply that the medicine acts against a vital function of the pathogen, a function that is not carried on by human cells. Examples of vital functions unique to pathogens and not present in human cells are the syntheses of the bacterial cell wall and of certain essential vitamins by microorganisms. They provide a scientific explanation for selective toxicity. For most of the medications discovered so far, however, the therapeutic ratio is not great, because at higher doses the medication in many cases interferes with some biochemical function of the human cell, for example, with the cell's ability to use sugars.

Some antimicrobial medicines kill microorganisms, while others prevent their growth only as long as the medicine is present. Medicines that kill bacteria are called **bactericidal,** while those that only inhibit their growth are called **bacteriostatic.** The similar terms "fungicidal," "fungistatic," "amebacidal," and "amebastatic" are occasionally used for fungi and amebae, respectively. "Static" medicines keep microorganisms from growing until they are killed or eliminated by host defenses.

As indicated in the preceding discussion, the medicines used against infectious agents come from a variety of sources. Some are plant or animal extracts, some are synthesized by chemists from simple chemical reagents, and some are antibiotics.

ANTIBIOTICS

Antibiotics are a diverse group of chemicals, produced by natural biosynthetic processes of microorganisms, that kill or inhibit the growth of other microbial species. Only a small percentage of them show selective toxicity and can safely be used as antimicrobial medicines for treating humans. In some instances the major portion of the antibiotic molecule is made by a microorganism, and the synthesis is then completed by chemists. In other instances scientists have learned how to synthesize the entire antibiotic in the laboratory. By convention, these partially or totally synthetic chemicals are nevertheless called antibiotics, since microorganisms can make them in whole or in part. Several antibiotics that are too toxic to use as antimicrobial medicines nevertheless are useful in treating certain cancers. Antibiotics used for antimicrobial chemotherapy come from three groups of microorganisms: actinomycetes (bacteria that grow in branching filaments), rod-shaped Gram-positive bacteria of the genus *Bacillus*, and molds. Indeed, almost all those used in antimicrobial chemotherapy come from only five genera: *Bacillus, Micromonospora, Streptomyces, Penicillium,* and *Cephalosporium.*

Most of the major groups of antibiotics were discovered before 1955 by exhaustive screening of thousands of cultures from natural sources. For example, cephalosporin was derived from a *Cephalosporium* mold isolated from the sea near a sewage outflow; streptomycin was derived from a *Streptomyces* actinomycete recovered from the throat of a chicken; and bacitracin came from a *Bacillus* culture from dirt in the wound of a girl named Tracy. Mutants of these cultures, either occurring spontaneously or induced by ultraviolet light, were found to produce chemical variants of the antibiotic as well as differences in yield. Other variants of the original antibiotic have been produced artificially by chemists. Selection of these variants has resulted in "families" of antibiotics, each resembling one of the original chemicals but having various useful properties for treating different kinds of infections.

Hundreds of tons and many millions of dollars' worth of antibiotics are produced each year. In the commercial production of an antibiotic, a carefully selected strain of the appropriate producing species (e.g., of *Penicillium* for penicillin) is inoculated into broth medium and incubated in huge vats. As soon as the maximum antibiotic concentration is reached, the drug is extracted from the medium and extensively purified. Its potency is then standardized by comparing its antimicrobial action with that of a reference sample of highly purified antibiotic. This comparison ensures that each capsule or pill that a patient takes has the correct amount of antibiotic activity in it. The antibiotic is next put into a form suitable for administration to patients (flavored, buffered, coated to protect against stomach acid, put into capsules, pressed into tablets, or mixed with substances to prolong its release into the blood).

New antibiotics must be evaluated extensively; for example, their spectrum of activity* is determined against a wide variety of pathogens to see which kinds of pathogens resist their action and which kinds are killed or have their growth inhibited.* Toxicity is determined by giving increasing doses to laboratory animals to see how much of the substance they can receive before injury occurs to tissues such as the kidneys, liver, bone marrow, or central nervous system. The antibiotic is then given to human volunteers to identify how it is broken down and excreted by the host. By measuring the concentration of the antibiotic in tissues, mother's milk, spinal fluid, blood, and other body fluids, the kinds of infections that can be reached by the drug, the doses needed, and how often it must be given can be determined.

MODES OF ACTION OF ANTIMICROBIAL MEDICINES

It is likely that most antimicrobial medicines will ultimately be shown to act by interfering with one or more essential enzymes within the cells of microorganisms. For many antimicrobial medicines it is known which enzymes are blocked,* while in other cases only the general type of biosynthesis (such as the production of proteins or bacterial cell wall) or cellular structure involved (such as cell membrane) is known.

---

*spectrum of activity—the range of different kinds of organisms against which the medicine is active

*inhibited—prevented

*blocked—prevented from functioning

TABLE 8–1 MODES OF ACTION OF SOME ANTIBACTERIAL ANTIBIOTICS

Antibiotic type	Mode of action
Penicillins	Interfere with bonding together of peptidoglycan cell wall layers; activate cell wall–destroying enzymes
Aminoglycosides	Interfere with protein synthesis at one or more stages by binding to the small subunit of bacterial ribosomes
Tetracyclines	Interfere with protein synthesis by binding to the small subunit of bacterial ribosomes and preventing attachment of tRNA to the ribosomes
Erythromycin, lincomycins, chloramphenicol	Interfere with protein synthesis by various mechanisms that involve attaching to the large ribosomal subunits
Polymyxins	Damage the cytoplasmic membrane
Rifamycins	Interfere with RNA synthesis by reacting with RNA polymerase

One mechanism by which enzyme action is blocked is by competitive inhibition, discussed in Chapter 4. The sulfa drugs are competitive inhibitors of folic acid synthesis, for example. The modes of action of several antibiotics are discussed in Chapters 2 and 5 and are summarized in Table 8–1.

SENSITIVITY TESTING* OF MICROORGANISMS

For some pathogenic species, susceptibility to therapeutic agents is predictable (for example, penicillin G almost always kills pneumococci, and chloroquine is generally active against the malaria-causing *P. vivax* protozoa). With other species (such as *S. aureus*), there is no way of knowing which antimicrobial agent is likely to be effective in a given case. In treating infections, it has often been the practice to try one medicine after another until a favorable response is observed or, if the infection is very serious, to give several antimicrobial agents together. Both approaches are undesirable, because with each unnecessary medicine there are unnecessary risks of toxic or allergic effects and undesirable alterations of the normal flora. *The cornerstone of rational treatment of infectious diseases thus rests with choosing the antimicrobial agent most likely to act against the offending pathogen but against as few other cells as possible.* Practical methods exist for determining the susceptibility of the more rapidly growing bacterial pathogens, and some of the basic concepts involved are discussed in the following paragraphs.

Malaria, page 565

Normal flora, page 209

Quantification of Susceptibility

A series of decreasing concentrations of the antimicrobial medicine is prepared in a suitable growth medium, and a suspension of the microorganism is added to test tubes containing each concentration. After a period of in-

*sensitivity testing—use of laboratory procedures to find out whether an antimicrobial medicine is active against a given microorganism

cubation, the concentrations in which visible growth has not appeared are determined (Fig. 8–2). The lowest concentration capable of preventing growth is called the **minimal inhibitory concentration (MIC).** Cultures containing concentrations of the antimicrobial agent that show no growth of the microorganism can be subcultured onto medium lacking antimicrobial medicines to see whether viable organisms remain. In this way, the **minimal bactericidal concentration (MBC)** can also be determined. The organism is said to be "sensitive" to the lowest concentration that inhibits growth. If the concentration required to kill the organisms is only two to four times as much as the inhibitory concentration, the antimicrobial agent is said to be bactericidal; if higher concentrations are required, it is bacteriostatic.

A   A series of tubes containing decreasing concentrations of antimicrobial agent.

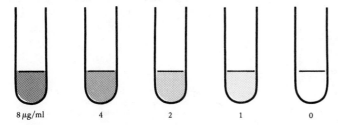

B   Addition of an invisible inoculum doubles the volume.

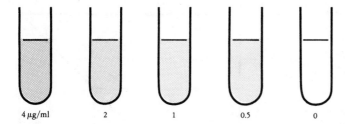

C   Appearance of growth in the more dilute solutions following incubation.

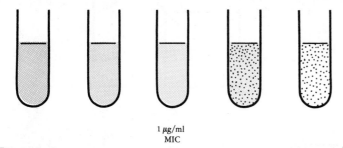

Figure 8–2
Minimal inhibitory concentration (MIC). In this example the MIC is 1.0 μg/ml. The tube containing medium without antibiotic is a growth control. In actual practice, an antibiotic control is also included (organism of known MIC).

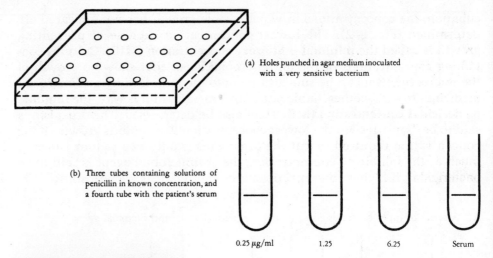

(a) Holes punched in agar medium inoculated with a very sensitive bacterium

(b) Three tubes containing solutions of penicillin in known concentration, and a fourth tube with the patient's serum

0.25 μg/ml   1.25   6.25   Serum

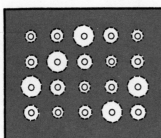

(c) The holes have been filled with fluid from the tubes and the culture plate incubated overnight; the sensitive bacterium grew everywhere except for areas around the holes in the agar

(d) The diameters of the areas of inhibited growth are averaged for each solution and plotted semilogarithmically; the diameter D of the zone of inhibition produced by the serum is used to determine the concentration C of antibiotic in the serum

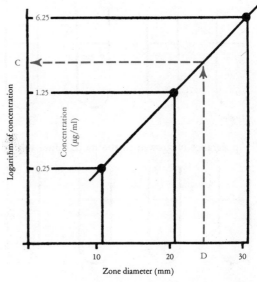

Figure 8–3
Assay of penicillin in a patient's serum.

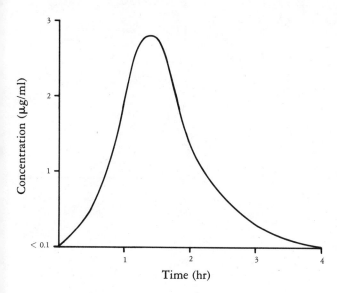

Figure 8–4
Concentration of penicillin in a patient's blood at different times after an oral dose.

Assays of Antimicrobial Agents

Merely knowing the MIC and MBC is not enough to tell whether a drug will be effective in treating an infection. It is also necessary to know whether these concentrations are likely to occur in a patient's infected tissues. To determine this, doctors give each new drug to human subjects in doses known to be safe and nontoxic. Samples of the blood, urine, and other body fluids are then collected at different time intervals following administration of the drug. A very sensitive organism is then used to measure the amount of antimicrobial drug present in the fluids. A culture of the organism is spread over the surface of agar medium. Holes are then punched out of the agar, and some are filled with fluid containing known concentrations of the drug to be assayed, while others are filled with the body fluid being tested. Following incubation, zones of inhibition form around the agar wells, their diameter depending on the concentration of antibiotic present. By measuring the zone sizes and plotting them against the corresponding concentrations, a curve relating zone size to concentration is obtained, from which the concentration of the antimicrobial medicine in the body fluid can be read (Fig. 8–3). Thus, concentrations present at different times following administration of a drug can be determined, as shown in Figure 8–4.

The Meaning of "Sensitive"

Although often used in the quantitative sense,[1] the word "sensitive" is also commonly used in a qualitative sense to describe organisms susceptible to concentrations of antimicrobial drugs known to occur in the blood of pa-

---

[1]See "Quantification of Susceptibility," above.

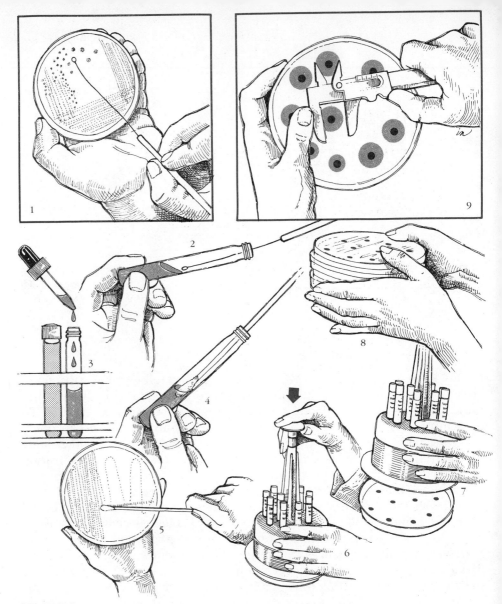

Figure 8–5
Steps involved in the Kirby-Bauer procedure for sensitivity testing using filter paper discs impregnated with antimicrobial medicines. (1) A microbiological loop is used to remove a colony from a culture of infected material. (2) The colony is transferred to a tube of liquid broth and incubated for a short time. (3) The growth of the bacteria causes the fluid to become turbid; this turbidity is adjusted to that of a standard turbidity by adding more sterile fluid. (4) A sterile swab is dipped into the culture. (5) The swab saturated with a suspension of bacteria is used to inoculate the entire surface of a Petri dish containing the sensitivity-testing medium. (6 and 7) A mechanical dispenser is used to deposit filter paper discs containing different antimicrobial medicines. (8) The Petri dishes are incubated overnight at 35° to 37°C. (9) On the next day the diameters of the inhibited growth around the discs are measured. Sensitivity and resistance to the different antimicrobial medicines is determined by comparing the size of the diameters with tables that give the range of sizes for sensitive and resistant bacteria.

Figure 8–6
Dr. William M. M. Kirby (standing) and Dr. Alfred Bauer demonstrating the technique for the sensitivity test that bears their name. (Courtesy of C. Iden Roberts.)

tients under treatment. The word "resistant" is used for microorganisms requiring substantially higher concentrations. Thus, an organism with an MIC of only 20 micrograms* per ml of polymyxin would nevertheless be called "resistant," since blood levels of this drug are usually lower than 5 micrograms per ml. Microorganisms requiring inhibitory concentrations on the borderline between sensitive and resistant are often called "intermediate."

Distinguishing Sensitive and Resistant Organisms

Automated and miniaturized equipment is now available to determine MIC's, but it is expensive. Therefore, most medical laboratories determine qualitative susceptibilities of bacterial pathogens by using filter paper discs impregnated with antimicrobial agents (Figs. 8–5 and 8–6). Thus, to determine whether the *S. aureus* strain isolated from a boil is sensitive to penicillin, a culture of the organism is spread over the surface of agar medium,

S. aureus, page 295

---

*microgram (μg)—one thousandth of a milligram

and an antibiotic disc containing a precise amount of penicillin is placed on top of it. Following incubation, the zone of inhibition of growth is measured, and if it is large enough, the organism is called *sensitive*. To determine whether the inhibition zone is large enough, it is necessary to refer to tables of values for each drug. These tables are prepared by determining MIC's and inhibition zone sizes simultaneously for bacteria of a wide range of susceptibility. In this way, the inhibition zone diameters are correlated with the MIC's, and knowing the expected blood levels of each drug, correlations between zone diameters and qualitative sensitivity can also be determined. Since antimicrobial agents diffuse through agar at different rates, there are different standards for each drug.

LIMITATIONS ON THE VALUE OF ANTIMICROBIAL AGENTS

Selection of Resistant Variants

During the first few years after the introduction of sulfa drugs and penicillin, there was great hope that such agents would soon eliminate most infectious diseases. Yet today, resistance limits the usefulness of all known antimicrobial medicines. For some organisms (such as *S. aureus*), penicillin-resistant strains were well established in nature before the introduction of penicillin, although at a low frequency. Heavy use of penicillin (measured in tons per year) soon fostered their selection, so that by 1950 more than 50 percent of the strains of staphylococci causing infections were resistant. Such strains do not readily disappear when usage of an antibiotic is discontinued, since they are well adapted for survival.

Clones* of sensitive bacteria almost always contain rare mutants resistant to antimicrobial medicines. Under certain conditions of sustained heavy usage of an antimicrobial agent, selection of a mutant may occur, giving rise to a resistant strain. Generally such mutant strains are not well adapted for survival under usual conditions and can maintain their dominance only with the help of heavy use of an antimicrobial medicine. If use of the medicine is stopped, they lose their competitive advantage and disappear. This is true of chloramphenicol-resistant *S. aureus*, for example.

Transfer of genetic information from resistant to sensitive strains of bacteria by transduction and conjugation can also markedly increase the incidence of resistance to antimicrobial medications. Transfer of the genetic information necessary for resistance occurs infrequently, but the presence of the appropriate antimicrobial medicine gives a marked selective advantage to the recipient bacterium. Moreover, resistance factors (R factors) transferred by conjugation, and perhaps by transformation and transduction, may contain genes for resistance to several medications. Thus, a person receiving a single antibiotic risks having a sensitive pathogen acquire resistance to several antibiotics simultaneously.

R factors, page 135

---

*clones—families of cells derived from a single parental cell by repeated divisions

Finally, the establishment of resistant strains in a community is markedly influenced by conditions that allow their spread from person to person and by their ability to colonize people. Thus, overcrowding, mobility of the human population, ventilation of houses and offices, sanitation, antibiotic usage, and virulence of pathogens all may play important roles. The administration of an antimicrobial medicine can markedly increase the chance that a person will be colonized by extraneous* antibiotic-resistant strains, presumably because sensitive strains among the normal body bacteria that would otherwise compete with the resistant forms are suppressed.

Nosocomial infections, page 610

Some of the mechanisms by which microorganisms resist the action of antimicrobial medicines are now known, and many studies of these mechanisms are being carried out in the hope of devising means to circumvent them. Some microorganisms resist the action of antimicrobial medicines by degrading them. Penicillinase, a microbial enzyme that can attack the penicillin molecule and destroy its activity, is one example. Resistance to other antibiotics occurs by the action of microbial enzymes that add a chemical group (such as acetate or phosphate) and thus cover up sites on the antibiotic molecule that would normally react with the microorganism. Resistance to other antimicrobial medicines is based on changes in the permeability of the cell wall or cell membrane that prevent entry of the antimicrobial agent. Finally, some bacteria resist sulfa drugs by increasing the production of *para*-aminobenzoic acid (PABA), thus preventing the competitive inhibition of growth produced by sulfa drugs in sensitive cells (see Chapter 4).

Other Limitations on Antimicrobial Effectiveness

Antimicrobial medicines only prevent the growth or increase the death rate of microorganisms. They have no effect on microbial toxins that have already been released, and, in some instances, by killing microbes they actually enhance the release of endotoxin, with life-threatening consequences. Moreover, the effectiveness of antibiotics is influenced by their ability to diffuse to the site of the infection. For example, aminoglycoside antibiotics (such as streptomycin) may fail to cure meningitis because they cross the blood-brain barrier poorly. Conditions at the site of infection may not be favorable for antibiotic action. Penicillin kills streptococci in a few hours in broth medium, but many days are required for the same concentration to kill them if the cocci are enclosed in a blood clot. The acid of urine interferes with the activity of streptomycin, and polymyxins are neutralized by nucleic acids in pus. For the most part, antimicrobial agents are of little value in eliminating infectious agents in abscesses or other sites where the organisms are not actively dividing; therefore, abscesses must be drained for antimicrobial agents to be maximally effective.

Blood-brain barrier, page 514

---

*extraneous—coming from elsewhere than the person's body

TABLE 8–2 ANTIBACTERIAL MEDICINES

Medicine	Principal pathogens used against	Examples of diseases
Penicillins		
Penicillin G (by injection)	*Neisseria gonorrhoeae* and *N. meningitidis*; *Bacillus anthracis*; species of *Streptococcus*, *Clostridium*, *Listeria*, *Bacteroides*, *Borrelia*, *Actinomyces*, and *Treponema*	Strep throat, gonorrhea, meningitis, anthrax, gas gangrene, endocarditis, pneumonia, blood infection, syphilis
Penicillin V (by mouth)	*Streptococcus pyogenes*, *S. pneumoniae*, *Staphylococcus aureus* (penicillinase nonproducing)	Strep throat, pneumonia, boils
Dicloxacillin (by mouth)	*Staphylococcus aureus* (penicillinase producing)	Boils, wound infections
Methicillin (by injection)	*Staphylococcus aureus* (penicillinase producing)	Pneumonia, bone and blood infections
Ampicillin (by mouth or injection)	Some *Streptococcus*, *Proteus*, *Salmonella*, *Escherichia*, *Haemophilus*, *Listeria* strains	Blood and kidney infections, meningitis
Ticarcillin (by injection)	*Pseudomonas aeruginosa*	Blood and wound infections
Chloramphenicol	*Salmonella typhi*, *Haemophilus influenzae*, *Bacteroides* species	Cases of typhoid fever or meningitis untreatable with ampicillin because of allergy or bacterial resistance
Tetracyclines	*Neisseria gonorrhoeae*, *Brucella* species, *Vibrio cholerae*, *Chlamydia psittaci*, *C. trachomatis*, *Mycoplasma pneumoniae*	Gonorrhea, brucellosis, cholera, pneumonia, genital infections, eye infections
Erythromycin	*Corynebacterium diphtheriae*, *Bordetella pertussis*, *Campylobacter fetus*, *Legionella pneumophila*, *Chlamydia trachomatis*, *Mycoplasma pneumoniae*	Diphtheria, whooping cough, blood and intestinal infections, eye infections, pneumonia
Lincomycins		
Clindamycin	*Bacteroides* species	Blood and wound infections
Sulfa Drugs		
Sulfisoxazole with trimethoprim	*Escherichia coli* *Shigella* species, *Nocardia*	Bladder infections Dysentery, pneumonia
Isoniazid (usually with ethambutol and rifampin)	*Mycobacterium tuberculosis*	Tuberculosis
Sulfones (usually dapsone plus rifampin)	*Mycobacterium leprae*	Leprosy
Metronidazole	*Bacteroides* species, *Haemophilus vaginalis*	Blood infections, vaginitis
Aminoglycosides		
Tobramycin	*Pseudomonas aeruginosa*, *Enterobacter* and *Klebsiella* species	Wound and blood infections, pneumonia
Gentamicin	*Serratia marcescens*, *Acinetobacter* species	Wound and blood infections, pneumonia
Streptomycin	*Yersinia pestis*	Plague
Amikacin	*Proteus* species	Blood and kidney infections

Finally, good treatment is almost totally lacking for some microorganisms, particularly the protozoa that cause African sleeping sickness and American trypanosomiasis. There is also increasing pessimism about whether the continued discovery of new antimicrobials can keep pace with the development of resistance to antimicrobial medications.

ANTIBACTERIAL MEDICINES

Some of the principal antibacterial medicines are discussed below, and some of their important characteristics are listed in Table 8–2.

Synthetic Medicines

Sulfa Drugs. Sulfanilamide, the substance responsible for the antimicrobial action of the red dye Prontosil, was the first sulfa drug to be introduced for widespread medical use. The impact on many infectious diseases was extremely gratifying. For example, by 1937, death rates for pneumonia and meningitis had fallen to about half of their previous levels. Yet sulfanilamide has a serious drawback: It is insoluble in acid urine and tends to precipitate in and damage the kidneys. However, the sulfanilamide molecule is easily modified chemically, and soon a number of other sulfa drugs with better solubility were produced. Figure 8–7 compares sulfanilamide with sulfisoxazole, a newer, highly soluble sulfa drug.

As mentioned earlier, sulfa drugs block folic acid synthesis and thereby interfere with cellular production of purines, pyrimidines, and other substances needed by the cell. Generally, their antibacterial action is bacteriostatic. They are active against both Gram-positive and Gram-negative bacteria, but sensitivity tests must be performed because many strains of bacteria are resistant. Sulfa drugs are well absorbed from the gastrointestinal tract. They are very widely used to treat bladder infections, most of

Competitive inhibition, page 88

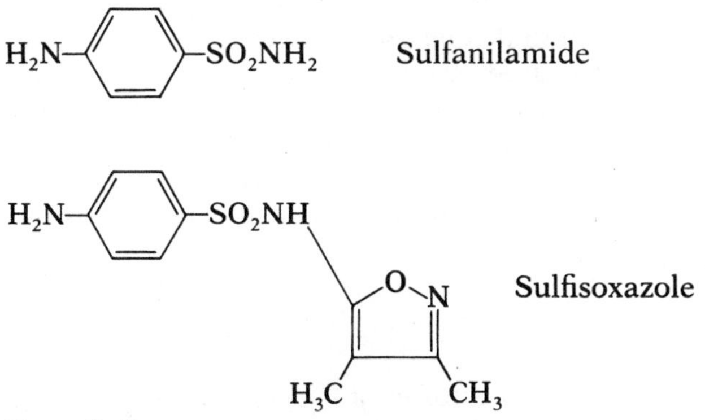

H_2N—⟨⟩—SO_2NH_2 Sulfanilamide

H_2N—⟨⟩—SO_2NH

H_3C CH_3 Sulfisoxazole

Figure 8–7
Structural formulas of sulfanilamide and sulfisoxazole. The latter is much more soluble under acid conditions and therefore less likely to precipitate in the kidneys.

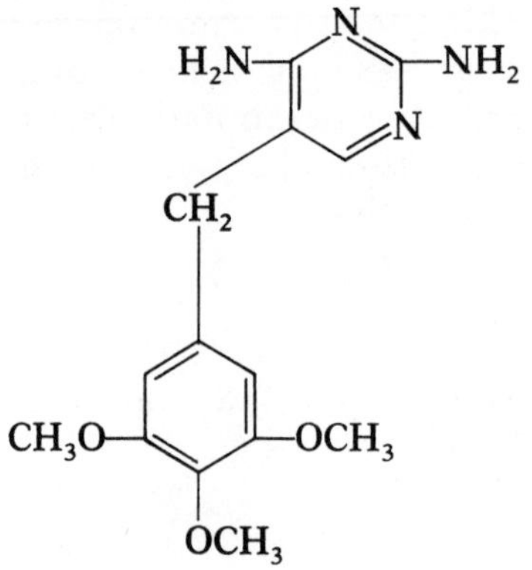

Figure 8—8

Trimethoprim. Trimethoprim given with a sulfa drug enhances the effectiveness of both medicines.

which are caused by *E. coli*. About 5 percent of the patients who receive sulfa drugs experience side effects, mostly allergic reactions involving fever and skin rash. More serious are occasional effects on the blood or bone marrow, which occur in about 0.1 percent of cases.

Trimethoprim. Trimethoprim (Fig. 8–8) also blocks folic acid–dependent metabolism, but at a different step than sulfa drugs. Trimethoprim therefore markedly increases the effectiveness of sulfa drugs, and the two are often given together. The combination has a very wide spectrum of activity against Gram-positive and Gram-negative bacteria and is useful in treating infections resistant to other medicines. Side effects are similar to those that occur with sulfa drugs alone.

Other Synthetic Antibacterial Medicines. The structures of *para*-amino-salicylic acid (PASA), isoniazid (INH), ethambutol, pyrazinamide, ethion-amide, and dapsone are shown in Figure 8–9. Like sulfa drugs and trimeth-oprim, these medicines block enzymatic reactions in bacterial cells. For example, dapsone and PASA, like sulfa drugs, are competitive inhibitors of PABA metabolism. On the other hand, INH competes with vitamin B_6 (pyr-idoxine). These medicines have essentially one use: the treatment of diseases caused by mycobacteria. Dapsone is used against *Mycobacterium leprae* (the cause of leprosy), and the others are used against *M. tuberculosis* (the cause of tuberculosis). The most common serious side effects are liver damage with INH and pyrazinamide and blindness with ethambutol. Patients are watched carefully for evidence of these effects, and the medicines are stopped at the first indication that they may be occurring.

Penicillins

Sir Alexander Fleming's filtrates of broth cultures of the penicillin-producing mold proved to have insufficient activity for treating infections, and it was not until a decade after his discovery that sufficiently potent samples of

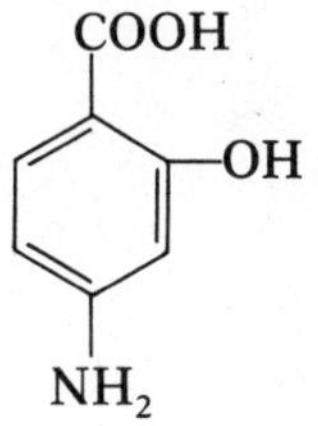

PASA

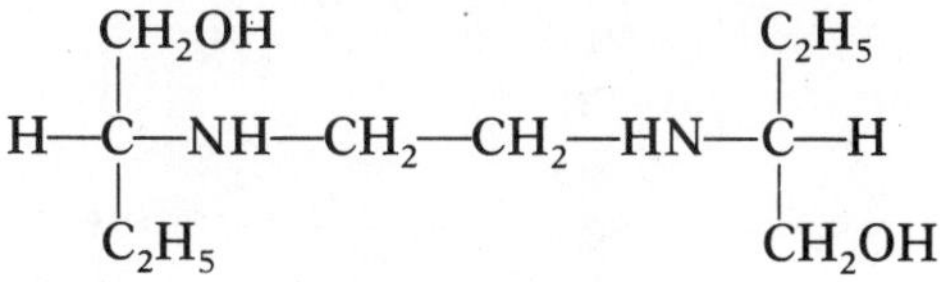

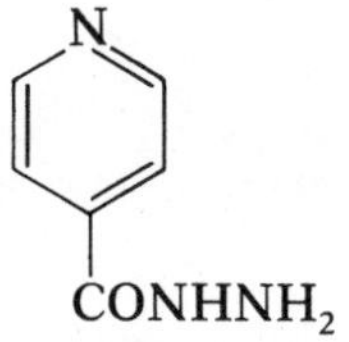

INH

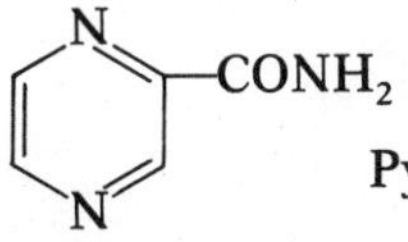

Pyrazinamide

Ethambutol

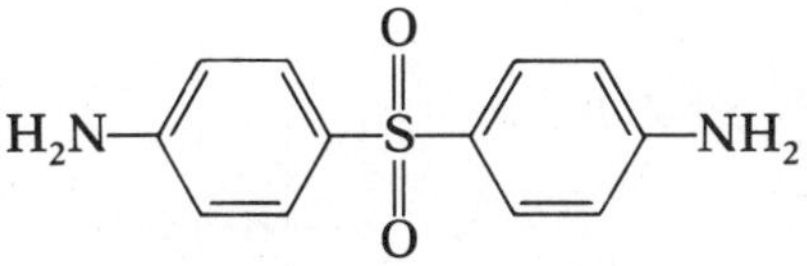

Dapsone (a sulfone)

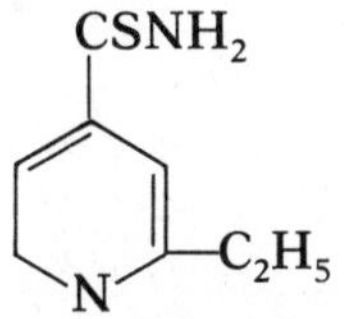

Ethionamide

Figure 8–9
Synthetic medicines used against mycobacteria.

penicillin were prepared. World War II spurred British and American workers to determine the chemical structure of penicillin and to develop means for its large-scale production. Several different penicillins were found in the mold cultures, and these were designated F, G, X, K, O, and so on. Penicillin G (or benzyl penicillin; see Fig. 8–10) was found to be most suitable for treating infections.

Penicillin G is a safe and effective antibiotic for systemic* therapy, and today it remains the drug of choice for susceptible organisms. However, penicillin G has an important drawback: It is unstable in acid solutions and therefore cannot consistently survive stomach acid and be absorbed after swallowing. Moreover, it is effective only against certain Gram-positive bacteria (such as *Streptococcus pyogenes)* and a few Gram-negative species (such

---

*systemic—pertaining to the whole body

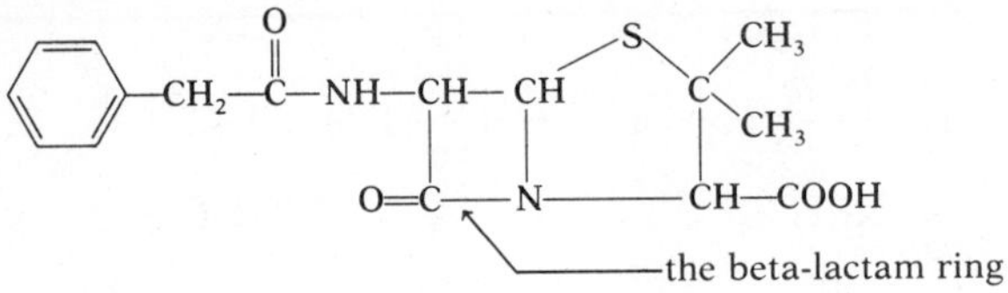

Penicillin G

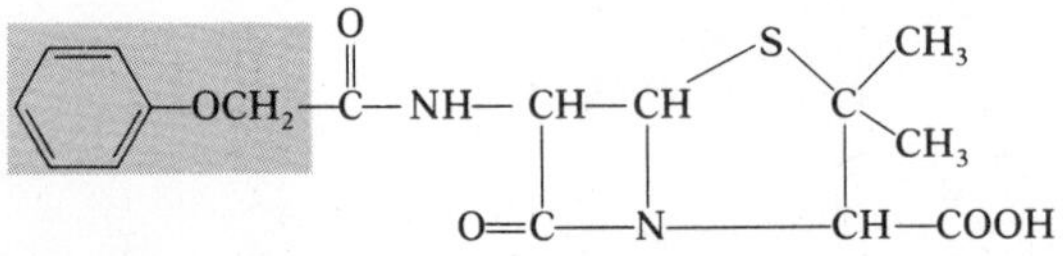

Penicillin V
(acid-resistant)

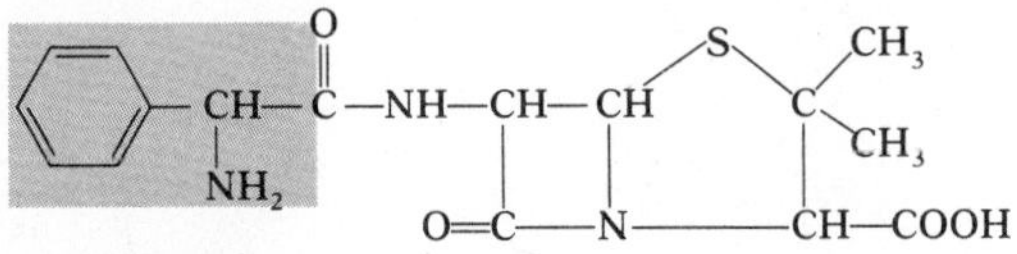

Ampicillin
(broadened spectrum
acid-resistant)

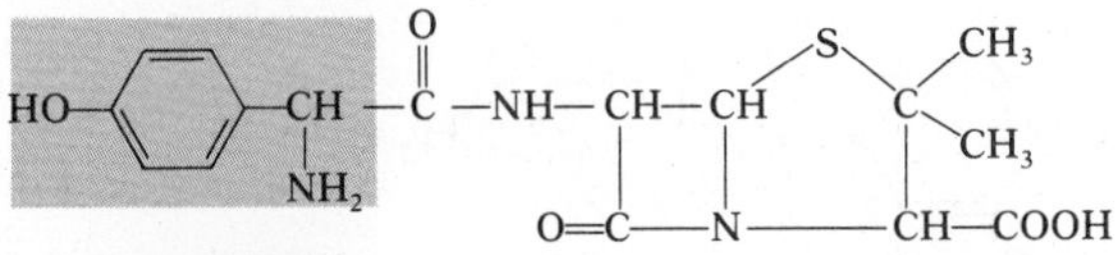

Amoxicillin
(like ampicillin
but more active
and requires less
frequent doses)

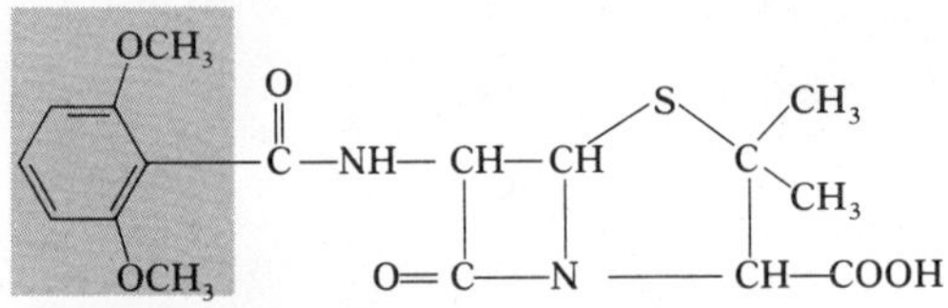

Methicillin
(penicillinase-resistant)

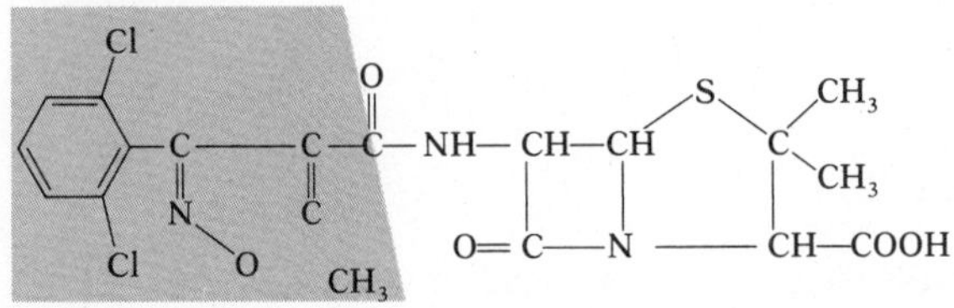

Dicloxacillin
(acid- and penicillinase-resistant)

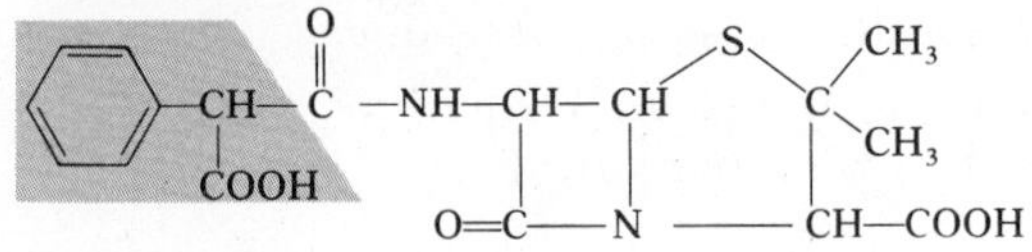

Carbenicillin
(broadened activity against
Gram-negative rods)

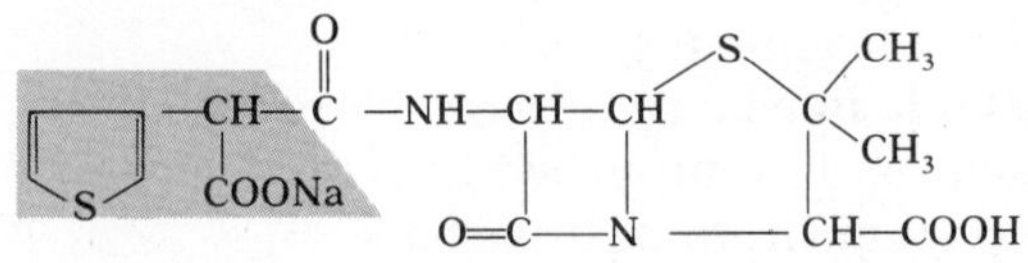

Ticarcillin
(like carbenicillin but 2
to 4 times more active
against *Pseudomonas*)

Figure 8–10
Some members of the penicillin family. Shaded areas indicate the modifications responsible for the changes in properties.

as *Neisseria meningitidis*). It is completely ineffective against bacteria that produce penicillinase, the enzyme that destroys the drug. The development of penicillin V, a relatively acid-stable compound, only partially overcame these difficulties.

In 1959, English scientists discovered that by altering the chemical structure of penicillin they could obtain derivatives of the antibiotic that were resistant to penicillinase. Subsequently, other derivatives were selected for use in treating infections by Gram-negative rods. Carbenicillin and ticarcillin, for example, are effective against many strains of *Pseudomonas aeruginosa*, a species resistant to most other penicillins.

Because their action is directed strictly against bacterial cell walls, penicillins are almost completely nontoxic. However, penicillins occasionally cause death when administered to people who are allergic to them. (An estimated 300 to 500 such deaths occur per year in the United States.) For this reason, doctors and nurses always ask patients about penicillin allergy before giving these medicines. In the past, some people probably became hypersensitive* to penicillin because they were unknowingly exposed to it in vaccines or in milk from cows under treatment with penicillin, but strict controls now make this possibility less likely.

Mode of action, page 31

Another drawback of the penicillins is their instability in aqueous solutions. Once the dry powder is dissolved it must be used promptly to avoid loss of antibacterial activity. Even water condensing from the air may decrease the activity of penicillin powders that have been refrigerated.

Widespread and often inappropriate use of penicillins has resulted in a selective increase in penicillin-resistant strains of bacteria as the penicillin-sensitive strains were killed. Some penicillin-resistant bacteria even grow in penicillin solutions and can cause infections when such contaminated medicines are administered. As is the case for most medicines taken orally, food may interfere greatly with the absorption of penicillins.

Cephalosporins

A strain of mold of the genus *Cephalosporium* was found to produce antibiotic substances somewhat similar in structure to penicillin and having the same mode of action. One of these substances, cephalosporin C, could be modified chemically to produce useful antibiotics. Some of them are susceptible to stomach acid and must be given by injection; others can be given orally. Like the penicillins, the cephalosporins are bactericidal. They are active against many Gram-positive and Gram-negative pathogens. Their chemical structure is sufficiently different from that of the penicillins to make them relatively resistant to penicillinase and to permit their administration to some people who are allergic to the penicillins.

---

*hypersensitive—allergic

Tetracyclines

Protein synthesis, pages
117–118

Unlike the penicillins and cephalosporins, which are of fungal origin, the tetracyclines are produced by certain strains of the actinomycete genus *Streptomyces*. They are bacteriostatic antibiotics that act by interfering with bacterial protein synthesis. A family of useful tetracylines has been produced through selection of appropriate producing strains of the organism and by chemical alteration. This family includes oxytetracycline, doxycycline, demeclocycline, minocycline, and tetracycline. All have a very similar spectrum of activity but vary in their stability, toxicity, and affinity for blood proteins. Tetracyclines are often referred to as "broad-spectrum" antibiotics because they show activity against many Gram-positive and Gram-negative species. The term "broad-spectrum," however, may be misleading because of the high incidence of tetracycline resistance among various genera of bacterial pathogens. The tetracyclines are usually given orally and reach relatively low levels of antimicrobial activity in body fluids.

Tetracyclines have been widely and indiscriminately used in animal feeds, in treating infections of humans when the etiological* agent could not be readily determined, and in efforts to prevent a secondary bacterial infection* in people suffering from colds, influenza, measles, heart failure, or surgical wounds. It is now known that the use of tetracyclines to prevent secondary infection commonly *increases* the chance of infection, probably because these drugs interfere with the protective effect of the normal bacteria of the body. Thus, tetracyclines may foster colonization or overgrowth by resistant microorganisms, including *Staphylococcus aureus*, *Candida albicans*, *Pseudomonas aeruginosa*, and species of *Proteus* and *Klebsiella*. The extensive use of tetracyclines has resulted in an increasing percentage of resistant strains of *Streptococcus pyogenes*, *S. pneumoniae*, *Neisseria gonorrhoeae*, and *Bacteroides* species. The common side effects of these drugs include diarrhea, yeast vaginitis, and discoloration of the teeth in infants and children.

Chloramphenicol

Originally isolated from *Streptomyces venezuelae*, chloramphenicol is now synthesized by chemical methods. Like the tetracyclines, it has a broad spectrum of antimicrobial activity, is generally bacteriostatic, and acts by interfering with protein synthesis. Unlike tetracycline, it readily diffuses into cells and spinal fluid to act on bacterial pathogens beyond the reach of other antibiotics. However, it is toxic at high doses, mainly because it interferes with the normal development of erythrocytes (red blood cells). Much more serious, however, is a reaction that occurs in about 1 of every 40,000 patients who receive chloramphenicol. This complication—aplastic anemia—is characterized by the inability of the body to form leukocytes (white blood cells)

---

*etiological—causative

*secondary infection—invasion by another microorganism or virus after the host has been weakened by the first one

and erythrocytes and often ends in death from infection or from the development of leukemia. For this reason, the use of chloramphenicol is generally reserved, and its use is restricted to life-threatening infections for which equally effective alternative treatment is not available. Nevertheless, chloramphenicol is used extensively in poor countries because of its low price. During the years of peak chloramphenicol use in the United States, fewer than half the people with aplastic anemia had received the antibiotic, showing that substances besides chloramphenicol can cause aplastic anemia.

Aminoglycosides

The aminoglycosides are classified as a group because of similarities in their chemical structure. These antibiotics are bactericidal and share many properties, but they do not necessarily kill the same bacteria. Streptomycin, the first of the aminoglycosides, was found after screening 10,000 cultures of soil bacteria. The producing strain was called *Streptomyces griseus*. Streptomycin is a bactericidal antibiotic that is active against many Gram-positive and Gram-negative bacteria and is one of the few antibiotics that is active against *Mycobacterium tuberculosis*. Streptomycin has little activity against anaerobic bacteria or under acid conditions. Because it is poorly absorbed from the gastrointestinal tract, it must be given by injection. Unfortunately, the usefulness of this promising drug was impaired soon after its discovery because of the high frequency of very resistant mutants among many bacterial pathogens. Streptomycin damages the kidneys and the nervous apparatus for body equilibrium, but the most feared toxic effect of the drug is irreversible deafness. Fortunately this is rare, but up to 15 percent of those receiving the medicine for more than a week have a slight but measurable loss of hearing. The severe toxic effects of streptomycin result when prolonged high concentrations of the drug are present in the bloodstream. This situation is especially likely to occur in people who have kidney diseases that interfere with normal excretion of the drug in the urine.

Neomycin, another aminoglycoside antibiotic, is too toxic to give by injection. However, it can be given safely by mouth because, like other aminoglycosides, it is very poorly absorbed, and not enough gets into the bloodstream to cause a toxic reaction. The value of orally administered neomycin is that it markedly reduces the numbers of intestinal bacteria. Neomycin is therefore administered by surgeons preparing to operate on a patient's intestine, since following oral neomycin treatment there are few bacteria left to cause infection of the surgical wound. In addition to its use before intestinal surgery, there are other uses for oral neomycin. For example, killing the intestinal bacteria with neomycin causes about a 25 percent decrease in blood cholesterol, and this is thought to decrease the chance of heart attacks in certain individuals who have too much cholesterol in their blood. Furthermore, since intestinal bacteria are also responsible for ammonia production, another use of neomycin treatment is to decrease the high levels of blood ammonia that occur in people with severe liver disease.

Gentamicin, tobramycin, and amikacin—three newer aminoglycosides—

are active against many Gram-negative bacteria that have acquired resistance to the older medications. However, resistant strains have already appeared in several areas of the United States. Although these aminoglycosides are also active against many Gram-positive bacteria, other less toxic antibiotics are generally available and are used instead.

Other Antibacterial Antibiotics

Erythromycin. Erythromycin is active primarily against Gram-positive bacteria; it is generally bacteriostatic and is readily absorbed after oral administration. The major use of erythromycin is for treatment of infections in people who are allergic to penicillins. Serious side effects are rare, although some forms of erythromycin can cause a mild liver injury.

Lincomycins. Lincomycin and clindamycin are chemically distinct from other antibiotics, but they have an antimicrobial spectrum similar to that of erythromycin. Their action is generally bacteriostatic. Although staphylococci resistant to erythromycin may initially be sensitive to lincomycins, resistant forms quickly arise on exposure to the latter drugs. Certain lincomycins have an important role in the treatment of infections caused by anaerobic Gram-negative rods of the genus *Bacteroides.* An uncommon but potentially serious side effect of treatment with lincomycins is the development of pseudomembranous colitis, an intestinal disease caused by heavy growth of *Clostridium difficile,* a toxin-producing Gram-positive rod that is resistant to lincomycins.

Polypeptide Antibiotics. Polypeptide antibiotics include bacitracin and polymyxin. They are very poorly absorbed after oral administration. Both bacitracin and polymyxin can be administered by injection, but because of their toxicity this usage is reserved for the treatment of infections caused by bacteria resistant to other antimicrobial drugs. Usually bacitracin and polymyxin are applied to body surfaces for treating superficial infections; however, this practice may lead to an allergic skin reaction to the antibiotic. Bacitracin is a bactericidal antibiotic produced by a strain of *Bacillus subtilis* and is effective in treating infections by Gram-positive bacteria such as *Staphylococcus aureus.* The polymyxins are produced by another species of *Bacillus* and, in contrast to bacitracin, are active against Gram-negative organisms. Polymyxins can be used to treat serious infections caused by *Pseudomonas aeruginosa* but share with bacitracin the property of poor diffusion into infected tissue. In addition, the polymyxins are readily inactivated by pus and dead tissue. Safer and more effective medicines generally are available.

Rifamycins. The rifamycins, discovered in 1957 in cultures of *Streptomyces mediterranei,* were found unsuitable for therapeutic use, but rifampin, a derivative, has shown significant activity against many Gram-positive and Gram-negative organisms as well as mycobacteria. Rifampin can be admin-

Mode of action, page 113

istered orally, and little toxicity has been noted. Its main defect has been that mutants highly resistant to it commonly arise. The rifamycins are chemically unrelated to other antibiotic families.

MEDICINES THAT ACT AGAINST EUKARYOTIC CELLS

A few medicines have proved useful in treating eukaryotic pathogens and certain malignancies (cancers and leukemias). For example, the antibiotic amphotericin is effective in some serious fungal infections. Quinine, a plant extract, is useful in treating malaria, a protozoan disease. Vaginitis caused by the protozoan *Trichomonas vaginalis* is treated with a synthetic antimicrobial agent called metronidazole. The antibiotic daunorubicin is useful in treating some leukemias. Further details on medications useful against eukaryotic cells are given in Appendix VI. Most have low therapeutic ratios.

ANTIVIRAL MEDICINES

Since viruses use cellular metabolic machinery, one might suspect that it would be difficult to find antiviral medications with an acceptably high therapeutic ratio. Indeed, so far very few medicines that are of proven benefit in the chemotherapy of viral illness have been found. Nevertheless, scientists have demonstrated that for at least some viruses there are enzymes essential for viral reproduction that are unique to the virus (and thus not required or found in normal body cells). On this fact is based the hope that, in the future, substances will be found that block the viral enzymes without causing a significant effect in the host cells. It is hoped that such substances would provide a way of treating viral illnesses effectively.

Antiviral medicines currently available to physicians include amantadine, vidarabine, idoxuridine, and methisazone. These medications have had only a minor impact on preventing or treating human viral diseases. Their modes of action are summarized in Table 8–3.

TABLE 8–3   MODES OF ACTION OF ANTIVIRAL MEDICINES

Medicine	Mode of action
Amantadine	Prevents uncoating of viral nucleic acid inside the host cell
Vidarabine	Inhibits viral DNA synthesis by reacting with viral DNA polymerase
Methisazone	Prevents final assembly of viruses reproducing inside the host cell
Interferon	Induces enzymes that interfere with translation of viral mRNA

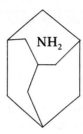

Figure 8—11
Amantadine, an antiviral medicine.

Amantadine

The amantadine molecule has an unusual shape and a low molecular weight (Fig. 8–11). Its usefulness is limited to the prevention of influenza A; it is active against all the strains tested. Use of this drug reduces the number of people who become ill with influenza by 50 to 100 percent when compared with groups of untreated individuals who received equal exposure to the virus. Amantadine is not effective against influenza B, however. There is also evidence that the medicine can decrease the severity of an established influenza A infection, especially if given soon after symptoms begin, although the response is not usually striking. Vaccination against influenza is generally considered preferable to taking amantadine to prevent the disease. However, since the medicine does not interfere with immunity, it can be given to afford protection while waiting for the vaccine to take effect.

About 1 of 20 healthy adults who take amantadine have side effects, usually confusion, dizziness, and insomnia. More rarely, serious side effects such as severe emotional disorders and heart failure can occur. The effect of the medicine on developing human fetuses and on infants is unknown. Therefore it is usually not given to expectant mothers or to nursing mothers, since it is excreted in breast milk.

Amantadine acts by blocking the uncoating of virus particles and the release of their nucleic acid in the host cell. The medicine may interfere with viral penetration into the cell, although definite scientific proof of this theory is lacking.

Vidarabine

Vidarabine (also called adenine arabinoside) is a purine nucleoside obtained from cultures of *Streptomyces antibioticus*. Chemically, it resembles the building blocks of DNA and RNA. It is similar to idoxuridine, an older and more toxic antiviral medication used to treat eye infections caused by herpes simplex (cold sore) viruses.

Nucleic acids,
Appendix I

Vidarabine acts by inhibiting viral DNA synthesis more readily than most cell DNA synthesis. This effect probably occurs when vidarabine or one of its metabolites reacts with viral DNA polymerase, the enzyme responsible for linking the subunits of DNA together. Thus, the polymerase is rendered unable to produce normal DNA.

So far vidarabine is of proven value in only one situation: encephalitis caused by herpes simplex virus. The death rate from this disease can be reduced from about 70 to 30 percent if the medication is given intravenously early in the course of the illness. Vidarabine can also be used as an eye ointment for treatment of corneal infections by the same virus. In test tube studies, vidarabine is active against certain other DNA viruses of the pox and the herpes groups.

No toxic side effects have been observed in the majority of patients treated with vidarabine, although a few have had skin rashes of gastrointestinal upsets. Tremors,* painful skin sensations, poor coordination, and other effects on the nervous system have also been observed in some patients. Finally, laboratory studies indicate that the medicine could possibly, but rarely, cause cancers, and it therefore should not be used in minor viral infections that would resolve without treatment.

Methisazone

Methisazone (also widely known by the trade name of Marboran) is one of a group of chemicals called thiosemicarbazones. The structural formula is shown in Figure 8–12.

The medicine acts within the host cell to prevent the final assembly of viral particles. Thus, only immature, noninfectious virus is produced.

Methisazone, which is taken by mouth, was shown in the 1960's to be markedly effective in preventing smallpox when given to people within one or two days of exposure to the disease. The medicine may also have been effective in treatment of mild smallpox, especially when given early in the illness. Methisazone is probably similarly effective in treating some diseases caused by other pox viruses and thus continues to have a therapeutic role now that smallpox has been eradicated. The medicine might also have an important place if for some reason smallpox should again appear, for example, through accidental escape of the virus from a laboratory. Since vaccination is no longer practiced, many people would be highly susceptible.

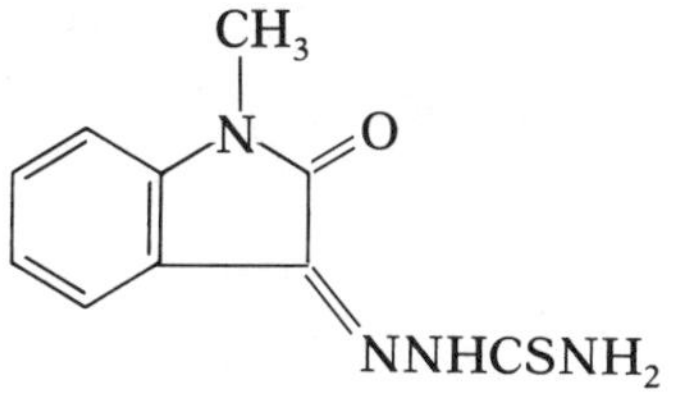

Figure 8–12
Methisazone, an antiviral medicine.

---

*tremors—involuntary shaking of parts of a person's body

Severe nausea and vomiting occur occasionally and are the major side effects of methisazone.

Interferons—Experimental Antiviral Substances from Human Cells

Interferons are antiviral glycoproteins* produced by virus-infected cells. These substances render normal noninfected cells resistant to a wide variety of unrelated viruses; in other words, they act nonspecifically against viruses. Humans produce several kinds of interferons that differ in their physical properties* and effects on body cells. Interferons have been very difficult to obtain in quantity in pure form, but new techniques will soon make this much easier.

Interferons act by attaching to certain sites on the surface of cells. The cells respond to the attachment of interferons by producing enzymes that prevent translation of viral messenger RNA. Viral protein synthesis is thus blocked. The enzymes produced by the cell prevent translation by at least two different mechanisms. They inactivate a substance needed to start translation, and they break down viral messenger RNA.

Besides acting to prevent reproduction of viruses, the interferons (or possibly unknown substances mixed with them) cause heightened activity of some of the cells responsible for immunity. This may explain the activity interferons show against certain cancers.

So far, interferons are available only for experimentation, and it is not yet known how they will be useful. Available information indicates that they should be effective against most of the important human viruses. Favorable effects are most likely to be obtained if interferons are given before or shortly after infection and continued for a number of days. They are usually given by injection, but experiments have shown that an interferon nasal spray is effective in decreasing the severity of colds.

Fever, nausea, vomiting, and temporary suppression of the ability of the bone marrow to make blood cells occur with the injection of interferon. It is not yet clear whether these effects are due to impurities in the medicine or to the interferon itself.

Other Antiviral Medications

Acyclovir is a promising experimental antiviral medicine. Like vidarabine and idoxuridine, acyclovir is a nucleoside derivative,* but acyclovir is much more active against herpes simplex virus, and there is scientific evidence

---

*glycoproteins—proteins with sugar molecules attached

*physical properties—for example, the ability to withstand heat or acid

*derivative—a molecule made from (derived from) another molecule by chemical alteration

that it may be useful in treating genital herpes. Early claims for many other substances, however, including vitamin C and lysine, have not been substantiated by scientific studies. Other chemicals, including substances [such as poly(I):poly(C)] that cause release of the patient's own interferon, have so far proved to be too toxic or of little effect in the treatment of humans.

Genital herpes, page 507

SUMMARY

Antimicrobial and antiviral treatment depend on finding medicines effective against the pathogen but not harmful to the host's normal cells. Such medicines are said to exhibit **selective toxicity.** The **therapeutic ratio** of a medicine is the highest dose a patient can tolerate without toxic effects, divided by the dose required to control a microbial infection. Antimicrobial medicines come from a variety of sources, such as plant and animal extracts, chemical reagents, and antibiotics. New antibiotics are evaluated for their spectrum of activity, toxicity, and amount needed to control an infection. The modes of action of antimicrobial medicines include interference with synthesis of the cell wall, interference with protein synthesis in a variety of ways, damage to the cytoplasmic membrane, and interference with nucleic acid synthesis. In treating infections, one chooses the antimicrobial agent most likely to act against the offending pathogen but against as few other cells as possible. Laboratory tests are used to determine how sensitive the pathogens are to antimicrobial medicines and whether these medicines will act against the pathogen in the human body. Resistant strains of pathogens have become common as a result of the overuse of antimicrobial medicines, and therefore these medicines should be used only when necessary. Limitations on the effectiveness of antimicrobial medicines include their ineffectiveness against microbial toxins already released, nondiffusion to the site of infection, and inactivation by acids of the body or materials such as pus that occur in infected tissue.

The principal groups of antibacterial medicines include the bactericidal penicillins, cephalosporins, and aminoglycosides and the bacteriostatic tetracyclines, sulfa drugs, erythromycin, lincomycins, and chloramphenicol.

Very few nontoxic medications have been found useful against eukaryotic cells or viruses. Amphotericin is of principal importance in serious fungal infections, while medicines as diverse as quinine (plant extract) and metronidazole (synthetic) have selected usefulness against protozoa. Amantadine, methisazone, and vidarabine have very narrow uses against certain viruses, but an experimental substance called interferon shows promise for the future. A very few highly toxic antibiotics have proved valuable in treating certain cancers.

SELF-QUIZ

1. Antibiotics can act against bacteria by all of the following means *except*
    a. interfering with protein synthesis
    b. interfering with cell wall synthesis
    c. damaging the cytoplasmic membrane
    d. attacking the mitochondria
    e. blocking nucleic acid synthesis
2. Sources of antimicrobial medicines include
    a. actinomycetes
    b. molds
    c. plants
    d. a, b, and c
    e. none of the above
3. Resistance of bacteria to antibiotics can be due to
    a. enzymatic destruction of the antibiotic
    b. impermeability of the cell wall to the antibiotic
    c. addition of a chemical group to an antibiotic molecule
    d. all of the above
    e. none of the above
4. Knowing the minimum inhibitory concentration of an antimicrobial medicine against an infecting microorganism would allow you to predict the response to treatment with that medicine *if* you also knew
    a. the minimum bactericidal concentration
    b. the concentration of the medicine in body fluids, achieved during treatment
    c. the name of the infecting species
    d. the therapeutic ratio of the antimicrobial medicine
    e. the pathways for excretion and inactivation of the medicine
5. All the following limit the value of antimicrobial medicines *except*
    a. increasing prevalence of resistant strains of pathogens
    b. production of toxins (poisons) by many pathogens
    c. inability of the medicines to diffuse to the site of infection
    d. chemical alterations that allow the medicines to resist microbial enzymes
    e. nongrowing organisms are often unaffected by antimicrobial medicines
6. All of the following are antibiotics *except*
    a. ampicillin
    b. sulfanilamide
    c. doxycycline
    d. amikacin
    e. chloramphenicol
7. In treating infections it is wise to
    a. use many medicines simultaneously to be sure that one will act against the pathogen

b. choose the medicine that is most likely to act against the pathogen but against as few other cells as possible

c. wait several days before starting treatment to see if the patient's body will fight off the infection

d. use a "broad-spectrum" antibiotic to suppress the normal body bacteria as much as possible

e. change medicines daily to avoid developing resistance

QUESTIONS FOR DISCUSSION

1. Would you expect a bacteriostatic antibiotic given along with penicillin to increase, decrease, or have no effect on its bactericidal action?

2. Does the therapeutic ratio differ for different pathogens?

FURTHER READING

Baldry, P. E.: *The Battle Against Bacteria.* New York: Cambridge University Press, 1965.

Black, C. D., Popovich, N. G., and Black, M. D.: "Drug Interactions in the G.I. Tract." *American Journal of Nursing* (September 1977).

Franklin, T. J., and Snow, G. A.: *Biochemistry of Antimicrobial Action,* 2nd ed. New York: John Wiley & Sons, 1975. Brief lucid book giving much information on modes of action and mechanisms of resistance.

Gause, G. F.: *The Search for New Antibiotics—Problems and Perspectives.* New Haven, Conn.: Yale University Press, 1960.

Goodman, L. S., and Gilman, A., eds.: *The Pharmacological Basis of Medical Therapeutics,* 5th ed. New York: Macmillan, 1975.

Hare, R.: *The Birth of Penicillin and the Disarming of Microbes.* London: George Allen and Unwin, 1970. Fascinating account of the discovery of penicillin.

McAdams, C. W.: "Interferon, the Penicillin of the Future?" *American Journal of Nursing* (April 1980).

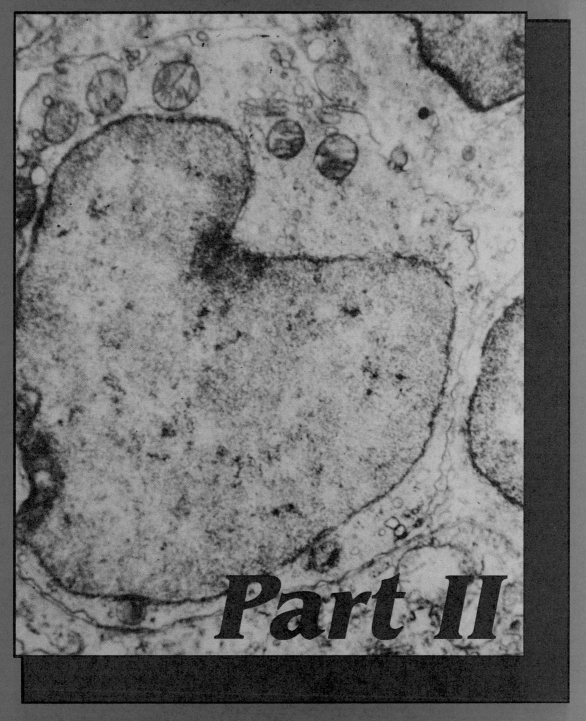

Part II

INTERACTIONS BETWEEN
MICROORGANISMS AND
THEIR HUMAN HOSTS

Chapter 9

HOST-PARASITE RELATIONSHIPS AND NONSPECIFIC HOST DEFENSES

In the premature infants' ward of a large teaching hospital, a baby recently died of an infection caused by a mixture of common organisms present in and on all the other patients and staff in the hospital. In pus swabbed from the baby's infected ear, the laboratory identified *Pseudomonas aeruginosa*, resistant to many antibiotics, and also bacteria of the genus *Klebsiella*. The baby died of this mixed infection, in spite of all possible medical attention. A similar situation also occurs with many patients who have leukemia and certain other cancers; they usually do not die as a direct effect of the cancer but rather from infections caused by organisms of the normal flora. For example, the yeast *Candida albicans* probably colonizes every individual, usually without causing serious disease, yet this yeast is one of the most frequent causes of death of leukemia victims.

OBJECTIVES

To know
1. At least two organisms normally found in each of the following locations: nose, throat, vagina, and colon.
2. In general terms, how innate, nonspecific host defenses differ from specific acquired immunity.
3. The major cells that participate in host defense.
4. The principal events that occur during the inflammatory response.
5. The mode of action of at least three important nonspecific antimicrobial factors.
6. At least four mechanisms of pathogenicity employed by bacteria.
7. How the bacterial cause of disease is established.

THE NORMAL FLORA

Each one of us carries huge numbers of microorganisms, both on body surfaces and within the body. Some viruses may even be found within eggs and sperm cells and so will be present from the time of conception. Other viruses and some bacteria are able to cross the placenta and enter the fetus in the uterus. The majority of organisms found in or on the body, however, are acquired after birth. These are **parasites** that gain benefit from their human host.

From the time of birth, the infant is fertile ground for colonization by a variety of microbial parasites. During passage through the mother's birth canal, the child encounters the organisms growing there. In some instances these microbes can cause disease in the baby, even though they may not harm the mother. Other organisms may simply colonize the child and grow without harming it. During the first few weeks of life, the infant usually becomes colonized with an abundant flora from the mother and others who are in close contact with it, and a **normal flora** of microorganisms is established. This normal flora stays with the human host throughout life, changing constantly in response to changes in the host and in the host's environment (Fig. 9–1).

Some species of bacteria and fungi are universally found among the normal flora (Fig. 9–2). Other organisms are commonly present in a majority of people tested. The flora varies considerably with the different areas of the body. Some of the reasons for this variation will become apparent later.

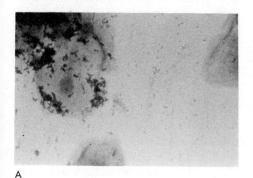

A

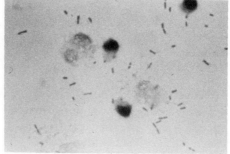

B

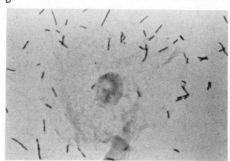

C

Figure 9–1
(a) Normal flora in the mouth. Gram stain. (× 1000) (b) Normal flora of the urinary tract (large cells are leukocytes). Gram stain. (c) Normal flora of the vagina. Gram stain. (× 1000) (Photographs courtesy of M. F. Lampe.)

TERMINOLOGY

The terminology used to describe host-parasite interactions is sometimes confusing. Most of the organisms of the normal flora do not cause disease under the conditions in which they are usually encountered. A great variety of host and microbial factors interact to determine the effect of organisms on the human body, whether these organisms are established members of the normal flora or new arrivals from the environment or from other hosts. **Colonization** simply implies establishment of microorganisms on a body surface. If the microbe breaches the surface, enters body tissues, and multiplies, **infection** is said to have occurred. An infection that causes noticeable impairment of body function is called an **infectious disease.**

A **pathogen** is any disease-producing microorganism or virus. **Pathogenicity** is the ability to cause disease. An **opportunist** is an organism that is able to cause disease only in hosts with impaired defense mechanisms, which might result, for example, from wounds, from alcoholism, or for many other reasons. The terms "opportunist" and "pathogen" should not be used in the absolute sense. *Pathogenicity always depends on a combination of factors involving both the host and the organism.* Table 9–1 gives some examples of bacteria that are commonly pathogenic.

Virulence refers to properties of microbes that enhance their pathogenicity. These properties are responsible for varying degrees of pathogenicity of different strains within a species. For example, pneumococci that have a capsule are often virulent, but those lacking a capsule are avirulent and have little pathogenic potential. The word "virulent" is also used in a quantitative sense, indicating the degree of pathogenicity. Thus, highly virulent microbes cause more severe disease than less virulent ones or require fewer organisms to establish infection.

HOST DEFENSES

Defenses may be specific or nonspecific. In general, inborn or **innate responses** that occur without previous exposure to the inciting* agent are nonspecific, whereas the **acquired response** to an agent is highly specific. For example, the formation of pus is an innate, nonspecific response, part of the inflammatory response to be discussed later. This response occurs upon exposure to many different agents, including certain kinds of bacteria, fungi, and nonliving materials. The acquired antibody response involves antibody molecules that are specific for the inciting agent. This latter acquired specific response is what is often meant by "the immune response." However, in the broader sense, nonspecific innate protection is also part of immunity. This

---

*inciting—causing a response

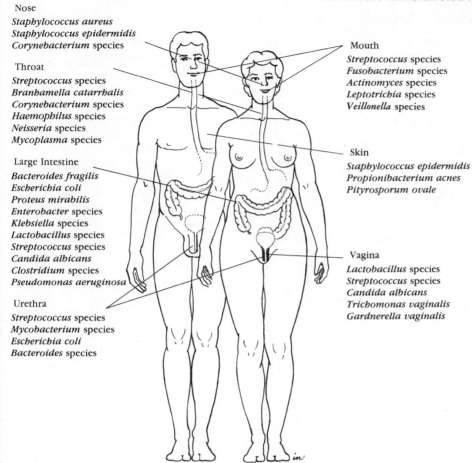

Nose
Staphylococcus aureus
Staphylococcus epidermidis
Corynebacterium species

Throat
Streptococcus species
Branhamella catarrhalis
Corynebacterium species
Haemophilus species
Neisseria species
Mycoplasma species

Large Intestine
Bacteroides fragilis
Escherichia coli
Proteus mirabilis
Enterobacter species
Klebsiella species
Lactobacillus species
Streptococcus species
Candida albicans
Clostridium species
Pseudomonas aeruginosa

Urethra
Streptococcus species
Mycobacterium species
Escherichia coli
Bacteroides species

Mouth
Streptococcus species
Fusobacterium species
Actinomyces species
Leptotrichia species
Veillonella species

Skin
Staphylococcus epidermidis
Propionibacterium acnes
Pityrosporum ovale

Vagina
Lactobacillus species
Streptococcus species
Candida albicans
Trichomonas vaginalis
Gardnerella vaginalis

Figure 9–2
Location of many of the organisms that are part of the normal flora on the male and the female human body.

chapter will consider nonspecific mechanisms of immunity, or host defense. Specific, acquired immunity will be discussed in subsequent chapters.

Cells that Participate in Host Defense

The white blood cells, or **leukocytes,** are the cells primarily responsible for defense. There are several kinds of leukocytes (Fig. 9–3) and subsets of the different kinds: each subset carries out special functions (Table 9–2). Notice in the table the three major kinds of leukocytes: granulocytes, mononuclear phagocytes, and lymphocytes. These leukocytes will be the basis of much of our discussion about both nonspecific and specific immune responses.

Granulocytes. The granulocytes are so named because of their character-istic cytoplasmic granules, which differ from one subset of granulocytes to

TABLE 9–1 EXAMPLES OF PATHOGENIC BACTERIA

Pathogenic in normal people[a]	Commonly opportunistic[b]	Occasionally opportunistic[c]
Bacillus anthracis	Actinomyces israelii	Acinetobacterium calcoaceticus
Bordetella pertussis	Bacteroides fragilis	Bacillus cereus
Borrelia recurrentis	Clostridium perfringens	Eikenella corrodens
Brucella abortus	Clostridium tetani	Enterobacter aerogenes
Chlamydia psittaci	Escherichia coli	Flavobacterium meningosepticum
Clostridium botulinum	Haemophilus influenzae	Propionibacterium acnes
Corynebacterium diphtheriae	Klebsiella pneumoniae	Pseudomonas cepacia
Coxiella burnetii	Neisseria meningitidis	Staphylococcus epidermidis
Escherichia coli[d]	Pasteurella multocida	Streptococcus agalactiae
Francisella tularensis	Proteus mirabilis	Streptococcus faecalis
Leptospira interrogans	Pseudomonas aeruginosa	
Mycobacterium tuberculosis	Serratia marcescens	
Mycoplasma pneumoniae	Staphylococcus aureus	
Neisseria gonorrhoeae	Streptobacillus moniliformis	
Salmonella typhi	Streptococcus pneumoniae	
Shigella dysenteriae		
Streptococcus pyogenes		
Treponema pallidum		
Vibrio cholerae		
Vibrio parahaemolyticus		
Yersinia pestis		

[a]Strains of these species cause disease in normal, nonimmune people.
[b]Commonly pathogenic only in cases of burns, wounds, viral infections, alcoholism, or similar conditions of impaired resistance.
[c]Commonly pathogenic only in cases of immaturity, birth trauma, defective immunity, large infecting dose, or when other factors overwhelmingly favor the microbe.
[d]Certain strains.

another. The names of the various granulocytes reflect the staining properties of the cells when a certain mixture of dyes is used. The **basophils** retain the basic dyes, and their granules stain a dark purplish blue. The **eosinophil** granules take on the bright red of the eosin dye. The **neutrophils** are neutral in their staining. Color plate 17 shows the appearance of the cells in a stained smear of human blood.

Most of the neutrophilic granulocytes are unique in that their nuclei consist of several lobes each. Thus, they are often called **polymorphonuclear neutrophils (PMN's)**, or "polys," or simply neutrophils. They account for more than half of the leukocytes in normal human blood (Table 9–2), and an increase in their number is associated with certain diseases, especially acute infections. As they accumulate in diseased areas, they make up a large portion of **pus.**

Neutrophils are essential for human survival, since they play a vital role in defense by engulfing or **phagocytizing** and destroying foreign or other

Phagocytosis by amebae, page 393

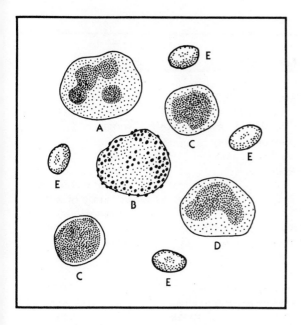

Figure 9—3
Cells of human blood. (a, b, c, and d) Leukocytes (white blood cells). (a) Polymorphonuclear neutrophil. (b) Basophil. (c) Lymphocyte. (d) Monocyte. (e) Erythrocyte (red blood cell).

unwanted materials. Although many types of cells can perform phagocytosis to a limited extent, only the polymorphonuclear neutrophils and cells of the monocyte-macrophage series are highly efficient phagocytes, often called the "professional phagocytes."

Mononuclear Phagocytes. Monocytes are also shown in color plate 17. These represent the fairly immature form of the macrophage cell line, and are found in normal circulating blood in small numbers (see Table 9–2). As is the case for all leukocytes, monocytes are produced in the bone marrow and released into the circulation. From the blood, monocytes can migrate into the tissues and there mature into various forms of tissue macrophages (also called histiocytes). This series of cells makes up the **mononuclear phagocyte system,** formerly known as the reticuloendothelial system (Fig. 9–4). Mononuclear phagocytes are found in virtually all tissues, but especially large accumulations of macrophages are located in the liver, the lungs and bronchial tissues, and the peritoneal cavity. Macrophages are also abundant in the spleen and lymph nodes, where they participate along with lymphoid cells in immunological responses.

Lymphocytes. In some cases, **lymphocytes** may resemble monocytes to a degree that makes it impossible to tell them apart on the basis of their morphology. Most lymphocytes are smaller and have a round nucleus rather than the indented nucleus seen in the monocytes (color plate 17). Small lymphocytes also contain less cytoplasm than monocytes, often just a narrow pale-staining rim around the darkly stained lymphocyte nucleus. As the lymphocytes differentiate, they become larger and contain more cytoplasm, so

TABLE 9–2   SETS AND SUBSETS OF HUMAN LEUKOCYTES

Cell type (% in blood)	Morphology*	Location in body	Functions
Granulocytes			
Polymorphonuclear neutrophils (PMN's, polys) (55–70%)	Lobed nucleus; granules in cytoplasm; ameboid appearance	Account for part of the leukocytes in the circulation; few in tissues except during inflammation	Phagocytosis and digestion of engulfed materials
Eosinophils (2–4%)	Large eosinophilic granules; nonsegmented or bilobed nucleus	As above	Participate in inflammation and immunity of some parasites
Basophils; mast cells (0–1%)	Lobed nucleus; large granules in cytoplasm containing histamine	Basophils in circulation; mast cells present in most tissues	Release histamine and other mediators of inflammation
Mononuclear phagocytes, monocytes, macrophages	Single nucleus; abundant cytoplasm	Macrophages present in all tissues and in lining of vessels; monocytes are less mature circulating forms	Phagocytosis and digestion of engulfed materials; can participate in killing foreign cells that are not engulfed
Lymphocytes	Single nucleus; little cytoplasm	In lymphoid tissues (such as lymph nodes, spleen, thymus, appendix, tonsils); also in the circulation	Participate in immunlogical responses
Plasma cells	Single nucleus pushed to one side of ovoid cell; cytoplasm packed with ribosomes	In lymphoid tissues	Antibody synthesis; derived from lymphocytes after antibody stimulation

*morphology—form and structure

that small, medium, and large lymphocytes can be distinguished on the basis of size. However, most of the lymphocytes in normal blood are small lymphocytes.

In addition to circulating in the blood, lymphocytes are the principal cells in lymph and in a vital network known as the **lymphoid system.** The lymphoid tissue of this system is distributed throughout the body in strategic locations (Fig. 9–5). The bone marrow, thymus, spleen, and lymph nodes are important lymphoid organs, and collections of lymphoid cells are also abundant beneath the mucous membranes in the respiratory and gastrointestinal tracts. During specific immune responses, subsets of lymphocytes interact in complex ways not yet fully understood.

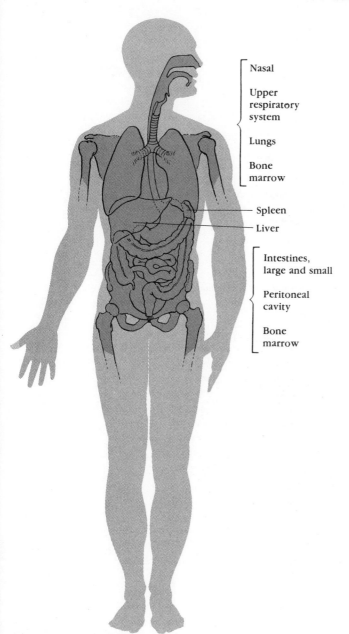

Nasal

Upper
respiratory
system

Lungs

Bone
marrow

Spleen

Liver

Intestines,
large and small

Peritoneal
cavity

Bone
marrow

Figure 9–4
Location of the cells in the mononuclear phagocyte system. Note that large accumulations
of macrophages are located in the liver, lungs, spleen, and peritoneal cavity.

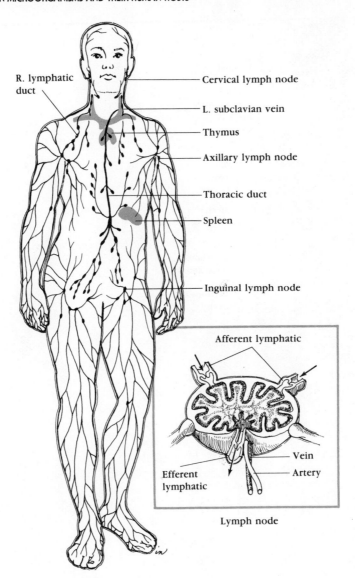

R. lymphatic duct

Cervical lymph node

L. subclavian vein

Thymus

Axillary lymph node

Thoracic duct

Spleen

Inguinal lymph node

Afferent lymphatic

Vein

Efferent lymphatic

Artery

Lymph node

Figure 9–5
Distribution of the lymphoid tissues.

Inflammation

Once a parasite has penetrated through external barriers such as the skin or mucous membranes and has entered the tissues, the first host response is a nonspecific reaction to injury called **inflammation** (Fig. 9–6). The same sequence of events occurs in response to any injury, whether caused by invading bacteria, burns, trauma, or any other stimuli. The extent of each event, however, varies depending on the nature of the insult. Table 9–3 lists the general sequence of events that occur during inflammation. There is an initial release of chemical mediators of inflammation at the site of injury.

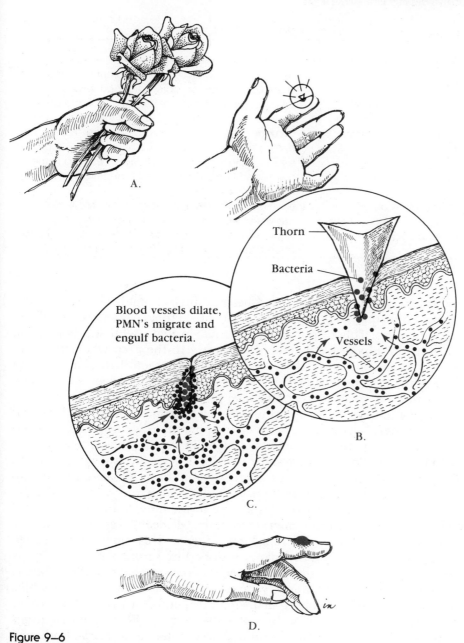

Figure 9–6
The inflammatory response, a nonspecific reaction to injury, is initiated by an actual injury. In this example, a rose thorn penetrates the skin (a) and causes inflammation (b). A microscopic view of the area shows, diagrammatically, the release of chemicals from injured mast cells and the introduction of bacteria into the wounded area. (c) Released chemicals cause the small blood vessels to dilate; PMN's migrate through the vessels and engulf the bacteria. (d) The signs of inflammation are redness, heat, swelling, and pain in the affected area.

TABLE 9–3   PRINCIPAL EVENTS OF THE INFLAMMATORY PROCESS

Event	Effects
Tissue injury	Release of kinins, prostaglandins, histamines, and other chemicals that act on adjacent blood vessels
Blood vessels dilate and show increased permeability to plasma which may clot	*Swelling* of the tissues results from the leakage of plasma; *elevated temperature* of the region may occur as a result of increased blood flow through the dilated vessels; *redness* may appear for the same reason; *pain* may result from increased fluid in the tissues and from direct effect of mediators on sensory nerve endings
Circulating white blood cells adhere to the walls of the altered blood vessels	The white blood cells migrate chemotactically through the vessel walls and to the area of injury; they are responsible for phagocytosis of foreign material and tissue debris, and for initiating antibody production

These substances lead to increased permeability of small blood vessels in that area, allowing the fluid portion of the blood (plasma) and some cells to enter the tissues. This response is protective, in that useful substances and cells from the blood are delivered promptly to the area where they are needed. The side effects of these events are so characteristic that they are known as the famous "cardinal signs" of inflammation, recognized by physicians since the earliest days of medical practice and known to everyone as indicators of injury. The four cardinal signs are **redness, heat, swelling,** and **pain.** A fifth sign, **loss of function,** is often apparent as well. The redness and heat result from increased amounts of blood in the area due to the dilation of blood vessels. The swelling and pain result from the leakage of plasma and other blood constituents into the tissue and from the action of inflammatory substances on nerve endings in the local area.

During inflammation, leukocytes adhere to the inner walls of the altered blood vessels. The leukocytes migrate from the permeable vessels in response to chemicals that attract different kinds of cells. This movement is called **chemotaxis.** The first kind of leukocyte to be lured from the circulation is the polymorphonuclear neutrophil (PMN). Microscopic examination of areas of inflammation early after injury reveals a predominance of neutrophils. Some stimuli lead to prolonged neutrophil responses, whereas others cause only brief responses. In any event, after the influx of PMN's, macrophages and some lymphocytes are attracted and accumulate in the area. Both neutrophils and macrophages actively phagocytize any foreign materials they can, often aided by **opsonins.** Opsonins are substances that cause increased phagocytosis, and they include both nonspecific materials, such as components of **complement,** and specific **antibodies,** both of which will be discussed in later sections.

Acute inflammation occurs soon after injury and lasts only a short time. An example is an infected wound that becomes acutely inflamed within a

TABLE 9–4 SOME IMPORTANT NONSPECIFIC ANTIMICROBIAL FACTORS

Antimicrobial or antiviral factor or system	Chemical nature	Source	Effects
Lysozyme	Protein (enzyme)	Most body fluids; also within phagocytes	Destroys bacterial cell walls
Beta-lysin	Protein	Serum, leukocytes	Attacks cytoplasmic membrane; active against Gram-positive bacteria
Peroxidase–hydrogen peroxide cofactor (halide ions or other system)	Protein (peroxidase enzymes)	Leukocytes, saliva, and other sources of peroxidases	Kills a variety of microorganisms; important killing mechanism in saliva and within neutrophils
Interferon	Protein	Bone marrow, spleen, macrophages	Interferes with the multiplication of viruses by causing the formation of antiviral protein
Complement system	Many distinct proteins	Produced by macrophages and other host cells	Proteins act in special sequence to produce effects such as chemotaxis, opsonization, and cell lysis

few days. The area fills with pus that drains within a week or so, after which the area heals. Neutrophils predominate in these lesions. Chronic inflammation, however, may continue for months or even longer and is characterized by fewer neutrophils and more mononuclear cells in the lesions over time.

Nonspecific Antimicrobial Factors

The nonspecific* protective effects of certain physical barriers to infection cannot be overemphasized. These barriers include the skin and mucous membranes and the mucus and other secretions that cover and constantly flush many body surfaces such as the eyes, nose, and mouth. (Some of these defense mechanisms will be considered in chapters dealing with the organ systems.) Table 9–4 lists some chemicals that are nonspecific antimicrobial substances. These are normally present in the blood and often in other body fluids and are nonspecific with respect to the invading organisms. Thus, **lysozyme** acts nonspecifically against a multitude of different bacteria, because it is an enzyme that can degrade bacterial cell walls. Similarly, **beta-lysin** acts against many different species and strains of Gram-positive bacteria and not just against a specific invading strain. **Interferons** are also

---

*nonspecific—reacting against any foreign material, living or nonliving

Interferons, page 347

Complement, page 245

nonspecific in their action against many different kinds of viruses (see Chapter 16). **Peroxidases** found in neutrophils, saliva, and other body fluids act together with hydrogen peroxide and certain ions to kill many kinds of organisms very effectively. **Complement system** is a complex set of interacting proteins that aid in the phagocytosis of a foreign substance.

Other Host Mechanisms of Defense

In addition to many nonspecific defenses, the human host responds with specific antibody and cell-mediated responses, as described in the next chapters. These greatly augment some of the nonspecific defenses. For example, antibodies specific for invading pathogens may opsonize them and significantly increase their phagocytosis and subsequent destruction.

Overall protection involves carefully controlled, complex interactions of many specific and nonspecific host defenses. If one part of the system fails, there are often other mechanisms that can compensate for the defect.

MECHANISMS OF PATHOGENESIS

Even though humans are efficient at protecting themselves, infections can still occur because microorganisms have become equally adept at evading and overcoming host defenses. A first step in becoming established often involves adhering to and colonizing the surface of the host or penetrating the host's exterior. Even large numbers of microbes cannot infect unless they have ways of penetrating the tissues and multiplying there. Mucosal tissue can chemically attract some motile bacteria to move toward it and thus helps to promote interactions between these bacteria and the mucosa.

Adherence

Many bacteria that live on mucous membranes can adhere selectively to the areas they colonize. This ability to adhere prevents them from being swept away by secretions and determines the areas that will be infected. For example, the gonococci that cause gonorrhea can survive well in many tissues when artificially introduced, but in the human host the usual sites of infection are limited to a particular type of epithelial cell layer to which these bacteria adhere selectively. Viruses must adhere to specific receptors on host cells in order to enter the cells and replicate within. Even some fungi use adherence mechanisms. The yeast *Candida albicans*, a common cause of oral and vaginal infections, has been shown to adhere selectively to cells from the areas it colonizes and often infects.

Pili, pages 45–46

Antiphagocytic Mechanisms

A necessary step in invasion is avoidance of engulfment and destruction by phagocytes. As mentioned previously, some bacteria have solved this problem by acquiring the ability to produce capsules that inhibit phagocytosis.

Probably the best-known example of this is the pneumococcus, a species of bacteria that commonly causes pneumonia but may also cause many other kinds of infections. Certain strains of pneumococci produce polysaccharide capsules, and capsule production has been shown to be directly related to the virulence of these strains. Other strains that are similar but lack the capacity to form capsules are avirulent. Many bacteria besides pneumococci also use this strategy. Others produce surface components other than capsules that can inhibit phagocytosis. For instance, the streptococci that cause the familiar "strep throat" have surface proteins, called M proteins, that are antiphagocytic and are important virulence factors for these bacteria.

M proteins, page 299

Interference With Host Immune Responses and the Production of Allergic States

Some infectious agents are able to suppress normal immune responses and thereby avoid destruction. Others have the ability to cause allergic responses that are harmful to the host and contribute to the disease process. These will be discussed in later chapters.

Production of Toxins and Enzymes

Microorganisms of many kinds—bacteria, fungi, algae, and others—may produce either enzymes or poisonous substances called **toxins** that injure the host. Often toxins are the direct cause of disease, but sometimes they are merely contributing factors.

Bacterial toxins have been studied extensively. Many of them fall into one of two major categories: **exotoxins** or **endotoxins** (Table 9–5). Others may have some of the characteristics of both categories or may be unique. Since

TABLE 9–5 IMPORTANT PROPERTIES OF BACTERIAL TOXINS

Property	Exotoxins	Endotoxins
Bacterial source	Gram-positive and -negative species	Gram-negative species only
Location in bacterium	Synthesized in cytoplasm and released from cell	Component of the cell wall
Chemical nature	Protein	Lipopolysaccharide containing Lipid A
Ability to form toxoid*	Present	Absent
Stability	Generally heat-labile,* 60–100°C for 30 min	Heat stable
Action	Each has distinctive effect	All have the same effect; fever and circulatory system damage

*toxoid—a chemically modified toxin, used as a vaccine
*heat-labile—destroyed by heating

some of the toxins have a role in the pathogenesis of major diseases, it is necessary to understand what these toxins are and how they function.

Exotoxins. Some bacterial exotoxins are among the most powerful poisons known, being lethal in extremely small amounts. Many exotoxins are known to be coded for by plasmids or by bacterial viruses called phages. The most powerful exotoxins are produced by the Gram-positive *Corynebacterium diphtheriae*, the cause of diphtheria; *Clostridium tetani*, the cause of tetanus (lockjaw); and *Clostridium botulinum*, the cause of botulism, a type of paralysis. The causative bacteria of diphtheria and tetanus have little tendency to invade their host, but their exotoxins, absorbed from a localized area of infection, are responsible for the symptoms of the disease. The bacterium responsible for botulism grows in food and releases its toxins there. In most cases the organism does not infect the host.

Endotoxins. As indicated in Table 9–5, endotoxins are actually a part of the cell walls of Gram-negative bacteria. The actions of the endotoxins are very similar, regardless of which Gram-negative bacterium produced the toxin. The principal toxic effects are fever and damage to the circulatory system. Most Gram-negative organisms, both pathogens and nonpathogens, possess endotoxins, in contrast to the few bacterial species that produce exotoxins. Endotoxins generally play a contributing rather than a primary role in pathogenesis.

Enzymes. Metabolic products of microbes, particularly enzymes, also contribute to the disease-producing potential of many organisms. Various bacteria can synthesize and release a tremendous array of degradative enzymes capable of breaking down many kinds of host tissues. Often the role of such enzymes in pathogenesis is not proved, but it is reasonable to assume that they do contribute. As we have seen, disease generally results from the interaction of many different mechanisms. Thus, a toxin may damage and kill cells in the area of toxin release, and then enzymes can degrade adjacent cells more readily. It is worth noting that some of the toxins are, in fact, enzymes and exert their toxic effects through enzyme activity. Table 9–6 lists some important microbial enzymes capable of contributing to the pathogenesis of infectious diseases.

THE ESTABLISHMENT OF INFECTIOUS DISEASES

Since all normal humans are colonized by large numbers of organisms living on their skin and mucous membranes, and since organisms may actually invade and infect without causing disease, it is necessary to establish some guidelines that will allow an organism to be identified as the cause of a disease. This was done during the last century by Koch and his co-workers. The guidelines are known as Koch's Postulates. Table 9–7 presents Koch's Postulates, which must be fulfilled if a microorganism is to be positively identified as the causative agent of a disease.

TABLE 9–6 SOME BACTERIAL EXTRACELLULAR PRODUCTS THAT MAY CONTRIBUTE TO VIRULENCE

Product	Function	Example of producing bacterium
Phospholipase	Breaks down lipid components of host cell membranes	*Clostridium perfringens*
Coagulase	Clots plasma	*Staphylococcus aureus*
Collagenase	Breaks down collagen, a tissue fiber	*Clostridium perfringens*
Hyaluronidase	Breaks down hyaluronic acid, a substance that binds cells together, thereby permitting spread of bacteria through tissues	*Streptococcus pyogenes*
Deoxyribonuclease	Breaks down DNA	*Staphylococcus aureus*
Leukocidin	Kills leukocytes	*Staphylococcus aureus*

All of us are **carriers** of potentially dangerous organisms that could cause disease if introduced into unusual locations or in very large numbers or into hosts with immune deficiencies. In other words, all of us carry many opportunists in our normal flora. In addition, some people carry highly pathogenic microbes and shed them into the environment. This may occur following disease caused by that organism, when enough of an immune response is induced to overcome the disease but not enough to rid the body completely of the pathogen. Thus, the host is well and lives in harmony with the pathogen but is able to disseminate it to others. "Typhoid Mary" was notorious because she carried the causative agent of typhoid fever and disseminated the bacteria to dozens of people she contacted. Unsuspecting carriers are currently responsible for much of the spread of gonorrhea, a serious and growing problem of public health. Carriers are also largely responsible for epidemics of meningitis caused by Gram-negative cocci called meningococci. Using special culture techniques, researchers have shown that about 15 percent of the population normally carries meningococci, but under the crowded and stressful conditions in which meningococcal epidemics occur, almost everyone present carries the organisms but only a small proportion actually develop the disease.

In some instances, microorganisms may live in the host tissues without indicating their presence. They are not shed to infect others, and they may actually benefit the host by helping to maintain a state of immunity. These microorganisms are said to be **latent.** The only problem is that if the immune capabilities of the host decrease, these latent microorganisms may reemerge

TABLE 9–7 KOCH'S POSTULATES FOR DETERMINING CAUSATIVE AGENTS OF DISEASE

1. The suspected microorganism must be present in every case of the disease.
2. The microorganism must be grown in pure culture from the diseased host.
3. The same disease must be reproduced when a pure culture of the organism is inoculated into a healthy susceptible host.
4. The microorganism must then be recovered from the experimentally infected host.

and cause disease. A good example is the virus that causes chickenpox. After infection and disease caused by this virus, a latent state is usually established. Years later, under appropriate circumstances, the virus may emerge again to cause a quite different disease called herpes zoster, more commonly called "shingles" (see Chapter 16). Another common example of latency occurs with the bacilli that cause tuberculosis. Most people exposed to sufficient doses of these tubercle bacilli become infected but do not develop tuberculosis. Instead they develop some resistance to reinfection with the bacilli, and the infecting organisms become latent. A few of them survive and are maintained in the tissues. In old age or under conditions of stress or immune deficiency, these latent tubercle bacilli are able to overcome host resistance and cause disease. This is called "reactivation tuberculosis."

Reactivation tuberculosis, page 447

Obviously, the outcome of any infection depends on a complex interaction between aggressive mechanisms of the invading organism and defensive mechanisms of the host. Often a state of truce is attained that can be overcome by changes in either the host or the parasite.

SUMMARY

Many microorganisms and viruses colonize normal humans and constitute the normal flora. The species normally found in different areas depend upon the microenvironments of various body locations.

Host defenses are either specific, resulting from acquired immunity, or nonspecific, resulting from innate or inborn capabilities of the host. Leukocytes that are important in host defense include granulocytes of three kinds (neutrophils, basophils, and eosinophils), lymphocytes, and monocytes. Neutrophils and monocytes (macrophages) are "professional phagocytes" that can engulf and destroy a variety of foreign materials, including many bacteria. Monocytes (macrophages) make up the mononuclear phagocyte system (reticuloendothelial system), which removes unwanted substances from the circulation. Lymphocytes, the major cells in lymphoid tissues, participate principally in specific immune responses.

Inflammation is a nonspecific reaction to injury and is characterized by redness, heat, swelling, and pain in inflamed areas. It results from the actions of locally released chemical mediators that cause enlargement and increased permeability of small blood vessels.

Many antimicrobial factors act nonspecifically. These include components of the complement system, lysozyme, peroxidases, interferons, and others.

Bacteria and other microorganisms employ many means of colonizing and invading tissues. Some virulence factors important for the pathogenicity of microorganisms include surface components that encourage adherence; capsules and other forms of antiphagocytic protection; exotoxins, endotoxins, and other kinds of toxins; and enzymes that are toxic or can degrade host tissues.

Koch's Postulates must be fulfilled to establish that an organism is the causative agent of a disease. Potential pathogens can be carried by healthy carriers, who may shed the pathogens or in whom the organisms may become latent. The pathogenic properties of the organism and the protective mechanisms of the host are equally important in determining the outcome of any host-parasite interaction.

SELF-QUIZ

1. Species of *Bacteroides* are
   a. among the normal flora of the mouth
   b. among the normal flora of the throat
   c. among the normal flora of the genitalia
   d. all of the above
2. Factors important in nonspecific, innate immunity include all of the following *except*
   a. interferon
   b. complement
   c. antibodies
   d. lysozyme
   e. peroxidase
3. Granules containing histamine are characteristic of
   a. neutrophils
   b. basophils
   c. eosinophils
   d. lymphocytes
   e. macrophages
4. Bacterial endotoxins are characteristically
   a. destroyed by mild heating
   b. part of the bacterial capsules
   c. capable of causing fever
   d. proteins
   e. enzymes
5. Which one of the following statements about bacterial exotoxins is *false*?
   a. They often are proteins.
   b. They may be produced by Gram-positive bacteria.
   c. They may be produced by Gram-negative bacteria.
   d. They are generally stable to heat (100°C, 5 minutes).
   e. They affect specific tissues.

QUESTIONS FOR DISCUSSION

1. Is *E. coli* a pathogen? Explain.
2. Does having an infection mean that an infectious disease is present? Why or why not?
3. What is the most important factor in host defense? Explain.

FURTHER READING

Burnet, M., and White, D. B.: *Natural History of Infectious Disease,* 4th ed. London: Cambridge University Press, 1972. An old favorite with good background material and many historical examples.

Kluger, M., Jr.: "The Evolution and Adaptive Value of Fever." *American Scientist* (January-February 1978), pp. 38–43.

Mim, C. A.: *The Pathogenesis of Infectious Disease.* New York: Academic Press, 1976.

Pike, J. E.: "Prostaglandins." *Scientific American* (November 1971).

Rosebury, T.: *Microorganisms Indigenous to Man.* New York: McGraw-Hill Book Co., 1962.

Smith, H.: "The Determinants of Microbial Pathogenicity." In Norris, J. R., and Richmond, M. H. (eds.): *Essays in Microbiology.* New York: John Wiley & Sons, 1978, pp. 13/1–13/32.

Chapter 10

IMMUNOLOGY: ANTIGENS, ANTIBODIES, AND CELL-MEDIATED IMMUNITY

On June 2, 1881, at Pouilly le Fort in France, a large crowd gathered at the farm of a veterinarian named Rossignol. They had come to see Rossignol ridicule and disprove the theories of Louis Pasteur, who claimed that sheep could be protected against anthrax, a serious disease, by injecting the animals with cultures of modified, or attenuated,* anthrax-causing organisms. Rossignol had injected living, attenuated anthrax organisms into 24 sheep, four cows, and a goat, and some days later had injected live, virulent anthrax bacilli into the same animals and also into the same number of control animals. It soon became apparent to the gathered crowd that Pasteur's theories were not absurd; all the immunized animals remained healthy! The nonimmunized control animals, however, did not fare well; 21 of the 24 sheep were dead in the morning; the other three sheep died during the day as the crowd watched, and the four cows were also ill with anthrax. Pasteur was highly praised, and very soon immunization of sheep against anthrax was widely practiced. These and similar results in animals paved the way for widespread use of immunization in humans.

---

*attenuated—modified so as to be incapable of causing disease

OBJECTIVES

To know

1. Some of the applications, in addition to immunization against infectious diseases, that fall within the scope of immunology.
2. How to diagram the molecules of antigen and antibody and indicate how they interact.
3. What haptens are and how they function.
4. Some of the functions of the different classes of immunoglobulins.
5. The events that occur during antibody production in both the primary and secondary responses.
6. How antigen-antibody interactions are used in the diagnosis of disease.
7. What the complement system is and what its functions are.
8. How cell-mediated immunity differs from antibody-mediated immunity.
9. The mechanisms of cell-mediated immunity operative against bacteria, viruses, and foreign tissues, such as tumors and transplants.
10. The four major characteristics of immune responses that are common to antibody-mediated and cell-mediated immunity.

THE SCOPE OF IMMUNOLOGY

The science of immunology (the study of immunity) developed from studies of the mechanisms of immunity to infectious diseases, but its scope has enlarged tremendously during this century. At present, immunology is of major importance not only in many aspects of medical practice but also in a wide variety of fields unrelated to medicine.

One familiar application of immunology is the use of immunological tests as an aid in the laboratory diagnosis of many diseases, both infectious and noninfectious. The "blood tests" used in diagnosing syphilis and infectious mononucleosis are examples of such tests. More sophisticated techniques such as **radioimmunoassays** are frequently used to measure extremely small quantities of materials such as hormones in the blood, and are also helpful in diagnosing and treating many diseases. A multitude of other immunological tests are also widely used.

Blood transfusions are often essential, life-saving procedures made possible by an understanding of the complex immunology involved. For example, transfusions of blood or blood components can aid in many surgical procedures. They also permit the replacement of erythrocytes in seriously anemic* patients and replacement of clotting factors in hemophiliacs.* Successful blood transfusions were not regularly possible until the early part of

---

*anemic—deficient in red blood cells

*hemophiliacs—individuals with a blood coagulation disorder, usually inherited this century, when Karl Landsteiner and his colleagues discovered the major blood groups. As more has been learned about the immunology of blood, transfusions have become increasingly safe and effective.

Transplantation of kidneys has become a routine medical procedure now that immunological responses to transplanted tissues are partially understood. It is probable that further advances in understanding will soon lead to success with transplantation of organs other than kidneys.

Major advances have been made in overcoming infectious diseases, not only by immunization but also by antimicrobial therapy. Many diseases, however, have not been controlled and may even be increasing in incidence. Cancer is one such group of diseases. But **immune responses** to cancer occur regularly, and many efforts are being made to unravel the intricacies of immunity to cancers and to use immune mechanisms in their prevention and treatment.

Even though immunity often protects against bacteria, viruses, fungi, and even chemicals in the environment, the same mechanisms that protect may under other circumstances be harmful to the host who is producing the immune response. Such harmful immune responses are called **hypersensitivity responses** or **allergic reactions.** Thus, allergies are other forms of immunological responses and play a large part in causing many diseases.

Immunological principles and techniques are used in areas far removed from the health sciences. For instance, in anthropology and criminology, immunological tests are often helpful in proving whether blood is of human origin. These tests can even be used to establish the group of humans the blood came from.

This chapter and Chapter 11 will explore the basic principles of immunology. We shall see that the study of immunology reveals much about the ways in which human hosts interact with the environment. During the immune response, cells multiply and differentiate.* In fact, the study of mechanisms of these responses is leading to a better understanding of the ways in which many kinds of cells and tissues* develop.

ANTIGENS AND ANTIBODIES

Antigens and antibodies are the materials that interact during specific immune responses. (Chapter 9 dealt with the nonspecific immune responses.) Most people are familiar with the term "antibodies" and associate them with protection against disease. This concept is only partially correct. The words antigen and antibody must be defined in terms of one another. An **antigen** (Ag) is a substance that causes the host to produce specific antibodies against it. An **antibody** (Ab) is a protein produced by the body in response to an antigen and is able to combine specifically with that antigen.

---

*differentiate—to modify or develop in such a way that the cells perform specialized functions

*tissues—groups of cells that associate into structural subunits and perform specific functions

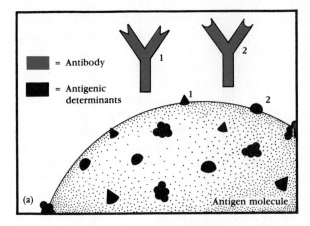

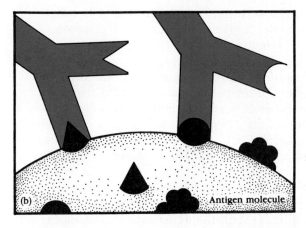

Figure 10–1

A schematic representation of antibody and antigen molecules. Each of the antibody molecules has two antibody reaction sites that are complementary to specific antigenic determinant sites. Note that a single antibody molecule has reaction sites for only one specific antigenic determinant. On the other hand, the antigen molecule has many antigenic determinant sites.

Figure 10–1 illustrates some of the properties of antigen and antibody molecules. Note that the antibody molecules have inverted sites on either arm of their Y-shape. The two sites on each antibody molecule have the same shape. These sites are the **antibody reaction sites** on the antibody molecule and represent the areas that combine with antigens. Typically an antibody molecule has two reaction sites, although some antibody molecules may have as many as ten.

The surface of the antigen molecule contains specific chemical groups known as the **antigenic determinants** that combine with antibodies. An antigen molecule has many such antigenic determinants.

Figure 10–1 shows that there is a close "lock-and-key" fit between antigenic determinant 1 and antibody 1, and between antigenic determinant 2 and antibody 2. Thus antibody 1 cannot combine with antigenic determinant 2 because it would not fit well. This need for a close fit accounts for the specificity a particular antibody molecule exhibits for one antigenic determinant and, conversely, the specificity that an antigenic determinant has for an antibody molecule.

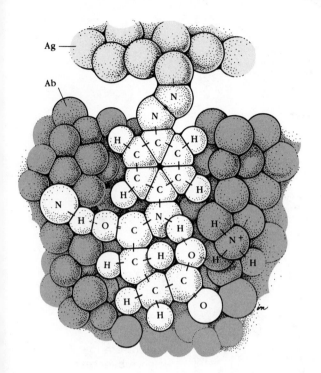

Figure 10–2
A schematic drawing showing the kinds of bonding forces between the antigen and the antibody molecules. Note that many of the bonds are weak hydrogen bonds. Thus, many such bonds are needed to hold the antibody to the antigen. Further note how the molecule of antigen "fits into" the antibody molecule to bring the atoms into close proximity.

The need for a close fit between the antigenic determinant and the antibody reaction site is understandable when the nature of the bonding between the two molecules is examined (Fig. 10–2). Weak forces such as hydrogen bonding hold the antigen and antibody molecules together. A large number of weak bonds are required for this purpose, and these bonds function only if the molecules are close to each other.

The Nature of Antigens and Haptens

Most antigens are large and chemically complex molecules. Substances having low molecular weights are usually not antigenic. Antigens may be proteins or polysaccharides; lipids rarely make good antigens. Another important characteristic of antigens is that they are usually foreign to the host that forms the antibodies to them. If this were not so, an individual would form antibodies against his own body constituents with the potential of producing tissue damage. Under unusual circumstances this does occur, resulting in **autoimmune diseases.**

Although low molecular weight substances by themselves are not often antigenic, they can serve as **haptens** (Fig. 10–3). A hapten is a substance that can react with specific antibodies but cannot cause the production of antibodies unless it is chemically combined with a large carrier molecule. For example, penicillin is a low molecular weight substance that is not antigenic by itself; however, it is destroyed by enzymes in the body. Some of these

Autoimmune diseases, page 274

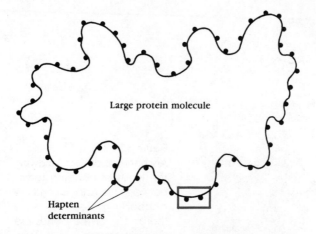

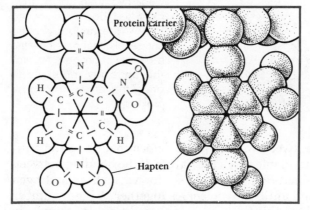

Figure 10–3
Haptens are low molecular weight substances that can act as antigens only if they are combined with a larger molecule. This diagram shows a large protein molecule acting as the carrier for a hapten.

penicillin breakdown products can combine with some of the large protein molecules of the host to form an antigenic hapten carrier complex. Antibodies are produced to this complex. The ability of many small molecules to act as haptens greatly increases the possibility of developing immunity or allergies, or both, to foreign substances.

The Nature of Antibodies

Antibodies, like other proteins, are made up of chains of amino acids. Each chain is coded for by a particular gene or genes. All antibodies belong to a group of proteins called **globulins** and are somewhat similar in structure. Since antibodies are involved in the immune response they are called **immunoglobulins** (Ig).

Five different classes of human immunoglobulins have been described, and these are referred to as IgG, IgM, IgD, IgA, and IgE. Each class differs from the others in the kinds of polypeptide chains* it has in its molecule.

---

*polypeptide chains—proteins, or chains of amino acids, joined together by peptide bonds

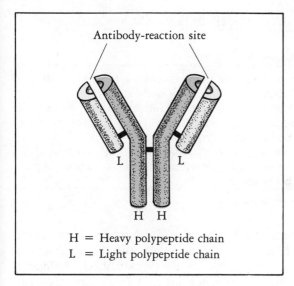

Antibody-reaction site

H = Heavy polypeptide chain
L = Light polypeptide chain

Figure 10–4
A molecule of human immuno-globulin G is made up of two heavy polypeptide chains and two light polypeptide chains. There are two antibody reaction sites on each molecule where combination with specific anti-gen can occur.

Immunoglobulins of the most abundant class, IgG, have been studied extensively. Figure 10–4 diagrams the structure of the IgG molecule. The molecule is Y-shaped and has two identical halves. Each half consists of a heavy (H) and a light (L) polypeptide chain held together by disulfide bonds.* The H and L chains of the immunoglobulin have both variable and constant parts (Fig. 10–5). The constant area of the molecule is virtually the same for all antibodies of the same class. The variable part of the antibody molecule

---

*disulfide bonds—covalent bonds between two sulfur atoms in different amino acids of a protein

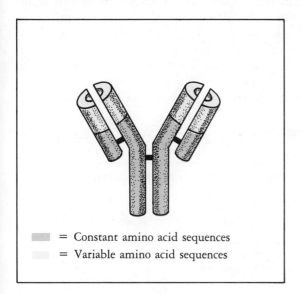

▨ = Constant amino acid sequences
▨ = Variable amino acid sequences

Figure 10–5
Variable amino acid sequences at the ends of both heavy and light polypeptide chains cooper-ate to make up the antibody reaction sites. The variability of these sequences accounts for the wide range of different antibody specificities.

TABLE 10–1 FUNCTIONS OF THE VARIOUS CLASSES OF HUMAN IMMUNOGLOBULINS[a]

Class	Percent of total antibody	Functions
IgG	80–85%	Protection within the body; causes lysis or removal of antigens, neutralizes viruses and toxins
IgM	5–10%	Protection within the body, as with IgG; especially effective in agglutinating antigens and in activating complement
IgA	10%	Protection of mucous membranes
IgE	0.05%	Aids in the expulsion of some multicellular parasites
IgD	1–3%	Functions during the development and maturation of the antibody response

[a]All classes of immunoglobulins can combine with specific antigen.

is at the outer end of each arm of the Y, the part of the antibody molecule that reacts with the antigenic determinant site. This antibody reaction site has a unique amino acid sequence that is specific for that antibody molecule.

Table 10–1 lists some of the important functions and Table 10–2 lists the properties of the different classes of immunoglobulins. IgG accounts for 80 percent of the antibodies found in the blood, lymph, and tissue fluids. It is the major circulating antibody. IgM is about five times the size of the IgG molecule (Fig. 10–6) and is usually the first antibody produced in response to an invading microorganism. Antibodies of the class IgA are secreted across mucous membranes and are important defense mechanisms in saliva, milk, and mucus. IgD is present in small amounts in the blood, and IgE is present in even smaller quantities. Although IgE is barely detectable in the blood, it is important because it is responsible for many allergies.

Only one class of immunoglobulin molecules, IgG, can cross the human placenta from the mother to the fetus. Although human fetuses can produce antibodies, under normal circumstances they are protected by maternal antibodies and do not synthesize their own. After birth, the maternal antibodies gradually disappear from the blood over a period of a few months as the infant begins to produce its own immunoglobulins (Fig. 10–7). By about six months of age, the child's own antibodies are responsible for most of the protection observed.

The ability of antibodies to recognize and combine with antigens is one

TABLE 10–2 IMPORTANT PROPERTIES OF HUMAN IMMUNOGLOBULIN CLASSES[a]

Class	Properties
IgG	Specific attachment to phagocytes; complement-fixation; ability to cross the placenta
IgM	Complement-fixation
IgA	Secretion into saliva, milk, mucus, and other external secretions
IgE	Specific attachment to mast cells[b] and basophils

[a]All of these properties depend upon the structure of the constant region of the immunoglobulin molecules.
[b]Mast cells are large cells that are found primarily in connective tissue.

Figure 10–6
A schematic representation of the IgM molecule. Note that there are five Y-shaped units bonded together. Each Y-shaped unit is similar in shape and size to one IgG molecule.

of their important functions. However, recognition alone often does not result in the desired effect of removing, neutralizing, or destroying the antigen. For these functions to be carried out, the constant regions (Fig. 10–5) of the Y-shaped immunoglobulin molecules play important roles (Table 10–2). Some antibodies attach to host cells by these constant regions and carry out special functions while attached to the cells. For example, IgG molecules that have combined with antigen attach to phagocytes (Fig. 10–8) and act as opsonins,* thereby increasing phagocytosis of the antigen. IgE molecules attach to basophils and similar granule-containing cells and, following com-

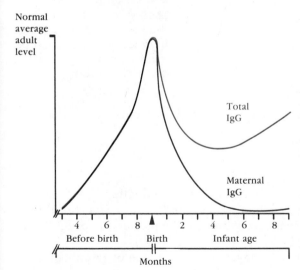

Figure 10–7
Immunoglobulin G levels in the fetus and infant. During gestation, maternal IgG crosses the placenta to the fetus. Normally the fetus does not make appreciable quantities of immunoglobulin but depends on the maternal antibodies that are transferred passively. After birth, the infant begins to produce immunoglobulins to replace the maternal antibodies, which gradually decay over a period of about six months. Usually by about three to six months, most of the antibodies present are those produced by the infant.

*opsonins—substances that cause increased phagocytosis

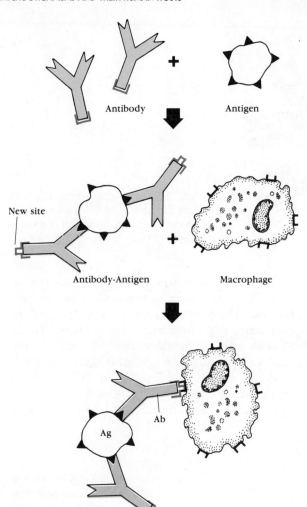

Antibody + Antigen

New site

Antibody-Antigen + Macrophage

Ab

Ag

Figure 10–8
After the antibody molecules coat the antigen molecule with antigen, they use the constant region of the molecule to attach themselves to the macrophages to aid in phagocytosis. Note that this diagram is not drawn to scale. The macrophages would be much larger than the antigen and antibody complex.

bination with antigen, cause release of the granule contents and subsequent allergic reactions.

Antibody Production

The sequence of events occurring during antibody production is complex, is carefully controlled, and varies depending on many factors. One variable is the nature of the antigen. Some antigens induce the production of IgM, and other antigens first induce IgM and later IgG (Fig. 10–9).

At least three cell types—macrophages and two kinds of lymphocytes—participate in producing most antibodies. One of the most illuminating discoveries of recent years was the finding that lymphocytes develop, or differentiate, into one of two populations called T and B lymphocytes (Fig. 10–10).

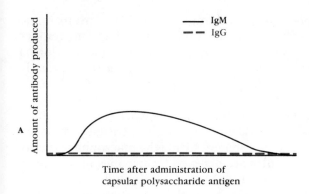

—— IgM
– – IgG

A

Amount of antibody produced

Time after administration of
capsular polysaccharide antigen

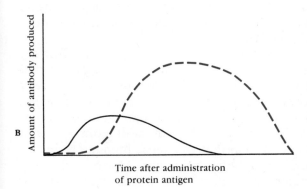

B

Amount of antibody produced

Time after administration
of protein antigen

Figure 10–9
The graphs in this figure show how different types of antigens cause different kinds of immunoglobulin molecules to be produced. Graph A shows that after the administration of a capsular polysaccharide antigen, only IgM is produced. Graph B shows that after the administration of a protein antigen, first IgM is produced and then IgG is produced in much larger quantities.

Both populations develop from precursor cells* formed in the bone marrow. Precursor lymphocytes that differentiate and mature within the thymus gland are known as **T lymphocytes.** Precursor lymphocytes that differentiate and mature within a structure known as the bursa fabricius in birds are known as **B lymphocytes.** Humans do not have a bursa fabricius and it is not clear exactly where in humans the B cells do differentiate and mature. It is thought that B cell development occurs within the bone marrow. Cells of the B lymphocyte line further differentiate into **plasma cells** and, once they are stimulated by antigen, synthesize and secrete large amounts of antibody. Some T lymphocytes are involved in cell-mediated immunity (discussed below), while others known as **T helper cells** cooperate with B cells to stimulate efficient antibody production. However, the T cells do not secrete any antibody. Lymphocytes of the T cell series are important in the regulation of many immune responses. One subset called **suppressor cells** serves to check or suppress certain responses (Fig. 10–11).

In humans, lymphoid tissues are widespread throughout the body and include the lymph nodes, spleen, tonsils, and others (see Fig. 9–5). The principal cells of these tissues, the lymphocytes, also circulate in the blood and

Distribution of lymphoid tissue, page 216

*precursor cells—cells that subsequently differentiate into more specialized ones

lymph. In addition to lymphocytes, the lymphoid organs contain many phagocytic macrophages, which can engulf foreign materials and remove them from the circulation. Macrophages cooperate with lymphocytes during the immune response in several ways. The lymphoid tissues are strategically located to protect the body against invasion by foreign agents via virtually any route.

When a foreign antigen enters the body, it soon reaches the circulation, is channeled through various lymphoid tissues, and is eventually removed from the circulation, usually by macrophages. Inside the macrophage, many kinds of large particles and molecules are digested by enzymes into smaller antigenic molecules, which can reach potential antibody-forming cells. Macrophages themselves lack the capacity to synthesize antibodies, but they are often important in the antibody response because they digest antigens into highly immunogenic fragments and allow these fragments to contact populations of responsive lymphocytes.

Among the many millions of lymphocytes in the body, only a few can respond to any given antigen. The antigenic determinant seeking a lymphocyte that can respond is rather like a person shopping for gloves; to get a good fit, it is necessary to select the proper size and style from a wide variety of gloves in stock. Just as there are many different people in a population, requiring many different sizes of gloves, so are there many different antigenic determinants—an estimated one million that can cause immune responses in a normal person. This specificity requires millions of different clones of lymphocytes of different subpopulations, each having a different recognition site for antigen. Antigenic determinants can stimulate only those lymphoid

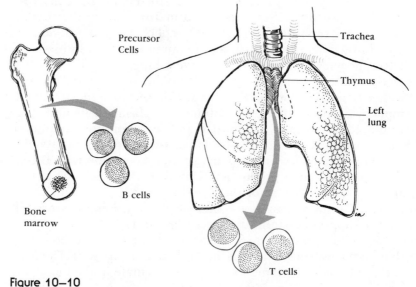

Figure 10–10
Precursor cells, originally from the bone marrow, differentiate under the influence of the bone marrow into B cells or under the influence of the thymus into T cells.

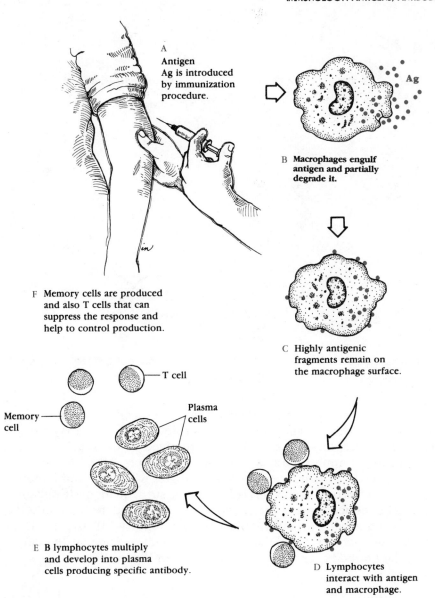

A
Antigen
Ag is introduced
by immunization
procedure.

B **Macrophages engulf
antigen and partially
degrade it.**

C **Highly antigenic
fragments remain on
the macrophage surface.**

F **Memory cells are produced
and also T cells that can
suppress the response and
help to control production.**

T cell

Memory
cell

Plasma
cells

E **B lymphocytes multiply
and develop into plasma
cells producing specific antibody.**

D **Lymphocytes
interact with antigen
and macrophage.**

Figure 10–11

Antibody production. Antigen is introduced by infection, immunization, or other means. The antigen is engulfed by macrophages, which partially degrade it, leaving highly antigenic fragments. These antigenic fragments interact with B lymphocytes that recognize the antigenic determinants and also with T cells that help in the antibody response (T helper cells). The B cells multiply and develop into a clone of plasma cells producing antibody specific for the antigen. Memory cells are also formed, which remain inactive in the body over long periods until the antigen is again available to stimulate them, inducing the memory response. A population of T cells also acts to suppress and control antibody production.

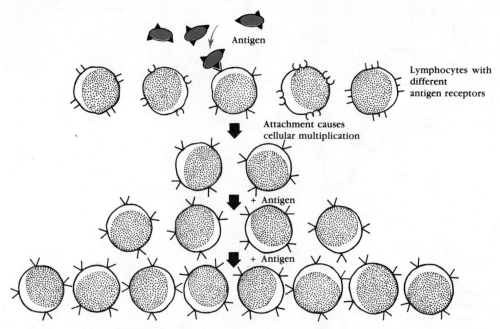

Figure 10–12

Clonal selection. Antigen selects lymphocytes with receptors for that antigen, causing cellular multiplication of those particular lymphocytes.

cells that recognize and respond to them by virtue of specific antigen-binding receptors fixed to their surface membranes (Fig. 10–12). On a B cell, the antigen-binding receptors are antibodies synthesized by the B cell that carries them; each B lymphocyte has many such receptors, all with the same specificity. Following the interaction of an antigen with its specific immunoglobulin receptors on a group of B cells, and the subsequent multiplication of these B cells induced by the interaction, the B cells differentiate into the highly efficient antibody-producing plasma cells (Fig. 10–13). These plasma cells synthesize and secrete the bulk of the antibodies. Steps in the development of the antibody response are diagramed in Figure 10–11.

Antigen-Antibody Interactions

Antibody-mediated immunity refers to the protection provided by antibodies, for example, immunity to tetanus and diphtheria following immunization. In vivo,* antibodies interact with antigens in many ways to lead to protection, or sometimes to allergy. Many of the same antigen-antibody reactions occur in vitro* as well, thus permitting the development of useful

---

*in vivo—in a living organism

*in vitro—in an artificial environment; for example, in a test tube or culture dish

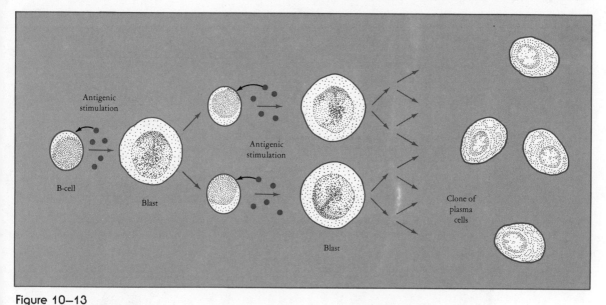

Figure 10–13
The postulated development of the immune response. Stem cells from the bone marrow become B cells. Upon stimulation with specific antigen, B cells enlarge to become blast cells, which divide to form two small lymphocytes. The process is repeated if antigen is available. During proliferation the B cells differentiate, finally yielding a clone of plasma cells synthesizing antibody molecules.

laboratory tests. For example, the concentration of antibody in a serum or other body fluid can be estimated by making dilutions of the fluid and testing it with antigen. The greatest dilution that gives a positive test for antibody activity represents the **titer** of the fluid. Therefore, the titer of a fluid is a semiquantitative measure of the antibodies it contains. **Serum,** which is the liquid portion that remains after blood has clotted, is easy to obtain and a good source of antibodies, so it is often tested. The term **serology** has arisen to mean the study of antibody reactions in vitro.

Serological Techniques

An increase in the amount of specific antibodies, or a *rise in titer*, in the serum is often useful for diagnosing infectious diseases. Early during infection, the titer of specific antibodies is usually very low or nonexistent; however, ten days or two weeks later the titer commonly increases as a result of antigenic stimulation by the infecting organisms. For instance, in the case of suspected typhoid fever, paired serum samples are taken as soon as possible during the illness and again approximately one to two weeks later. An appreciable (fourfold or greater) increase, or rise in titer, of antibodies specific for the typhoid organism indicates that the illness is indeed typhoid fever.

(a) Killed *Treponema pallidum* spirochetes, fixed to the slide, are incubated with the subject's serum. If specific antibodies are present they combine with the spirochetes.

(b) The slide is washed to remove excess serum, and fluorescent-labeled anti-human gamma globulin (anti-HGG) antiserum is added. The anti-HGG antibodies combine with any human antibodies already on the spirochetes.

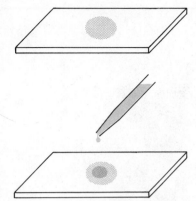

(c) Following incubation, the slide is washed and examined microscopically using ultraviolet light. The *T. pallidum* organisms coated with antibodies fluoresce and can be visualized readily.

T. pallidum +
specific antibodies

Labeled
anti-HGG

Fluorescing
T. pallidum

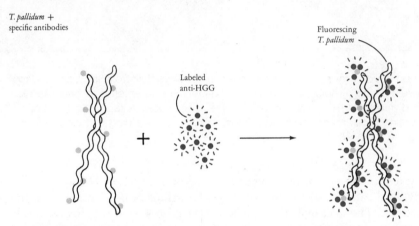

Figure 10–14
The fluorescent antibody test. This test employs indirect immunofluorescence to demonstrate humoral antibodies against *Treponema pallidum.*

Various tests have been developed for measuring quantities of antibodies present or for determining the location of antibodies or antigens in tissues, by use of antigen or antibody reagents* labeled with easily detected substances such as radioisotopes.* Radioimmunoassays are among the most sensitive of all immunological tests. Substances that fluoresce when exposed to ultraviolet light are frequently used as labels that make antigens or antibodies visible (Fig. 10–14).

*reagents—substances chosen for use in a chemical process on account of their known biological or chemical activity

*radioisotopes—radioactive isotopes of a chemical element; for example, carbon-14 is a radioisotope of carbon-12

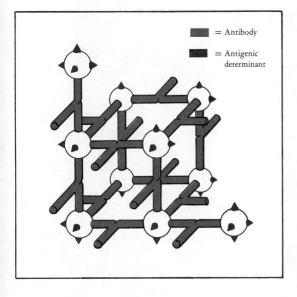

= Antibody

= Antigenic
  determinant

Figure 10–15
Lattice formation by antigen and antibodies leads to precipitation. Bivalent antibodies combine with multivalent antigens, linking the two in an insoluble complex lattice formation that precipitates.

The Principle of Lattice Formation

Many antibodies have the ability to cross-link antigen molecules into large lattice formations, as illustrated in Figure 10–15. If the antigen is a soluble material, the molecules will be precipitated in large lattices and a visible precipitate will form. If the antigen is particulate, for example bacteria or erythrocytes, the particles will be agglutinated into a visible mass (Fig. 10–16).

Precipitation and agglutination tests are regularly used and have many

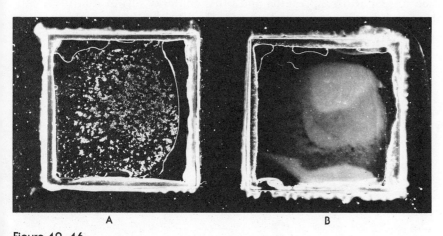

A B

Figure 10–16
Agglutination of erythrocytes. (a) Agglutinated red blood cells and antibody. (b) Non-agglutinated red blood cells (control).

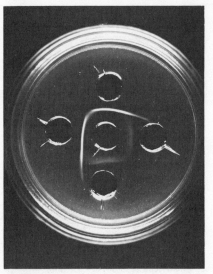
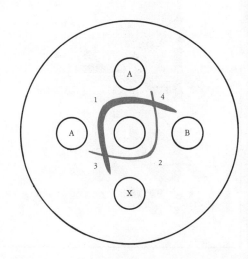

Figure 10–17
Ouchterlony double diffusion in agar. Antigen is placed in wells A, B, and X cut in the agar and antibody in the center. Each diffuses through the agar, and a line of precipitate is formed where the two meet in proper proportions. This method can be used to determine whether an unknown substance x is identical to or shares antigenic determinants with a known substance. (Photograph courtesy of M. Tam.)

practical applications. There are many clever variations of these tests, especially of the precipitation tests. These reactions can be carried out in a gel by allowing antigen and antibody to diffuse through the gel and to combine to form visible lines of precipitation (Fig. 10–17). Electrophoresis, a technique used for separating proteins and other materials from mixtures, can be combined with diffusion-in-gel; this method, called immunoelectrophoresis, gives better results with mixtures of antigens, as shown in Figure 10–18.

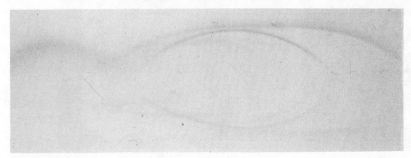

Figure 10–18
Immunoelectrophoresis. Human serum was subjected to an electric current. The various serum proteins present migrated to different areas, depending on their electric charges and other properties. Following electrophoresis, rabbit antiserum against human serum was placed on the slide and the two diffused. More than 30 different proteins in serum, each represented by a distinct line, can be distinguished by this method.

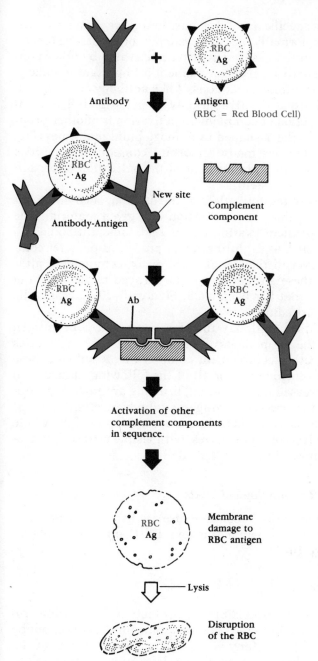

Figure 10–19
The activation of complement by combination of antigen with antibody starts a series of reactions that can end with lysis of cells bearing the antigen.

Complement-Fixation Reactions

Erythrocytes, bacteria, or other cells are sometimes **lysed,** or disrupted, as the result of an interaction between antigenic components of the cell, specific antibodies, and **complement** (C), a system of proteins found in the blood (Fig. 10–19). In contrast to antibodies, which are specific for a particular antigen, complement is not antigen-specific and takes part in a large number of immunological reactions. The normally inactive system in blood can be turned on or **activated** by certain antibodies. It can also be activated by other nonimmunological means. The activation of complement is important in inflammation, as well as in specific immune reactions. When antibody participates, the complement system is usually activated only after antibody has combined with antigen. The function of complement activation is another property of the constant region of the Y-shaped immunoglobulin molecules (Fig. 10–1). Complement activation has many important consequences besides cell lysis. Opsonization is one important result of the activation of certain components of the complement system.

Many in vitro tests measure the activation, or the **fixation,** of complement as an indirect measure of antibody concentration in a given material. The advantage of using a **complement-fixation test** is that some antibodies do not give a visible reaction, such as agglutination or precipitation, after combining with antigen. However, if they react with, or fix, complement after reacting with an antigen, these antibodies can be assayed by determining the amount of complement that has been fixed. Figure 10–20 diagrams the complement-fixation test procedure.

During complement-fixation, antibody that has combined with antigen then combines with one of the components of complement, triggering a series of reactions involving the other complement components. Results of these reactions occurring in vivo include any or all of the following: destruction of the antigen by lysis; increased inflammation, bringing antimicrobial substances and phagocytes to the area of the antigen-antibody reaction; removal of the antigen by phagocytosis; and others. In vitro, completion of the whole series of reactions leads to lysis of erythrocytes, with visible release of hemoglobin from the cells, as shown in color plate 18.

Some Commonly Used Immunological Tests

Some of the antigen-antibody tests in common use are listed in Table 10–3. Many of these are referred to elsewhere in this text. The Coombs antiglobulin test is diagramed in Figure 10–21.

CELL-MEDIATED IMMUNITY

Although antibodies are often responsible for protection, in reactions referred to as **antibody-mediated immune reactions** some immunity cannot be credited to antibody protection. Many years ago, around the turn of the century, it was discovered that serum from people or animals immune to diphtheria could be given to nonimmune people or animals, resulting in temporary protection against diphtheria. This was called **passive transfer** of immunity. It was shown to be caused by preformed, specific antibodies against diphtheria toxin that were present in the transferred serum (Fig. 10–22). However, when passive transfer of serum was used to try to protect

Positive complement–fixation reaction

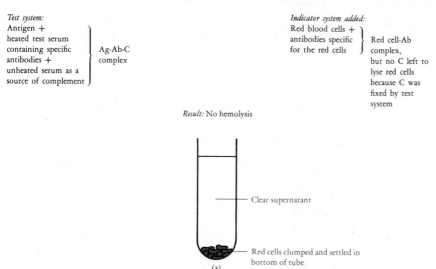

Test system:
Antigen +
heated test serum
containing specific
antibodies +
unheated serum as a
source of complement
} Ag-Ab-C complex

Indicator system added:
Red blood cells +
antibodies specific
for the red cells
} Red cell-Ab complex, but no C left to lyse red cells because C was fixed by test system

Result: No hemolysis

— Clear supernatant

— Red cells clumped and settled in bottom of tube

(a)

Negative complement–fixation reaction

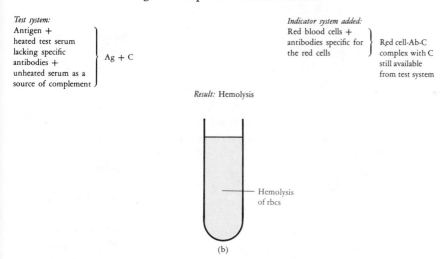

Test system:
Antigen +
heated test serum
lacking specific
antibodies +
unheated serum as a
source of complement
} Ag + C

Indicator system added:
Red blood cells +
antibodies specific for
the red cells
} Red cell-Ab-C complex with C still available from test system

Result: Hemolysis

— Hemolysis of rbcs

(b)

Figure 10–20
Complement-fixation test procedure.

against tuberculosis, the efforts failed, even though the serum contained specific antibodies against the causative organisms, the tubercle bacilli. Not until the 1940's was it found that a degree of immunity to tubercle bacilli could be passively transferred, not by giving serum antibodies but by trans-

TABLE 10–3   SOME COMMONLY USED IMMUNOLOGICAL TESTS

Test	Principle	Examples of applications
Agglutination	Cross-linking of particles of antigen by antibody molecules	Blood typing and cross-matching Tests to aid in diagnosis of bacterial infections such as typhoid fever, brucellosis, rickettsial infections, and many others
Precipitation	Cross-linking of soluble molecules of antigen by antibody	Precipitin-in-gel to measure amounts of serum immunoglobulins and presence of various classes of immunoglobulins Tests to aid in the diagnosis of infectious diseases such as viral hepatitis and streptococcal infections
Complement-fixation	Complement is fixed by antigen-antibody molecules and not available to cause visible lysis in an indicator system	Measurement of complement levels in blood Tests to aid in the diagnosing of infectious diseases such as infections with many viruses, rickettsiae, and other bacteria
Immunofluorescence assays	Antigen or antibody labeled with a fluorescent tag; reacts with the corresponding (nonfluorescing) antibody or antigen to give a visible reaction	An aid in the rapid diagnosis of certain infectious diseases such as infections caused by pneumococcus
Virus neutralizations	Specific antibodies neutralize viruses and prevent their entrance into and growth within cultured host cells	An aid in the diagnosis of infectious diseases such as poliomyelitis

ferring lymphocytes from an immune donor to a nonimmune recipient. This form of immunity results from the activities of lymphoid cells and is called **cell-mediated immunity.**

More recently, it has been shown that cell-mediated immunity involves T lymphocytes that respond specifically to antigen. These are called **sensitized T cells.** Both macrophages and specifically sensitized T lymphocytes are essential for the development of cell-mediated immunity. The participating lymphocytes carry cell-surface molecules that recognize the antigen to which they can respond, much as immunoglobulins on B cells recognize antigen to trigger antibody production. Following the interaction of lymphocytes with antigen, the lymphocytes are stimulated to multiply extensively (Fig. 10–23). In addition, some of the specific T lymphocytes are the principal producers of substances called **lymphokines,** which participate in cell-mediated immunity in a nonspecific manner (Table 10–4). Some of these substances act on macrophages or other cells to influence their functions. For example, some of the lymphokines convert inactive macrophages into very active cells that can inhibit or kill ingested microorganisms.

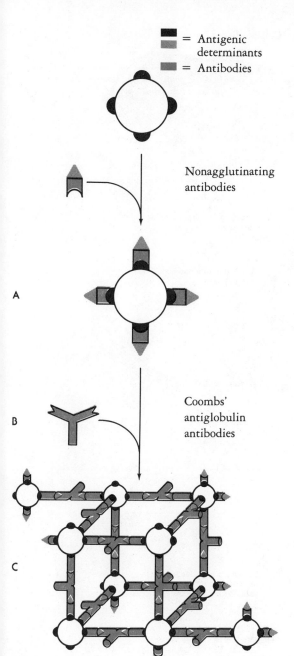

= Antigenic
determinants
= Antibodies

Nonagglutinating
antibodies

A

Coombs'
antiglobulin
antibodies

B

C

Figure 10–21
The Coombs antiglobulin test.
(a) Antibodies sometimes
combine with antigen in such
a way that agglutination or
precipitation cannot occur to
give a visible reaction. (b)
Specific antiglobulin antibodies can be used to link the antigen-antibody complexes together. (c) The result is lattice
formation and a visible reaction.

Passive Transfer   Cell-Mediated Immunity

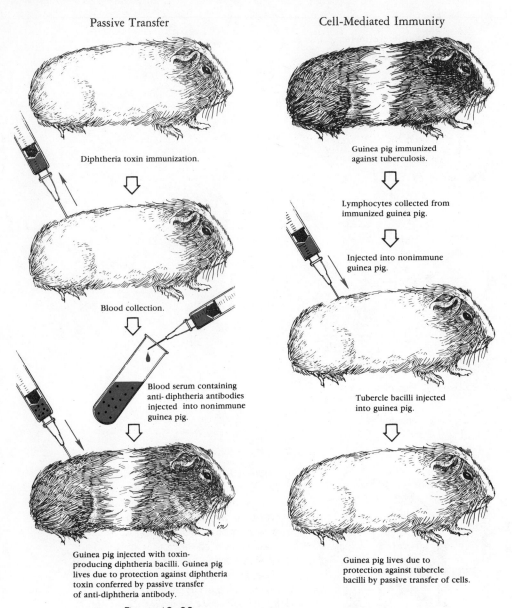

Diphtheria toxin immunization.

Guinea pig immunized
against tuberculosis.

Lymphocytes collected from
immunized guinea pig.

Blood collection.

Injected into nonimmune
guinea pig.

Blood serum containing
anti- diphtheria antibodies
injected into nonimmune
guinea pig.

Tubercle bacilli injected
into guinea pig.

Guinea pig injected with toxin-
producing diphtheria bacilli. Guinea pig
lives due to protection against diphtheria
toxin conferred by passive transfer
of anti-diphtheria antibody.

Guinea pig lives due to
protection against tubercle
bacilli by passive transfer of cells.

Figure 10–22
Passive transfer of immunity and cell-mediated immunity.

Antimicrobial Cell-Mediated Immunity

As indicated previously, cell-mediated immunity is involved in tuberculosis,
a disease caused by *Mycobacterium tuberculosis*, the tubercle bacillus. This
organism is engulfed by and multiplies within macrophages of the nonimmune host. Antibodies are not effective against the bacilli. Resistance to

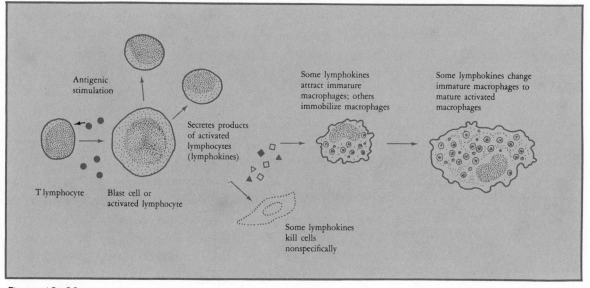

Figure 10—23
The production of lymphokines by antigen-stimulated sensitized T lymphocytes.

**TABLE 10—4   LYMPHOKINES (PRODUCTS OF SENSITIZED LYMPHOCTES):[a] PROBABLE
MEDIATORS OF CELL-MEDIATED IMMUNITY AND DELAYED HYPERSENSITIVITY**

Lymphokine	*Activity in vitro*	*Probable action in vivo*
Chemotactic factor	Attracts macrophages	Causes influx of macrophages into area where antigen is present
Migration-inhibitory factor	Inhibits migration of macrophages	Keeps macrophages immobilized in area near antigen
Lymphotoxin	Kills many different kinds of cells nonspecifically	Kills foreign cells; also kills cells of host, resulting in tissue damage
Macrophage-activating factor	Causes macrophages to become metabolically active and to synthesize many degradative enzymes	Results in macrophage activation; activated macrophages are able to kill foreign cells and to degrade ingested materials efficiently

[a]Lymphokines are synthesized and released by sensitized lymphocytes following stimulation by specific antigen. Only a few of the lymphokines are listed in this table.

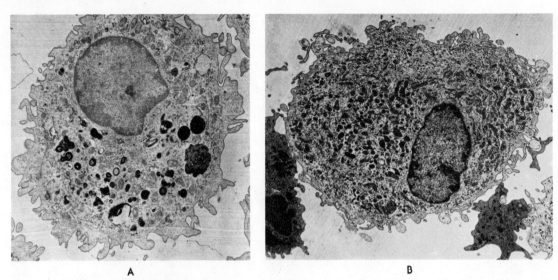

A   B

Figure 10–24

Normal and activated macrophages. (a) A macrophage from the lung of a normal nonimmunized rabbit. Few phagocytic vacuoles (dark areas) are present. (b) A macrophage from the lung of a rabbit 5 days after immunization with mycobacteria. The cell is highly activated, as evidenced by the abundant phagocytic vacuoles and lysosomes (dark areas) present in the cytoplasm. (Courtesy of Q. N. Myrvik and E. S. Leake.)

tuberculosis depends on cell-mediated immunity, in which activated macrophages phagocytize and destroy tubercle bacilli (Fig. 10–24).

During tuberculosis the following sequence of events usually occurs: Specifically sensitized T lymphocytes are stimulated by antigens of the infecting tubercle bacilli to produce lymphokines, which cause nonactivated macrophages to migrate to the infected area and to become immobilized and highly activated. These activated macrophages then kill or inhibit the growth of the intracellular tubercle bacilli that they have ingested, thereby limiting the infection. Furthermore, in tuberculosis and some other diseases, lymphokines, along with macrophages, can participate in **granuloma** formation, in which immune cells surround and wall off foreign materials that cannot be disposed of in other ways.

A similar sequence of events probably occurs in other infectious diseases caused by microorganisms. These include leprosy, brucellosis, some of the diseases caused by fungi, and many others.

Antiviral Cell-Mediated Immunity

All viruses must live within host cells; therefore, it is not surprising that cell-mediated immunity directed toward virus-infected cells is of primary importance in overcoming many viral infections. The way in which cell-mediated immunity operates against virus-infected cells differs considerably

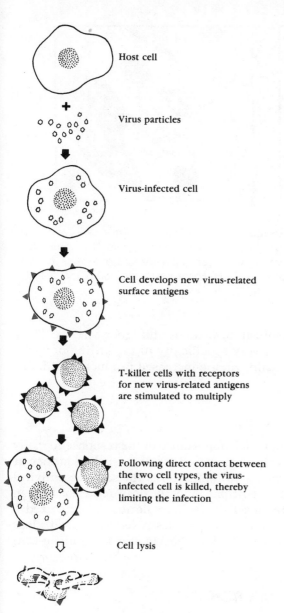

Host cell

Virus particles

Virus-infected cell

Cell develops new virus-related surface antigens

T-killer cells with receptors for new virus-related antigens are stimulated to multiply

Following direct contact between the two cell types, the virus-infected cell is killed, thereby limiting the infection

Cell lysis

Figure 10–25
Cell-mediated immunity against virus infections. Sensitized T lymphocytes can kill virus-infected cells following direct contact between the two cell types.

from the mechanisms of antibacterial and antifungal cell-mediated immunity. In a virus-infected cell, the virus directs the cell to maintain viral proteins on its surface. These viral proteins are then recognized by specific T lymphocytes known as **T killer cells.** The T killer cells are able to distinguish virus-infected cells from normal cells by these viral surface antigens. The T killer cells then destroy only the cells that are infected with the particular virus to which those T lymphocytes have responded (Fig. 10–25). Other T lymphocytes produce interferon, one of the lymphokines. The interferon acts

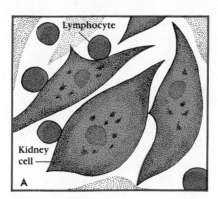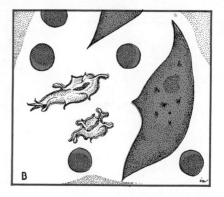

Figure 10–26
Cell-mediated immunity against tissues requires direct contact between sensitized T lymphocytes of the host and the cell to be destroyed. For example, cells of a transplanted kidney can be destroyed by direct contact with special populations of T cells. (a) Lymphocytes directly contact cells to be destroyed. (b) Following destruction of the cells, the lymphocytes are free to act again.

nonspecifically to inhibit the replication of many different viruses. Thus, cell-mediated immunity operates very specifically in the killing of virus-infected cells and also nonspecifically via the increased production of interferon.

Antitissue Cell-Mediated Immunity

Cell-mediated immunity is also of major importance in the response to many tumors (cancers) and to foreign cells in tissue transplants, such as kidney or skin grafts. This antitissue cell-mediated immunity is the same as antiviral cell-mediated immunity, and the principal mode of destruction of foreign cells is **contact killing** of the foreign cells by subpopulations of sensitized T lymphocytes (Fig. 10–26). For a foreign cell to be destroyed by a T cell, close contact must be established between the lymphocyte and the antigen-containing foreign cell.

PRIMARY AND SECONDARY IMMUNE RESPONSES

The immune response, whether it is mediated by cells or by antibodies, occurs much faster and more efficiently after a second exposure to an antigen than after the first exposure. After the first contact with antigen, the responding lymphocytes multiply extensively in the **primary response.** Upon subsequent exposures to the same antigen, the accelerated response that occurs is called the **secondary response,** or memory response. This response not only occurs more rapidly, but is also more intense than the primary response (Fig. 10–27). One reason for this is that some of the cells that multiply during a primary response survive as memory cells and increase the total popula-

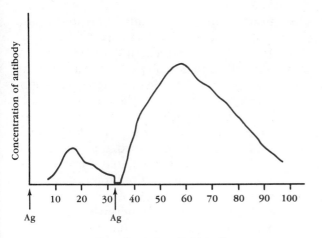

Concentration of antibody

10 20 30 40 50 60 70 80 90 100

Ag   Ag

Days after antigen (Ag) injection

Figure 10–27
A memory response is characteristic of antibody response. The first exposure to antigen causes a *primary response,* during which small amounts of antibodies are produced for a short period of time. Subsequent doses of the same antigen stimulate a *memory response,* during which much more antibody is produced over a longer period of time (brown line).

tion of cells capable of responding to the specific antigen. Figure 10–27 shows the difference in extent and timing of antibody production during the primary and secondary responses.

The secondary response is the basis of immunization methods used to prevent diseases. After being immunized, the host can give a prompt and accelerated memory response when exposed to the same antigen, either in the form of infecting organisms or as a "booster" dose of antigen in subsequent immunization procedures. Vaccines and immunization procedures are discussed in some detail in Chapter 12.

COMMON CHARACTERISTICS OF ANTIBODY-MEDIATED AND CELL-MEDIATED IMMUNITY

Of the many characteristics common to the various forms of immunity, several are worth special emphasis. These are recognition, amplification, memory, and control.

Recognition accounts for the specificity of acquired immune responses and for the ability to respond to foreign substances. In the case of antibody responses, recognition depends on the presence of immunoglobulin on B cell surfaces. In cell-mediated immunity, recognition depends on surface molecules on T cells that are apparently not immunoglobulins but can carry out similar recognition functions. In either event, recognition marks the antigen as foreign or different, but other steps must occur if the antigen is to be removed or destroyed.

Amplification is achieved in many ways during the antibody response. Complement activation is one important way. Each antigen-antibody complex that activates complement may activate hundreds of complement molecules that function in diverse ways. Some opsonize particles, making them more readily phagocytized. Other complement molecules lead directly to the disruption of foreign cells. Others function in the inflammatory response and aid in the delivery of antimicrobial substances and cells to areas where they are needed. Phagocytes are absolutely essential in the amplification of many antibody responses, since they are the cells that actually kill or destroy many of the foreign substances marked by antibodies. Without phagocytes, abundant supplies of specific antibodies may be ineffective. The amplification of cell-mediated responses is carried out by activities of certain substances produced by antigen-stimulated T lymphocytes. The mechanisms by which this takes place are not fully understood, but it is obvious that amplification is equally important in cell-mediated and antibody-mediated immunity.

Memory responses are also vital for all forms of acquired immunity. They are efficient; it is not necessary to maintain high levels of a particular antibody if that antibody can be produced at top efficiency when the need arises. This is what happens, for example, in the case of tetanus toxin. Following immunization in childhood, the antibody levels decline and may be insignificant; however, memory cells remain in the circulation for many years. If an infected wound becomes the source of tetanus toxin, the first molecules of toxin call forth a vigorous and effective response, even as long as ten years after the primary immunization series.

Such powerful and potentially dangerous events as antibody production, complement activation by antibodies, cell killing by lymphocytes, and increased phagocytosis must obviously be under strict and careful control. Some of the mechanisms of control are currently the focus of major research efforts.

SUMMARY

Immunological responses can be protective, causing immunity, or they can be harmful, causing hypersensitivity or allergic reactions. Frequently, both immunity and hypersensitivity occur simultaneously. The responses may be mediated either by antibodies or by cells.

Antibodies are protein immunoglobulins that react specifically with the antigens that induced their production, or with closely related molecules. Antigens and antibodies react by means of many weak bonds, so the reaction is reversible.

Proteins, polysaccharides, and some other large molecules are usually good antigens if they are foreign to the individual making the response. However, even small molecules called haptens, which are not antigenic by themselves, can be made antigenic by attaching them to large carrier molecules.

The five classes of antibody or immunoglobulin molecules differ in their structure and function. They are called IgG, IgA, IgM, IgD, and IgE.

Complex cellular interactions between lymphocytes and macrophages occur during antibody production, ending in the synthesis of large amounts of antibody by plasma cells. Antigen selects the clones of cells that can respond specifically.

A rise in titer of antibodies is useful in diagnosing certain diseases. Antigen-antibody reactions often used diagnostically include agglutination, precipitation, complement-fixation, radioimmunoassay, and many others.

Cell-mediated immunity is carried out primarily by T lymphocytes, with the aid of macrophages. It functions in different ways against bacteria, viruses, and foreign cells.

Characteristics common to antibody-mediated and cell-mediated immunity include recognition, amplification, memory, and control.

SELF-QUIZ

1. The antibodies that can cross the placenta from mother to fetus are of the class
   a. IgG
   b. IgA
   c. IgM
   d. IgD
   e. IgE
2. Antibodies of the class IgM have all of the following properties *except* that of
   a. protection by causing lysis or removal of antigens
   b. being the largest of the immunoglobulins
   c. fixing complement efficiently
   d. being of major importance in mucous secretions
   e. agglutinating antigens efficiently
3. Plasma cells
   a. come from T lymphocytes
   b. belong to the mononuclear phagocyte system
   c. are "professional phagocytes"
   d. produce most antibody molecules
   e. produce only IgG antibody molecules
4. Immunity against viruses may involve
   a. antibodies in the circulation
   b. cell-mediated immunity
   c. antibodies in secretions
   d. interferon
   e. all of the above
5. Paired samples of serum were taken from a patient during a disease with fever and again two weeks later. The first sample had antibody titers of less than 10 against *Brucella abortus* (the cause of brucellosis), and the later sample's titer was 160. It is probable that the patient
   a. had brucellosis
   b. did not have brucellosis, since the titer was less than 200

c. should be tested again, since the titer was less than 200
d. had a false positive reaction
e. had a false negative reaction

QUESTIONS FOR DISCUSSION

1. From the information given in this chapter, deduce several ways in which the immune response can be controlled.
2. Many attempts have been made to immunize against certain diseases, without success. What are some reasons why immunization procedures might not lead to permanent immunity?
3. Lipids are often not very good antigens, yet it is possible to cause immunological responses against them. How might this be done?
4. How could deficiencies in macrophage function affect specific immune responses?
5. Complement is a system that functions nonspecifically, yet certain defects in complement function interfere with protection by specific antibodies. How can this be explained?

FURTHER READING

Cooper, M. D., and Lawton, A. R.: "The Development of the Immune System." *Scientific American* (November 1974).

Golub, E. S.: *The Cellular Basis of the Immune Response.* Sunderland, Mass.: Sinauer Associates, Inc., 1977. An overview of the biology of the immune response with an emphasis on cell interactions and the regulation of immune responses.

Jerne, N. K.: "The Immune System." *Scientific American* (July 1973).

Kabat, E. A.: *Structural Concepts in Immunology and Immunochemistry.* 2nd ed. New York: Holt, Rinehart and Winston, 1976. A sophisticated immunochemical approach toward understanding basic immunology.

Muller-Everhard, H.: "Chemistry and Function of the Complement System." *Hospital Practice 12*:33–43 (1977).

Raff, M. C.: "Cell Surface Immunology." *Scientific American* (May 1976).

Rose, N. R. and Friedman, H.: *Manual of Clinical Immunology.* Washington, D.C.: American Society for Microbiology, 1976. A comprehensive manual dealing with applications of immunology for the detection and analysis of a wide variety of diseases.

Talmage, D. W.: "Recognition and Memory in the Cells of the Immune System." *American Scientist* (March/April 1979).

Thaler, M. A., Klausner, R. D., and Cohen, H. J.: *Medical Immunology.* Philadelphia: J. B. Lippincott Co., 1977. An easy-to-read approach to immunology for those interested in its medical application. The diagrams are very good and helpful.

Chapter 11

IMMUNOLOGY: HYPERSENSITIVITIES AND TRANSPLANTATION

In the early part of the twentieth century, Dr. Prausnitz, a German physician, observed that foods could cause symptoms of allergy in some individuals; in fact, he himself suffered from hives whenever he ate fish. He thought substances in the blood were responsible for these symptoms. To prove this hypothesis, he had the help of a medical student named Kustner, who was not allergic to fish or any other foods. Prausnitz first injected an extract of fish into Kustner's skin, without any visible effect. Next he injected some of his own blood serum into Kustner's skin, and a day or so later injected the fish extract into the same area. This time Kustner immediately had hives develop in the area of injection. This was the first recorded passive transfer of allergy, and it was evidence that some substance in blood serum played an important role in this type of allergy. Although it was suspected that serum antibodies were responsible, nearly half a century was to pass before the exact nature of those antibodies was discovered.

In this way, it became apparent that immunological responses are not always of immediate benefit; some of them may actually harm the host and may contribute to the pathogenesis of a variety of diseases. The term **hypersensitivity** has been applied to immunological reactions that cause tissue damage or malfunction in the host. Another word for hypersensitivity is **allergy.** Even though the direct effect of allergic reactions is discomfort or harm, the overall effect may still be to the benefit of the host, for example, by contributing to and increasing the inflammatory response.

259

OBJECTIVES

To know
1. The four major categories of hypersensitivities.
2. The mode of action of IgE-mediated hypersensitivities and how they are treated.
3. About other kinds of immediate hypersensitivities in addition to the IgE-mediated type.
4. What cells are responsible for delayed hypersensitivity, how they function, and some examples of delayed hypersensitivity reactions.
5. The role of immunity in the rejection of tissue transplants and in immune responses to tumors.
6. The major features of the ABO and Rh blood group systems.
7. About Rh disease (hemolytic disease of the newborn) and how it is prevented.
8. The causes of autoimmune diseases.

MAJOR TYPES OF HYPERSENSITIVITIES

Hypersensitivities are usually classified into two broad categories: **immediate** and **delayed.** In either type a state of sensitivity must exist before a reaction will occur. During the first exposure to an antigen, no allergic symptoms appear. Immediate reactions occur rapidly after exposure of a sensitized individual to a specific antigen, and reach their maximal effects within minutes to a few hours, depending on the type of reaction. Delayed reactions, however, are not apparent at all until about 6 to 12 hours after exposure to antigen and reach their peak two to three days later. Immediate hypersensitivity reactions are caused by antibodies, whereas those of delayed hypersensitivity result from the activities of *cells* rather than antibodies. As a rule, blood serum collected from a person with an immediate-type allergy to a specific antigen and transferred to a normal, nonallergic person will make the recipient temporarily allergic to that antigen, as shown by Prausnitz and Kustner in the experiment described earlier. This kind of passive transfer, known as the Prausnitz-Kustner or P-K reaction, results from the transfer of specific antibodies in the serum. Delayed hypersensitivities, on the other hand, can be transferred with living lymphoid cells from an allergic individual but not with serum; therefore, these lymphoid cells, specifically sensitized lymphocytes, are responsible for this type of allergy. Delayed hypersensitivity can also often be transferred from one person to another by an *extract* of sensitized lymphocytes which is called **transfer factor.** Some of the characteristics of the four major categories of hypersensitivities are outlined in Table 11–1.

Immediate Hypersensitivities

Immediate hypersensitivities are classified into three categories on the basis of the mechanisms responsible for each: IgE-mediated, cytotoxic, and immune complex–mediated reactions. An understanding of the mechanisms of

TABLE 11–1 SOME CHARACTERISTICS OF THE MAJOR TYPES OF HYPERSENSITIVITIES

Type	Time of reaction, following challenge of sensitized individual with antigen	Mediated by activities of	Examples
Immediate			
IgE-mediated	Immediate, within seconds or minutes; fades by an hour or several hours	IgE antibodies, fixed to mast cells in tissues and to basophils	Hayfever, hives, asthma, anaphylactic shock
Cytotoxic	Immediate, within seconds or minutes	Antibodies and complement,[a] reacting with antigen that is a part of a cell	Blood transfusion reaction with mismatched blood
Immune complex	Immediate, within minutes to few hours	Antibodies and complement,[a] reacting with soluble antigens that are not part of a cell	Serum sickness, Arthus-type reactions, farmer's lung
Delayed			
Cell-mediated	Delayed; first visible at about 6 hr, peaks at 24 to 48 hr; gradually declines over a period of days	Specifically sensitized lymphoid cells	Positive tuberculin skin test; poison ivy rash

[a]Complement may not always be necessary.

each of these categories is important in order to know how to deal with each condition. The body's immune response to any antigen is usually complex, involving a mixture of protective and hypersensitivity responses. However, most allergic reactions are predominantly one or another of the four types listed in Table 11–1.

Cytotoxic, as the name implies, refers to allergic reactions in which cells are damaged or destroyed. This can occur in a number of ways; antibody and complement lysis is a common mechanism. An example of this kind of allergic reaction is the damage that occurs because of a mismatched blood transfusion.* This kind of hypersensitivity reaction will be considered later in our discussion of blood transfusions.

Immune complex–mediated reactions are allergic reactions caused by antigen and antibody bound together in so-called immune complexes. These molecular complexes can be deposited in the walls of blood vessels, where they activate complement, thereby attracting neutrophils to the area. The damage in this kind of reaction is often a result of enzyme release from the many neutrophils attracted to the reaction site. The mechanisms of immune complex–mediated reactions are complicated and will not be discussed further here. Instead, the IgE-mediated reaction of immediate hypersensitivity will be considered in greater detail.

---

*mismatched blood transfusion—case in which the blood of the donor is incompatible with that of the recipient because of antibody-antigen reactions

IgE-Mediated Allergies

The manifestations of IgE-mediated allergy, seen for example in hayfever, hives, and some forms of asthma, are familiar to all. These are collectively called **anaphylactic reactions.** Very severe, systemic reactions of this type are called anaphylactic shock.

Hayfever (allergic rhinitis) occurs when a sensitized person inhales a specific antigen. The itching, tearing eyes, sneezing, and runny nose that result are all evidence of extensive irritation of the upper respiratory tract. Hives (urticaria) is a skin disease characterized by itchy red bumps that are actually fluid-filled lesions on the skin, called wheals. Often the offending antigen is eaten, as was the case with Prausnitz and fish. Someone who is allergic to strawberries may experience an outbreak of hives soon after eating even a very small quantity of the fruit. Eczema is another skin condition that is often but not always caused by IgE-mediated reactions.

Asthma is a fairly common and sometimes serious affliction, which also may result from various causes; however, IgE-mediated reactions are a frequent cause. Asthma is characterized by wheezing, coughing, and difficulty in breathing, particularly in exhaling.

Generalized (systemic) anaphylaxis, or **anaphylactic shock,** is an important manifestation of IgE-mediated allergy. It may be fatal, but fortunately it is rare. Generalized anaphylaxis is similar to other IgE-mediated reactions, but instead of being localized to the skin or to the respiratory tract it occurs systemically throughout the body. Breathing difficulties occur as a result of swelling and tissue changes in the bronchi, and other internal tissues are also affected. Death often occurs from a combination of causes, including generalized leakage of fluid from the blood vessels into the tissues, producing shock. The symptoms and signs of generalized anaphylaxis become apparent within seconds or minutes after exposure to the causative antigen and include itching and flushing of the skin, difficulty in breathing, and collapse as a result of shock.

Bee stings and other insect stings are a frequent cause of generalized anaphylactic reactions (Fig. 11–1), but in the United States today penicillin is the substance most often responsible. A metabolic product of penicillin acts as a hapten and combines with proteins of the host; the resulting antigen causes an immune response, making the patient allergic. Penicillin is often injected, so that a large quantity of the antibiotic rapidly reaches the circulation. If the recipient has previously become sensitized to penicillin, an immediate reaction can occur systemically and may be fatal within minutes. Although systemic anaphylaxis is extremely rare in patients receiving penicillin, less severe allergic reactions to penicillin are not uncommon and are estimated to occur in as many as 10 percent of the recipients of this antibiotic. Because of the danger of systemic anaphylaxis, any person who has had an allergic reaction to penicillin, or to any other material, should avoid subsequent treatment with the reaction-inducing agent.

IgE immunoglobulins were not discovered until 1965, more than 40 years after Prausnitz's experiment and years after the other classes of immuno-

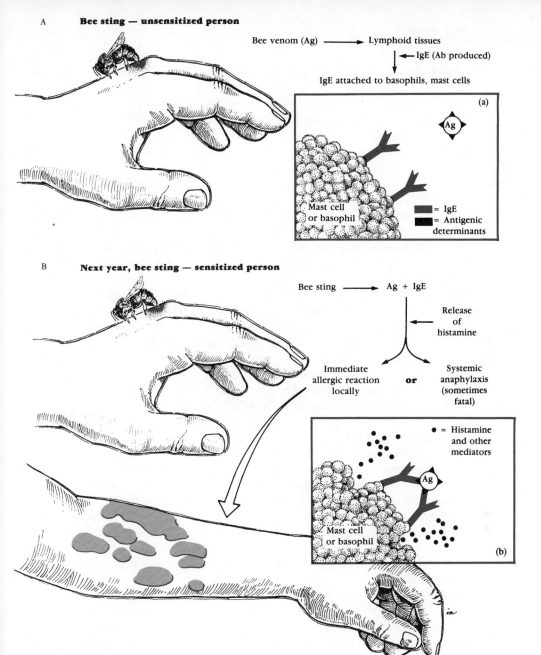

A   Bee sting — unsensitized person

Bee venom (Ag) $\longrightarrow$ Lymphoid tissues

$\longleftarrow$ IgE (Ab produced)

IgE attached to basophils, mast cells

(a)

Ag

Mast cell
or basophil

= IgE
= Antigenic
  determinants

B   Next year, bee sting — sensitized person

Bee sting $\longrightarrow$ Ag + IgE

$\longleftarrow$ Release
of
histamine

Immediate
allergic reaction
locally

or

Systemic
anaphylaxis
(sometimes
fatal)

= Histamine
and other
mediators

Ag

Mast cell
or basophil

(b)

Figure 11–1
Bee sting, an example of an immediate hypersensitivity reaction. Allergic reactions mediated by IgE can result from sensitization to the venom of bee stings or other insect stings. As in any allergy, a period of sensitization is necessary during which the immunological response occurs. (A) In this case, IgE is produced and attaches to mast cells and basophils (a). (B) When a sensitized individual is again stung by a bee, the venom antigen reacts with cell-bound IgE (b), resulting in the release of chemical mediators, the substances that actually cause reaction. Symptoms of the reaction may occur locally at the site of the sting, or in highly sensitized persons the symptoms may occur systemically, producing systemic or generalized anaphylaxis.

globulins had been found. One reason is that IgE is present in serum in such very small quantities, compared with IgG. The amount of IgE present in normal blood is about one fifty-thousandth of the amount of IgG present! In fact, most of the IgE is attached to basophils and similar cells called **mast cells,** which are found in most parts of the body. These cells are particularly abundant in the nose, lungs, and skin, which is one reason why IgE-mediated reactions usually occur in these locations.

Mast cells and basophils, page 213

Mast cells and basophils, shown in Figure 9–3, contain many large granules packed with histamine and other substances. Figure 11–1a indicates that IgE attaches to mast cells and basophils by the constant region of the molecule, in much the same way that other immunoglobulins attach to phagocytes. Once attached, the IgE molecule can survive for many weeks with its antibody-reaction sites available to react with antigen. For example, a person with an allergy to ragweed pollen will have many IgE antibodies specific for the pollen fixed to mast cells and basophils throughout the body and especially in the upper respiratory tract. This is because the pollens enter through the respiratory tract and stimulate the production of IgE by lymphoid cells in the adjacent lymphoid tissues. Many of the IgE molecules attach to cells near the site of synthesis, and relatively few enter the circulation and move to other areas. When antigen is not present, these IgE antibodies are harmless, but when the specific ragweed pollen is inhaled, it combines rapidly with the cell-fixed IgE. Within seconds the antigen-IgE reaction leads to the release of histamine and other substances from the mast cell and basophil granules (Fig. 11–1b). These chemical substances cause tissue damage and are responsible for the attack of hay fever that follows inhalation of the allergy-inducing antigen, or **allergen,** by a sensitized host. If instead of being inhaled the allergen were injected into a highly allergic host, systemic anaphylaxis might occur.

Histamine is the principal mediator of hay fever and hives; therefore, antihistamine drugs are very effective in treating these allergic diseases. The other substances released by mast cells are mainly responsible for the symptoms of asthma, and consequently antihistamines are of no value for treating asthma.

Skin tests are often performed to determine the substances to which a person is allergic. In these tests, extremely small amounts of antigen are injected or introduced directly into the skin to induce a reaction called **cutaneous* anaphylaxis.** If the subject is sensitized, or allergic, to that substance, a wheal forms within seconds or minutes at the site of the antigen injection. The reaction reaches a peak in 20 to 30 minutes and fades within a few hours. Great care must be taken during skin testing to avoid systemic anaphylaxis in highly sensitized persons. The tests are always carried out under careful supervision by the physician and with epinephrine (a drug that counteracts anaphylaxis) and other appropriate medical supplies at hand.

---

*cutaneous—of, or pertaining to, the skin

In view of the dangers that can follow exposure to allergens, it may seem strange that one of the common ways of preventing severe allergies involves injecting increasing amounts of a specific allergen to which an individual is sensitized. This procedure is called **immunotherapy** (or desensitization). Its purpose is to make the person less allergic to the offending antigen. The dose of antigen is gradually increased over a period of months. Often the patient reacts less to the antigen as treatment continues. The reasons for the success of this form of therapy are not fully understood. It is known that antibodies (blocking antibodies) of classes other than IgE are produced; such antibodies of the class IgG may block the reaction by combining with the allergen before it can react with IgE on the cells. There is also evidence that suppressor cells may be induced by this treatment and may suppress the IgE response.

Since all normal people produce IgE, why doesn't everyone suffer from allergies? The reasons are not known, but members of some families tend to develop IgE-mediated allergies much more readily than others. Current research is focused on the genetics of these responses and the mechanisms of immunological control that govern these reactions.

Delayed Hypersensitivity (Cell-Mediated)

Common examples of delayed hypersensitivity are reactions to poison ivy and positive reactions to the tuberculin skin test. The latter test is probably familiar to most people. It involves the introduction of very small quantities of protein antigens of tubercle bacilli into the skin. Instead of giving an immediate reaction, as described previously, the person with delayed hypersensitivity shows no reaction to the antigens for at least six hours or usually about a day. Then the area of injection becomes reddened and gradually thickened, or indurated, reaching a peak reaction at about two to three days. There is no formation of wheals, as seen in IgE reaction sites.

A positive tuberculin skin test does not indicate that the tested person has active tuberculosis. It simply means that the person has had enough exposure to tubercle bacilli or related organisms to become allergic to their antigens. The test is useful as a diagnostic aid and in epidemiological studies. Similar tests for delayed hypersensitivity are used for studying many other diseases.

In delayed hypersensitivity, the T lymphocytes are responsible for recognition of antigen and initiation of the response, but macrophages also play essential roles. Sensitized T cells encounter the antigen and are stimulated to produce substances that participate in the reaction. Some of these substances attract macrophages, some immobilize macrophages at the site, and others act to increase phagocytosis and the power to kill intracellular microorganisms. The accumulation of cells and fibrin* is largely responsible for the thickening at the reaction site.

---

*fibrin—protein formed from fibrinogen by action of thrombin in the clotting of blood

Another kind of cell-mediated hypersensitivity reaction of particular interest is **contact hypersensitivity.** A familiar example is the poison ivy reaction (Fig. 11–2), but in fact contact hypersensitivity is much more widespread. It is one of the major problems in occupational and industrial medicine. Many chemicals used in industry are haptens that can combine with host proteins and lead to sensitization. Because of their repeated exposure, industrial workers are especially likely to become sensitized. In contact hypersensitivity, the delayed reaction begins hours after contacting the antigen and develops more slowly than the tuberculin-type reaction, usually reaching a peak at about three days. The rash of poison ivy is characteristic of these lesions. Some of the most common allergens in contact hypersensitivity include chromium salts (used in tanning leather), nickel (in costume jewelry and watch bands), and certain dyes.

The skin test performed to detect contact hypersensitivity is the patch test, in which suspected allergens are applied to the skin under a patch of adhesive bandage for three days. A positive reaction, peaking at three days, resembles the rash of poison ivy, with itching and blisters in the skin.

SPECIAL PROBLEMS INVOLVING HYPERSENSITIVITY STATES

Having examined some of the basic mechanisms of immediate-type and delayed-type hypersensitivities, we can now discuss some special problems that involve different kinds of allergic responses. It is of interest to consider the balance between the protective and harmful aspects of immunological responses in these circumstances.

Transplantation

Successful organ transplantation depends largely on overcoming or bypassing the body's very efficient mechanisms for rejecting or destroying foreign materials. Since people who need transplants, such as those with kidney failure, usually are not lucky enough to have an identical twin, most human tissue grafts (primary grafts) are transplanted from a normal donor to a genetically different human recipient, not previously exposed to the donor's antigens. The immunological response causing rejection of primary grafts results principally from the action of sensitized T lymphocytes (T killer cells). Killing of the grafted tissue cells depends on a direct lymphocyte-to-graft-cell contact. Upon contact with the specific graft cell antigen, the T killer cells cause damage to the membranes of the graft cells by releasing cytotoxin and other substances.

Contact killing, page 254

It is possible to take cells from patients and from prospective donors of grafts and test them in the laboratory for major antigenic differences. The tissues are typed in a manner similar to blood typing, after which the donor and recipient can be matched as well as possible to minimize antigenic differences. The greater the antigenic differences, the stronger the rejection

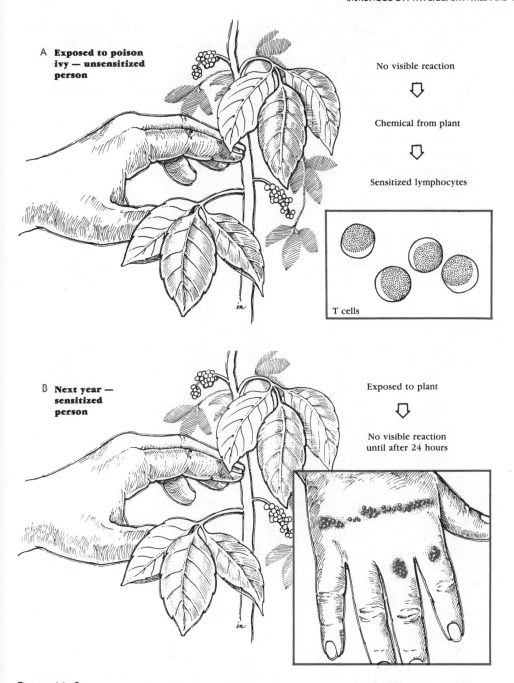

A **Exposed to poison ivy — unsensitized person**

No visible reaction

⇩

Chemical from plant

⇩

Sensitized lymphocytes

T cells

B **Next year — sensitized person**

Exposed to plant

⇩

No visible reaction until after 24 hours

Figure 11–2

Poison ivy, an example of a delayed hypersensitivity reaction. Note that there is a difference between this reaction and the bee sting reaction (Fig. 11–1). Immediate hypersensitivity reactions involve basophils and mast cells, whereas the delayed hypersensitivity reactions involve T lymphocytes.

will be; therefore, close family members are the most suitable donors. Even with fairly good matches, **immunosuppressive agents** must be used to decrease responses to the remaining differences. Immunosuppression results in an overall immune suppression, often leading to one of the major complications of transplantation—infection. It is difficult to adjust the dosage of immunosuppressive agents so that the graft rejection is suppressed but enough protective immunity against infection remains.

Much research is directed toward understanding the control of immune responses. If it were possible to suppress responses to transplanted tissues selectively, while leaving immune responses otherwise intact, the problems of graft rejection would be solved.

Immunity to Tumors

The same mechanisms that cause hypersensitivity reactions under some conditions are of great benefit to the host under other circumstances. For example, some of the mechanisms of delayed hypersensitivity also function importantly in protection against tumors (cancers). Immune responses to tumors are tremendously varied, although some generalizations can be made. It is well established that tumor cells have unique antigens (tumor antigens) that can stimulate an immune response, even though the tumor cells arise from the host's own tissues. As in transplantation rejection, a cell-mediated immune response occurs, and tumor cells are killed following direct contact with immune (sensitized) cells. This kind of lymphocyte killing of tumor cells is demonstrable in the laboratory (Fig. 11–3). It is possible that this sort of protective response eliminates many cells that would otherwise grow into detectable tumors.

There are several possible explanations for the occurrence of tumors in spite of an active immune response against them. First of all, tumor cells grow rapidly and may simply outgrow and overwhelm the capacity of immune cells to destroy them. Furthermore, tumor cells usually have a coating substance on their surfaces, somewhat like the capsules of bacteria. This substance inhibits their effective contact with immune cells. Alternatively, blocking factors may block cell killing (Fig. 11–4). The blocking factors are either specific antibodies, antigen-antibody complexes, or other antigen-specific factors. In addition, many patients with tumors have been shown to have T suppressor cells that specifically prevent an immunological reaction from occurring.

T suppressor cells, page 238

To complicate the matter further, it is known that macrophages and certain other cells besides immune lymphocytes can also kill tumor cells. For example, it has been shown that activated macrophages have a selective toxicity for tumor cells over normal cells. Still, the T lymphocyte seems to be of primary importance in immunity to tumors, aided and abetted by a variety of other mechanisms.

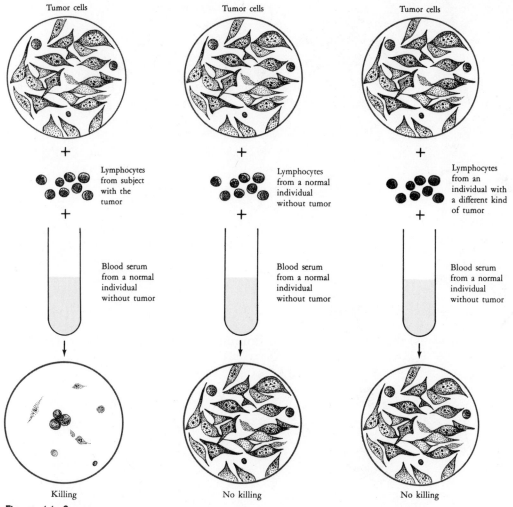

Figure 11—3
Lymphocytes from a tumor-bearing subject can kill specific tumor cells in culture.

Contributions to Infectious Diseases

Some microorganisms can induce a high degree of hypersensitivity in most human hosts. This ability contributes substantially to the pathogenesis of many infectious diseases, because many of the symptoms of infectious diseases stem from allergic reactions induced by the invading microbes. Hypersensitivity reactions are often useful in assessing the response of the host to an infectious agent or in diagnosing infectious diseases.

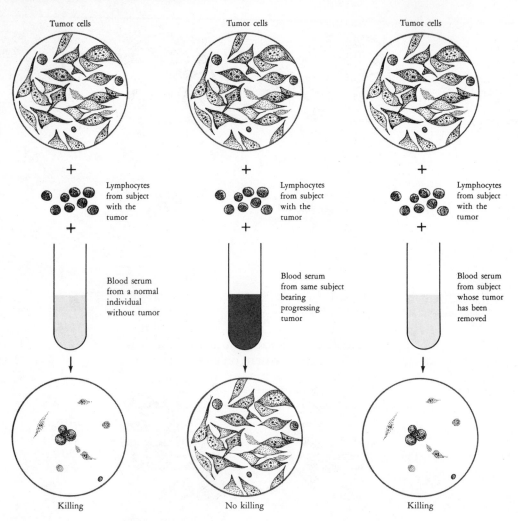

Figure 11—4
Serum from a tumor-bearing subject may interfere with tumor-cell killing by that subject's lymphocytes.

Blood Grouping and Blood Transfusion

Blood transfusion is a form of tissue transplantation, with foreign cells being transferred to a recipient of the same species; however, the cells in blood are free and fully exposed to both antibodies and complement in plasma. For this and other reasons, reactions to blood transfusions are usually cytotoxic reactions caused by specific antibody and complement, rather than cell-mediated reactions.

The ABO blood-group system concerns the red blood cell antigens that cause the strongest and most common transfusion reactions. Understanding the ABO system has made blood transfusions feasible. It is important to remember that there are hundreds of different antigens in blood plasma and hundreds more on the surfaces of erythrocytes and leukocytes; in fact, there are dozens of minor blood group systems of red blood cell antigens in addition to the major ABO system. The Rhesus (Rh) system is the most important minor blood group system. Fortunately, many of the minor antigens do not cause a strong antibody response, or the antigens are relatively rare; otherwise blood transfusions would not be practical.

Because of the complexities of blood group antigens, the following discussion is limited to explanations of the most important systems, the ABO and the Rh systems.

The ABO System

In the ABO system, the presence or absence of only two antigens, called A and B, accounts for the four major blood groups, as shown in Table 11–2. Antigens A and B, carbohydrate molecules having specific structural properties, are present in large quantities on the surfaces of erythrocytes of the appropriate types. Group O cells lack both antigen A and antigen B; group A cells have antigen A but lack antigen B; group B cells have antigen B but lack antigen A; and group AB red blood cells possess both the A and B antigens. As expected, when antigens A and B are present on a person's cells, antibodies are not made against them—that could be disastrous for the host! What is not expected and is unique to this blood group system is that people who lack A and B antigens on their erythrocytes normally have antibodies against the A and B antigens. These are called **natural antibodies,** meaning antibodies that are formed without deliberate antigenic stimulation. They are of the IgM class. In this case, antigens very similar to A and B occur in many foods, dusts, bacteria, and other substances to which people are frequently exposed, and it is assumed that these stimulate the production of antibodies that will react with A and B antigens on erythrocytes.

Table 11–2 includes a summary of the plasma antibodies normally present in humans of each of the four major ABO blood groups. There are some

TABLE 11–2 SOME CHARACTERISTICS OF THE ABO BLOOD GROUP SYSTEM

Blood type	Antigen(s) present on the erythrocytes	Antibodies normally present in the plasma	Percent in a mixed Caucasian population[a]
O	Neither A nor B	Both anti-A and anti-B	45
A	A	Anti-B	41
B	B	Anti-A	10
AB	Both A and B	Neither anti-A nor anti-B	4

[a]The incidence of ABO blood groups may vary greatly between genetically different populations.

subgroups and exceptions to the simplified information in Table 11–2, but for the most part this is a less complex blood-group system than many others, including the Rh system.

Transfusion reactions are usually the result of destruction of the donor's erythrocytes by the recipient's antibodies. When a pint or unit of blood is transfused, the donor plasma is immediately diluted by the 10 to 12 pints of blood in the adult recipient; therefore, the transferred donor blood must contain an extremely large amount of antibodies to cause a detectable reaction. Donor erythrocytes, on the other hand, are constantly bathed in the recipient's plasma and therefore have a high risk of being promptly destroyed by the recipient's antibodies and complement. Thus, before transfusions are given, compatibility tests (in addition to ABO and Rh blood typing) between the bloods of a donor and a recipient must be conducted to minimize the possibility of a transfusion reaction.

The Rh (Rhesus) System

The Rh (Rhesus) system is another important blood group system. Although this system is complex and many antigens are involved, only the strongest one, called the **D** or simply the **Rh antigen,** will be discussed. If the Rh antigen is present on the erythrocytes, a person is Rh-positive; if it is lacking, a person is Rh-negative.

The Rh-positive person can receive either Rh-positive or Rh-negative blood, because the Rh-positive blood is compatible and the Rh-negative erythrocytes lack the Rh antigen altogether. It is the Rh-negative adult who may experience an Rh blood transfusion reaction as a result of being immunized against the Rh antigen, either from transfusions, transplantation, or, in the case of women, pregnancies that can introduce this antigen from the fetus. In contrast to the ABO system, Rh-negative individuals do not have anti-Rh antibodies until exposed to Rh-positive cells.

Anti-Rh antibodies formed by a pregnant woman may damage her offspring. The disease that results is called **hemolytic disease of the newborn,** or simply Rh disease (Fig. 11–5). While an Rh-negative mother is carrying an Rh-positive fetus, a few fetal cells may pass across the placenta to the mother but usually not enough to stimulate a primary antibody response. At birth, however, the baby's blood cells often enter the mother's circulation in sufficient numbers to cause a vigorous immune response. Abortion procedures are also very likely to cause large numbers of blood cells to pass from the fetus into the mother. The anti-Rh antibodies formed by the Rh-negative mother cause her no harm because they cannot damage her cells, which lack Rh antigens. However, problems may arise for any subsequent Rh-positive fetus that the mother may carry. With such a fetus, the few Rh-positive cells that cross the placenta from the fetus into the mother's circulation may be enough to stimulate a secondary immune response. Large amounts of anti-Rh antibodies of the class IgG would then be produced. These antibodies can cross the placenta to the fetus and cause extensive

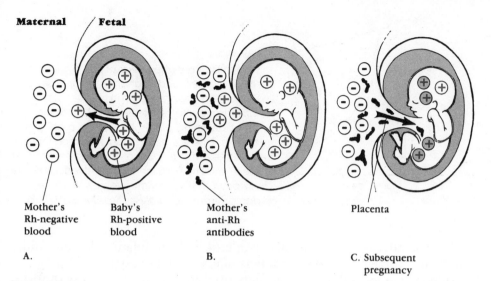

Maternal / **Fetal**

Mother's
Rh-negative
blood

Baby's
Rh-positive
blood

Mother's
anti-Rh
antibodies

Placenta

A.

B.

C. Subsequent
pregnancy

Figure 11–5

Events leading to Rh disease (hemolytic disease of the newborn). (a) Fetal erythrocytes bearing the Rh antigen determinant enter the maternal circulation, and the Rh antigen determinants become available to stimulate an immune response. (b) Anti-Rh antibodies are formed. (c) During subsequent pregnancy with an Rh-positive fetus, the anti-Rh antibodies can cross the placenta, leading to destruction of fetal erythrocytes and Rh disease.

disease due to fetal erythrocyte damage. Although miscarriage and loss of the fetus can result, the damage from this phenomenon is often not apparent until very soon after birth, as indicated by the name hemolytic disease of the newborn.

The child whose Rh antigens are attacked by its mother's antibodies often survives until birth because harmful products of erythrocyte destruction can be eliminated from its system by certain enzymes, present in large amounts in the mother but only in very small amounts in the fetus and newborn. Soon after the exchange of materials between the circulations of mother and fetus is interrupted at birth, however, the child becomes acutely ill. Not only are the products of erythrocyte destruction toxic, but the baby is also seriously anemic and may not have enough erythrocytes to survive. In this critical situation, replacement transfusions are life-saving. The baby's blood is gradually withdrawn while fresh Rh-negative blood is transfused. The benefits of this are twofold: first, toxic products, maternal antibodies, and the infant's Rh-positive erythrocytes are all removed; and second, the infant's erythrocytes are replaced by Rh-negative erythrocytes, which cannot react with any remaining anti-Rh antibodies. The child does not produce large numbers of Rh-positive cells until the transfused cells age and are removed from the circulation, usually within a matter of weeks. By this time, the maternally derived anti-Rh antibodies have also decreased to a harmless level. An additional treatment for hemolytic disease of the newborn, used

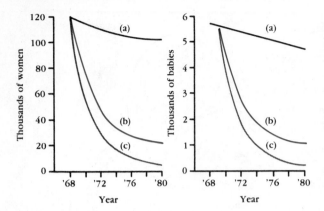

Figure 11–6
This graph illustrates the effects of prophylactic treatment of women and infants for Rh disease. Line (a) is the estimated reduction in number of women sensitized and infants affected by Rh disease as a result of declining birth rate with no prophylactic treatment. Line (b) is the actual reduction that has been achieved with 80% of Rh-negative women treated, and line (c) is the estimated reduction that could occur if 100% of the women were treated. (Adapted from *Hospital Practice*, June 1978.)

with some success, is ultraviolet irradiation of the infant. This procedure inactivates some of the toxic products of erythrocyte destruction.

An effective means is now routinely used to prevent Rh disease. Anti-Rh antibodies are given to susceptible (nonimmunized) Rh-negative mothers at or soon after the birth of an Rh-positive child, or following the abortion of an Rh-positive fetus. This procedure can prevent the mother from becoming immunized by the baby's Rh-positive cells and will thereby prevent Rh disease in her next child. The success of this preventive measure has been phenomenal, as indicated in Figure 11–6. Since surgical procedures associated with either spontaneous or induced abortions can also lead to sensitization, every Rh-negative woman not already sensitized who undergoes such procedures should also have an injection of anti-Rh antibodies within 72 hours of the abortion.

AUTOIMMUNE (AUTOALLERGIC) DISEASES

Important control mechanisms exist to prevent immune responses against components of one's own tissues (self components) under normal circumstances. There are some B cells in the body that can react against host (or self) tissues. These B cells are prevented from secreting antibody by the presence of T suppressor cells. When the T suppressor cells do not function properly, serious immune reactions against self substances can occur and can lead to the development of **autoimmune** or **autoallergic disease.** Autoimmune disease is said to be present when the individual's tissues are damaged by their interaction with specific antibodies or sensitized cells. It is often very difficult to prove that a disease is autoimmune in nature, however, since the evidence is usually circumstantial.

Among the proven examples of autoimmune diseases are some forms of hemolytic anemia and of thyroiditis. Many other diseases are suspected to be autoimmune in nature, including rheumatoid arthritis, a common inflammatory joint disease.

It has been suggested, although not proved, that autoimmune responses contribute to aging. It is known that T lymphocytes decrease in number and in their ability to function with age and that immune capabilities generally wane in old age. Because T lymphocytes are important in controlling and suppressing immune responses, it is possible that a decrease in T cell control with aging permits the development of immune responses against components of self.

SUMMARY

Immunological reactions that cause tissue damage or malfunction in the host are called hypersensitivity or allergic reactions. These may be either immediate or delayed reactions. The four major kinds of hypersensitivity reactions are the cell-mediated (delayed) type and three kinds of antibody-mediated (immediate) types: IgE-mediated, cytotoxic, and immune complex–mediated.

IgE-mediated allergic reactions occur promptly following the exposure of a sensitized person to antigen, as a result of antigen reacting with IgE attached to mast cells or basophils. In this antigen reaction, histamine and other mediator substances are released from mast cells and basophils, causing the symptoms. IgE hypersensitivity diseases include hives, hayfever, eczema, and asthma. Antihistamines and other drugs that counteract the mediator substances are used for treating reactions. Immunotherapy is used to help prevent development of the allergic state.

Cytotoxic immediate hypersensitivity reactions involve damage or death to cells bearing specific antigen. Damage is usually caused by specific antibody plus complement; however, complement is not always required.

In immune complex–mediated reactions, soluble antigens react with specific antibodies in such a way as to form complexes that interact with complement and neutrophils. The damage is often caused by enzymes released from the neutrophils.

Cell-mediated hypersensitivity reactions are delayed, reaching a peak one to three days after exposure to antigen. These reactions result from the activities of specifically sensitized lymphocytes. Delayed hypersensitivity is important in the pathogenesis of a number of diseases, and delayed hypersensitivity skin reactions are often helpful in diagnosis.

Special areas in which hypersensitivity reactions are important include blood transfusions, Rh disease, transplantation, and immunity to tumors.

SELF-QUIZ

1. Immediate allergic reactions caused by IgE antibodies
   a. reach a peak in six to eight hours
   b. require neutrophils for the reaction to occur

    c. are often caused by haptens used in industry

    d. are common in newborn infants

    e. involve the release of histamine from mast cells

2. Cytotoxic allergic reactions have all of the following characteristics *except*

    a. They occur immediately, within seconds to minutes after exposure to antigen.

    b. Cells are injured or destroyed.

    c. The causative antigen is soluble and is present in the circulation.

    d. Complement lysis of the cells may be responsible for the damage.

    e. Complement is not always necessary for the reaction to occur.

3. Immune complex–mediated hypersensitivity reactions

    a. reach a peak at 24 to 48 hours

    b. require the activity of sensitized T cells

    c. are best treated with antihistamine drugs

    d. involve the activation of complement

    e. require mast cells and basophils

4. Contact hypersensitivity reactions are

    a. severe, but fortunately rare

    b. exemplified by hives

    c. delayed, reaching a peak at about three days

    d. systemic

    e. treated effectively with antihistamines

5. Immunity to solid tumors, such as certain cancers, involves many of the same mechanisms as hypersensitivity reactions that are

    a. immediate

    b. cell-mediated

    c. mediated by IgE antibodies

    d. mediated by IgG antibodies and complement

    e. mediated by IgM antibodies and complement

QUESTIONS FOR DISCUSSION

1. A molecule of antigen must bridge two molecules of IgE on a mast cell or basophil in order to cause release of histamine and other mediator substances, thus causing a hypersensitivity reaction. Does this suggest ways that antigens could be changed chemically to block the reaction? Draw diagrams to show how this might be done.

2. Knowing that antigen-antibody and complement are required for immune complex–mediated hypersensitivity reactions, how might kidney disease caused by such reactions be diagnosed?

3. Hemolytic disease of the newborn can be caused by ABO incompatibility between mother and fetus, but it is much more frequently caused by Rh incompatibility. How might this be explained?

FURTHER READING

Allison, A. C., and Ferluga, J.: "How Lymphocytes Kill Tumor Cells." *New England Journal of Medicine, 295*:165–167 (1976). A brief review of evidence pertaining to the mechanisms of killing of tumor cells by lymphocytes.

Beer, A. E., and Billingham, R. E.: "The Embryo as a Transplant." *Scientific American* (April 1974).

Freda, V., Pollack, W., and Gorman, J.: "Rh Disease: How Near the End?" *Hospital Practice 13*:66–69 (1978).

Fudenberg, H. H., Stites, D. P., Caldwell, J. L., and Wells, J. V. (eds.): *Basic and Clinical Immunology*, 2nd ed. Los Altos, California: Lange Medical Publications, 1978. More than 50 authorities have contributed to this excellent book covering basic immunology, immunobiology, immunological laboratory tests, and various diseases considered in clinical immunology.

Gell, P. G. H., Coombs, R. R. A., and Lachmann, P. J. (eds.): *Clinical Aspects of Immunology*, 3rd ed. Philadelphia: F. A. Davis, 1975. An excellent, comprehensive book presenting many authors' views of the clinical aspects of immunology.

Old, L. J.: "Cancer Immunology." *Scientific American* (May 1977).

Patterson, R. (ed.): *Allergic Diseases*. Philadelphia: J. B. Lippincott Co., 1972. Some basic descriptions of immunological mechanisms are included, along with detailed clinical descriptions of allergic diseases.

Rose, N. R.: Autoimmune diseases. *Scientific American* (February 1981).

Chapter 12

VACCINES AND IMMUNIZATION PROCEDURES

Near the end of the eighteenth century, an English country doctor named Edward Jenner went to see a young dairymaid who was suffering from a pock-like rash. He thought the girl had smallpox, a common disease in those days, but she told him that was impossible because she had already had a similar disease of cattle called cowpox. Instead of shrugging this statement off as a bit of folklore, Jenner decided to investigate and determine whether there was any truth in this old belief. First, he injected material from smallpox lesions into people known to have had cowpox; the injection had little effect, convincing him that the cowpox had, indeed, protected these individuals. To make sure, he did an experiment that would probably never be done today. He deliberately gave a little boy cowpox, and later inoculated him with the smallpox material. The boy was protected against smallpox; in fact, he became the first person to be vaccinated intentionally. The terms **vaccinate** and **vaccine,** from the Latin word *vacca* meaning cow, were coined and are still used to honor Jenner for his work. Some people use "vaccine" to mean only a suspension of organisms or viruses used for immunizing, but the term is often more broadly applied to any immunizing preparation. Thus, in the broader sense, a vaccine may contain killed or altered (attenuated) microorganisms, or parts of organisms, or products of microbes such as exotoxins or toxoids.

Immunology has come a long way since the days of Jenner, and later Pasteur and Koch, when the major thrust of research was the development of vaccines. Nevertheless, effective immunization procedures are as essential as ever, and considerable research effort is still being expended to develop new and better vaccines. The effectiveness of current procedures is reflected by the decreased incidence of infectious diseases, lowered mortality in childhood, and increased life spans (Fig. 12–1).

OBJECTIVES

To know
1. Several factors that influence immunization.
2. The difference between active and passive immunity and how both forms of immunity are induced.
3. What sources of information are available to learn about the best current immunization procedures.
4. Some important hazards of immunization to be guarded against.
5. Some questions subjects should be asked before they are given vaccines or other immunizing preparations.

FACTORS THAT INFLUENCE IMMUNIZATION

Use of Attenuated or Killed Vaccines

The important principle of **attenuation** was discovered entirely by chance but was used to good advantage by Pasteur. He had been working with a bacterial disease of chickens called fowl cholera, and he neglected his bacterial cultures during a vacation period. Upon his return to the lab, he found that the aging cultures, when injected into chickens, failed to produce disease in the expected manner. Even more surprising was the finding that fresh virulent cultures of the same bacterial pathogen did not produce disease when later injected into the same chickens. Pasteur had unknowingly selected avirulent, or attenuated, mutant strains of the pathogen, which were capable of immunizing the chickens against reinfection with a fully virulent strain of the same organism. Attenuated organisms have mutations in the genes that are normally required for pathogenicity. Pasteur later applied the

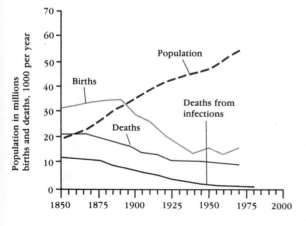

Figure 12–1
Representation of the birth rates and death rates in Great Britain from 1850 to the present, compared with population growth. Notice that the rate of deaths from infections decreased dramatically, especially during the twentieth century. This resulted in part from immunization programs but also from better sanitation and control measures, the use of antibiotics, and improved treatment of infectious diseases.

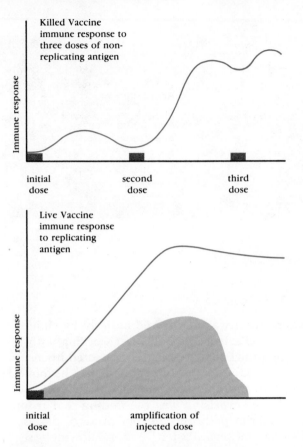

Figure 12–2
Killed versus living vaccine. Note that the living vaccine gives a higher level of antibody with only one dose of antigen. The antibody production is in response to a replicating vaccine.

principle of attenuation to other organisms and viruses, notably to the rabies virus, using living attenuated strains of the virus to induce immunity.

In addition to living attenuated organisms, suspensions of killed whole bacteria or viruses are often used as vaccines. Figure 12–2 shows one advantage of live attenuated vaccines over killed vaccines, namely, that live vaccines are replicating, so the dose of antigen administered is amplified. The result is a long-lasting, effective response, similar to that obtained after multiple injections of killed vaccine. In addition, whenever possible, it is preferable to obtain the antigen causing immunity rather than to use the whole organism. This purified material is then used to immunize against the pathogen. Pathogens contain many antigens, and only a few are likely to contribute to producing immunity. The others can cause trouble by producing hypersensitivity reactions. Some examples of immunizing preparations in common use are listed in Table 12–1.

Induction of Antibody Formation or Cell-Mediated Immunity

Antibody-mediated immunity, page 240

Cell-mediated immunity, pages 246–254

In choosing immunization procedures, it is essential to consider the mechanisms of immunity desired and especially whether cell-mediated or antibody-mediated mechanisms will be effective. For example, to immunize

TABLE 12–1   EXAMPLES OF PREPARATIONS USED TO IMMUNIZE

Disease	*Immunizing agent*	*Mode of preparation*
Poliomyelitis	Polio viruses of three types	Attenuated strains grown in human cell cultures
Rubella (German measles)	Rubella virus	Attenuated virus
Rubeola (measles)	Rubeola virus	Attenuated virus
Mumps	Mumps virus	Attenuated virus
Smallpox	Vaccinia virus	Attenuated virus that resembles variola virus of smallpox
Influenza	Influenza viruses	Killed viruses or viral components
Diphtheria	Toxoid of *Corynebacterium diphtheriae*	Toxin produced by bacterial cultures is purified and inactivated to toxoid
Tetanus	Toxoid of *Clostridium tetani*	Toxin produced by bacterial cultures is purified and inactivated to toxoid
Whooping cough	*Bordetella pertussis*	Killed suspension of bacteria
Typhoid fever	*Salmonella typhi*	Killed suspension of bacteria
Cholera	*Vibrio cholerae*	Killed suspension of bacteria
Tuberculosis	Bacille Calmette-Guérin (BCG)	Attenuated organisms that resemble tubercle bacilli, grown in culture
Meningococcal meningitis	Meningococcal polysaccharides, groups A and C	Purified capsular polysaccharides from cultured *Neisseria meningitidis* serogroups A and C
Pneumococcal pneumonia	Polyvalent pneumococcal polysaccharides	Purified capsular polysaccharides from cultures of *Streptococcus pneumoniae* serogroups that commonly cause pneumonia

against diphtheria or tetanus, diseases caused by bacterial exotoxins, the desired response is induction of specific antibodies (called **antitoxins**) that neutralize the toxins. However, in the case of tuberculosis, inducing specific antibodies is not effective, since the bacilli that cause tuberculosis live within host macrophages and can be attacked only by cell-mediated mechanisms. Effective immunization against tuberculosis requires cell-mediated immunity, which is stimulated by giving live attenuated bacilli.

Dead organisms or isolated microbial products, injected intramuscularly or subcutaneously in appropriate doses, tend to induce antibody production but not cell-mediated immunity. The induction of cell-mediated immunity, on the other hand, is favored by actual infection with attenuated, avirulent

TABLE 12–2   SOME ADVANTAGES AND DISADVANTAGES OF SEVERAL COMMONLY USED ADJUVANTS

Adjuvant preparation	Characteristics	Advantages	Disadvantages
Freund's complete adjuvant	Killed mycobacteria in water-in-oil emulsion	Excellent adjuvant; useful for increasing cell-mediated immunity	May cause severe inflammatory response, so not safe for human use
Freund's incomplete adjuvant	Water-in-oil emulsion without mycobacteria	Gives sustained and effective antibody responses	Can be used in humans but persists in tissues and may cause some inflammation
Adjuvant 65	Peanut oil with other lipids	Gives sustained and effective antibody responses; biodegradable and safe for human use	—
Aluminum compounds (alum)	Aluminum phosphate or aluminum hydroxide	Increases antibody responses; routinely used for humans	—
Bordetella pertussis	Killed bacteria	Causes increased antibody responses to toxoids given simultaneously	—

strains of organisms or by injection of antigens with special substances called **adjuvants**.

Adjuvants

Adjuvants act in a variety of ways to increase immune responses. Table 12–2 lists some of the advantages and disadvantages of a few commonly used adjuvants. The major advantage of all of these is that they increase the efficiency of the immune response when injected simultaneously with antigen. Some of the ways adjuvants have been shown to function include the following:

1.   Slow release of antigen from the injection site. Antigens can leak slowly from a water-in-oil emulsion or from the surfaces of alum particles deposited in the tissues.
2.   Stimulation and activation of macrophages, which are necessary to pre-

pare antigens so that they can stimulate potential antibody-producing cells. Macrophages also assist in other lymphocyte activities.
3. Direct effects on the T and B lymphocytes that function in the immune response.

Antigenic Competition

To avoid multiple injections, it would seem advantageous to give many antigens in a single dose, and this is often possible. A familiar example is the use of diphtheria, pertussis (whooping cough), and tetanus antigens in a single immunizing preparation—the DPT vaccine given to children. The killed pertussis organisms in the vaccine act as adjuvants and actually increase the response to the diphtheria and tetanus toxoids. It is not always possible to combine immunizing agents, however. One reason is that **competition of antigens** may occur. This is the opposite of the adjuvant effect observed when some antigens are combined. In antigenic competition, antibody responses to a given antigen are much lower than expected, or are completely inhibited, if the antigen is administered along with another antigen that can compete. This competition is not fully understood, but there is evidence that it can result from several different causes. One cause is the production of T cells that suppress the response to one of the antigens; another cause seems to be lack of enough receptors on macrophages to aid in all the responses. There may also be other limiting factors. In addition, it is known that T lymphocytes produce soluble substances that regulate normal immune responses; overproduction of such regulator substances could contribute to antigenic competition.

Viral Interference

Care must be taken to avoid a different kind of competition that can occur when living viruses are used to immunize, namely, **viral interference.** For example, it has been observed that simultaneous administration of equal amounts of the three types of polio viruses results in the production of antibodies to some, but not all, of the viruses. Thus, the vaccine must be adjusted to contain less of the interfering virus in order to obtain satisfactory responses to all three of the viruses. Interfering viruses inhibit the replication of a second virus and thereby prevent the antigenic stimulus normally provided by the replicating virus.

ACTIVE AND PASSIVE IMMUNIZATION

The use of vaccines and other immunizing preparations induces active immunity and represents **active immunization.** It is also possible to immunize passively, generally by the transfer of an antibody-containing immune serum to induce **passive immunization.** This kind of immunization is passive because those who receive the antiserum are protected by preformed antibod-

Passive immunity, pages 246–250

ies rather than by their own actively produced antibodies. Passive immunity does not last long, since the transferred antibodies decay within weeks or a few months. Active immunity, on the other hand, is long-lasting because of the production of memory cells that remain present for many years. The advantage of passive immunity is that large amounts of antibody can be provided very quickly without waiting for antibodies to be produced. This is especially important in counteracting certain exotoxins that are often rapidly fatal.

IgG antibodies, page 234

The newborn infant is protected by passive immunity, its antibodies being transported across the placenta from mother to fetus. Only IgG antibodies are transported. This is a **naturally acquired** passive immunity that lasts as long as the IgG persists, usually for three to six months. Passive immunity is **artificially acquired** by injecting either immune serum or the gamma globulin fraction of immune serum, which contains the antibodies. Although immune serum prepared by immunizing animals, such as horses or goats, works perfectly well against antigens and is sometimes used, it is much more desirable to use human immune serum or, better yet, human immune globulin. Serum from a foreign species contains many highly antigenic proteins that quickly sensitize the host and can give rise to serious or even fatal hypersensitivity reactions. The sera from different human individuals can also induce immune responses to human serum proteins, but in general these responses do not lead to dangerous allergic reactions. Some of the antibody preparations currently available, along with their uses, are listed in Table 12–3.

It is possible to transfer cell-mediated immunity passively in animals fairly readily by transferring lymphocytes collected from immunized donors. This can also be done in humans, but it is usually not practical, for a number of reasons. First, the transferred cells are genetically different unless donor and recipient are identical twins, so any transfer of immunity is transient, lasting only until the donor cells are rejected. Then, too, the injection of foreign lymphocytes sensitizes the recipient to human tissue antigens and should be avoided, in case the recipient ever needs blood transfusions or tissue transplants in the future.

CURRENT IMMUNIZATION RECOMMENDATIONS

Each country recommends schedules of immunization procedures that are based on conditions in that location. In the United States, the Public Health Service Advisory Committee on Immunization Practices sets the standards, which are constantly updated as conditions change. General guidelines are given here, but specific current information can be obtained from local health departments.

Table 12–4 indicates the usual immunization schedules recommended for children in the United States. Vaccination against smallpox is no longer recommended, since the disease has been officially declared eliminated and the chance of contracting the virus is remote. Moreover, the risks of immu-

TABLE 12–3 SOME AGENTS USED FOR THE PASSIVE TRANSFER OF IMMUNITY

Agent transferred	Effective against	Characteristics of the agent[a]
Diphtheria antitoxin	Diphtheria	Produced in horses, sheep, goats, or rabbits against diphtheria toxoid[b]
Tetanus antitoxin	Tetanus	Globulin from immunized humans is best; if not available, horse, sheep, or rabbit immune globulins are used
Trivalent antibotulinum antitoxin	Botulism	Prepared in animals against types A, B, and E toxins of *Clostridium botulinum*
Polyvalent antitoxin against several clostridia of gas gangrene	Gas gangrene	Prepared in animals against toxins of *Cl. perfringens*, *Cl. novyi*, *Cl. histolyticum*, and *Cl. septicum*
Immune serum globulin	Infectious hepatitis	Globulin fraction from large pools of adult human plasma containing antibodies against type A viral hepatitis; helps to prevent jaundice
Anti-HBs globulin	Serum hepatitis	Globulins from human plasma, rich in antibodies against small, surface antigen of hepatitis B virus
Zoster immune globulin	Varicella (chickenpox)	Prepared from pooled plasma of patients recovering from herpes zoster; used to prevent serious varicella infections in immunodeficient children
Vaccinia immune globulin	Smallpox and disseminated vaccinia	Prepared from blood of revaccinated servicemen
Human rabies-immune globulin	Rabies	Used in conjunction with active immunization after exposure to rabies virus

[a]Antitoxins and immune globulins are generally given as early as possible after exposure to disease-producing agents and, if possible, before disease develops.
[b]Diphtheria toxoid tends to induce allergic reactions in adult humans, so animals must be used to produce antitoxin.

nization, although they are small, far outweigh any expected benefits. Similarly, the Bacille Calmette-Guérin (BCG) vaccine for tuberculosis is not routinely recommended in the United States, because the chance of exposure to the disease is small at the present time and BCG causes a positive tuberculin skin test, thus removing a possible diagnostic tool. BCG immunization is widespread in many countries, however.

Travelers to foreign countries face different kinds of risks, depending on their itineraries, and the US Public Health Service issues annually a booklet

TABLE 12–4 IMMUNIZATION OF CHILDREN IN INFANCY

Age	Vaccine
2 months	DPT*,TOPV*
4 months	DPT, TOPV
6 months	DPT, TOPV (optional)
12 months	Tuberculin skin test
15 months	Measles, mumps, rubella
18 months	DPT, TOVP
4-6 years	DPT, TOVP
14-16 years	Td* adult-type
Every 10 years	Td adult-type

DPT = Diphtheria, Pertussis, Tetanus.
TOPV = Trivalent Oral Polio Vaccine (3 strains of polio I, II, III).
Td = Tetanus, diphtheria adult-type vaccine.
Adapted from 1977 Report of the Committee on Infectious Diseases, American Academy of Pediatrics.

called "Health Information for International Travel." This information may change from week to week, so it is essential to check with a physician or a local health department before making a trip abroad. The World Health Organization (WHO) also makes recommendations, which are incorporated into the USPHS information.

HAZARDS INVOLVED IN IMMUNIZATION

As the preceding sections have implied, hazards may be involved in immunization, but the benefits of protection usually far outweigh the dangers. However, the dangers are great in immune-deficient or immunosuppressed patients. Even attenuated vaccines, normally harmless, can cause serious infections in the immunologically incompetent individual who is unable to cope with infections. Patients known or suspected to have immune deficiencies of any kind generally should not receive live attenuated vaccines. The attenuated vaccinia virus, for example, has caused fatal infections in immunologically deficient children who have been given the virus to immunize them against smallpox.

In addition, live attenuated virus vaccines are usually not given to pregnant women, since there is a theoretical risk of danger to the fetus. Several viruses, notably rubella virus, can damage the fetus and so it is recommended that women who have not had rubella be immunized with rubella vaccine before the childbearing period.

An unusual complication of immunization was widely publicized during the influenza immunization campaign in 1976. In order to prevent an expected epidemic of "swine influenza," many people were given influenza vaccine. A very low percentage of immunized people, on the order of 0.001 percent, experienced a particular kind of paralytic disease that almost always disappeared with time. Extremely rare cases were fatal, however, and therefore instances of this disease appearing in immunized individuals were highly publicized. Since that time, progress has been made in preparing more purified vaccines and vaccines composed of subunits of the influenza virus, so this particular hazard is not likely to occur again during influenza immunizations.

Those who administer immunizing preparations of any kind must remember to obtain answers to these questions before giving the preparations:

1. Is the person known to be allergic to anything in the preparation? Some viral vaccines are grown in hens' eggs and therefore may contain egg proteins. Other proteins, preservatives, penicillin, and other antigenic materials are sometimes present.
2. Are any people in contact with the person to be immunized, such as family members or co-workers, known to have immune defects? Some live, attenuated vaccines can be transmitted to contacts and can seriously damage or even kill immunologically deficient hosts.
3. Is there any evidence of infection at the time of immunization?
4. Is the person to be immunized pregnant?

A "yes" answer to any of these questions either rules out immunization at that time or else demands careful consideration of the possible hazards involved.

FUTURE PROSPECTS

One prospect that seems close at hand is the total elimination of vaccination for smallpox. At present, immunization is required only for travelers to a few countries.

Strong research programs are now directed toward developing vaccines to protect against some of the venereal diseases that are epidemic, in particular, gonorrhea. These programs are hampered by a lack of basic knowledge about the mechanisms of pathogenicity and immunity operative against the causative gonococci. At present, it does not seem likely that a vaccine that could prevent gonorrhea will be developed in the near future. The same is true of syphilis, except that the likelihood of a vaccine against syphilis is even more remote than an antigonococcal vaccine, because the bacteria that cause syphilis cannot be grown in laboratory media, as gonococci can.

It is probable that vaccines will eventually be developed to protect special populations against organisms to which they may be vulnerable. Several such vaccines have already been developed; some of these are mentioned in Table 12–1. An example is the multivalent vaccine* against prevalent strains of pneumococci. These are quite useful for the limited population of elderly people and others at high risk of developing pneumococcal pneumonia. Vaccines against meningococci are also valuable for protecting military recruits

---

*multivalent vaccine—vaccine made from several strains of a disease-producing organism

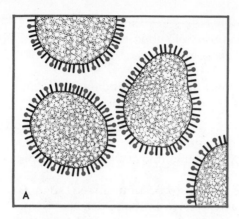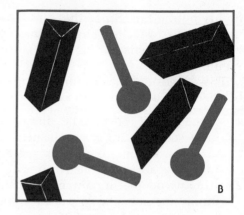

Figure 12–3

The components of influenza vaccines. (a) The complete virus particle of the influenza virus is often killed and used to prepare the vaccines. (b) Alternatively, the important protein antigens of the virus can be purified and given together with an adjuvant as a subunit vaccine to immunize against that particular strain of influenza virus. Two major antigens from the viral surfaces are rod-shaped hemagglutinin antigen and mushroom-shaped neuraminidase enzyme antigen.

against meningococcal meningitis. Other such vaccines are being studied and should become available in the near future.

A prospect of concern is that antibiotic resistance will develop in organisms that have already been routinely sensitive to antibiotic therapy. This has already occurred for many bacteria, for example, some strains of gonococci and pneumococci. This prospect makes the development of effective vaccines even more urgent.

Another prospect of particular concern is that people will become increasingly complacent about immunizations as the incidence of infectious diseases wanes, and will neglect to have their children immunized. This is already true to some extent; an increase in unprotected populations could lead to serious epidemics.

A more hopeful prospect is that the ability to manipulate genes by genetic engineering and cloning will lead to the preparation of highly specific subunit antigens for use in immunization (Fig. 12–3).

SUMMARY

Immunizing preparations include living, attenuated bacteria or viruses and vaccines made of killed microorganisms. Even better are preparations containing only antigenic subunits of the organisms, since other antigens capable of causing allergic reactions are thus avoided. A suitable route and schedule of immunization should be chosen to induce the desired response of either antibody-mediated immunity or cell-mediated immunity.

Adjuvants cause increased immune responses when given together with antigen. Antigenic competition has the opposite effect, causing a lowered response or no immune response to one or more of the antigens in a mixture. Viral interference also causes decreased responses to some of the viruses in a mixture.

Vaccines cause an artificially induced, active immunity that is usually long-lasting because of the production of memory cells. Immunity can be transferred passively by transferring preformed antibodies in immune serum or lymphocytes from immune donors to nonimmune recipients. Newborn babies and infants have naturally acquired passive protection from maternal IgG that crosses the placenta.

Current immunization recommendations are available from state health departments and from the Center for Disease Control (USA) and the World Health Organization. People administering vaccines must be aware of the hazards involved in immunization and ways to avoid ill effects from immunization.

Future prospects include, among others, the development of more effective subunit vaccines via genetic engineering and the cloning of genes.

SELF-QUIZ

1. Attenuated, living viruses are used to immunize against all of the following diseases except
   a. smallpox
   b. poliomyelitis
   c. rubella
   d. influenza
   e. measles
2. Killed bacteria of whooping cough (pertussis) in the diphtheria-pertussis-tetanus vaccine not only induce immunity against whooping cough, but they also
   a. are essential for a response against the toxoids in the preparation
   b. decrease the response to diphtheria toxoid
   c. decrease the response to tetanus toxoid
   d. decrease the response to both toxoids
   e. have an adjuvant effect
3. Adults are given booster doses of tetanus and diphtheria toxoids containing about the same amount of tetanus toxoid as the DPT vaccine used for children, but with only 10 to 25 percent as much diphtheria toxoid, because
   a. the diphtheria preparation is more toxic than the tetanus toxoid
   b. adults are less likely to get diphtheria than tetanus
   c. diphtheria is not dangerous in adults
   d. adults have more natural immunity against diphtheria than do children

    e. hypersensitivity reactions caused by antigen-antibody complexes are likely to occur if too much toxoid is given to a preimmunized person

4. All of the following are true of tetanus immune globulin (TIG) treatment *except* that it is composed of
    a. immunoglobulins
    b. is produced from human serum
    c. gives passive immunity
    d. causes lifelong immunity
    e. is used to treat nonimmunized individuals with wounds that might become infected with tetanus organisms

5. Babies are normally protected passively by maternal IgG for a period of
    a. 1–2 weeks
    b. 3–6 weeks
    c. 6–8 weeks
    d. 3–6 months
    e. 6–12 months

QUESTIONS FOR DISCUSSION

1. The vaccinia virus, formerly extensively used to immunize against smallpox, is not the same as the smallpox virus, yet it induces very effective immunity against smallpox. How can this be explained?
2. Suppose that a method for growing the spirochetes that cause syphilis was suddenly discovered. How might a vaccine against syphilis then be made and tested? What information is needed before a vaccine is developed?
3. How might genetic engineering and cloning of genes be used to make more effective immunizing preparations?
4. Immunization with diphtheria-pertussis-tetanus vaccine does not prevent diphtheria organisms from colonizing the throat. Why, then, is this vaccine effective against diphtheria?

FURTHER READING

Brachman, Philip S., et al.: *Health Information for International Travel 1980.* PHS Washington, D.C. U.S. Government Printing Office Publication No. (CDC) 80-8280. This is published as a supplement to the *Morbidity and Mortality Weekly Report.* The information is frequently updated by a weekly "Blue Sheet" sent out by the Bureau of Epidemiology, Center for Disease Control, Atlanta, Georgia. This publication gives up-to-date information on the vaccination requirements for all the countries in the world in addition to other information on vaccination.

Dick, G.: *Immunization.* London: Update Publications Ltd. 1978. Available from Update International, Inc., 2337 Lemoine Avenue, Fort Lee, New Jersey 07024. A simple, concise summary of immunization procedures used in the USA, the UK, and other countries, with chapters on immunization procedures in developing countries and other pertinent topics.

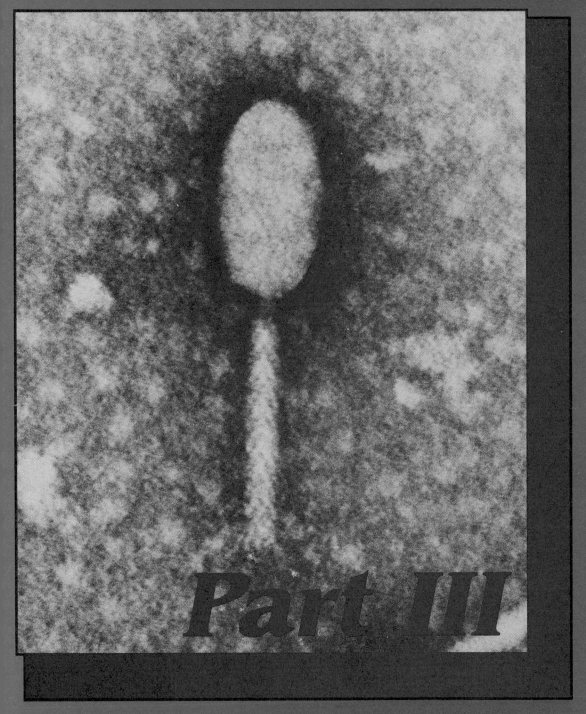

Part III

THE MICROBIAL WORLD

Chapter 13

GRAM-POSITIVE BACTERIA

In the morgue of City Hospital of Berlin, Germany, Dr. Hans Christian Gram was preparing slides from lung tissue of patients who had recently died of pneumonia. The year was 1884, not long after Koch had set forth his postulates for establishing the causative agents of diseases. Gram was trying to prove that the bacteria causing the pneumonia were present in the tissue of the diseased lung.

He looked up from his microscope and sighed. Again only some of the slide preparations had taken up the crystalline violet stain and retained it after treatment with iodine and alcohol. In these preparations, the bacteria were easy to distinguish from the nuclei of the human lung cells. But for some reason, when Gram tried to stain other preparations by the same method, the stain washed out in the alcohol rinse.

Gram reported his findings in a paper published in 1884. Although he was disappointed that many bacteria did not take up the stain, as early as 1886 a textbook by Flugge stated that "the method of Gram is mainly useful for the differential staining of bacteria in tissues and for the diagnostic differentiation of species." Others soon recognized that the Gram stain technique (as it came to be called) was useful in classifying bacteria as Gram-positive or Gram-negative. The staining technique made it possible to find many of the kinds of bacteria now known to infect body tissues, and the Gram stains have remained important tools for differentiating and classifying bacteria.

OBJECTIVES

To know
1. Some general characteristics of Gram-positive bacteria.
2. What is meant by "lactic-acid bacteria," what they look like, and where they might be found.
3. The names of some pyogenic Gram-positive cocci and their characteristics.
4. How to describe *Corynebacterium diphtheriae* and how it causes disease.
5. Two species of *Bacillus* that cause disease and how these species differ.
6. How to describe the pathogenic strains of *Clostridium,* and what diseases they cause.

GENERAL CHARACTERISTICS OF THE GRAM-POSITIVE BACTERIA

The Gram-positive bacteria are grouped together on the basis of their cell-wall structure, as reflected in the Gram stain reaction (see Appendix II). Most Gram-positive bacteria have cell walls consisting of multilayers of peptidoglycan.

Gram-positive cell wall structure, page 28

Gram-positive bacteria may be cylindrical (rod-shaped), or they may be spherical or oval. Some of the rods are spore-formers. Several of the cocci cause abundant production of pus by the infected host and hence are known as pyogenic (pus-causing) cocci. Some rods and cocci are notorious for producing dangerous, even lethal, exotoxins.*

Exotoxins, page 221

Several groups of Gram-positive bacteria known collectively as the **lactic acid bacteria** produce lactic acid as a major end-product of their fermentation of sugars. Among these are the non–spore-forming, rod-shaped lactobacilli, which are common among the microbial flora of the human oral cavity and genital mucous membranes. They are especially abundant in the normal flora of the vagina during the childbearing years and help the vagina to resist infection (Fig. 13–1). They rarely cause disease, although they may contribute to dental caries under certain conditions. Lactobacilli prefer acidic and relatively anaerobic conditions. In addition to their habitats in humans, they are often present in animals, in decomposing plant material, and in milk, yogurt, and other dairy products. They are important in making cheeses, pickles, and certain other foods (see Chapter 31).

The streptococci are also members of the lactic acid group of bacteria, but as the name implies, they are spherical or oval in shape. Many of them are harmless members of the normal flora, especially in the oral cavity, whereas others are highly pathogenic.

A majority of the Gram-positive bacteria are aerobic and facultatively anaerobic; however, a few genera are primarily anaerobic. These include the

---

*exotoxins—toxic substances secreted by the bacterial cell

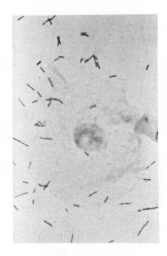

Figure 13–1
Normal flora of the vagina. Gram stain. (×1000) (Courtesy of M. F. Lampe.)

medically important clostridia, most of which are strictly anaerobic. The Gram-positive, spore-forming rods of the genus *Clostridium* (Fig. 13–2) are responsible for a number of serious human and animal diseases.

The formation of endospores characterizes not only clostridia but also aerobic Gram-positive rods of the genus *Bacillus* (Fig. 13–3). One species of this genus, *B. anthracis,* is the cause of the disease anthrax.

SOME IMPORTANT EXAMPLES OF PATHOGENIC GRAM-POSITIVE BACTERIA

The following Gram-positive bacteria are examples of those that cause disease in humans.

The Pyogenic Cocci

Pathogenic Gram-positive cocci of the genera *Staphylococcus* and *Streptococcus* are commonly pyogenic; i.e., pus formation is a major aspect of their effect on the human host.

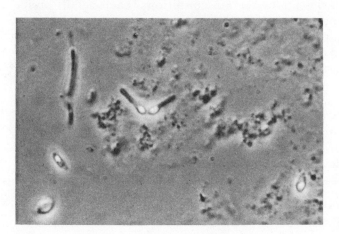

Figure 13–2
Clostridium botulinum from food. Phase. (×1000) (Courtesy of D. T. Maunder.)

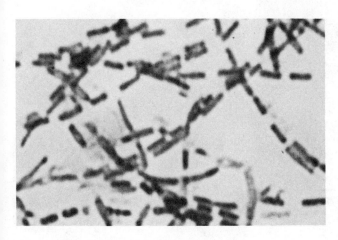

Figure 13—3
Bacillus anthracis. Spore stain.

Staphylococci. Normally, staphylococci live in the nose and sometimes on moist areas of the skin of humans and other animals without causing harm. However, staphylococci are frequently carried from those areas to other parts of the body. They can infect almost any tissue if host defenses are lowered. They are resistant to drying, a property that favors their transmission from one host to another. *Staphylococcus aureus* is a common cause of wound infections, boils, and other skin infections and also of abscesses in bones and other tissues. Moreover, certain strains of this organism are a common cause of food poisoning, and some apparently cause toxic shock syndrome.

In infected material and cultures, staphylococci are typically arranged in clusters, although single or paired cocci, or short chains, are seen under some conditions (Fig. 13–4). *Staphylococcus aureus* grows readily on the usual laboratory media such as blood agar plates, and well-developed colonies are present after 18 hours of incubation (color plate 19). These colonies are the color of thick cream, and most strains of the organism destroy eryth-

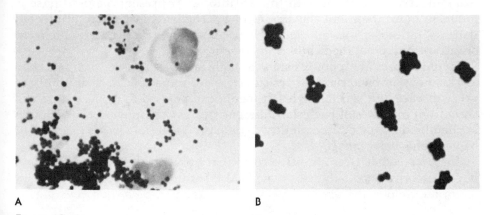

A   B

Figure 13—4
(a) *Staphylococcus aureus* from pus of a wound infection. (Courtesy of J. Portman.) (b) *Staphylococcus aureus* from culture. Gram stain. (Courtesy of J. Portman.)

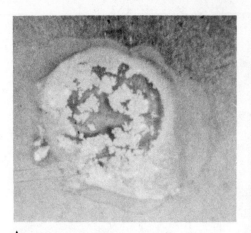 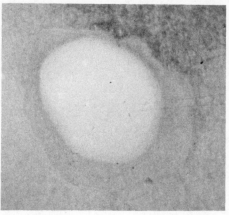

A   B

Figure 13–5
(a) *Staphylococcus aureus*, positive coagulase test. (b) *Staphylococcus* species, negative coagulase test. (Courtesy of C. E. Roberts.)

rocytes, causing a small zone of clearing (beta hemolysis) around their colonies on sheep blood agar. An important aid in identifying pathogenic staphylococci is the **coagulase test** (Fig. 13–5). Nonpathogenic staphylococci are generally coagulase-negative. Virtually all strains of *S. aureus* produce enzymes, called coagulases, that cause plasma to coagulate (clot), whereas none of the other medically important cocci do this under the conditions of the test. The ability of *S. aureus* to ferment mannitol is also an important feature in its identification. Besides coagulase, this organism produces numerous other cellular products; several are enzymes capable of destroying animal tissues. Others are toxins; one damages the skin, while another results in food poisoning. Of great importance is the enzyme **penicillinase,** produced by many strains, which destroys penicillin, thus protecting the bacterium from effective treatment with this antibiotic. At present, penicillinase is found in more than half the strains of *S. aureus* recovered from infected patients. Indeed, strains of *S. aureus* have appeared that are resistant to most of the antimicrobial medications in current use.

Staphylococci are more resistant to high salt (NaCl) concentrations than are many other bacteria. This resistance is not important in their ability to cause disease, but special high-salt media can be used to recover staphylococci from mixtures of bacteria. Different strains of staphylococci can often be identified by their susceptibility to different bacterial viruses (also called bacteriophages, or phages).

There are other Gram-positive cocci that grow on the human body and have little virulence*; therefore, medical laboratories are compelled to differentiate among the various species. For example, *Staphylococcus aureus*

---

*virulence—the relative capacity of a pathogen to overcome body defenses

must be differentiated from *S. epidermidis*, a coagulase-negative species almost universally present on the human skin, and from *Micrococcus* species, also common in the normal flora. The chief distinguishing characteristics are shown in Table 13–1. *Micrococcus* species and *S. epidermidis* are only occasional causes of infection, usually of the heart valves.

Streptococci. *Streptococcus pyogenes* is the cause of "strep throat," scarlet fever, wound infections, and infections of the skin, ear, lung, and other tissues. It may also be responsible for delayed effects occurring after the actual infection. Examples include glomerulonephritis (a kidney disease) and rheumatic fever (a heart disease). In infected material, *S. pyogenes* most commonly occurs as diplococci or in short chains, while in culture longer chains are characteristic (Fig. 13–6). *Streptococcus pyogenes* is easily cultivated on the usual laboratory media but prefers a medium more nearly resembling body tissue, enriched with blood and having an atmosphere with more carbon dioxide and water vapor than are present in air. Its colonies on blood agar, when touched with a bacteriological wire loop, tend to slide intact across the medium. Some strains are encapsulated. The majority of strains of *S. pyogenes* also show a zone of beta hemolysis around their colonies (color plate 20), where complete erythrocyte destruction and clearing of the medium have occurred due to the activity of two enzymes (called streptolysin O and streptolysin S) released from *S. pyogenes* cells. Cultivation of *S. pyogenes* mixed with other species, as from a culture of the throat, may result in an inhibition of the growth of *S. pyogenes*. Anaerobic culture conditions and selective media help to minimize this effect. Another feature of *S. py-*

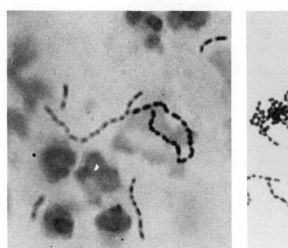 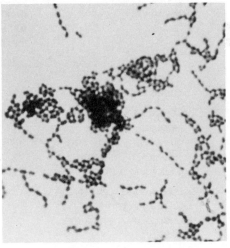

A   B

Figure 13–6

(a) *Streptococcus* species. Gram stain of a smear from culture. (Courtesy of J. Portman.)

(b) *Streptococcus* species. Gram stain from pus.

TABLE 13–1   CHARACTERISTICS OF SOME MEDICALLY IMPORTANT GRAM-POSITIVE BACTERIA

Genus and species	Morphological and cultural characteristics	Special properties	Importance in disease
Staphylococcus aureus	Medium-sized cocci, frequently in clusters; creamy, golden colonies; beta-hemolytic; grows in high-salt medium	Produces acid from mannitol; coagulase-positive; phage typing can be done; often resistant to anti-microbials	Causes pyogenic infections of many organs, e.g., skin, bone, bloodstream
Staphylococcus epidermidis	Medium-sized cocci 1 μm); in clusters, white colonies; grows on high-salt medium	Does not ferment mannitol; coagulase-negative	Part of normal flora but can be opportunistic
Micrococcus sp.	Large cocci in pairs and clusters	Is nonfermentative (does not produce acid from sugars anaerobically); coagulase-negative	Part of normal flora but can be opportunistic
Streptococcus pyogenes	Cocci, frequently in pairs and chains; cocci may be oval or round; small colonies with large zones of beta hemolysis	Growth inhibited by bacitracin; group A carbohydrate present; M proteins important antigens; susceptible to penicillin	Causes pyogenic infections of many tissues, "strep throat," scarlet fever, glomerulonephritis (kidney disease)
Streptococcus pneumoniae	Oval diplococci with prominent capsules; small colonies with zones of alpha hemolysis	Growth inhibited by optochin; organisms lysed by detergents; capsular polysaccharide antigens usually susceptible to penicillin therapy	A cause of pneumonia, ear infections, meningitis
Viridans group of streptococci	Small to medium-sized cocci in long or short chains or in pairs	Usually alpha-hemolytic	Common in normal flora; can be opportunistic
Corynebacterium diphtheriae	Rods, club-shaped with irregularly staining granules; gray or black colonies on tellurite-containing media	Exotoxin production, caused by bacteriophage; is responsible for diphtheria	Causes diphtheria; also infections of the skin and nose, especially in hot climates
Bacillus anthracis	Square-ended rods; tend to occur in chains; endospores near center of the rods	Aerobic	Causes anthrax
Bacillus cereus	Spore-forming, rod-shaped bacteria	Aerobic; forms endospores; some strains produce an exotoxin	Strains that produce an exotoxin cause food poisoning

Table continues on the opposite page

TABLE 13–1   CHARACTERISTICS OF SOME MEDICALLY IMPORTANT GRAM-POSITIVE BACTERIA (*Cont.*)

Genus and species	Morphological and cultural characteristics	Special properties	Importance in disease
Clostridium perfringens	Plump, nonmotile rod of variable length occurring singly or in chains	Anaerobic although it tolerates low levels of O_2; produces endospores; forms polysaccharide capsules	Causes gas gangrene; produces many toxins that contribute to the disease
Clostridium tetani	Round endospores at the end of rods; motile	Strictly anaerobic; has swarming growth	Causes "lockjaw" or tetanus by action of powerful neurotoxin
Clostridium botulinum	Endospores in the center of the rods; motile	Strictly anaerobic	Soil organism that can produce lethal exotoxins in improperly prepared foods

ogenes is that it is more likely to be inhibited by low concentrations of the antibiotic bacitracin than are other beta-hemolytic streptococci, and this property aids in its identification.

Some of the components of *S. pyogenes* are diagramed in Figure 13–7. Particularly important is the **M protein,** which differs from one strain to another and is therefore useful in distinguishing different strains. This M protein is antiphagocytic* and thus aids the invasiveness of the organism. Immunity to *S. pyogenes* depends on an antibody to the M protein of that strain. It is therefore "strain-specific," meaning that an antibody to one strain protects against infection with that strain only and does not protect against the other 50 or more strains of *S. pyogenes*. Some strains produce an exotoxin that causes scarlet fever.

The carbohydrate* antigens of streptococci are useful in grouping these cocci. For example, *Streptococcus pyogenes* possesses a type of carbohydrate referred to as the group A carbohydrate. *S. agalactiae*, a species of strepto-cocci with the group B carbohydrate, is a prominent cause of meningitis in newborn babies and is also responsible for mastitis in cows. Cultural char-acteristics also help to distinguish this organism from *S. pyogenes*. The co-lonial appearance differs, and *S. agalactiae* breaks down an organic material called hippurate, while *S. pyogenes* does not.

Streptococcus pneumoniae is a common cause of bacterial pneumonia, ear infections, and meningitis. Its most striking characteristic is a large capsule, which is generally necessary for disease production (Fig. 13–8a). There are about 80 different types of *S. pneumoniae*, as judged by their different cap-sular antigens, but relatively few of these types are commonly pathogenic. Also called the "pneumococci," *S. pneumoniae* generally occur as oval dip-

*antiphagocytic—prevents phagocytosis

*carbohydrate—compound containing principally carbon, hydrogen, and oxygen atoms in a 1:2:1 ratio

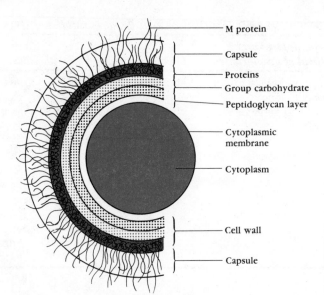

M protein
Capsule
Proteins
Group carbohydrate
Peptidoglycan layer
Cytoplasmic membrane
Cytoplasm
Cell wall
Capsule

Figure 13–7
Stylized cross-sectional drawing of some important components of *Streptococcus pyogenes*.

lococci with their adjacent sides flattened (Fig. 13–8b). Colonies of pneumococci grown aerobically on blood agar are surrounded by a zone of partial destruction of the erythrocytes, causing an area of greenish discoloration called **alpha hemolysis** (color plate 21). Unlike any of the many other alpha-hemolytic streptococci, pneumococci are strongly inhibited by optochin (a chemical relative of quinine) and are lysed by bile* and some other detergents. These properties are useful in identifying pneumococci.

Gram-Positive Rods

Corynebacteria. *Corynebacterium diphtheriae*, the cause of diphtheria, is a club-shaped, Gram-positive rod of variable size with a tendency to stain irregularly (Fig. 13–9). Its cytoplasm contains characteristically staining granules. Media useful for identifying *C. diphtheriae* include Loeffler's, which is composed largely of solidified serum, and medium containing potassium tellurite, a chemical that inhibits many other bacterial species. Tellurite forms a black compound within *C. diphtheriae*, causing the colonies to appear gray or black. Growth occurs aerobically on many media and is generally rapid, although some strains do not develop visible colonies for several days. Different colony types can be identified, a property that is useful in tracing epidemic spread of the organisms. Most, but not all, strains of *C. diphtheriae*

Exotoxins, page 221

secrete a powerful exotoxin that diffuses from the organism and is responsible for the seriousness of diphtheria infections. In many strains of *C. diphtheriae*, production of this toxin depends on the presence of a bacteriophage,

---

*bile—a secretion of the liver that aids in the digestion of fats

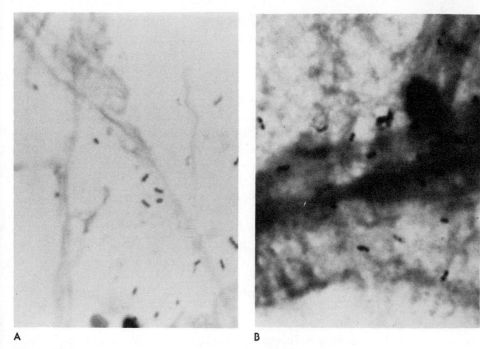

Figure 13–8

(a) *Streptococcus pneumoniae*, showing the capsules in the sputum of a patient with pneumonia. Gram stain. (×1000) (Courtesy of S. Eng.) (b) *Streptococcus pneumoniae* in pus from the middle ear. Gram stain. (×1200) (Courtesy of Leon J. Le Beau, Ph.D., University of Illinois Medical Center, Chicago.)

an example of **lysogenic conversion.** Bacteria that look similar to *C. diphtheriae* are common in the normal flora; these include *C. pseudodiphtheriticum* in the throat and the anaerobic *Propionibacterium acnes* on the skin.

Lysogenic conversion, pages 341–342

Bacillus *Species.* Members of the genus *Bacillus* are aerobic, spore-forming rods that are widespread in nature. *Bacillus anthracis* is the cause of anthrax, a serious disease of sheep, cattle, and sometimes humans. In this

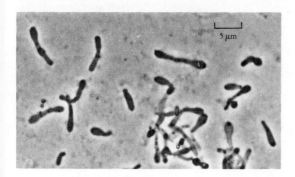

Figure 13–9

Phase-contrast photomicrograph of *Corynebacterium diphtheriae* after 18-hour incubation on Loeffler's medium. (Courtesy of J. T. Staley and J. P. Dalmasso.)

disease, an extensive ulcer often forms at the site of infection, or severe pneumonia develops if the organisms are inhaled. The dangerous nature of anthrax is due to toxins produced by the bacillus. *Bacillus anthracis* is a large Gram-positive rod that tends to be square-ended when seen in infected material (Fig. 13–10a). Like many other *Bacillus* species, its cells occur as short chains. It grows well either aerobically or facultatively anaerobically on many of the usual laboratory media enriched with blood. Stained preparations from cultures on such media reveal long chains; the individual bacteria contain single endospores near the center of their rod-shaped cells. The cells synthesize a unique capsule when incubated in carbon dioxide (Fig. 13–10b). Unlike most *Bacillus* species, *B. anthracis* has no flagella and therefore is nonmotile. It can also be distinguished from other species of its genus by its susceptibility to a bacteriophage. The endospores of *B. anthracis* are highly resistant to environmental conditions and remain fully infectious for years in soil, wool, hides, and other habitats. Cases of human anthrax occurred recently in the United States, contracted from spores on the leather of imported West Indian drums.

Of the many *Bacillus* species commonly present in the natural environment, one called *B. cereus* has caused dangerous infections of the eye when introduced via eye medications contaminated with the growing bacteria. Some strains of *B. cereus* produce an exotoxin responsible for a form of food poisoning.

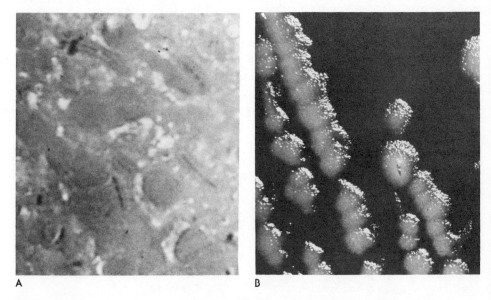

A   B

Figure 13–10
(a) *Bacillus anthracis* from infected tissue. Note capsule surrounding chains and square-ended rods. Gram stain. (× 1000) (Courtesy of Leon J. Le Beau, Ph.D., University of Illinois Medical Center, Chicago.) (b) *Bacillus anthracis* colonies showing the rough surface. (Courtesy of S. Eng.)

Clostridia. The clostridia are all anaerobic, although some tolerate low concentrations of oxygen, depending on the species. All the clostridia have the ability to produce endospores. Endospores, page 49

Clostridium perfringens is the cause of most cases of gas gangrene, although other species of clostridia may also cause this disease. Certain strains of *C. perfringens,* which frequently possess remarkably heat-resistant endospores, cause food poisoning, and others produce a severe, gangrenous infection of the intestine. *Clostridium perfringens* is an encapsulated, nonmotile rod that tolerates low concentrations of oxygen, in contrast to other clostridia that are strict anaerobes and are motile. It produces a large number of different toxins, as discussed in Chapter 26. These toxins probably contribute to disease production by degrading host tissues. The normal habitat of *C. perfringens* is the intestine of humans and other animals, and soil is commonly contaminated with its endospores.

Clostridium botulinum, a soil organism, is the cause of botulism, an often fatal disease characterized by severe paralysis. This motile organism is strictly anaerobic. Endospores are formed near the center of its cells (Fig. 13–2). A powerful exotoxin secreted by *C. botulinum* causes the symptoms of botulism, and its presence can be detected in blood to help diagnose the disease. Indeed, six types of *C. botulinum* occur, identifiable by differences in the toxins. In some types, it has been demonstrated that lysogenic conversion is involved in toxin production.

Clostridium tetani, a third species of *Clostridium,* is the cause of tetanus (lockjaw), a disease characterized by continuous muscle spasm resulting from the action of a neurotoxin* produced by the tetanus organisms. *Clostridium tetani* shows two striking features. The first of these is the spherical endospore that forms at the end of the bacillus (Fig. 13–11), in contrast to

---

*neurotoxin—a toxin that acts on the nervous system

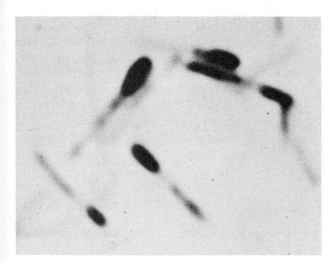

Figure 13–11
Clostridium tetani. Gram stain of anaerobic culture on blood agar of an exudate from a skin laceration contaminated with soil. This patient succumbed to clinical tetanus.

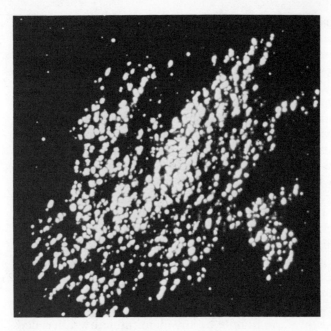

Figure 13–12
Clostridium tetani colonies on 48-hour blood agar plate. Colonies are swarming on the plate. (×48)

the usual oval endospore that forms near the center of the rod in the other pathogenic species of *Clostridium* (compare Figs. 13–2 and 13–11). The second feature is a swarming growth that spreads to cover the entire surface of solid media (Fig. 13–12). Final identification of *C. tetani* depends on identifying its toxin, which acts in a manner quite distinct from the toxin of *C. botulinum*.

SUMMARY

Gram-positive bacteria may be rods or cocci, aerobic or anaerobic. Some produce dangerous exotoxins, and some readily induce pus formation. The lactic-acid bacteria produce lactic acid as a major end-product of sugar fermentations. This group includes streptococci and various lactobacilli important in food production. Endospore-forming Gram-positive bacteria may be aerobic, of the genus *Bacillus*, or anaerobic, of the genus *Clostridium*.

Important pathogenic Gram-positive bacteria include, among others, members of the following genera: *Staphylococcus, Streptococcus, Corynebacterium, Bacillus,* and *Clostridium*.

SELF-QUIZ

1. All of the following are characteristic of some of the Gram-positive bacteria *except* that they

a. contain peptidoglycan in their cell walls
b. produce endotoxin
c. may ferment sugars to produce lactic acid or other acids
d. may form endospores
e. may produce exotoxins

2. The lactic-acid bacteria are all
   a. cocci
   b. rods
   c. pathogenic
   d. fermentative
   e. rare

3. Exotoxins are often important in disease production by all of the following Gram-positive bacteria *except*
   a. *Streptococcus pyogenes*
   b. *S. pneumoniae*
   c. *Clostridium tetani*
   d. *C. perfringens*
   e. *Corynebacterium diphtheriae*

4. Antibodies that protect against *Streptococcus pyogenes* are directed
   against the specific
   a. capsular antigens
   b. C carbohydrate
   c. M protein
   d. exotoxin
   e. pili

5. *Bacillus anthracis* is
   a. anaerobic
   b. susceptible to drying
   c. nonpathogenic
   d. characterized by round terminal spores
   e. likely to grow in chains in culture

QUESTIONS FOR DISCUSSION

1. A polyvalent vaccine has been developed to protect against pneumococcal pneumonia. What components of *Streptococcus pneumoniae* are likely
   to be in the vaccine? In other words, what substances of the pneumococci
   are most important in causing disease?

2. An individual may suffer from "strep throat" more than once but will
   rarely have more than one attack of scarlet fever, yet both diseases are
   caused by *Streptococcus pyogenes*. How is this explained?

3. Bacteria of the genus *Clostridium* frequently infect war wounds caused
   by gunshot or shrapnel. What are several reasons why these organisms
   are common in such wounds but uncommon in some others, for example,
   a wound caused by a kitchen knife?

FURTHER READING

Buchanan, R. E., and Gibbons, N. E., (eds.): *Bergey's Manual of Determinative Bacteriology*, 8th ed. Baltimore: Williams and Wilkins, 1974. A reference work that aids in identifying the various species of bacteria.

Finegold, S. M., et al.: *Scope Monograph on Anaerobic Infections.* Kalamazoo, Michigan: Upjohn Co., 1974. Many colored pictures.

Fox, G. E., et al.: "The Phylogeny of Prokaryotes." *Science 209*:457–465 (1980). A new approach to bacterial taxonomy.

Krause, R. M.: "Cell Wall Antigens of Gram-positive Bacteria and Their Biological Activities." In Schlessinger, D. (ed.). *Microbiology 1977*. Washington, D.C.: American Society for Microbiology, 1977, pp. 330–338.

Chapter 14

GRAM-NEGATIVE BACTERIA

It is a lovely summer day, just right for a picnic at the lake. You call a friend and agree to meet in a couple of hours. Each of you will bring something to eat. Potato salad is one of your favorite picnic foods so you make a large bowl of it.

All afternoon you eat and swim in the warm sun. You go home comfortably tired and fall asleep. After several hours, you are suddenly awakened by a stomach ache. Vomiting and diarrhea follow, and you are left feeling miserable. Who or what could have been the culprit that perpetrated this nastiness?

Who could guess that a microscopic organism, *Salmonella enteritidis,* a Gram-negative rod, could have been responsible? The potato salad that was left on the table in the warm sun had provided a rich medium for its rapid growth.

OBJECTIVES

To know
1. Some general characteristics of Gram-negative bacteria.
2. The difference between the two species of *Neisseria* that are of medical importance.
3. Which Gram-negative rods are primarily aerobic and which are facultative.
4. The similarities and differences between organisms of the genera *Francisella, Brucella, Yersinia, Bordetella,* and *Haemophilus.*
5. The major characteristics of the enterobacteria, some of the organisms included, and how they differ.
6. The differences between the strictly anaerobic genera *Bacteroides* and *Fusobacterium.*

GENERAL CHARACTERISTICS OF THE GRAM-NEGATIVE BACTERIA

The Gram-negative bacteria represent one of the most diverse and wide-spread groups of organisms in existence. Table 14–1 presents a summary of characteristics of some of the medically important Gram-negative bacteria.

Gram-negative bacteria have similarities in their cell walls that account for their staining properties. Recall that the Gram-negative cell wall contains a thin layer of peptidoglycan with additional exterior layers of components made up of fats, sugars, and proteins joined together to form large molecules (see Chapter 2). One component contains fats, or lipids, bonded to sugars in **lipopolysaccharides.** This material is of particular interest because the lipid portion, which is the same in most Gram-negative bacteria, has toxic properties. This is the cell-wall endotoxin discussed previously (Chapter 9). The endotoxins of the various Gram-negative bacteria all have the same effects on the host, namely, production of fever and damage to the circulatory system. Sometimes this results in fatal shock.

While endotoxin activity is a common characteristic of the Gram-negative bacteria, it is only one of the many mechanisms that these bacteria use to cause disease. Some species also produce potent exotoxins, and many species possess a large number of enzymes that aid in the invasion and degradation of host tissues. In addition, some Gram-negative bacteria are protected against phagocytosis to some extent by their ability to produce capsules or other antiphagocytic surface components.

As is true of Gram-positive organisms, virtually every kind of metabolic pattern can be found among the diverse Gram-negative bacteria, ranging from the aerobic pseudomonads through the facultative bacteria of the gastrointestinal tract (the enterobacteria) to the anerobic species, such as members of the genus *Bacteroides*.

Lipopolysaccharides, Appendix I

GRAM-NEGATIVE COCCI

The genus *Neisseria* includes most of the pathogenic Gram-negative cocci of importance to humans. Other members of the genus, common inhabitants of normal human mucous membranes, are rarely pathogenic.

Neisseria gonorrhoeae

Neisseria gonorrhoeae, the "gonococcus," causes gonorrhea. Like the other neisseriae, it is a Gram-negative diplococcus with adjacent sides flattened (Fig. 14–1). In material from infected patients, cells of *N. gonorrhoeae* characteristically occur within polymorphonuclear phagocytes (Fig. 14–2). They grow aerobically on specialized media such as chocolate agar (an agar medium that has a chocolate color because heated blood is added to it). Growth of the gonococci either requires or is enhanced by carbon dioxide. The organisms are oxidase-positive, meaning that their colonies give a positive test for the enzyme cytochrome C oxidase. Glucose is metabolized, yielding acid by-products. Disease-producing strains of *N. gonorrhoeae* generally possess

TABLE 14–1 CHARACTERISTICS OF SOME MEDICALLY IMPORTANT GRAM-NEGATIVE BACTERIA

Genus and species	Morphological and cultural characteristics	Special properties
Neisseria gonorrhoeae	Diplococci, often within pus cells Grows well on chocolate agar with increased CO_2 Oxidase-positive	Very susceptible to drying Adheres to host cells via pili Cause of gonorrhea
Neisseria meningitidis	As above	Types A, B, C determined by polysaccharide antigens Causes meningitis and bloodstream infections
Pseudomonas aeruginosa	Rods, motile by means of a single flagellum Respiratory metabolism Green pigment	Produces exotoxin and endotoxin Opportunistic Infects almost any location, especially burns, wounds, urinary tract
Acinetobacter calcoaceticus	Short rods or cocci in pairs Nonmotile Respiratory metabolism	Opportunist, especially in hospitalized patients
Francisella tularensis	Tiny rods Nonmotile Bipolar staining Require cystine	Infection transmitted directly from wild rodents, such as rabbits, or indirectly via arthropods, such as ticks Causes tularemia
Brucella abortus	Tiny rods Nonmotile May require serum, increased CO_2 Slow-growing	Infection transmitted from infected cattle or via contaminated milk Causes brucellosis (undulant fever)
Brucella melitensis	As above	Transmitted from infected goats or via goat's milk
Bordetella pertussis	Tiny rods Nonmotile Large capsule Bordet-Gengou medium used to isolate the organism	Causes pertussis (whooping cough)
Escherichia coli	Rods Motile Lactose-fermenting	Normal flora of the gastrointestinal tract Some strains cause diarrhea, urinary tract infection, meningitis Some strains produce exotoxins
Salmonella species	Rods Motile Non–lactose-fermenting	Species are recognized according to the disease produced: S. enteritidis causes self-limiting gastroenteritis S. typhi causes typhoid fever and other enteric fevers S. cholerae-suis causes bloodstream infections
Shigella species	Rods Nonmotile Non–lactose-fermenting	Shigella dysenteriae produces exotoxin, causes severe dysentery Other species cause less severe dysentery

Table continues on the following page

TABLE 14–1   CHARACTERISTICS OF SOME MEDICALLY IMPORTANT GRAM-NEGATIVE BACTERIA (*Cont.*)

Genus and species	Morphological and cultural characteristics	Special properties
Proteus species	Rods Motile Non–lactose-fermenting Objectionable odor	Some species very motile, with "swarming" growth Opportunist, can cause urinary tract and bloodstream infections
Yersinia pestis	Small rods Bipolar staining Grows slowly Prefers 28°C Nonmotile Non–lactose-fermenting	Lives in wild rodents Causes bubonic and pneumonic plague in human hosts Produces exotoxin
Klebsiella *Enterobacter* *Serratia*	Rods Usually lactose-fermenting	Opportunists
Vibrio species	Curved rods Motile	*V. cholerae* produces exotoxin, causes cholera Other species cause gastrointestinal disease
Haemophilus influenzae	Small rods Nonmotile Requires special growth factors	Causes meningitis Infects ear, respiratory tract, and epiglottis
Bacteroides species	Rods, variable shape Strictly anaerobic Nonmotile	Normal flora of gastrointestinal tract Causes abscesses and bloodstream infections
Fusobacterium species	Slender, pointed rods Strictly anaerobic Nonmotile	As above May also contribute to periodontal disease Usually susceptible to penicillin

pili (Fig. 14–1), a property related to the ability of these bacteria to attach specifically to the cells they colonize. The majority of strains are killed by cold or drying.

Neisseria meningitidis

Neisseria meningitidis, the meningococcus, is a prominent cause of meningitis and serious bloodstream infections. It is closely related to *N. gonorrhoeae* but is less demanding in its growth requirements. The two species can be distinguished from one another by antigenic differences and differences in certain metabolic patterns of sugar utilization. Although there are seven different antigenic types, identified by polysaccharide antigens on the surface of the cells, most infections are due to one of the three types A, B, or C.

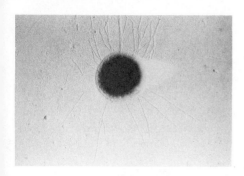

Figure 14-1
Electron micrograph of a piliated virulent *Neisseria gonorrhoeae*. (Courtesy of S. Falkow.)

GRAM-NEGATIVE RODS THAT GROW UNDER AEROBIC CONDITIONS

Many species of medically important Gram-negative rods can grow under aerobic conditions, and most of these are also able to grow in an anaerobic environment, meaning that they are facultative anaerobes. However, the metabolic patterns differ significantly from one group to another, as indicated below, and laboratories use this fact to help identify the different species and subspecies.

Pseudomonads

Pseudomonas aeruginosa is a common cause of hospital-acquired infections. It is motile by means of a single flagellum at the end of the cell (Fig. 14-3). The organisms grow readily and rapidly on many media at 37°C. Some strains of this organism can also grow in a variety of aqueous solutions, even in distilled water and some disinfectant solutions! The metabolism of *P. aeruginosa* is respiratory; i.e., the final electron acceptor in the breakdown of foods is inorganic, usually oxygen. However, *P. aeruginosa* can grow well anaerobically on many media that contain nitrate, which can substitute for oxygen as a final electron acceptor. In this reaction, nitrate is reduced by combining with electrons. Most strains of *P. aeruginosa* produce a greenish discoloration of the growth media due to the production of two pigments (color plate 22). Several toxins, including one especially powerful exotoxin,

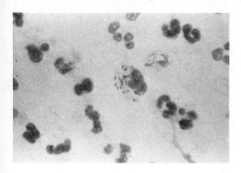

Figure 14-2
Neisseria gonorrhoeae urethral smear with intracellular organisms. Gram stain. (×1000) (Courtesy of M. F. Lampe.)

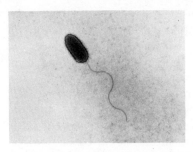

Figure 14–3
Pseudomonas aeruginosa showing a single polar flagellum. (×5250) (Courtesy of V. Chambers.)

are produced by some strains. Surprisingly, the exotoxin of *P. aeruginosa* has the same mode of action as diphtheria toxin. *Pseudomonas aeruginosa* is found in many environmental sources and in the normal human intestine. It is not highly invasive, usually infecting only individuals with impaired host defenses. Treatment is often difficult because most strains are resistant to many antimicrobial medicines.

Acinetobacters

Acinetobacter calcoaceticus, also known by its old name, *Herellea vaginicola*, is another organism with respiratory metabolism that tends to cause infections in hospitalized patients. These organisms, like the pseudomonads, occur widely and are of low invasiveness, but they can act as opportunists in hosts whose defenses are lowered by previous illness or weakened by surgical procedures. In many of its properties, however, *A. calcoaceticus* differs considerably from the pseudomonads: it occurs as short rods or cocci in pairs; it is nonmotile and oxidase-negative; it does not produce green pigment; and it does not reduce nitrate.

Francisella tularensis

This organism, the cause of tularemia, is also known by its old name, *Pasteurella tularensis*, and shows some similarities to *Yersinia pestis*, formerly called *Pasteurella pestis*. *Francisella tularensis* infects wild rabbits and numerous other species of wild animals, and is transmitted to human beings by direct contact with infected animals or via arthropods such as ticks. The organism is a tiny, nonmotile rod that sometimes stains more darkly at the ends than in the middle, a property called **bipolar staining**. It requires special media containing glucose, blood, and cystine for growth. The organism produces acid without gas from several sugars and can be identified by reaction with specific antibodies.

Brucella abortus

Six species of *Brucella*, including *B. abortus* and *B. melitensis*, can cause the disease brucellosis, also called "undulant fever" because of the recurring fever characteristic of the disease. *Brucella abortus* is a tiny, often very short, nonmotile rod that may require medium enriched with serum and glucose, as well as an increased carbon dioxide concentration, for growth. The organism is slow-growing in culture and difficult to isolate. Brucellosis is usually acquired by drinking unpasteurized milk from infected cattle or goats or by contact with infected animals. The different species of *Brucella* infect other animals, including pigs and dogs, as well as cattle and goats. Identification of the species of *Brucella* depends on biochemical tests, on the pattern of susceptibility to certain dyes and to bacteriophages, and on reactions with specific antibodies.

Bordetella pertussis

As the name implies, *B. pertussis* causes the disease pertussis, otherwise known as whooping cough. The organism resembles the brucellae in size, shape, and lack of motility, but it characteristically has a large capsule. A special medium called Bordet-Gengou medium, containing potato extract, glycerol, and a large amount of rabbit blood, is generally used to isolate the organism (Fig. 14–4). Different antigenic types of *B. pertussis* occur, and vaccines against pertussis must include the prevalent type if they are to be effective. *Bordetella pertussis* inhabits the human respiratory tract, and pertussis develops in most, but not all, of those infected with it.

Enterobacteria

The enterobacteria are facultatively anaerobic, Gram-negative rods, including species that are medically important, others that are plant pathogens, and many that are free-living, nonpathogenic organisms. The most important members of this group are mentioned below and in Table 14–1. *Escherichia coli* (color plate 23) is not only the most extensively studied member

Figure 14–4
Bordetella pertussis colonies on Bordet-Gengou agar. (1-2) (Courtesy of León J. Le Beau, Ph.D., University of Illinois Medical Center, Chicago.)

TABLE 14–2 PROPERTIES OF SOME ENTEROBACTERIA

Genus	Lactose fermentation	Motility	Tryptophan to indole	H_2S production	Other properties
Escherichia	+	+	+	−	Acid and gas from many sugars; metallic sheen on EMB[a] agar; some strains produce exotoxin
Shigella	−	−		−	Some produce exotoxin
Salmonella	−	+		+	
Yersinia	−	−		−	Produces exotoxin (Y. pestis)
Klebsiella	+	−		−	Capsules, mucoid colonies
Enterobacter	+	+		−	Lack sheen on EMB
Serratia	+	+			Red pigment produced
Proteus	−	+		+	Swarming in some species; distinctive odor; degrades urea to ammonia

[a]EMB = eosin methylene blue agar. (Note color plate 24.)
Note: These properties are not found in all species or strains of a given genus but are characteristic of many or medically important organisms in the genus.

of this group, but owing to its simplicity and the ease with which it can be grown, it is also probably the most thoroughly understood of all living creatures.

The different enterobacteria are distinguished partly by biochemical tests (Table 14–2) and partly by antigenic structure (Fig. 14–5). Note that the typical organism has three main antigenic components. These are chemically distinct, the cell wall (O antigen) being the lipopolysaccharide complex that also contains the endotoxin activity of the cell wall; the flagellar H antigen, a protein; and the capsular K antigen, a polysaccharide. Since these antigens may vary, not only between species but also from one strain to another within a species, there is the possibility of tremendous variety among the enterobacteria. The antigens help to identify the species and strain of organism.

Escherichia coli lives in the intestine of humans and other animals as part of the normal flora; however, within the species there are hundreds of different strains that vary in many ways, including their potential to produce diseases such as diarrhea and meningitis (especially in children) and urinary tract infections. All the strains have similar growth characteristics (Table 14–2), and lactose fermentation is the rule. Differences occur in certain metabolic capabilities, in antibiotic susceptibility, and in antigenic structure. In

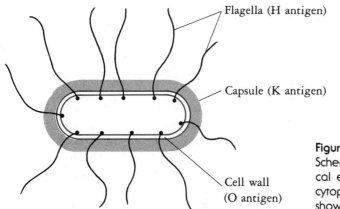

Flagella (H antigen)

Capsule (K antigen)

Cell wall
(O antigen)

Figure 14–5
Schematic drawing of a typical enterobacterial cell (the cytoplasmic membrane is not shown).

strains that cause diarrhea, at least three factors may play a role: toxin production, the ability to attach to the intestine, and the ability to invade intestinal-lining cells. Strains with a certain type of K antigen are likely to cause meningitis, and certain antigenic types occur more commonly as causes of urinary tract infections than other types. Some strains are a cause of the so-called "traveler's diarrhea."

Shigella species, also inhabitants of the intestine, are close relatives of *E. coli*, but they differ in being nonmotile and in failing, usually, to ferment lactose. *Shigella dysenteriae*, the most dangerous of this species, produces an exotoxin that acts on the intestine and central nervous system of experimental animals (and presumably of humans as well) and causes the disease called bacillary dysentery. Three other species are common causes of less severe dysentery: *S. flexneri*, *S. boydii*, and *S. sonnei*. All four species can invade cells of the intestinal lining, but only *S. dysenteriae* produces the exotoxin.

Salmonella species are also related to *E. coli*, but like the shigellae, these organisms do not ferment lactose. The three species currently recognized differ in their pathogenic potential.

The species *Salmonella enteritidis* actually consists of many hundreds of distinct strains that share the ability to produce gastroenteritis (diarrhea, vomiting, and often fever). These organisms inhabit the intestines of many warm- and cold-blooded vertebrate species, including chickens, turkeys, and pet turtles, all of which are common sources of human infections. The infections originating from foods constitute many cases of "food poisoning." Most infections with this species are short-lived and self-limited, not requiring treatment for the infection.

Salmonella typhi is a second species of the genus *Salmonella*, consisting of strains that cause enteric fevers such as typhoid fever. It differs biochemically and in its antigenic structure from the other species of salmonellae and has a characteristic capsular (K) antigen called the Vi antigen. Different strains of *S. typhi* can be distinguished from one another by their susceptibility to various bacteriophages.

Salmonella cholerae-suis represents the third species. These organisms typically cause septicemias, severe bloodstream infections.

Proteus mirabilis, like *E. coli*, is a member of the normal intestinal flora that can cause infections and disease if it reaches the urinary tract or the bloodstream. It is noted for its distinctive and objectionable odor and for its ability to swarm (color plate 25). Rather than forming discrete colonies, as most bacteria do, this organism can swarm in a filmy growth over the entire surface of the medium. Spreading from the site of inoculation occurs in a discontinuous fashion, presumably because the genes responsible for swarming are activated by conditions in the area of heavy growth and become inactive when the organisms reach fresh medium. Like shigellae and salmonellae, *Proteus mirabilis* is another of the Gram-negative rods that fail to ferment lactose.

Yersinia pestis, formerly called *Pasteurella pestis*, is a highly virulent organism that causes the notorious disease plague, in either the pneumonic or bubonic form. This organism resembles *Shigella* species in that it is nonmotile and does not ferment lactose, but it differs antigenically and in several other ways. It grows slowly and grows best at a temperature of 28°C, rather than the usual 35° to 37°C favored by many medically important bacteria. It shows bipolar staining, as does *Francisella tularensis*. It produces a potent toxin that contributes to its pathogenicity. Other species of *Yersinia* cause abdominal infections. *Yersinia pestis* lives in wild rodents such as rats and ground squirrels and is present in these animals in many parts of the United States.

Klebsiella, *Enterobacter*, and *Serratia* are, like *Proteus* species and most *E. coli*, enterobacteria capable of causing infections of the urinary tract, blood, and lungs in people with impaired host defenses. All have been responsible for serious infections acquired in hospitals from contaminated water and aqueous solutions. They commonly harbor R factors that are effectively transmitted to other enterobacteria in hosts receiving antibiotics.

Vibrios

The vibrios are curved, Gram-negative rods with flagella at the end of the rod. One of these, *Vibrio cholerae* (Fig. 14–6), produces an exotoxin that is important in causing human cholera. *Vibrio cholerae* is oxidase-positive, tolerates strongly alkaline conditions and a high salt concentration, and has distinctive antigens that identify it.

Another species, *V. parahemolyticus*, requires sodium chloride and is widespread in coastal waters. It is a prominent cause of gastrointestinal disease in humans who eat crabs and other seafoods. Recently, *V. parahemolyticus* has been shown to produce an exotoxin active against heart muscle.

Other Facultative Anaerobes

Other Gram-negative rods are also responsible for disease. One small, nonmotile rod, *Haemophilus influenzae*, is a common cause of meningitis, inflammation of the epiglottis (epiglottitis), and infections of the ear and res-

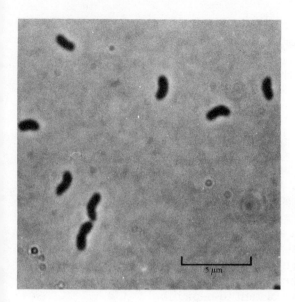

Figure 14–6
Vibrio cholerae, a curved bacterium, which causes cholera. Phase-contrast photomicrograph. (Courtesy of J. T. Staley and J. P. Dalmasso.)

piratory tract. Some of the properties of this and similar species of the genus *Haemophilus* are presented in Table 14–3. The species of this genus require certain growth factors that can be provided by components of blood or by other microorganisms. These growth requirements account for the phenomenon of **satellitism,** wherein *Haemophilus* colonies develop in the area immediately surrounding a colony of a different bacterium that can supply the required substance (Fig. 14–7). *Haemophilus influenzae* can be subdivided into several types on the basis of antigens found on its surface, type b being most often responsible for cases of severe illness.

TABLE 14–3 PROPERTIES OF A FEW SPECIES OF *HAEMOPHILUS*[a]

Species	Diseases produced	Requirement for X-factor[b]	V-factor[c]	Hemolysis on blood agar
H. influenzae	Meningitis; infections of ear, respiratory tract, and epiglottis	+	+	−
H. parainfluenzae	Respiratory infections; subacute bacterial endocarditis	−	+	−
H. aegyptius	Conjunctivitis (pink eye)	+	+	−
H. ducreyi	A venereal disease	+	−	+

[a]All species are small, nonmotile Gram-negative rods.
[b]X-factor (hemin) is a blood component, part of the enzyme catalase and also of the cytochromes. + = factor required.
[c]V-factor (NAD), also a blood component, is nicotinamide adenine dinucleotide. + = factor required.

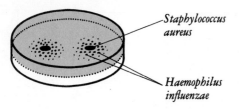

Staphylococcus aureus

Haemophilus influenzae

Figure 14–7
Colonies of *Haemophilus influenzae* growing as satellites around *Staphylococcus aureus*. The entire surface of the medium has been inoculated with *H. influenzae*, but growth occurs only adjacent to the staphylococcal colonies, which supply both X and V growth factors.

ANAEROBIC GRAM-NEGATIVE RODS

The anaerobic Gram-negative rods most often identified in normal flora and also as a cause of disease are of the genera *Bacteroides* and *Fusobacterium*.

Bacteroides fragilis, an organism of low virulence universally found in the intestine, can cause abscesses and bloodstream infections. It is nonmotile and, like the other bacteroides, tends to be variable in shape. It is a strictly anaerobic Gram-negative rod that grows rapidly at 37°C on a medium containing blood or serum. The normal habitat of *B. fragilis* is the large intestine, where it occurs in large numbers. Like most medically important Gram-negative rods, this organism is resistant to penicillin.

Fusobacterium nucleatum is another nonmotile, strict anaerobe, with a long, slender shape and pointed ends. It normally inhabits the mouth and upper respiratory tract. Fusobacteria differ from most Gram-negative rods in that they are unusually susceptible to penicillin.

Both *F. nucleatum* and *B. melaninogenicus* occasionally cause abscesses and bloodstream infections, and there is evidence that both may contribute to periodontal disease.

SUMMARY

All Gram-negative bacteria have a similar cell wall composition, which accounts for their staining characteristics. Many of these bacteria produce endotoxins that cause fever and damage to the circulatory system in the host.

The Gram-negative bacteria cover the full range of metabolic patterns. They range from aerobes to facultative anaerobes to strict anaerobic species.

The important human pathogens of the Gram-negative bacteria fall into the following groups: pseudomonads, acinetobacters, enterobacteria, and vibrios as well as members of the genera *Neisseria*, *Bacteroides*, and *Fusobacterium*.

SELF-QUIZ

1. The organism *Neisseria gonorrhoeae*
   a. is frequently found within pus cells
   b. occurs in short chains

    c. cannot utilize glucose
    d. is oxidase-negative
    e. is usually transmitted via contaminated clothing and dishes

2. Infections with *Pseudomonas aeruginosa* are feared because
    a. the pigments it produces are lethal
    b. the organism grows so rapidly under anaerobic conditions
    c. the organisms are usually resistant to so many antimicrobial drugs
    d. the organisms are so plentiful in the normal lung
    e. the organisms are so difficult to isolate

3. Any of the following organisms are likely to be the cause of hospital-acquired infections *except*
    a. *Brucella abortus*
    b. *Pseudomonas aeruginosa*
    c. *Acinetobacter calcoaceticus*
    d. *Klebsiella* species
    e. *Serratia* species

4. A cause of meningitis, *Haemophilus influenzae,*
    a. cannot be grown in culture
    b. grows best at 28°C
    c. will not grow on agar containing blood
    d. can also cause infections of the ear and epiglottis
    e. occurs only rarely

5. Species of *Shigella* and *Salmonella* are non–lactose-fermenting, but they can usually be distinguished from one another by the following property
    a. Gram-staining reaction
    b. motility
    c. presence of endotoxin
    d. presence of organisms in feces
    e. presence of organisms in fecally contaminated water

QUESTIONS FOR DISCUSSION

1. A case of gastroenteritis thought to be cholera occurs. How might this be identified as cholera or another disease? If it is cholera, how could the source of the disease be traced?

2. A woman had a severe disease with fever after skinning and cooking wild rabbits her son had shot. What is a probable diagnosis and what are the characteristics of the causative organisms?

3. Gram-negative rods are most often the cause of urinary tract infection, but *Shigella* is not found in urine. Why is this, and what organisms would be expected to cause urinary tract infections?

FURTHER READING

Sanderson, K. E.: "Genetic Relatedness in the Family Enterobacteriaceae." *Annual Review of Microbiology 30*:327–349 (1976).

Schlessinger, D. (ed.): *Microbiology*. Washington, D.C.: American Society for Microbiology, 1977. Large sections with many authors on *Pseudomonas aeruginosa* and on a variety of Gram-negative endotoxins.

Williams, F. D., and Schwarzhoff, R.: "Nature of the Swarming Phenomenon in Proteus." *Annual Review of Microbiology 32*:101–122 (1978). A review of the theories of why and how *Proteus* organisms swarm. Excellent pictures.

Chapter 15

OTHER BACTERIA OF MEDICAL IMPORTANCE

On an isolated, flat peninsula formed by volcanic action thousands of years ago is the city of Kalaupapa, Molakai, Hawaii. More than a hundred years ago, a leper colony was established there. Even today the only access to that part of Molokai is by airplane or by foot down a steep cliff.

Leprosy is caused by the bacterium *Mycobacterium leprae*. Not until recent years has the medical community come to some understanding of this disease and been able to control it with the drug sulfone. Previously, people who had leprosy were isolated from the rest of the community in leper colonies such as the one at Kalaupapa.

Father Damien, a Belgian Roman Catholic priest, came to Kalaupapa in 1864. After viewing the appalling conditions that existed there, he devoted the rest of his life to caring for the sick and improving their living conditions. He contracted the disease himself and died in 1889. Now the people who live in Kalaupapa live there by choice. All have or have had leprosy now under control. No new people are admitted to the colony, because leprosy can be controlled, and the affected person no longer needs to be isolated from family, friends, and community.

OBJECTIVES

To know
1. The three different species of spirochetes that cause human diseases, which ones infect other animals, and how this is important in their transmission.
2. Two major disease-producing mycobacteria and how the lipids of mycobacteria influence their properties.
3. The relationship of the actinomycetes to other microorganisms, the kinds of diseases the actinomycetes cause, and what other medically important function some of them, in particular the streptomycetes, have.
4. The important characteristics of the mycoplasmas.
5. The obligately parasitic bacteria and what diseases they cause.

Although many bacteria can be classified as either Gram-positive or Gram-negative on the basis of their cell wall properties, other bacteria do not fit so neatly into this kind of scheme (see Appendix VII). Some have waxy cell walls that the Gram-stain dyes do not penetrate; some lack a cell wall altogether; and others have unique, flexible cell walls. Some of these bacteria that are of medical importance will be considered in this chapter.

SPIRAL-SHAPED BACTERIA

The spirochetes are spiral-shaped because of their flexible cell walls. Other spiral-shaped bacteria include the spirilla and a group with gliding motility. Organisms of the latter group live in salt or fresh water, and none of them is known to be of medical importance. Spirochetes and spirilla may also live in water and be nonpathogenic; however, some species, especially some of the spirochetes, are pathogenic.

The spirochetes move by means of unique structures called **axial filaments** (Fig. 15–1). Electron microscopy has revealed that each axial filament is actually composed of fibrils, identical in structure to flagella, originating near each end of the organism. The fibrils extend toward each other between two layers composing the cell wall and apparently overlap in the midregion of the cell. The layer of cell wall material covering the filament is called the **sheath.** As with flagella, the molecular basis for movement with axial filaments is not yet clear. Some of the smaller spirochetes are so slender that they cannot be seen by ordinary light microscopy, and dark-field or phase-contrast microscopy must be used (Fig. 15–2). Many spirochetes have been difficult or impossible to culture in vitro.

Several spirochetes cause diseases. Under certain abnormal conditions, spirochetes of the normal human flora can cooperate with anaerobic bacilli called **fusobacteria** to cause fusospirochetal disease in the mouth, lung, or other tissues.

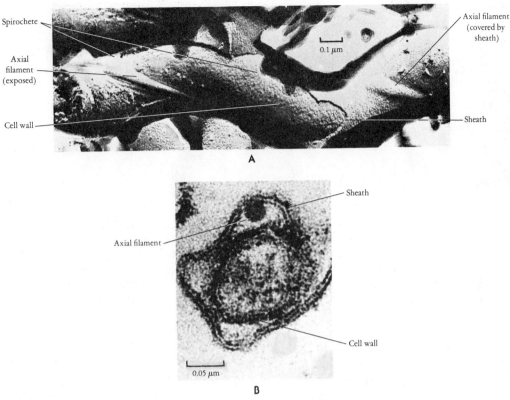

Figure 15–1
Electron micrographs showing the axial filament of a spirochete of the genus *Leptospira*.
(a) Longitudinal view. (b) Cross section. (Courtesy of R. K. Nauman, S. C. Holt, and C. D.
Cox, *J. Bacteriol.*, 98:264, 1969.)

Treponema pallidum

Treponema pallidum is the spirochete of greatest medical importance because
it is the pathogen that causes syphilis. It is a slender organism with tightly
wound coils and ranges up to 20 μm in length (Fig. 15–3). Dark-field microscopy is usually used to see *T. pallidum* and to observe its characteristic
rotational and flexing motions. Study of the organism is difficult because it
cannot be cultured in vitro and must be grown in laboratory rabbits. This
procedure is cumbersome and expensive, therefore, tremendous efforts are
being made to find more suitable means of cultivating these spirochetes.
Although the organisms can be maintained in vitro and their metabolism
studied, they either do not multiply in culture or do so to a very limited
extent. There are indications from the in vitro studies that *T. pallidum*, long
thought to be a strict anaerobe, is probably aerobic and that its metabolism
is inhibited under absolutely anaerobic conditions. Normally *T. pallidum* is
a parasite only of humans. Species similar in appearance to it live in the
mouth without causing disease.

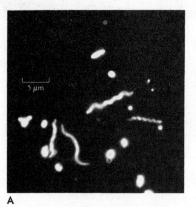
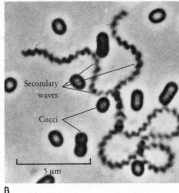
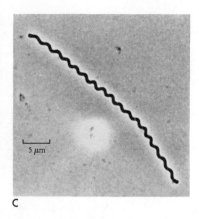

A   B   C

Figure 15–2
Spirochetes. (a) Dark-field photomicrograph of spirochetes from a monkey's mouth. (Courtesy of B. L. Williams.) (b) Phase-contrast photomicrograph of a very long free-living spirochete showing secondary waves. Note the size relative to the cocci, also present in the photograph. (Courtesy of J. T. Staley and J. P. Dalmasso.) (c) Although helical, this organism is not a spirochete; it is a member of a genus of multicellular gliding bacteria and lacks an axial filament. (Phase-contrast photomicrograph courtesy of J. T. Staley and J. P. Dalmasso.)

Borrelia recurrentis

Borrelia recurrentis (Fig. 15–4), a cause of relapsing fever, is a Gram-negative organism with loosely wound spirals that are thicker and easier to see than those of *T. pallidum*. Some strains of *B. recurrentis* have been grown anaerobically in complex media enriched with serum or other animal proteins. This organism is a parasite of humans and of the louse, *Pediculus humanis*. In the United States, relapsing fever is also caused by other species of *Borrelia* that are tick-borne.

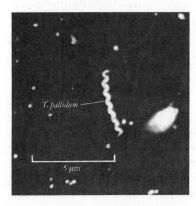

Figure 15–3
Dark-field photomicrograph of *Treponema pallidum*, the cause of syphilis. (Courtesy of F. Schoenknecht and P. Perine.)

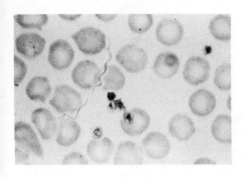

Figure 15—4
Borrelia recurrentis. Peripheral blood smear from a patient with relapsing fever. Giemsa stain.

Leptospira interrogans

Leptospira interrogans (so named because its shape resembles a question mark) causes leptospirosis. This organism, like *T. pallidum,* is a slender spirochete with tightly wound coils, best observed by dark-field microscopy. However, *L. interrogans* grows well aerobically in liquid media containing serum, and frequently the organism bends near the end to give the appearance of a hook. There are a large number of different antigenic types of *L. interrogans,* and various types normally live in the urinary systems of dogs, rats, cattle, and other animals. The "triple vaccine" given routinely to dogs is designed to protect them against leptospirosis (disease caused by infection with leptospira) and also against distemper and hepatitis. Human leptospirosis is usually a mild, self-limiting disease characterized by fever, but severe forms involving the eye, nervous system, and liver sometimes occur.

THE MYCOBACTERIA

The mycobacteria are irregularly shaped rods with large quantities of lipids in their cell walls. They grow aerobically, and their high lipid content often causes them to form a film or layer of growth over the surface of liquid media. The lipids also account for the characteristic staining properties of the mycobacteria. Stains do not ordinarily penetrate the waxy, lipid-containing cell wall, so heat or other methods must be used to cause the dyes to penetrate. Once the cell walls are stained, they are highly resistant to decolorization, even by acid alcohol that removes stains from most bacteria; therefore, the mycobacteria and a few other related organisms are known as the **acid-fast bacteria.** Some of the mycobacteria cause serious diseases in humans, including tuberculosis, Hansen's disease (leprosy), and others.

Mycobacterium tuberculosis

Mycobacterium tuberculosis, or the tubercle bacillus, is the cause of human tuberculosis. It is a slender, beaded-appearing, acid-fast rod (Fig. 15–5). Because *M. tuberculosis* resists killing by strong acids and alkalis and also by

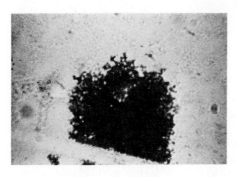

Figure 15-5
Acid-fast colony in blood culture of a patient with untreated pulmonary tuberculosis. Triple stain. (Courtesy of Mattman and Oakes.)

certain antibiotics and disinfectants that kill many other bacteria, the tubercle bacilli can be recovered from mixtures of organisms. One method commonly used involves collecting sputum from a tuberculous patient, mixing it with a high concentration of sodium hydroxide, and incubating it for about 15 minutes. This treatment destroys most bacteria and partially degrades the mucus in the sputum. It does not kill the mycobacteria, which can then be cultured. *M. tuberculosis* has simple growth requirements, but it nevertheless often fails to grow on ordinary laboratory media because these media commonly contain substances that inhibit its growth. At best, growth is slow; the generation time is about 12 hours compared with approximately 20 minutes for *E. coli*, a rapidly growing bacterium. Unless the medium on which *M. tuberculosis* is growing contains a wetting agent,* the organisms tend to adhere strongly to each other. Under moist conditions of growth they will form long tangled ropes, or cords, of cells attached side to side (color plate 26). Other characteristics that help to identify this species include their production of excess niacin (a B vitamin readily detected in the laboratory), their ability to reduce nitrate, and their susceptibility to certain antimicrobial agents.

Mycobacterium leprae

Mycobacterium leprae, the cause of Hansen's disease (leprosy), is morphologically indistinguishable from *M. tuberculosis*, but *M. leprae* has not been grown on artificial media. It can infect armadillos and can be cultured in the footpads of specially prepared mice. Growth is very slow, with a generation time of 12 days or more in mice.

In addition to *M. tuberculosis* and *M. leprae*, several other species of mycobacteria can cause human infections.

THE ACTINOMYCETES

The branching, filamentous, Gram-positive actinomycetes include bacteria of the genera *Actinomyces*, *Nocardia*, and *Streptomyces*. These organisms are

*wetting agent—a substance that allows water to adhere to the surface of other material

related to the mycobacteria, and some of the species of *Actinomyces* and *Nocardia*, like the mycobacteria, are acid-fast. In nature, the actinomycetes are responsible for much of the decomposition of organic materials in soil and are the source of the peculiar odor unique to soil. Several species of *Streptomyces* are important producers of antibiotics. Because these organisms form spores and filaments similar to fungal spores and hyphae, they were once considered to be fungi. However, it is now well established that they are bacteria and possess cell wall peptidoglycans and other bacterial characteristics. This is an important point, because it means that penicillin and certain other antibacterial antibiotics can be used to treat infections caused by actinomycetes.

Nocardia asteroides

Nocardia asteroides is an aerobic actinomycete of the soil that can produce serious lung or generalized infections, especially in people with debilitating diseases. After about four days' growth in culture, the filaments that are produced, often acid-fast, fragment into rod-shaped bacteria. Each of these bacteria can then develop into another colony of the organism.

Actinomyces israelii

Actinomyces israelii is an anaerobic actinomycete that causes actinomycosis, a chronic, deep-tissue infection of the face and other parts of the body. This organism, like *N. asteroides*, produces non–spore-forming filaments that fragment (Fig. 15–6). However, *A. israelii* is generally not acid-fast and is not primarily a soil organism. It normally lives in the alimentary and upper respiratory tracts of humans. Other similar but less pathogenic species of *Actinomyces* can be distinguished from *A. israelii* by their antigens and by their metabolic products.

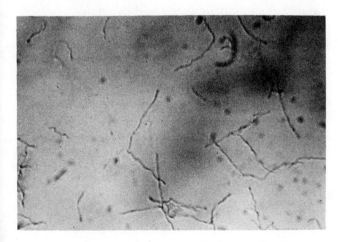

Figure 15–6
Actinomyces israelii. Wet mount. Interference phase microscopy. (×1000) (Courtesy of Leon J. Le Beau, Ph.D., University of Illinois Medical Center, Chicago.)

BACTERIA WITHOUT CELL WALLS

Organisms of the genus *Mycoplasma* lack cell walls. As a result, they are plastic, easily deformed, and can pass through filters that retain most other bacteria. Nevertheless, different species have characteristic shapes (Fig. 15–7). Mycoplasmas are also unusual bacteria in that their cytoplasmic membranes contain sterols, which stabilize the membranes and protect the cells against osmotic lysis.* Colonies of mycoplasmas are often so small that a lens or dissecting microscope is required to see them well. The organisms grow down into the agar, often producing a dense central zone surrounded by a flat translucent area. The whole colony resembles a fried egg (color plate 27). Most species require a medium containing serum as a source of the sterols the bacteria need but cannot synthesize. Most strains of mycoplasmas also require vitamins and a number of precursors of proteins and nucleic acids. They may be either aerobic or anaerobic and inhabit diverse environments, depending on the species. Some live in soil, others in sewage, and still others are parasitic on plants and animals.

Although mycoplasmas are responsible for a number of diseases of animals, only two species—*M. pneumoniae* and *M. hominis*—have been proved to cause disease in humans. Of the two, *M. pneumoniae* is of much greater importance, since it is a common cause of bacterial pneumonia. Mycoplasmal pneumonia is not typical of most bacterial pneumonias, and the causative bacteria, as mentioned previously, are small and easily deformed so they are able to pass through filters that retain most bacteria. For these reasons, this kind of pneumonia was long thought to be caused by viruses and is still known as primary atypical pneumonia.

Mycoplasma pneumoniae organisms are readily cultured, producing tiny colonies in about ten days. They grow best aerobically and produce abundant quantities of hydrogen peroxide. Erythrocytes in the medium are hemolysed. The surface of *M. pneumoniae* contains areas that attach firmly to receptors on cells lining the respiratory tract, permitting the mycoplasmas to colonize and infect the lung. Other species of *Mycoplasma* are part of the normal flora of human mucous membranes but show little or no pathogenicity.

OBLIGATELY PARASITIC BACTERIA

Although most of the bacteria considered so far have the ability to parasitize humans or other animals, they can also grow in the absence of living cells. Bacteria of the genera *Rickettsia*, *Coxiella*, and *Chlamydia*, however, generally lack some of the metabolic capabilities necessary for an independent existence and are obligately parasitic.

The Rickettsiae

The rickettsiae (Fig. 15–8), including the genera *Rickettsia* and *Coxiella*, cause a variety of generalized infections or pneumonia (Table 15–1). All of these

---

*osmotic lysis—the disruption or disintegration of a cell caused by osmosis

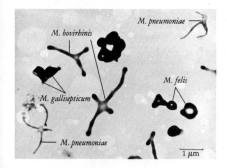

Figure 15–7
Electron micrograph showing the morphology of different species of *Mycoplasma*. (Courtesy of E. Boatman.)

agents are short, nonmotile rods, transmitted by arthropod vectors such as insects and arachnids. In general, rickettsiae do not survive well in the environment. The exception is *Coxiella burnetii,* which is resistant enough to drying and heat to be transmitted by air and by food products. Rickettsiae can be cultivated in embryonated eggs or tissue cell cultures. Some species stimulate the production of antibodies that happen to cross-react with antigens of some *Proteus* species. The presence of such antibodies can be used diagnostically.

The Chlamydiae

Chlamydiae (Fig. 15–9) differ from rickettsiae in that they are spherical rather than rod-shaped and do not require arthropod vectors for transmission. Two species are recognized. *Chlamydia trachomatis* strains are responsible for trachoma (the leading cause of blindness in the world), inclusion conjunctivitis, urethritis, pneumonia in infants, and a venereal disease. Within host cells, *C. trachomatis* produces compact areas of growth called **inclusion bodies,** which can be stained with iodine or certain dyes. *Chlamydia psittaci* causes a lung infection called ornithosis or psittacosis. It differs from

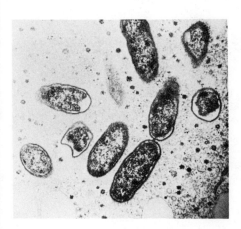

Figure 15–8
Electron micrograph of *Rickettsia rickettsii* in cytoplasm of Malpighian tube cell of *Dermacentor andersoni.* Rickettsia at upper right under binary fission. (Courtesy of W. Burgdorfer.)

TABLE 15–1   SOME CHARACTERISTICS OF RICKETTSIAL DISEASES

Disease	Causative species	Principal hosts	Vectors
Typhus	*Rickettsia prowazekii*	Humans	Lice
Scrub typhus	*R. tsutsugamushi*	Rodents	Mites
Rocky Mountain spotted fever[a]	*R. rickettsii*	Rodents, dogs	Ticks
Rickettsial pox	*R. akari*	Mice	Mites
Q fever	*Coxiella burnetii*	Rodents, cattle, goats, sheep	Ticks[b]

[a]Additional spotted fevers are caused by other rickettsiae.
[b]*C. burnetii* may also be transmitted via air, milk, and other means, in addition to arthropods.

C. trachomatis in failing to develop inclusion bodies. Individual organisms of *C. psittaci* can be seen by using appropriate staining and light microscopy. A variety of mammals and birds can be infected by various strains of *C. psittaci.*

LEGIONELLA PNEUMOPHILA

Legionella pneumophila (Fig. 15–10), a bacterium that causes pneumonia, is one member of a newly recognized group of similar bacteria. This organism was isolated from patients with the well-publicized Legionnaire's disease, an epidemic of pneumonia that occurred in 1976, largely among those attending an American Legion convention in Philadelphia. Subsequent epidemiological "detective work," using cultural and serological techniques, revealed that this rod-shaped bacterium is a widespread cause of bacterial pneumonias, particularly in smokers and the elderly. It can be cultivated aerobically on an agar medium containing charcoal, yeast extract, cystine, and an iron compound.

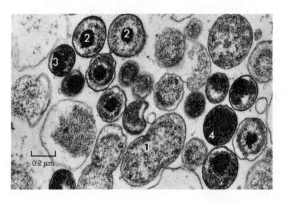

Figure 15–9
Chlamydiae growing in a tissue cell culture. The numbers indicate development from dividing form (1) to mature infectious bacterium (4). (Courtesy of R. R. Friis, *J. Bacteriol.,* 110:706, 1972.)

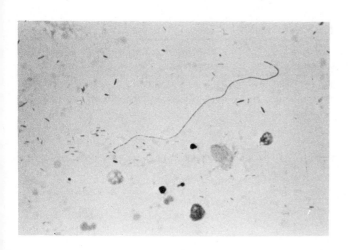

Figure 15–10
Legionella pneumophila agent in yolk sac of embryonated egg as stained by the Gimenez method. (Courtesy of McDade.)

SUMMARY

A variety of bacteria that can cause disease are categorized primarily by special characteristics other than their Gram-staining reaction. Some of them are not readily seen by light microscopy, for example, the spirochetes. Others, such as the mycobacteria, have lipid-containing cell walls that stain best with acid-fast staining techniques. Others lack cell walls or are obligate intracellular parasites. Some of the medically important bacteria of such groups are briefly characterized in Table 15–2.

TABLE 15–2   CHARACTERISTICS OF SOME OTHER MEDICALLY IMPORTANT BACTERIA

Genus and species	Morphological and cultural characteristics	Special properties
Treponema pallidum	Very slender, tightly coiled spirals Motile via axial filament Not cultured in vitro	Visible by dark-field microscopy but not by Gram staining Highly susceptible to cold and drying Usually spread by venereal contact Can cross placenta Cause of syphilis
Borrelia species	Gram-negative, loosely wound spirals Some can grow anaerobically on rich media	Transmitted via lice or ticks Cause of relapsing fever
Leptospira interrogans	Slender, tightly coiled spirals Grows aerobically in liquid media containing serum	Visible by dark-field microscopy Lives in urinary tract of animals (rodents, cattle, dogs, etc.) Transmitted to humans via urine

Table continues on the following page

TABLE 15–2 CHARACTERISTICS OF SOME OTHER MEDICALLY IMPORTANT BACTERIA (*Cont.*)

Genus and species	Morphological and cultural characteristics	Special properties
Mycobacterium tuberculosis	Slender, acid-fast rods High lipid content Grows slowly on special media Produces excess niacin Reduces nitrate	Causes tuberculosis of lungs and other tissues and tuberculosis meningitis
Mycobacterium leprae	Slender, acid-fast rods High lipid content Not grown in cultures in vitro	Cause of Hansen's disease (leprosy)
Nocardia asteroides	Gram-positive filaments and rods May be acid-fast Aerobic Easily cultured	Soil bacterium Causes lung or generalized infections, usually in immunologically compromised hosts
Actinomyces israelii	Gram-positive filaments and rods Anaerobic Easily cultured	Part of the normal flora Can cause chronic deep-tissue infections, usually in damaged tissues
Mycoplasma pneumoniae	Lacks cell walls Facultative Grows in rich media containing serum	A cause of human pneumonia Other species of *Mycoplasma* cause diseases of animals
Rickettsia species	Short, Gram-negative rods Nonmotile Obligate parasites, grow only within cells	Various species cause typhus, spotted fevers Transmitted via arthropods such as lice and ticks
Coxiella burnetii	As for *Rickettsia*	Transmitted via air, milk, other food products Cause of Q fever
Chlamydia trachomatis	Gram-negative Nonmotile Obligate parasites	Causes the eye disease trachoma, a venereal disease, and other infections Transmitted by contact with other humans
Chlamydia psittaci	As for *C. trachomatis*	Causes lung disease (ornithosis or psittacosis) Transmitted by birds
Legionella pneumophila	Gram-negative rods Difficult to culture	A cause of pneumonia

SELF-QUIZ

1. *Treponema pallidum* is able to
   a. move by means of flagella
   b. grow on chocolate agar with increased CO_2
   c. grow best under strictly anaerobic conditions
   d. grow in laboratory mice
   e. cause a sexually transmitted disease

2. All of the following are true of *Borrelia* species except that they
   a. are Gram-negative spirals
   b. can be a cause of venereal disease
   c. may grow anaerobically in rich media
   d. may live in lice
   e. may live in ticks
3. The mycobacteria are
   a. Gram-negative
   b. tiny cocci
   c. acid-fast
   d. anaerobic
   e. rapidly growing organisms
4. Actinomycosis is a disease that
   a. is caused by a fungus
   b. is caused by an aerobe
   c. results from infection with a soil organism
   d. generally involves the urinary tract
   e. can be treated with penicillin
5. All of the following are true of rickettsial organisms *except* that they
   a. grow well on very rich media under anaerobic conditions
   b. include organisms of the genera *Rickettsia* and *Coxiella*
   c. are short, nonmotile rods
   d. are frequently transmitted by arthropods
   e. usually do not survive well in the environment, in air, or in water

QUESTIONS FOR DISCUSSION

1. What steps might be taken to try to culture *Treponema pallidum* or *Mycobacterium leprae?*
2. It is unusual that contaminated urine is a source of infection, but this occurs in leptospirosis. How can this be explained?
3. Organisms of the genus *Streptomyces* are the source of a number of antibiotics. How might one go about looking for new antibiotics produced by *Streptomyces?*

FURTHER READING

Johnson, R. C.: *The Biology of Parasitic Spirochetes.* New York: Academic Press, 1976.
Maniloff, J.: "Molecular Biology of Mycoplasma." In Schlessinger, D. (ed.): *Microbiology 1978.* Washington, D.C.: American Society for Microbiology, 1978, pp. 390–393. One of a series of seminar papers on the mycoplasmas; all the papers contribute to the knowledge of this group of bacteria.
Marshall, E.: "Hospitals Harbor a Built-In Disease Source." *Science* 210:745–749 (1980). Concerns tracking down the source of *Legionella pneumophila* infection in a hospital.

Chapter 16

VIRUSES

Illnesses that we now know are caused by viruses have existed for thousands of years. In 1100 BC, Rameses V of Egypt died of a disease that left pockmarks on his mummified face—probably smallpox. It is ironic that the first and most devastating viral disease ever recorded is apparently the first viral disease to be eradicated. About a dozen years ago, the worldwide incidence of smallpox was about ten million cases annually. However, no cases have been reported anywhere in the world since 1978, and there is hope that this disease, which killed countless millions and left scarred the faces of such notable people as George Washington, will never be seen again. Nevertheless, many other viral infections afflict people. It is estimated that every American suffers from two to six viral infections a year and has as many as 200 during his or her lifetime.

A modern viral disease that is not given its proper respect is influenza ("flu"). In 1918–1919, influenza put one fifth of the human race to bed and killed more than 20 million people. Its effects have been noted since the twelfth century, and more than 45 epidemics have been noted by historians since the fifteenth century. For all its deadliness, influenza does not conjure up the terror of such diseases as cholera, polio, and smallpox. Most people do not fear influenza, which is generally accepted as an inescapable nuisance. One unique feature of the influenza virus is that it undergoes major (and minor) changes in its antigenic structure; these changes leave people without much antibody protection against the new types. Epidemics result at regular intervals. Thus, after the pandemic* Spanish flu in 1918–1919, another type arose in 1957—the Asian flu, which originated in China. In the United States alone, more than 69,000 people died of this disease. In 1968–1969, the virus underwent another major change in its antigenic composition, and the so-called Hong Kong flu killed almost 34,000

*pandemic—a worldwide epidemic

people in the United States. In early 1976, an influenza virus strain similar to the one responsible for the 1918–1919 pandemic was found in several Army recruits at Fort Dix, New Jersey. This finding prompted a nationwide effort to immunize the most susceptible people in the population. The swine flu vaccination program was initiated, but the expected epidemic never developed. However, many scientists fully expect that the next epidemic will be caused by an organism with properties of swine influenza. If this occurs before 1982, there will be 115 million doses of swine influenza vaccine available for use. If preventive methods as effective as those against smallpox or polio can be developed for influenza, the public will no longer have to accept influenza as a disease from which there is no escape.

OBJECTIVES

To know
1. Three kinds of interaction that occur between bacteriophages and bacteria.
2. The major components and structure of viruses.
3. The life cycle of a virulent bacteriophage and a temperate bacteriophage.
4. The significance of lysogenic conversion to medical microbiology.
5. How the mechanism of specialized transduction works.
6. The four types of interaction of animal viruses with their host cells, and some examples of these.
7. The basis for cell transformation by Epstein-Barr (EB) virus.

GENERAL DESCRIPTION

Viruses are infectious particles that differ from cells in several important ways. Each virus particle or **virion** contains a *single* type of nucleic acid (DNA or RNA), whereas cells always contain both kinds of nucleic acid. The nucleic acid is surrounded by a protein coat or **capsid** and often by an outer layer or envelope that contains carbohydrates and lipids. The presence of either RNA or DNA but not both is an important property that distinguishes viruses from cells. Thus, there are DNA viruses and RNA viruses. The nucleic acid of a virus may occur as double-stranded DNA, single-stranded DNA, double-stranded RNA, or single-stranded RNA. Like cells, the main function of viruses is to reproduce themselves. However, viruses lack the large number of enzymes and cell structures that are responsible for generating energy and synthesizing many macromolecules, such as protein and nucleic acid. They use the enzymes and structures of the cells they infect. Thus, outside the cells, viruses are inert* particles. The few enzymes that viruses possess

Viruses, pages 15–16

---

*inert—inactive

336

THE MICROBIAL WORLD

TABLE 16–1   COMPARISON OF VIRUSES AND CELLULAR MICROORGANISMS

Viruses	Cells
Have RNA *or* DNA, never both	Have both DNA and RNA
Reproduce only inside living cells	Reproduce in the absence of living cells
Nucleic acid and protein replicate independently of one another	Nucleic acids and protein replicate together as a single unit
Contain very few, if any, enzymes	Contain many enzymes

are generally related to the entry of the virion into the host cell. A comparison of viruses and cells is given in Table 16–1.

Because viruses consist of only a few structures, generally a protein coat surrounding a nucleic acid molecule (Fig. 16–1), they are much smaller than most bacteria (Fig. 16–2). Because of their small size, most viruses can be seen adequately only by electron microscopy. As shown in Figure 16–1, viruses can have a variety of shapes. Indeed, electron microscopic techniques have made it possible to see in detail the few simple structures of virus particles (Fig. 16–3).

All forms of life are infected by viruses. However, most information has been gained about the viruses that infect bacteria, the **bacteriophages** ("eaters of bacteria"), often abbreviated **phages.** Generally, these viruses serve as excellent models for the viruses that infect animal cells, so we will focus on the bacteriophages first and then consider infections of animal cells.

Electron microscopy, Appendix II

GENERAL FEATURES OF BACTERIOPHAGE–BACTERIAL INTERACTIONS

The events that take place in cells infected by bacteriophages differ depending on the type of bacteriophage involved. Some bacteriophages multiply and lyse the host cell, releasing newly formed virions. These bacteriophages are **virulent** and are exemplified by the so-called "T" phages. The infection is a **lytic** one. Other bacteriophages multiply and their progeny are released from the cell without killing it. In still other cases, the viral nucleic acid becomes integrated into the DNA of the cell and multiplies as the DNA of the cell multiplies. In these instances, the cell actually acquires new chromosomal genes in the form of the inserted* viral DNA. These three types of interaction are shown in Figure 16–4. Our focus will be on the ways viruses interact with the cells they infect. These interactions can be divided into a number of well-defined steps. The following discussion concerns the bacteriophages that contain double-stranded DNA.

LYTIC INFECTIONS (Fig. 16–5)

Step 1. Adsorption

The first step in the interaction of viruses with host cells is the attachment of the virus to the cell surface (Fig. 16–5). This is called **adsorption.** Bacte-

*inserted—incorporated into the chromosome

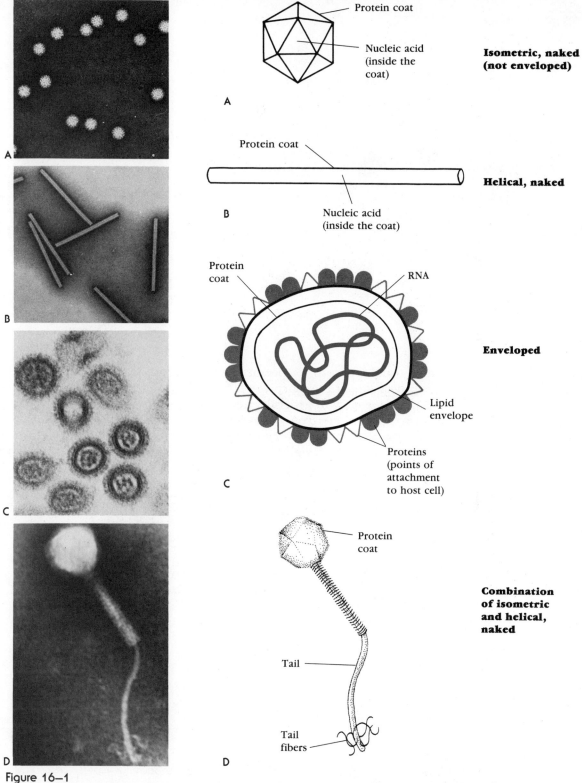

A   **Isometric, naked (not enveloped)**

Protein coat

Nucleic acid (inside the coat)

B   **Helical, naked**

Protein coat

Nucleic acid (inside the coat)

C   **Enveloped**

Protein coat

RNA

Lipid envelope

Proteins (points of attachment to host cell)

D   **Combination of isometric and helical, naked**

Protein coat

Tail

Tail fibers

Figure 16—1

Virus morphology. (a) Bushy stunt virus. (×260,000) (b) Tobacco mosaic virus. (×140,000) (c) Influenza virus. (×200,000) (d) Marine bacteriophage. (×275,000) (a and b courtesy of R. Williams, c and d courtesy of E. Boatman.)

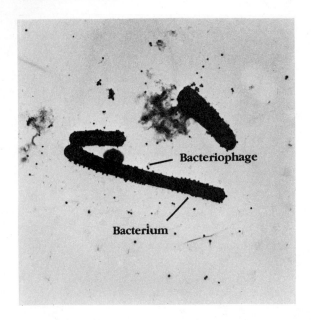

Figure 16–2
Bacteriophages attached to bacteria. Note their relative sizes. (Courtesy of D. Birdsell.)

Pili, pages 45–46

Flagella, pages 43–44

riophages usually adsorb to specific sites on the cell wall, but some adsorb only to pili or flagella. Adsorption is a very specific interaction, and any given bacteriophage will generally adsorb to the surface of only one species of bacterium, and often to only a few of the strains that compose the species. This fact has proved to be very useful in distinguishing between closely related bacteria and especially between strains of bacteria of the same spe-

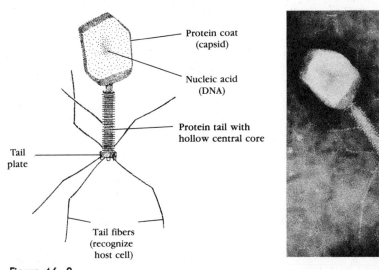

Protein coat (capsid)

Nucleic acid (DNA)

Protein tail with hollow central core

Tail plate

Tail fibers (recognize host cell)

Figure 16–3
The bacteriophage T4, one of the most complex of the bacterial viruses. ($\times 270,000$) (Courtesy of R. Williams.)

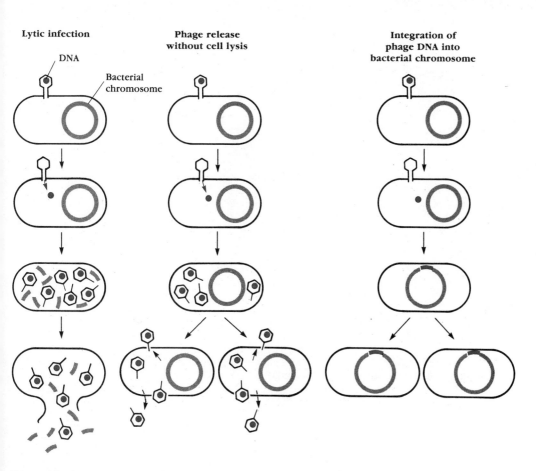

Figure 16–4
Three types of bacteriophage-bacterial interactions.

cies. The technique of using viruses to distinguish among bacterial strains is termed **bacteriophage typing** (see Chapter 29).

Step 2. Entrance

Once the virion becomes firmly attached to the surface of the cell, the nucleic acid penetrates the cell wall and membrane and enters the cytoplasm. The capsid remains on the outside; its job of protecting the nucleic acid from harmful substances in the environment is finished. Bacteriophages differ in this respect from animal viruses. In the latter case, the entire virus particle enters the cell.

Step 3. Multiplication

Within a few minutes after the phage DNA enters the cell, leaving its protein coat on the outside, all the bacterial DNA is degraded. Therefore, the only

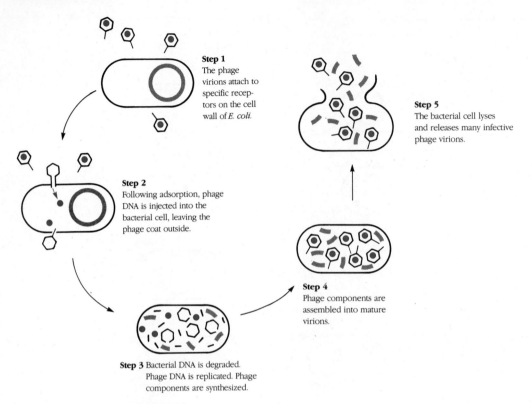

Step 1
The phage virions attach to specific receptors on the cell wall of *E. coli.*

Step 2
Following adsorption, phage DNA is injected into the bacterial cell, leaving the phage coat outside.

Step 3 Bacterial DNA is degraded. Phage DNA is replicated. Phage components are synthesized.

Step 4
Phage components are assembled into mature virions.

Step 5
The bacterial cell lyses and releases many infective phage virions.

Figure 16–5
The sequence of events in a lytic infection.

functioning genes left in the cell are in the viral DNA. In this way the bacteriophage DNA directs the operations of the bacterial cell to its own purpose—the synthesis of more bacteriophages. The bacteriophage DNA contains genes that code for the mRNAs and proteins necessary for phage replication. These include the enzyme that degrades the host DNA, all the proteins of the capsid and other bacteriophage structures, and the many enzymes required for the synthesis of bacteriophage DNA. Note that unlike the multiplication of bacteria, in which the protein and DNA are produced in a synchronized* manner, bacteriophage multiplication involves the separate synthesis of protein and DNA. This independent production of macromolecules is typical of all virus infections. The bacteriophage uses many of the host enzymes as well as other cellular components, such as the ribosomes and the energy-generating machinery, to synthesize its own macromolecules. This explains why phages, like all viruses, are considered to be *obligate intracellular parasites.*

Protein synthesis, pages 115–116

*synchronized—occurring at the same time

Step 4. Maturation

During this stage the nucleic acid replicates, and many of its genes code for the proteins that constitute the outside structure of the phage. When sufficient nucleic acid, protein coats, and the other components of the phage have been synthesized, the various components combine in a specific sequence to form intact bacteriophages. The combination of all the components to form complete bacteriophages is termed **maturation.**

Step 5. Release

The last step in the life cycle involves the synthesis of a bacteriophage **lysozyme** inside the infected cell. This enzyme degrades the bacterial cell wall, and the cell lyses. About 100 bacteriophages per infected cell are released into the medium, where they are then able to infect other uninfected cells. Since the virus is able to multiply only within a living cell, once it lyses the cell in which it is living it becomes "inert" and is able to function only by infecting another cell.

Lysozyme, page 33

INTEGRATION OF BACTERIOPHAGE DNA INTO BACTERIAL DNA

Lysogenic Infection (Fig. 16–6)

Some bacteriophages, exemplified by the bacteriophage lambda, inject their DNA into bacteria, leaving their protein coat behind just like the "T" bacteriophages. However, rather than replicating, the DNA *integrates* into a specific site in the DNA of the host. The phage DNA in this state is termed a **prophage,** and the bacterial cell carrying a prophage is called a **lysogenic cell.** The bacteriophage that is able to establish a lysogenic infection is termed a **temperate bacteriophage.** Once integrated, the bacteriophage DNA is a part of the bacterial chromosome and multiplies as the bacterial genes replicate (Fig. 16–6). The genes coding for the structural proteins* of the phage are repressed, and these proteins are not synthesized.

However, in rare situations, the viral DNA may be excised from the bacterial DNA and enter the cytoplasm. Once this happens, the genes coding for structural proteins are no longer repressed, so the proteins are synthesized. The life cycle of the bacteriophage continues in a fashion identical to that of the "T" bacteriophages.

Cells containing a prophage multiply and behave almost as though they did not contain viral genes. However, the presence of the prophage often confers new, readily recognizable properties on the bacterial cell. For example, certain prophages of *Corynebacterium diphtheriae* are responsible for synthesizing the exotoxin that causes the symptoms of diphtheria. Streptococci can synthesize the toxin associated with scarlet fever only if they are

---

*structural proteins—proteins that form part of the virus, such as those of the capsid

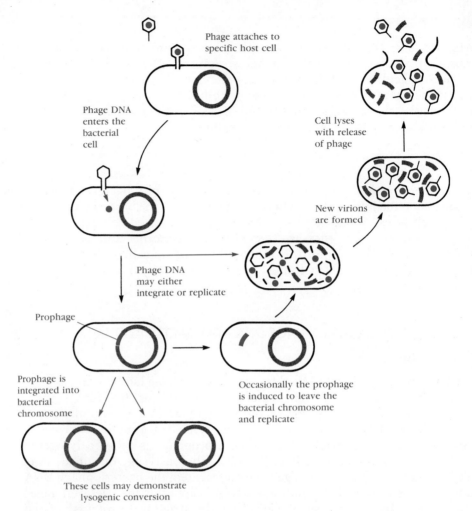

Phage attaches to
specific host cell

Phage DNA
enters the
bacterial
cell

Cell lyses
with release
of phage

New virions
are formed

Phage DNA
may either
integrate or replicate

Prophage

Prophage is
integrated into
bacterial
chromosome

Occasionally the prophage
is induced to leave the
bacterial chromosome
and replicate

These cells may demonstrate
lysogenic conversion

Figure 16—6
The sequence of events in a lysogenic infection.

lysogenic for a particular bacteriophage. Certain strains of *Clostridium bot-ulinum* synthesize botulinum toxin only if they are lysogenic for certain phages. The bacteriophages carry the genes for the synthesis of the toxin. The phenomenon in which a prophage confers on the bacterium the ability to synthesize new materials is termed **lysogenic conversion.**

When prophage DNA is excised from the bacterial chromosome, it can take a piece of the bacterial chromosome with it (just as the excision of the F particle leads to formation of F') (Fig. 16–7). This piece of DNA can then be encapsulated* as part of the bacteriophage and can be transferred when

---

*encapsulated—enclosed by the protein coat of the virus

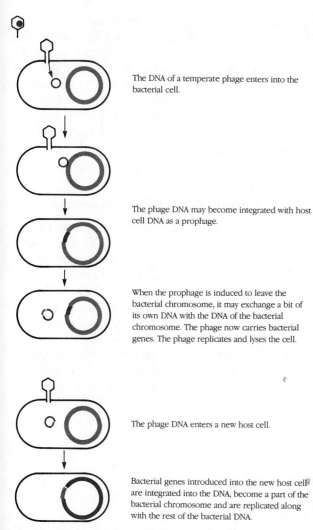

The DNA of a temperate phage enters into the bacterial cell.

The phage DNA may become integrated with host cell DNA as a prophage.

When the prophage is induced to leave the bacterial chromosome, it may exchange a bit of its own DNA with the DNA of the bacterial chromosome. The phage now carries bacterial genes. The phage replicates and lyses the cell.

The phage DNA enters a new host cell.

Bacterial genes introduced into the new host cell are integrated into the DNA, become a part of the bacterial chromosome and are replicated along with the rest of the bacterial DNA.

Figure 16–7
Specialized transduction.

other bacteria are infected by the phage carrying bacterial genes. This process is known as gene transfer by **transduction.** Since only bacterial DNA that lies adjacent to the site of integration of the phage DNA can be transferred, this form of transduction is termed **specialized transduction,** in contrast to **generalized transduction,** in which *any* bacterial gene can be transferred.

Generalized transduction, pages 143–144

BACTERIOPHAGES THAT COEXIST WITH THEIR HOST CELL

Since viruses are inert particles unless they are within living cells, it is advantageous for viruses not to destroy these cells but rather to coexist with them. An example of such coexistence is **lysogeny.** Several other bacterio-phages coexist in a different way. They invade in the same manner as the "T" bacteriophages and utilize the host machinery for their own multipli-cation. However, they do not destroy the bacterial DNA or take over the host cell metabolism completely; rather, they allow the cell to carry out its own life processes. Instead of lysing the cell, the virus particles are continuously extruded* from the cell, which results in some damage to the cell surface. The cells are not killed, although they do not reproduce as vigorously as uninfected cells because some of their energy is diverted to the production of bacteriophages (Fig. 16–8).

VIRUS—HOST CELL INTERACTIONS—ACUTE INFECTIONS (Fig. 16–9)

The typical acute* virus infection in humans leads to the death of the in-fected cell within 12 to 48 hours and results in symptoms of disease. How-ever, in many instances so few cells are destroyed that the infections are subclinical* or inapparent. These infections can be shown to have occurred because antibodies against the virus can be detected in the blood. For ex-ample, only about 10 percent of the population infected with poliovirus show any symptoms involving the central nervous system. Mumps produces symp-toms in only about 50 percent of the people that the virus infects.

The previously described bacteriophage-bacterium association serves as an excellent model for the interaction of viruses and animal cells at the molecular level. However, there are some important additional aspects of the virus–animal cell interaction that are important to consider. (Some of the methods used to study viruses are discussed in Appendix II.)

Adsorption

Animal viruses attach to the host cells by means of specific attachment sites on the virion and corresponding receptor sites on the host cell surface. Whether or not viruses can cause disease most often depends on whether the viruses can attach to the cells of the particular animal species (Fig. 16–9). Sometimes some tissues of a particular species possess the receptors and therefore can be infected, whereas other tissues in the same species do not possess the receptors and so are not susceptible to infection. For example, poliovirus infects only cells of primates* because nonprimate cells do not synthesize the appropriate receptors. In humans, not all cells are infected, since only certain cells have receptors. The susceptible cells include those in the nasopharynx and gut and certain motor nerve cells* of the spinal cord.

---

*extruded—forced through the cell surface without lysing the cell

*acute—short-lived, as opposed to chronic

*subclinical—without symptoms

*primates—the group of mammals that includes monkeys and humans

*motor nerve cells—nerve cells responsible for controlling muscles

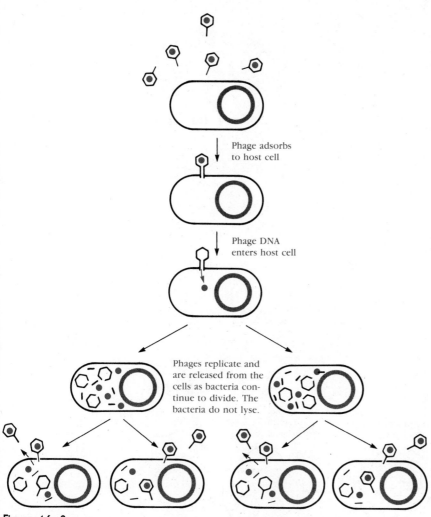

Figure 16—8
Life cycle of bacteriophages that are extruded from their host cell.

Infection of these spinal cord cells causes poliomyelitis, also called infantile paralysis. Some animal viruses can attack a wide range of hosts; other viruses, especially measles and pox viruses, can infect a very broad range of human tissues.

Entrance

In animal systems, complete virions are engulfed into animal cells by phagocytosis, and the uncoating of the envelope and protein coat from the nucleic acid occurs either inside a vacuole* or in the cytoplasm. Some animal viruses

---

*vacuole—a membrane-enclosed cavity in the cytoplasm

In the figure:

Phage adsorbs to host cell

Phage DNA enters host cell

Phages replicate and are released from the cells as bacteria continue to divide. The bacteria do not lyse.

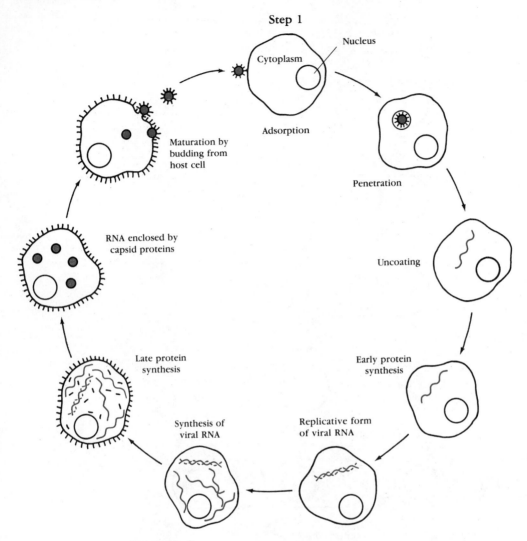

Figure 16–9
The replication of an envelope virus, the influenza virus. Note that the nucleic acid in this virus is RNA rather than DNA.

contain more than one coat or layer, and the process of uncoating may require several steps.

Replication and Maturation

As is the case in phage multiplication, the nucleic acid of the viruses codes only for the synthesis of viral proteins. Some of these proteins are concerned with nucleic acid synthesis and are produced early in the infection process. The structural proteins are synthesized later. As in all cases of virus multiplication, the nucleic acids and proteins are synthesized separately but come together as part of the maturation process to form a complete virion.

Release

Many animal viruses are enclosed by a lipoprotein envelope (Fig. 16–1), which is acquired by the virus at the time of its release from the cell. This envelope is composed of the membrane of the host cell, which coats the virions as they are released (Fig. 16–9). These viruses generally do not kill the cell. Other viruses lyse the cell without acquiring the membrane envelope. These are often released by cell wall disruption, which results in the death of the cell. These viruses have their counterparts in bacteriophages.

VIRAL DAMAGE TO ANIMAL CELLS

In many cases the ultimate result of viral infection of an animal cell is the death of the cell. Cell damage is not just caused by the formation of vast numbers of viral particles (100,000 per cell in the case of poliovirus); the effects of the proteins coded by virus genes on normal cellular processes may be of even greater importance. For example, viral proteins commonly change the permeability of one or more of the cellular membranes. Furthermore, some viral proteins can specifically inhibit host DNA, RNA, or protein synthesis. Infected cells may also develop abnormalities in their chromosome structure.

INTERFERON

A very important response of some virus-infected cells is the synthesis of **interferons,** proteins that interfere with the replication of viruses. This protein is synthesized as a result of viral infection but is coded by the DNA of the infected host. Interferon is released from infected cells and protects neighboring uninfected cells. It acts intracellularly to inhibit viral replication, in contrast to antibody, which acts extracellularly to inactivate virus particles and thereby prevent their entry into cells. Normal cell functions are unaffected by interferon. Most viruses cause the synthesis of interferon, and this response affects other concurrent or subsequent viral infections. The stimulus for the cell to produce interferon appears to be double-stranded RNA synthesized by the virus. Chemically synthesized RNA also has this effect. Because this natural defense mechanism may be artificially induced by administering chemically synthesized RNA, there exists a future possibility for the short-term prevention of viral illness. Indeed, laboratory animals have been protected against some viral diseases by giving them synthetic RNA before or shortly after infection. At this time, extensive clinical testing at medical centers across the United States is aimed at determining whether interferon will be an effective chemotherapeutic agent in the treatment of different types of cancer.

Interferon, pages 219–220

VIRUS—HOST CELL INTERACTIONS—NONACUTE INFECTIONS

In the past, virology has emphasized acute viral infections and the viral agents that cause them. It is now known that the interaction of the same virus with the same host can produce variable outcomes, which may result in a number of different clinical diseases. Thus, viruses that cause acute disease under some circumstances may also cause a prolonged or chronic infection with clinical symptoms quite distinct from those of the acute clinical course. To a certain degree, the clinical course depends on the immunological response of the host cells as well as on variability within viral strains. The long-term interactions of viruses with their host cells can be divided somewhat arbitrarily into three categories: latent infections, chronic infections, and slow virus infections.

Latent Infections

In this situation, symptoms can come and go repeatedly. When the symptoms are absent, the complete virus often cannot be detected. A good example is the herpes simplex virus, the cause of cold sores. The first infection usually occurs in childhood. The virus then infects certain ganglia,* where it remains without causing symptoms. When the virus is activated by such agents as fever and sunburn, it once again causes cold sores. When these have healed, the virus and host cells again exist in apparent harmony until the next recurrence of the disease. The varicella-zoster virus, another herpesvirus, is the cause of chickenpox in childhood. It can remain latent for years and then produce the disease called shingles (herpes zoster) in other children and adults.

It is not known why the latent virus cannot be detected. Perhaps there are no complete virus particles. Alternatively, there may be some viruses present but too few to be measured.

Chronic Infections

In this situation, the virus can be shown to be present even though disease symptoms are not apparent. Infected individuals may act as carriers. Disease may later develop as a result of interaction between the virus and the cells of the host. An example of a chronic infection is lymphocytic choriomeningitis. When mice are infected with the causative virus, the response may range from no symptoms of disease to a severe and acute illness resulting in death within a week.

Slow Virus Infections

Very closely related to latent and chronic infections are slow virus infections. The adjective "slow" refers to the course of the disease and not to the virus.

---

*ganglia—groups of nerve cells

In fact, the same virus may be capable of causing both slow and acute infections. These insidious disorders are characterized by a progressive degeneration to a usually fatal outcome. In recent years it has become clear that certain neurological diseases of both humans and animals result from virus infection. One of these disorders, kuru, was transmitted to cannibalistic women and children of New Guinea as they prepared the brains of dead relatives for cooking. Other degenerative diseases, such as multiple sclerosis and diabetes, may ultimately be shown to be slow virus diseases, although at present there is little evidence to support this theory.

CLASSIFICATION OF VIRUSES

The taxonomy* of viruses has been and continues to be a problem. The ideal goal is to classify viruses into groups based on their evolutionary relationships. However, very little is known about such relationships among viruses, so classification schemes are primarily of practical value for researchers in the field. A committee of virologists has recommended that viruses be subdivided into families and genera, but at this time such a classification scheme is not widely used. Viruses are grouped and named for a variety of reasons, many of which are trivial, including the disease they produce (polioviruses cause poliomyelitis); their mode of transmission (arboviruses require an arthropod vector); the tissue from which they were isolated (adenoviruses were first isolated from adenoids and tonsils); and the geographical site from which they were first isolated (Coxsackie viruses were first isolated in Coxsackie, New York). Appendix VIII lists the important groups of animal viruses and some of their characteristics. The characteristics that are important in classification include the following details of virus structure:

1. The nature of the nucleic acid—DNA or RNA—and whether the nucleic acid is double-stranded or single-stranded. Double-stranded RNA viruses are rare.
2. The shape of the virus particle (see Fig. 16–1).
3. Whether or not an envelope is present (see Fig. 16–1).
4. The size of the virion. They range in size from 18 to 300 nanometers.*

VIRUSES AND CANCER IN HUMANS

For more than 70 years, it has been known that certain viruses can cause cancer* in animals. Tumor-causing viruses have been found in frogs, mice, guinea pigs, rabbits, cats, dogs, and New World monkeys.* Thus, it seems reasonable to expect that viruses will be found to cause at least some cancers

*taxonomy—the science of the classification of organisms

*nanometer (nm)—1 billionth of a meter; 10^{-9} meter

*cancer—group of abnormally growing cells that can spread from their site of origin; also termed malignant tumor

*New World monkeys—monkeys found in Central and South America

in humans. However, even though tremendous efforts have been made and billions of dollars have been spent, there is no absolute proof yet that viruses do indeed cause human cancer.

There is now a strong indication that a type of herpesvirus, called Epstein-Barr (EB) virus after the two scientists who did much of the work on it, may cause a cancer called Burkitt's lymphoma.* The Epstein-Barr virus, like its relatives herpes simplex and varicella-zoster, can infect cells and then go "underground," losing its viral identity and becoming latent. The viral DNA persists in the cells it infects, the B lymphocytes, in two forms. One form is a plasmid, a circular double-stranded DNA molecule that multiplies independently of the chromosome. The other form is the viral DNA integrated into the chromosome of the host cell, where it is replicated along with the rest of the host chromosomes (Fig. 16–10). The latent viruses may become activated in a few cells in the population, and virus particles are then released that can reach the saliva and be spread to other individuals. These virus particles, which are released from cells, are of two types. One type, the **lytic** virus, causes infected cells to synthesize more viruses, similar to the system described for the "T" bacteriophages or animal viruses that cause acute infections. The other type is a **transforming** virus that infects B lymphocytes and causes them to multiply indefinitely when they are grown in tissue culture.* This transformation* of lymphocytes apparently is caused by EB-virus DNA that has become inserted into the chromosome of the lymphocyte. Thus, the virus in its latent state is analogous to the bacteriophage lambda when it lysogenizes its host bacterium. Like lambda, the EB-virus DNA multiplies as the host chromosomes multiply. Also like lambda, the viral DNA can become excised from the host chromosome, resulting in the synthesis of virus particles that lyse the cell.

Although the evidence is highly suggestive that EB virus causes Burkitt's lymphoma, many intriguing questions remain to be answered. Apparently, a high percentage of people around the world are infected with the EB virus, but only people in certain parts of Africa ever develop Burkitt's lymphoma. Factors other than infection with the virus must be required for lymphoma development. The EB virus is the causative agent of infectious mononucleosis, yet epidemiological studies have indicated that people who have had infectious mononucleosis do *not* have a higher probability of contracting Burkitt's lymphoma later in life. Almost everyone becomes infected with the EB virus, but very few ever develop infectious mononucleosis.

VIRUSES AND CANCER IN ANIMALS

Viruses that cause tumors in animals generally transform the cells of the animal when these cells are grown in tissue culture (Fig. 16–11). The nucleic

---

*lymphoma—cancer that arises in the lymph nodes

*tissue culture—culture of animal or plant tissues that grow in laboratory glassware in an enriched medium

*transformation—modification of the properties of cells; not to be confused with DNA-mediated transformation in bacteria

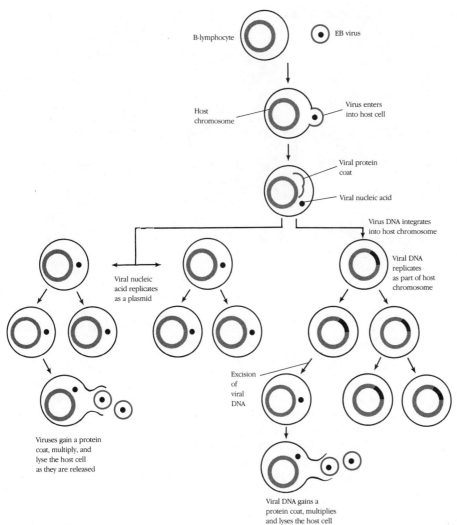

B-lymphocyte

EB virus

Host
chromosome

Virus enters
into host cell

Viral protein
coat

Viral nucleic acid

Virus DNA integrates
into host chromosome

Viral nucleic
acid replicates
as a plasmid

Viral DNA
replicates
as part of host
chromosome

Excision
of
viral
DNA

Viruses gain a protein
coat, multiply, and
lyse the host cell
as they are released

Viral DNA gains a
protein coat, multiplies
and lyses the host cell

Figure 16–10
The replication of the Epstein-Barr virus.

acid of these transforming viruses becomes integrated into the DNA of the
cells they transform. This integrated viral DNA transforms the host cells,
perhaps by coding for enzymes that alter key proteins in the cell. However,
many transforming viruses contain RNA rather than DNA as their genetic
material, and the question arose as to how these viruses transformed the
cells they infected. The question was answered by two Americans, David
Baltimore and Howard Temin. They demonstrated that RNA viruses have
an unusual enzyme, **reverse transcriptase,** which copies the viral RNA into
a complementary strand of DNA. This DNA is then integrated into the DNA
of the host cell.

Exactly how the integrated viral DNA transforms the cell is one of the

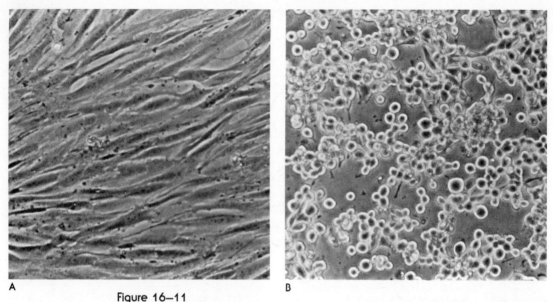

A   B

Figure 16–11
(a) Normal NRK cells (from the rat) growing in tissue cells. (b) NRK cells infected with avian sarcoma virus. This tumor-inducing virus transforms the cells and alters their properties in a number of ways. (a and b courtesy of W. Carter.)

major questions in tumor biology. Evidence is accumulating that the integrated DNA codes for an enzyme or enzymes, which modify key proteins located near the surface of the cell. Thus, some viruses that cause cancer in animals and probably in humans appear to have properties similar to those of the temperate phages. They can lose their viral identity by becoming a part of the host chromosome. However, in this state the viral DNA can code for proteins that cause profound changes in the host cell.

SUMMARY

Viruses are obligate intracellular parasites that require a living host cell in order to replicate. Each virus particle, or virion, consists of either DNA or RNA surrounded by a protein coat. The virion contains few, if any, enzymes and no ribosomes or other organelles required for replication. Consequently, viruses depend on host cells to supply these needs.

Virus replication is unique. Most viruses adsorb to specific receptors on host cells. Penetration of viral nucleic acid into the host cell follows. The various viral components replicate independently of one another and then assemble into complete virions. Mature virions are released from the cell.

The DNA of some viruses can integrate into the DNA of the host cell. In this state the viral DNA replicates as the host DNA replicates. The viral DNA may code for proteins that confer unusual abilities on the host cell, such as the ability to code for toxin synthesis.

Although thus far no virus has been shown definitely to cause human cancer, a herpesvirus—Epstein-Barr virus—which causes infectious mononucleosis is a likely cause of the cancer called Burkitt's lymphoma.

SELF-QUIZ

1. All viruses have one property in common:
   a. They are extracellular parasites.
   b. Only their nucleic acid enters infected cells.
   c. They always kill the cells they infect.
   d. They exist as plasmids inside the infected cell.
   e. The nucleic acid multiplies independently of the protein.
2. Viruses differ from cells in all of the following ways *except:*
   a. A virion has only one kind of nucleic acid; a cell has two.
   b. Viruses are inert outside a living cell; cells grow in the absence of other cells.
   c. Viral nucleic acid and viral protein replicate independently of one another; in cells, they multiply together.
   d. Some viruses code for the enzyme reverse transcriptase; bacteria have no such enzyme.
   e. Viruses do not require ribosomes for protein synthesis; cells do.
3. Specialized transduction is associated with
   a. temperate bacteriophages
   b. virulent bacteriophages
   c. DNA transformation
   d. lysogenic conversion
   e. "T" bacteriophages
4. The most common reason for a cell being resistant to viral infection is that
   a. the cell does not have the receptors to which the virus can adsorb
   b. the proper enzymes may not be present inside the cell
   c. the virus fails to mature properly
   d. the viral DNA is not transcribed by host enzymes
   e. the complete virion is not released by the cell
5. The Epstein-Barr (EB) virus is closely associated with
   a. smallpox
   b. influenza
   c. measles
   d. infectious mononucleosis
   e. polio

QUESTIONS FOR DISCUSSION

1. What explanations might account for latent viral infections?
2. If a virus is recovered from a tumor, what evidence would you want before you could be certain that the virus caused the tumor?

FURTHER READING

Burke, D.: "The Status of Interferon." *Scientific American* (April 1977).

Croce, C., and Kaprowski, H.: "The Genetics of Human Cancer." *Scientific American* (February 1978).

Fenner, F., et al.: *The Biology of Animal Viruses*, 2nd ed. New York: Academic Press, 1974. Authoritative two-volume reference on viruses of warm-blooded animals, including their molecular biology, classification, and ecology and the pathogenesis of the diseases they cause.

Fenner, F., and White, D. O.: *Medical Virology*, 2nd ed. New York: Academic Press, 1976. Concise, up-to-date text covering most aspects of medically important viruses and viral diseases, including persistent infections and viroids.

Henle, W., Henle, G., and Lennette, E.: "The Epstein-Barr Virus." *Scientific American* (July 1979).

Holland, J. J.: "Slow, Inapparent, and Recurrent Viruses." *Scientific American* (February 1974).

Horne, R. W.: "The Structure of Viruses." *Scientific American* (January 1963).

Kaplan, M., and Webster, R.: "The Epidemiology of Influenza." *Scientific American* (December 1977).

Kellenberger, E.: "The Genetic Control of the Shape of a Virus." *Scientific American* (December 1966).

Luria, S., Darnell, J., Jr., Baltimore, D., and Campbell, A.: *General Virology*, 3rd ed. New York: John Wiley and Sons, 1978. A very popular text on virology, covering bacteriophages, animal virology, and a little on plant and insect viruses. The coverage emphasizes molecular aspects of virology.

Rafferty, K. A., Jr.: "Herpes Viruses and Cancer." *Scientific American* (October 1973).

Spector, D. H., and Baltimore, D.: "The Molecular Biology of Poliovirus." *Scientific American* (May 1975).

Watson, J. D.: *Molecular Biology of the Gene*, 3rd ed. Menlo Park, Calif.: W. A. Benjamin, 1976. The most authoritative and current text in molecular biology. Easy to read, with good illustrations and a glossary.

Wood, W. B., and Edgar, R. S.: "Building a Bacterial Virus." *Scientific American* (July 1967).

Chapter 17

THE ALGAE—AN INTRODUCTION TO THE EUKARYOTES

About 300 years ago, a Dutch merchant named Antony van Leeuwenhoek had a hobby of grinding glass lenses. He became quite an expert and actually made the first simple microscope, which could magnify images to almost 300 times their original size. In 1674 he wrote to the Royal Society of London, describing what he had seen in a drop of lake water by using his primitive microscope. He reported that he saw

> . . . very many little animalcules, whereof some were roundish, while others a bit bigger consisted of an oval. On these last, I saw two little legs near the head, and two little fins at the hindmost end of the body. Others were somewhat longer than an oval, and these were slow a-moving, and few in number. These animalcules had divers colours, some being whitish and transparent; others with green and very glittering little scales, others again were green in the middle, and before and behind white; others yet were ashen grey. And the motion of most of these animalcules in the water was so swift, and so various, upwards, downwards, and round about, that 'twas wonderful to see. . . .

This letter represents the first known description of what we now know to be eukaryotic single-celled organisms, the algae and protozoa so abundant in lake water.

OBJECTIVES

To know
1. The basic structure of eukaryotic cells.
2. The major groups of eukaryotic organisms whose members include single-celled organisms.
3. Some important characteristics of algae.
4. Some major groups of algae.
5. Several symbiotic relationships that occur between algae and other kinds of organisms.
6. Some practical uses of algae.

Prokaryotes—the bacteria—are microscopic, and their cells are simply constructed. All organisms other than the bacteria are made up of eukaryotic cells that have a more complex construction, sharing many properties in common with the bacteria but differing in some important ways. Some eukaryotic organisms are also single cells: the algae, fungi, and protozoa. These are often microscopic, but some of them are quite large. For example, the seaweeds are plant-sized algae, some of the kelps reaching 50 feet or more in length, long enough to span the width of a football field. Among the fungi, mushrooms, and tree brackets are familiar examples of large organisms, and some of the protozoa also can be seen with the unaided eye.

The dividing lines between different groups of eukaryotes are not always clear. There is often lively argument and disagreement about the group in which a particular organism belongs. Some important differences between the groups will become apparent in the following chapters.

CELLULAR STRUCTURE AND FUNCTION OF EUKARYOTES

The cellular structure of all eukaryotic cells is basically the same, whether they are single-celled organisms or part of multicellular plants or animals. Thus the cells of microscopic algae, fungi, and protozoa are constructed in much the same way as the cells of humans. Figure 17–1 shows an electron micrograph of a typical eukaryotic cell of the green alga *Chlamydomonas*, along with a diagram to indicate the important organelles.* Figure 2–1 shows an electron micrograph of a typical bacterial cell for comparison with the eukaryotic cell. As is the case with prokaryotic cells, eukaryotic cells must be constructed in order to separate the contents of the cell from its surroundings, to store and reproduce the genetic information of the cell, to synthesize cellular components, and to generate, store, and utilize energy-rich compounds. Since eukaryotic cells carry out more metabolic processes

Figure 2–1, page 23

---

*organelles—characteristic structures, each of which performs a specific function in a cell

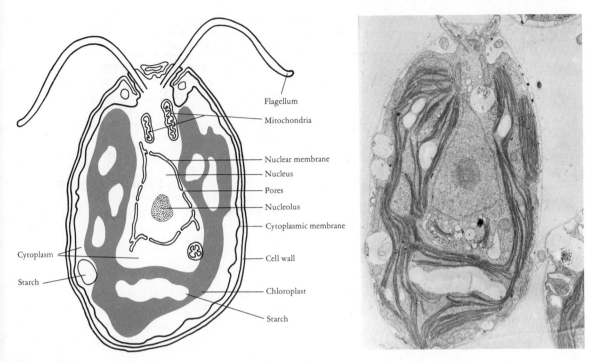

Figure 17–1

Chlamydomonas, a representative green alga. (Photograph courtesy of V. Goodenough; V. Johnson and K. Porter, *J. Cell Biol.,* 38:403, 1968.)

and contain more genetic information than prokaryotic cells, the eukaryotes are considerably larger, as well as being more complex in structure. The important features of eukaryotic cells are described briefly in this chapter. Table 17–1 lists the major components of eukaryotic cells.

Cell Walls and External Materials

Although a great many eukaryotic cells, including those of humans, lack cell walls, many eukaryotes, including the algae and fungi, have characteristic cell walls quite different from those of bacteria. The cell walls of algae contain cellulose and are similar in composition to those of plants. Fungal cell walls are composed largely of polymers of glucose, mannose, or related sugars; some of these polymers (called glucans, mannans, and chitin) occur in an arrangement that when viewed by electron microscopy resembles thatching.

The functions of cell walls are the same in eukaryotic and prokaryotic cells; the walls maintain the shape of the cells and retain the highly concentrated cellular contents. Unfortunately, there are at present no antibiotics useful for attacking fungal cell walls. Penicillin and other antibiotics that act on bacterial cell walls have no effect on fungal cells, which lack the

TABLE 17–1 MAJOR COMPONENTS OF EUKARYOTIC CELLS

Component	Structure and features
Cell wall	Often absent in eukaryotic cells except for plants and protists; when present, composed of cellulose or other polymers of glucose, mannose, or related sugars
Cytoplasmic membrane	Phospholipid double membrane, rich in sterols
Genetic material	DNA organized into pairs of chromosomes and enclosed by a nuclear membrane in a true nucleus
Protein-synthesizing apparatus	Large ribosomes in long strings or attached to membrane material in the cytoplasm
Energy-producing apparatus	Mitochondria or chloroplasts, containing enzymes
Organelles of movement	Flagella or cilia with nine-plus-two arrangement of microtubules

Peptidoglycan, pages 28–29

bacterial peptidoglycan. It is reasonable to assume that antibiotics may eventually be found that are capable of attacking fungal cell walls. There are enzymes available that specifically degrade chemical linkages in fungal cell walls; these enzymes are useful for laboratory studies of the fungi but cannot be used for treating fungal diseases.

Capsules and other layers outside the cell wall or external cell membrane occur in eukaryotic cells, but not to the extent that they do in bacteria. Only one fungal species of medical importance, *Cryptococcus neoformans*, has a prominent polysaccharide capsule, which is essential for its virulence (Fig. 17–2). External polysaccharide materials are visible in electron micrographs of many fungi and protozoa, but most of these are not well characterized. Some of them function in the attachment of the organisms to other cells or materials they colonize.

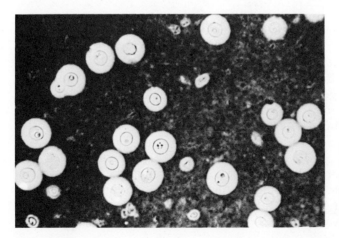

Figure 17–2
Cryptococcus neoformans, India ink preparation showing the polysaccharide capsule. (Courtesy of N. Goodman.)

Cytoplasmic Membranes

Both eukaryotic and prokaryotic cells are enclosed by similar double-layered, protein- and lipid-containing semipermeable cytoplasmic membranes; however, these differ in that those of eukaryotic cells are rich in sterols, a kind of fat molecule that gives them greater strength. Only a few bacteria, those lacking cell walls, have sterol-rich membranes.

Flagella and Cilia

Motile eukaryotic cells usually move by means of either flagella or the shorter cilia. Both cilia and flagella have the same basic structure shown in Figure 17–3, with two central hollow fibrils or microtubules surrounded by nine pairs of similar tubules, all encased within a single membrane. These are notably different from the simpler prokaryotic flagella, which are made up of long filaments of protein.

Cytoplasmic Streaming

The cytoplasm of eukaryotic cells moves about in a manner called **cytoplasmic streaming** that allows for the distribution of materials inside the cell. Bacteria, being much smaller, do not need such a mechanism for moving materials inside the cell, and they lack cytoplasmic streaming.

Mitochondria and Chloroplasts

Within the cytoplasm of nonphotosynthetic eukaryotic cells are found mitochondria, organelles concerned with the generation of energy (see Fig. 2–21). Chloroplasts carry out similar functions in photosynthetic eukaryotic organisms. Mitochondria, which are about the size of bacterial cells, contain DNA, small bacterial-type ribosomes, and many enzymes attached to the inner membrane or free in the fluid inside the mitochondrion (Fig. 17–4). The mitochondria resemble bacteria not only in size but also in that they

Figure 2–21, page 42

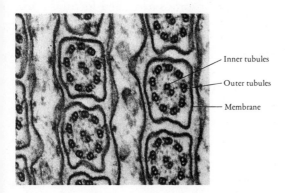

Inner tubules

Outer tubules

Membrane

Figure 17–3
Cross section of eukaryotic flagella from the protozoa *Trichonympha*. Each flagellum consists of nine pairs of outer tubules, two inner tubules (9 + 2 arrangement), and an enclosing membrane. (×65,000) (Courtesy of A. W. Grimstone.)

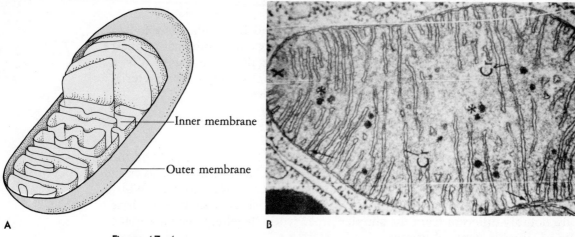

A   B

Figure 17–4

(a) Diagrammatic representation of a mitochondrion. The enzymes are located in two positions. Some are in the fluid region inside and others are attached to the inner membrane, which because of its convolutions has an extraordinarily large surface area. (b) Electron micrograph of a mitochondrion. (Courtesy of J. T. Staley.)

elongate and divide as bacteria do, and their division is independent of cellular division. Thus, many investigators think that bacteria engulfed by eukaryotic cells long ago established a mutually beneficial relationship with the cells and evolved into present-day organelles. This theory has not been proved, however, and is highly controversial. Regardless of their origin, mitochondria are areas of active energy production in eukaryotic cells. Chloroplasts are in many respects similar to mitochondria, but they contain the enzymes and pigments required for photosynthesis (Fig. 17–1).

Ribosomes

Protein synthesis in eukaryotic cells is performed either by long strings of ribosomes held together by molecules of messenger RNA or on the surface of folded membranes with many rows of ribosomes arranged for mass production. Cells that synthesize and secrete large amounts of proteins have their ribosomes positioned in assembly-line fashion on the membranes. Examples are the plasma cells of humans, cells that synthesize and secrete large quantities of protein antibody molecules. The eukaryotic ribosomes are larger than prokaryotic ribosomes. Antibiotics that interfere with protein synthesis as performed by prokaryotic ribosomes are not effective against eukaryotic cells.

True Nuclei

Within eukaryotic cells, but separated from the cytoplasm by a double-layered membrane of its own, is the nucleus. As shown in Figure 17–1, the nuclear membrane has pores that permit materials from the nucleus and the cytoplasm to mingle. Whereas bacteria have their DNA loosely organized into a circular double strand in the cytoplasm, eukaryotic cells have their DNA organized into chromosomes enclosed by the nuclear membrane, and thus the organelle is called a **true nucleus.** Inside this true nucleus, chromosomes occur in pairs, each chromosome made of DNA in a complex with special proteins. Eukaryotic cells may have as few as one pair of chromosomes, as in some of the yeasts, or as many as 23 pairs for a total of 46 chromosomes, as in humans.

Reproduction

The reproduction of eukaryotic cells may occur by either sexual or asexual means. During asexual reproduction, the chromosomes are duplicated and separated into two identical sets by the process called mitosis,* ensuring that each daughter cell gets the same genetic information. During sexual reproduction, the chromosome pairs separate, providing each sex cell (gamete) with only half the usual number of chromosomes, that is, one of each pair, instead of two. Recombination of male and female gametes from different parents gives rise to exchange of genetic information and the production of new individuals. Most of the eukaryotic organisms use both sexual and asexual reproduction at different times.

MAJOR GROUPS OF EUKARYOTIC ORGANISMS IN THE MICROBIAL WORLD

Fungi

The fungi are eukaryotic organisms that lack chlorophyll and are generally nonmotile. Examples are the yeasts, molds, mushrooms, smuts and rusts of grains, and many others, some of which cause disease in humans. Chapter 18 is devoted to this group of eukaryotes.

Protozoa

The protozoa also lack chlorophyll, but most of them are motile. Examples include amebae, paramecia, the parasitic organisms that cause malaria and sleeping sickness in humans, and many others. Chapter 19 is devoted to the protozoa.

Algae

The algae are the chlorophyll-containing eukaryotes. These are photosynthetic eukaryotic cells as contrasted with the prokaryotic photosynthetic cells of photobacteria (Table 17–2). At present, ten major divisions of eu-

---

*mitosis—a process of chromosome duplication and cell division which ensures that each daughter cell will have the same genetic make-up as the parent cell.

TABLE 17–2 SOME COMPARISONS BETWEEN ALGAE AND PHOTOBACTERIA

Algae	*Anaerobic Photobacteria*
Eukaryotic	Prokaryotic
Contain chlorophyll a and other photosynthetic pigments	Contain different pigments
Aerobic	Anaerobic
Produce oxygen as a product of photosynthesis	Do not produce oxygen during photosynthesis

karyotic algae are distinguished by their flagella, by their pigments, and by other biochemical and structural characteristics; however, the classification of algae is controversial, and several different classification schemes are accepted. Undoubtedly, these will change as more information is gathered about these organisms.

Certain pigments are essential for photosynthesis, and a large variety of these are found in algae, including green and yellow pigments. The red and brown algae are so named because of characteristic pigments found in these groups.

Morphology. Algae may be single-celled, or they may be groups of cells collected into colonies or plantlike forms. Color plate 28 illustrates some relatively simple microscopic forms of algae, and more complex macroscopic forms are shown in Figure 17–5 and color plate 29.

Many of the multicellular algae possess holdfasts* that allow them to cling to rocks or other fixed structures in the water. Another morphological feature common to many of the multicellular algae is the presence of gas-containing bladders that help the algae to float in a position suitable for using sunlight. The bladder of the large kelp *Nereocystis luetkeana* (Fig. 17–6) can be 4 to 6 inches long. The gases in the cavity of the float and stalk are air and carbon monoxide (color plate 30).

Microscopic forms of algae also have unique structures, as indicated by the exquisitely formed diatoms shown in Figure 17–7. The diatoms have hard, silicon-containing cell walls constructed in a way that resembles round Petri dishes (centric) or covered boxes (pennates), as illustrated in Figure 17–7. The construction of a diatom covering is diagramed in Figure 17–8. Very large single-celled green algae of the genus *Acetabularia* are illustrated in Figure 17–9. These single cells may grow as tall as 10 cm; thus, they have been very useful in studying the functions of the nucleus and cytoplasmic components of the cell, as indicated in the figure legend.

Physiology. The algae are aerobic photosynthetic organisms that use simple inorganic compounds. In the ocean, factors that limit their growth in-

---

*holdfasts—rootlike structures

A   B

Figure 17—5

Laminaria saccharina (a) and *Sargassum muticum* (b) are examples of some of the large macroscopic species of algae found in the waters of Puget Sound. Note in (b) the small gas bladders that keep this alga floating so as to take full advantage of the sun. (Courtesy of M. Nester.)

clude nitrates, phosphates, silicates, and essential trace constituents. Most important is the availability of sunlight.

All the algae contain chlorophyll a, and other photosynthetic pigments present in the cells usually contribute to the colors of the algae, which range from greens to browns to reds. Pigment constitution is one of the major criteria for classifying the algae, allowing them to be placed into major groups, the phyla. Other physiological characteristics of each group include storage products and the nature of the cell walls.

The pigment constitution determines the wavelengths of light that can be used for algal growth and hence the depths at which algae will grow underwater. Since most light does not penetrate below about 150 feet, algae are usually found in water less than 150 feet deep, and the largest numbers of marine algae are found above the 30-foot depth. However, the latitude and the clarity of the water both influence the extent to which light can penetrate, and in some warm waters a few of the red algae have been found growing

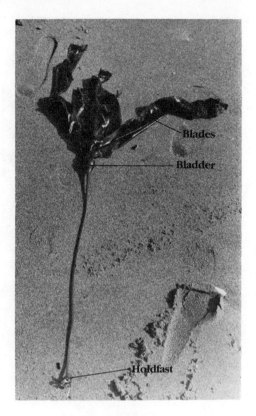

Figure 17—6
Nereocystis luetkeana, the bladder kelp.
This alga has a large bladder filled with
carbon monoxide and air that keeps the
blades of the alga floating on the surface
of the water to expose these surfaces to
sunlight so that photosynthesis can be carried out. (Courtesy of M. Nester.)

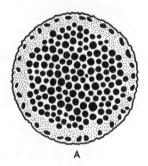

Figure 17—7
Diatoms. (a) Centric. (b) Pennate, with a central
slit (raphe) through which mucilage is extruded.
Diatoms with a raphe exhibit a characteristic
gliding movement.

Figure 17—8

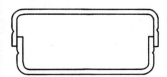

The construction of a diatom frustule. The hard coverings of diatoms (frustules) resemble covered boxes or dishes. When the cell divides, each daughter cell retains half of the parent frustule and subsequently forms another corresponding half.

as deep as 780 feet. With increasing depth, pressure becomes a limiting factor, but the quantity and quality of transmitted light are of primary importance.

Reproduction. Reproduction of algae occurs via the formation of either sexual or asexual spores that can develop into new organisms. The spores may be motile cells without cell walls or wall-enclosed nonmotile cells. Most algae can also undergo vegetative reproduction by cell division or fragmentation, without producing specialized spores.

The modes of sexual reproduction in the algae are many and varied. Some examples are included in the following discussion of specific organisms.

Properties of Groups. Some of the groups of algae overlap with groups of other eukaryotic organisms, but in general the algae are unique because of their ability to carry out photosynthesis, and the photosynthetic pigments are characteristic of the algal groups.

The **green algae** are found in fresh and salt water, in soil, and on plants. Although their chlorophylls predominate to give them a green color, they also contain other photosynthetic pigments called carotenes and xanthophylls. Many of them are similar in appearance to plants, and they even form starch as a storage product, as plants do. No wonder many people refer to them as primitive plants! Most of the green algae are nonmotile, but among the microscopic green algae are motile organisms of the genus *Chlamydomonas* (shown in Fig. 17–1) that have proved extremely useful in genetic studies.

The **brown algae** contain xanthophyll and carotene pigments in sufficient quantities to overwhelm the green chlorophyll and give these algae their brown color. These organisms are multicellular and are usually visible seaweeds. The alga with a gas bladder pictured in Figure 17–6 is an example of a brown alga. Other examples are the *Sargassum* species. Members of the Sargassum family are common along the coasts of Florida and the West Indies. Masses of these seaweeds are swept by storms and currents into the relatively calm center of the Atlantic Ocean and grow in dense masses in the fabled Sargasso Sea. Sexual reproduction is common in these and other brown algae.

Most of the **red algae** also live in the sea, but some are found in fresh water. These algae possess an abundance of red pigments. They store a unique carbohydrate product similar to starch. Their sexual reproduction may be fairly complex. Male and female cells are produced by separate

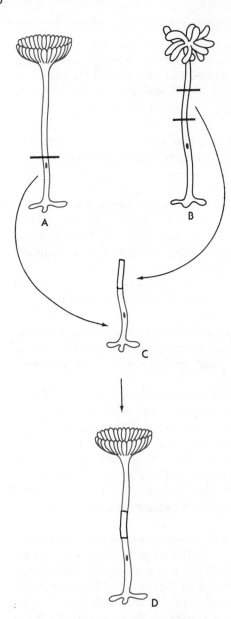

Figure 17–9
Cells of different species of *Aceta-bularia* (a, b) can be grafted together (c). The cap formed by the grafted alga is characteristic of the species from which the nucleus was derived (d). Experiments such as these have shown that the nucleus determines both the ability to regenerate cellular constituents and the nature of the regenerated cell components.

organisms; the nonmotile male cells break off and are carried passively by water currents to the still-attached female cells. Once an egg is fertilized, the resulting zygote produces a number of spore sacs, each containing a single spore.

Dinoflagellates are also primarily marine organisms, plentiful among the phytoplankton.* Most of them are simple unicellular organisms that gen-

*phytoplankton—the floating and swimming algae and prokaryotic organisms of lakes and oceans

erally reproduce asexually. People who live near the sea are familiar not only with the phytoplankton but also with dinoflagellates that cause "red tide" and with the paralytic shellfish poisoning that can be a consequence. The poisoning is caused by a nerve toxin produced by some species of the dinoflagellate *Gonyaulax* and certain other algae. At times when the weather is warm, these algae proliferate excessively and actually color the water red. Molluscs, such as clams and mussels, ingest the toxin-producing algae and concentrate the toxin in their tissues. Eating seafood that is heavily contaminated with the toxin can lead to paralytic shellfish poisoning, which may be fatal. For this reason, beaches are often closed to shellfish collection during periods of red tide.

The diatoms and coralline algae are included in a separate group that comprises algae of both fresh and salt water and some that live in soil and on plants. The **diatoms** belong to one class within this group. The **coralline algae,** belonging to another class of the same group, contain calcium in their cell walls. A third class includes algae that greatly resemble protozoan amebae.

An even closer relationship to the protozoa is observed in flagellated algae of still another group represented by the genus *Euglena*. Many biologists regard these organisms as protozoa capable of photosynthesis, and indeed the cells have many characteristics of protozoa. They lack a rigid cell wall, and they usually divide asexually by binary fission. They are truly a borderline group of organisms, with properties of both algae and protozoa.

Symbiotic Algae. Several algae exist in symbiotic relationships* with other organisms. The coralline algae live together with the corals and are the primary producers of nutrients used to build the coral reefs. By using carbon dioxide for their growth, they help to precipitate calcium in the reefs. Calcified algae first appeared more than 100 million years ago, and they have been important in the formation of limestone.

Lichens are organisms composed of algae and fungi living together in a symbiotic relationship (color plate 31). The algae are the primary producers, utilizing sunlight and carbon dioxide to make carbohydrates and gaining protection and some nutrients from their fungal partners. Lichens are an important source of food in extreme northern climates; for example, the lichen known as reindeer moss supports herds of reindeer. Other lichens are used for human consumption.

Algae can also live in symbiotic relationships with sea animals. The large sea anemone, common in tide pools, gets its green color and many nutrients from algae that live within it.

Some Uses of Algae. Apart from their essential role as primary producers in the food chain, algae are also useful as foods for human consumption.

*symbiotic relationship—intimate relationship between members of two different species under circumstances in which each can benefit from the other

Some of the seaweeds are widely enjoyed, especially by the Japanese, who use a yellow-brown seaweed to make the versatile kombu. This can be cooked and eaten with meat or rice or in soup; it can be served as a vegetable, sweetened as a dessert, or steeped to make a drink. The Japanese also use brown "fir-needle" algae, or matsuma, as food. Some red algae called red laver are collected, dried, and used as food by many different peoples of the Pacific Rim nations, including the Chinese, Chinese-Americans, and American Indians.

Iodine is a nutrient abundant in algae. In former years, large brown algae were used commercially as a source of iodine; however, there are now cheaper ways to obtain iodine in large amounts.

Algae are still the source of agar and sodium alginate, both prepared commercially on a large scale. The alginate, a storage product of brown algae, is used as a stabilizer and to make smooth emulsions in manufacturing many products such as cosmetics, sauces, and ice cream. In fact, 100 gallons of commercial ice cream require nearly a pound of alginate to ensure its creamy, smooth consistency! Fortunately, kelp are efficient producers, supplying as much as 44 pounds of alginate per ton of kelp. Also extensively used as an emulsifier and thickening agent is carrageenan, or Irish moss, from red algae.

The hard, silica-containing coverings of dead diatoms have formed large deposits of diatomaceous earth, used for making filters and as a heat insulator, since it will resist temperatures of 600°C or higher.

SUMMARY

Eukaryotic organisms, some of whose members are single-celled, include algae, fungi, and protozoa. All of these share the typical eukaryotic cell structure. Fungi have cell walls containing polysaccharides such as glucans, mannans, and chitin, and algae have cellulose in their cell walls. Protozoa lack cell walls. All these organisms have similar cytoplasmic membranes, ribosomes, and either mitochondria or chloroplasts concerned with energy production. Motile eukaryotic organisms generally move by means of flagella or cilia, both of which have the typical microtubule structure of nine pairs plus two central fibrils. All these organisms have true nuclei, and they generally use both asexual and sexual means of reproduction.

The fungi and the protozoa are further discussed in separate chapters. The algae are aerobic, chlorophyll-containing photosynthetic organisms, both macroscopic and microscopic. Some algae have evolved symbiotic partnerships with other organisms, including corals, fungi, and the sea anemone. Algae are primary producers in the food chain, and some of the large, multicellular algae are used directly as food. Algae are the source of agar, alginate, and other products.

SELF-QUIZ

1. Eukaryotic cell structure is characteristic of all the following organisms *except*
   a. cyanobacteria
   b. fungi
   c. algae
   d. protozoa
   e. humans
2. Among the eukaryotes, a prominent capsule
   a. is usually present
   b. is usually made of protein
   c. is present in most algal species
   d. is found in many strains of *Cryptococcus neoformans*
   e. never occurs
3. Mitochondria are generally
   a. present in bacterial cells
   b. concerned with cellular reproduction
   c. the site of energy production
   d. about the same size as bacterial plasmids
   e. present in the photobacteria
4. Ribosomes of eukaryotic cells
   a. are the target of action for many antibiotics
   b. are smaller than bacterial ribosomes
   c. function in energy production
   d. are associated with messenger RNA
   e. are usually found free in the nucleus
5. Important properties of the algae include all the following *except*
   a. anaerobic
   b. photosynthetic
   c. contain chlorophyll a
   d. often, but not always, nonmotile
   e. used for producing agar

QUESTIONS FOR DISCUSSION

1. What are some characteristics of algae that help to explain why the algae are not generally pathogenic for humans?
2. Considering the eukaryotic structure of fungi and protozoa, why is it difficult to find antibiotics effective in treating diseases caused by fungi and protozoa?
3. Knowing some basic differences between photosynthetic bacteria and photosynthetic algae, which would you expect to be more efficient in producing energy and oxygen, and why?

FURTHER READING

Carson, R. L.: *The Sea Around Us.* New York: New American Library, 1950. A marine biologist presents a fascinating story of the sea and its inhabitants.

Chapman, V. J., and Chapman, D. J.: *The Algae,* 2d ed. New York: Macmillan, 1973. An excellent introduction to the algae.

Gibor, A.: "Acetabularia: A Useful Giant Cell" *Scientific American* (November 1966).

Isaacs, J. D.: "The Nature of Oceanic Life." *Scientific American* (March 1969).

Prescott, G. W.: *The Algae: A Review.* Boston: Houghton Mifflin, 1968.

Round, F. E.: *The Biology of the Algae.* London: Edward Arnold, 1965. A scholarly, general account of the biology of algae and cyanobacteria.

Shilo, M.: "Formation and Mode of Action of Algal Toxins." *Bacteriological Reviews,* *31:*180–193 (1967). An in-depth review of algal toxins.

Taylor, D. L.: "Algal Symbionts of Invertebrates." *Annual Review of Microbiology,* 27:171–187 (1973). Emphasizes the biology of the algal partner in many symbiotic relationships.

Chapter 18

THE FUNGI

One hot, dusty summer not long ago, a group of scientists and college students worked long and hard in archeological digs near Chico, California. They sifted dirt and looked for additional relics of an early Indian settlement that had been located on that site. As the weeks passed, one person after another became ill with a respiratory disease, until finally a majority of these previously healthy individuals had the infection known in California as San Joaquin Valley fever, or coccidioidomycosis. A fungus growing in the soil is the cause of this illness. How could a mold, adapted to life in dry, dusty soil, enter and live within human lungs? What capabilities might these and other fungi have in their natural environments?

OBJECTIVES

To know
1. The basic structures of yeasts and molds.
2. What is meant by dimorphic fungi and some important examples of dimorphic fungi.
3. The structure of some fungal reproductive spores.
4. The meaning of mycosis.
5. The major kinds of mycoses, according to the tissues involved, and examples of each kind.
6. Examples of toxin and allergic effects of fungi.
7. Important practical functions and uses of fungi.

On the whole, fungi are tremendously useful and beneficial in the living world, despite their having some adverse effects. The more than 200,000 known species of fungi represent one of the most versatile and diverse groups of organisms that exist. They are eukaryotic, lack chlorophyll, and live either as saprophytes* on dead organic materials or as parasites* on living matter. They are abundant in soil, on vegetation, and in many other habitats. Their roles in the production of breads, cheeses, beers, and wines are well appreciated. Some of the fungi, such as mushrooms, serve directly as food, whereas others may spoil food. Fungal abilities to decompose organic materials keep us from being inundated by rubbish. On the other hand, their actions as plant pathogens have changed the course of history on more than one occasion. A striking example is the potato blight, caused by the fungus *Phytophythora infestans*, that ruined the Irish economy in the 1840's. Within a decade, the population of Ireland decreased by three million—one million died, and two million emigrated. The development of fungal products as antibiotics has had an even more profound effect on world history.

The many and diverse genera of fungi can be placed in five classes, based on the kinds of reproductive spores produced or on the lack of sexual spores. These classes and some of their characteristics are summarized in Table 18–1.

MORPHOLOGY

The forms of fungi are varied, ranging from single-celled yeasts or unicellular spores to multicellular molds, mushrooms, tree brackets, and others. The cells are enclosed by cell walls of chitin, glucans, mannans, and other polysaccharide constituents mixed with proteins.

Molds

Molds consist of long, branched tubular filaments called **hyphae** (singular: hypha) that grow in masses of **mycelia** (singular: mycelium). The cottony molds familiar on bread or fruit (color plate 32a) represent the visible mycelia of mold colonies, but the filaments also grow down into the food, making the molds more extensive than they appear. At first glance, the hyphae resemble some of the bacterial filaments, but close inspection shows that hyphae are much wider than bacterial filaments and are often divided into cellular compartments, each containing one or more true nuclei characteristic of eukaryotic cells (Fig. 18–1). Even those fungi without compartmented hyphae contain many nuclei in each hypha. Reproductive spores are produced on the mycelia in large enough numbers to color the mold colonies. For example, black bread mold (*Rhizopus*) gets its color from black spores that form on the white, cottony mycelia, and some species of *Penicillium* mold (color plate 32b) are colored green by huge numbers of mycelial spores.

---

*saprophytes—organisms that live on dead organic material

*parasites—organisms that live on or in another organism and gain benefit at the expense of the host

TABLE 18–1 SOME CHARACTERISTICS OF THE CLASSES OF FUNGI

Classes	Characteristics
Zygomycetes	Hyphae not divided into separate cells Sexual reproduction by oomycetes Spores are flagellated and have no cell walls Cell walls contain cellulose but no chitin Includes downy mildew on grapes and white rust on cabbages and other plants
Zoomycetes	Hyphae not divided into separate cells Cell walls composed of chitin Sexual reproduction by zygospores Asexual reproduction is much more common than sexual reproduction Includes black bread mold *Rhizopus*
Ascomycetes	Hyphae divided into separate cells Sexual spores formed in a sac called an ascus Asexual reproduction common, with formation of conidia spores Includes yeasts, morels, truffles, and many common molds
Basidiomycetes	Hyphae divided into separate cells Sexual spores formed on a club-shaped structure called a basidium Includes mushrooms, smuts and rusts of plants
Deuteromycetes (Fungi Imperfecti)	Hyphae divided into separate cells Cannot be classified by their sexual spores because their sexual stages are unknown Includes many fungi of medical importance

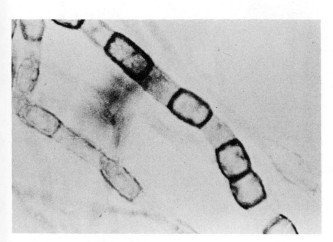

Figure 18–1
The septated hyphae and arthrospores of *Coccidioides immitis.*

Yeasts

Yeasts are unicellular fungi that reproduce asexually by budding (Fig. 18–2) or fission. The buds that separate from the mother cells are called blasto-spores, and each can grow into a new fungal organism. Yeast colonies resemble some bacterial colonies. They are often smooth and creamy in consistency.

Dimorphic Fungi

The versatile fungi are not always limited to existence as a single form, either yeast or mold; many of them, especially those capable of causing diseases in humans, can grow in more than one form, depending upon environmental conditions. They can change from one to the other form in response to temperature, nutrients, and other environmental shifts. These organisms are called **dimorphic fungi**. Examples are shown in Figure 18–3. The spores of certain fungi that grow in the soil as molds are readily carried in the air and can be inhaled; in the different environment of the lung, the spores develop into characteristic tissue forms.

METABOLISM AND GROWTH

Since fungi lack chlorophyll, they cannot carry out photosynthesis, but they produce many degradative enzymes and thus can utilize a tremendous variety of organic substances for growth and energy. They thrive on organic wastes and in soils enriched by organic wastes, and by so doing they play a vital role in the recycling of materials. Fungi exist that can degrade virtually any organic material, even leather and the proteins of skin, hair, and nails.

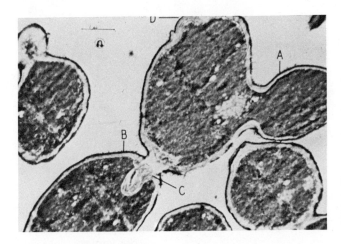

Figure 18–2
Electron micrograph of a longitudinal section through budding yeast cells.

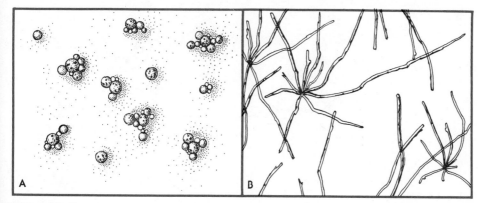

Figure 18—3
An example of dimorphism in fungi. This organism can exist in more than one form.

Some fungi are halophilic* and can live in high-salt media, for example, on cured hams or in brine; others can live in media having a high sugar content. The high sugar content in preserves or jams discourages or prevents bacterial growth, but it cannot stop the growth of some of the molds. Fungi grow either aerobically or as facultative anaerobes, and they can grow over a wide temperature range, many of them preferring room (or even cooler) temperatures. As anyone who leaves food in the refrigerator for long periods of time knows, some fungi thrive at refrigerator temperatures. Yet other species are thermophilic* and grow well at temperatures higher than 45°C. Compost and hay often contain large numbers of thermophilic fungi.

When a fungal spore reaches a suitable substrate* in a favorable environment, it germinates and begins to grow (Fig. 18–4). For example, an airborne spore of a *Penicillium* species may reach a ripe orange, rich in sugars and other organic materials. Perhaps it finds an invisible break in the orange peel, and there it germinates, undaunted by the acidity of the environment. Indeed, most fungi prefer acid conditions for growth. A filamentous hypha pushes from the single-celled spore. As the hypha grows, it branches and forms other hyphae, which grow together in a tangled mass. Some of the hyphae push down into the orange, others grow up into the air (aerial hyphae), and reproductive spores are formed on these aerial hyphae. Enzymes secreted by the mycelium degrade the pulp of the orange and even the orange peel into nutrients that can be absorbed and utilized by the fungus. After a time, mature spores are released from the mycelium and carried by air currents to other susceptible substrates.

Similarly, the yeasts responsible for producing wines from grape juice are found in the waxy film of microbes growing on fresh grapes. There they compete with other yeasts and molds and with bacteria. When juice is

---

*halophilic—capable of growing in a high-salt (NaCl) medium

*thermophilic—capable of growing at high temperatures

*substrate—surface on which an organism will grow

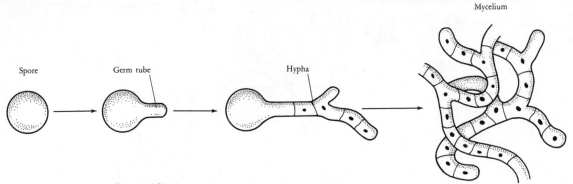

Figure 18—4
Spores of fungi germinate by forming a projection from the side of the cell (a germ tube), which elongates to form hyphae. As the hyphae continue to grow, they form a tangled mass called a mycelium.

pressed from the grapes under favorable conditions, the appropriate *Saccharomyces* yeast begins to bud and grow, using sugar and other nutrients available in the grape juice. Of course, one of the by-products of this growth is ethyl alcohol, but many other more subtle metabolic by-products contribute to the taste of the wine.

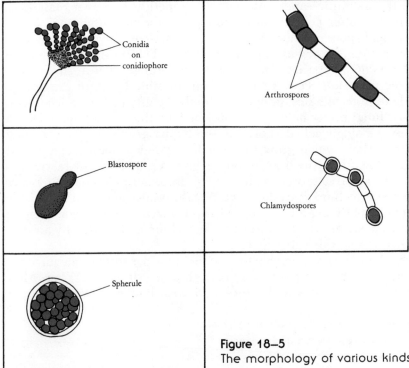

Figure 18—5
The morphology of various kinds of fungal spores.

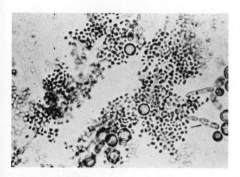

Figure 18–6
Candida albicans showing the chlamy-dospores (large round circles), which are highly resistant to adverse conditions.

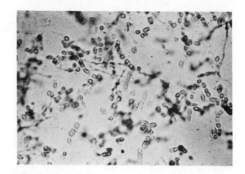

Figure 18–7
Arthrospores of *Coccidioides immitis* (the asexual spores).

REPRODUCTION

Fungi reproduce by either sexual or asexual reproductive spores (Fig. 18–5). The spores formed by sexual processes, following the union of two fungal cells and the exchange of genes, are often carried in unique structures designed to disseminate the spores at an appropriate time. Both sexual and asexual spores and single-celled and multicellular spores can occur on the same fungi. The characteristic spores are often essential in identifying fungal species, and sexual spores are important determinants for categorizing fungi into classes.

Some of the spores are also resistant to unfavorable conditions, such as lack of nutrients or adverse temperatures. However, these are reproductive spores, and they are not nearly as resistant as the endospores formed by certain bacteria. Among the asexual spores of fungi particularly adapted as resting spores are chlamydospores, formed by thickening and rounding of cells in the hyphae (Fig. 18–6).

Other asexual spores are blastospores of budding yeasts, described previously, and arthrospores, produced by the breaking apart of cells in the hyphae, as illustrated for *Coccidioides immitis* (Fig. 18–7). Arthrospores of *C. immitis* infected the archeology students and led to the production of San Joaquin Valley fever, as described in the chapter opening.

Conidia are spores that form at the tip of specialized branches of the hyphae, rather than being formed directly from hyphal cells as are the chlamydospores, arthrospores, and blastospores. Some conidia are single-celled and small and are called microconidia; other conidia, large and multicellular, are called macroconidia (Fig. 18–8).

MEDICALLY IMPORTANT FUNGI

Fungi cause diseases called **mycoses,** and often the name of the disease begins with the name of the causative fungus. Thus, histoplasmosis, a disease of worldwide distribution, is a mycosis caused by the fungus *Histoplasma cap-*

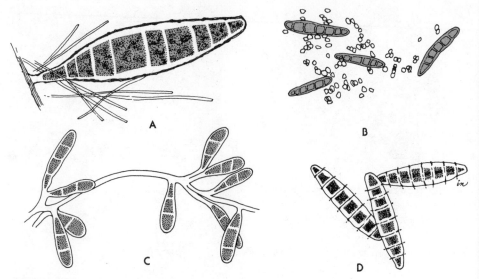

Figure 18–8
Some examples of microconidia and macroconidia.

sulatum. Similarly coccidioidomycosis is a mycosis caused by *Coccidioides immitis*, a fungus unique in certain areas in the American Southwest. Diseases caused by the yeast *Candida albicans* are called candidosis by the British; however, Americans and many others break the general rule and call these diseases candidiasis. By either name, diseases resulting from *C. albicans* are among the most common mycoses. Other mycoses are listed in Table 18–2.

Notice that the mycoses described in Table 18–2 fall into two general groups, according to the tissues involved: the systemic mycoses, in which fungi infect deep tissues and organs, often systemically in many locations; and the intermediate mycoses, in which the infection is usually limited to the respiratory tract or the skin and subcutaneous* tissues. These two categories often overlap. A third category of mycoses, in which the fungi infect only the hair, skin, or nails of humans (or the horns of some animals as well), is known as the superficial mycoses (Table 18–3).

Systemic and Intermediate Mycoses

Some fungi, such as *Histoplasma capsulatum* and *Coccidioides immitis*, cause infection in most normal, nonimmune people who are exposed to them. In areas where *C. immitis* grows as a mold in the soil, more than 90 percent of the resident population will give evidence of having been infected with *C. immitis*. Only about half of these individuals have any symptoms of disease as a result of infection, and a very small percentage will suffer from a severe

---

*subcutaneous—below the skin

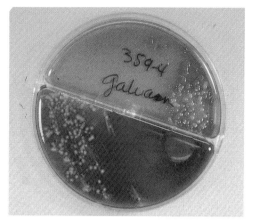

COLOR PLATE 25 A culture of *Proteus morganii*, similar to *Proteus mirabilis*, shows swarming growth on blood agar and the growth of individual colonies on MacConkey agar. (Courtesy of S. Eng.)

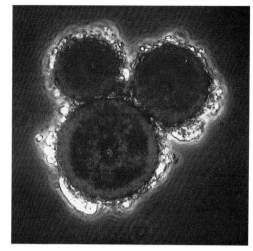

A

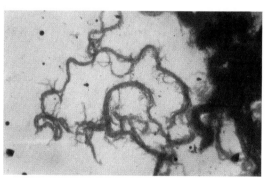

COLOR PLATE 26 Cording in *Mycobacterium tuberculosis*. (×1000) (Courtesy of C. E. Roberts.)

B

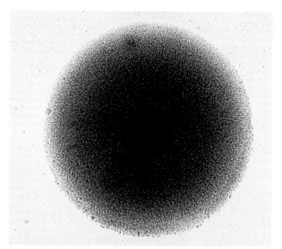

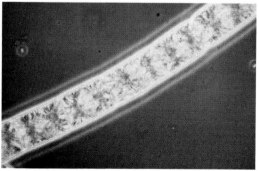

C

COLOR PLATE 28 Examples of microscopic algae. **A.** *Mycanthococcus,* **B.** *Nostoc.* **C.** *Zygnema.* (Courtesy of J. T. Staley.)

COLOR PLATE 27 Classic "fried-egg" appearance of a single colony of *Mycoplasma arthritidis* after 10 days' growth on agar. Crystal violet/methylene blue stain. (Courtesy of Gabridge.)

A

B

COLOR PLATE 29 **A.** Various macroscopic algae found on the beaches of Puget Sound, Washington. Note the variety of colors and shapes of the different species of algae. **B.** The alga *Fucus distichus* found on the rocks near the high-tide level in Puget Sound, Washington. (**A** and **B** courtesy of M. Nester.)

A

B

C

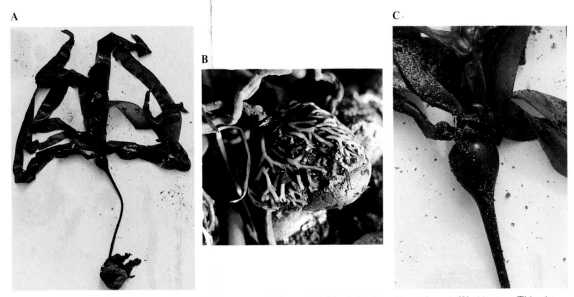

COLOR PLATE 30 **A.** *Nereocystis luetkeana* (bladder kelp) from Puget Sound, Washington. This picture shows an entire young kelp in early spring. **B.** The holdfast of an old bladder kelp. Note how strongly it is attached to the rock. These kelp are annuals, meaning that they grow to their enormous size in just one season. **C.** A close-up photograph of the bladder of a bladder kelp. The bladder, filled with air and carbon monoxide, holds the blades near the surface of the water so that the kelp can utilize the sun in photosynthesis. (**A, B,** and **C** courtesy of M. Nester.)

A

B

COLOR PLATE 31 **A.** Lichens found growing on the branches of a cherry tree. (Courtesy of M. Nester.) **B.** Lichens found in Alaska. Note the variety of shapes and colors. (Courtesy of E. Morgan.)

A

B

COLOR PLATE 32 **A.** *Rhizopus* mold growing on bread. The asexual spores are the black dots at the ends of the colorless branches (mycelia). **B.** A *Penicillium*-type mold on a crust of bread. The asexual spores at the ends of the mycelium are green and give a green appearance to the area infected by the mold. (**A** and **B** courtesy of M. Nester.)

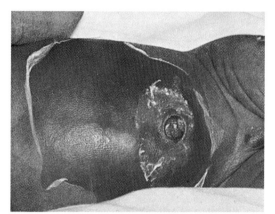

COLOR PLATE 33 Staphylococcal scalded skin syndrome (SSSS).

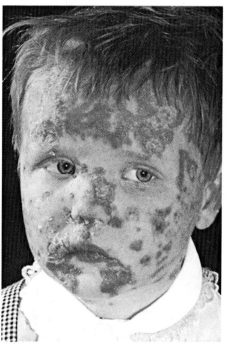

COLOR PLATE 34 Streptococcal impetigo.

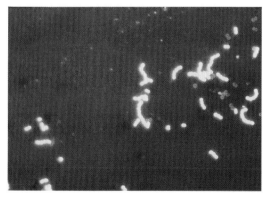

COLOR PLATE 35 Chains of *Streptococcus pyogenes* from a fluid medium culture. The organisms have been allowed to react with specific antibodies, which have been dyed with fluorescein to make them visible. (Courtesy of C. E. Roberts.)

Note: Color plates 33, 34, and 36 are courtesy of Korting, G. W.: *Hautkrankheiten bei Kindern und Jugendlichen.* Copyright 1969 and 1972 by F. K. Schattauer Verlag GmbH, Stuttgart.

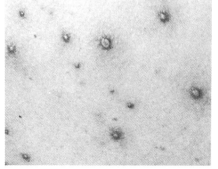

COLOR PLATE 36 Varicella (chickenpox).

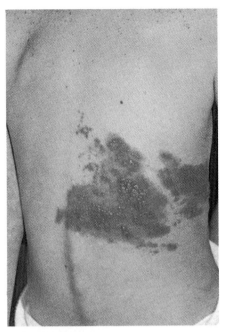

COLOR PLATE 37   Herpes zoster.

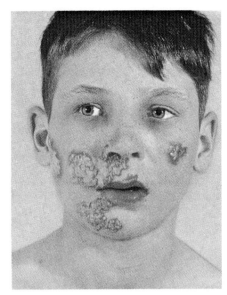

COLOR PLATE 38   Herpes simplex.

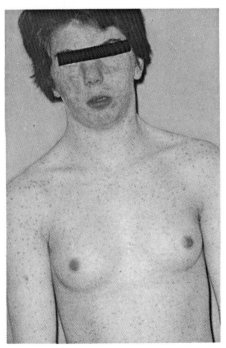

COLOR PLATE 39   Rubeola (measles rash).

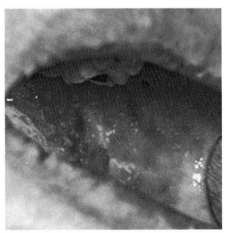

COLOR PLATE 40   Involvement of the mucous membrane of the mouth in rubeola. The white spots surrounded by redness are known as Koplik spots.

Note: Color plates 37, 38, 39, and 40 are courtesy of Korting, G. W.: *Hautkrankheiten bei Kindern und Jugendlichen.* Copyright 1969 and 1972 by F. K. Schattauer Verlag GmbH, Stuttgart.

TABLE 18–2 CHARACTERISTICS OF SOME SYSTEMIC AND INTERMEDIATE MYCOSES

Mycosis	Type of disease	Causative fungus	Characteristics
Coccidioidomycosis	Systemic[a]	Coccidioides immitis	Often asymptomatic, usually respiratory infection, rarely generalized Dimorphic fungus, limited to certain geographic areas, especially in SW United States
Histoplasmosis	Systemic	Histoplasma capsulatum	As above, but geographic location of fungus is widespread
Blastomycosis (North American)	Systemic	Blastomyces dermatitidis	Skin, lung, or generalized infections Dimorphic fungus, widespread Disease common in Africa, as well as North America
Cryptococcosis	Systemic	Cryptococcus neoformans	Asymptomatic or mild respiratory infection Occasionally causes severe meningitis and disseminated infections Yeast, widespread Forms prominent capsules when growing in the human body
Candidiasis (candidosis)	Intermediate[b] or systemic (opportunistic)	Candida species, especially C. albicans	Mucous membrane infections most common, esp. in vagina and oral cavity; skin infections also common Disseminated, generalized infections in immunologically deficient subjects

Table continues on the following page

TABLE 18–2   CHARACTERISTICS OF SOME SYSTEMIC AND INTERMEDIATE MYCOSES
(Cont.)

Mycosis	Type of disease	Causative fungus	Characteristics
Sporotrichosis	Intermediate[b]	*Sporothrix schenckii*	Chain of subcutaneous ulcers along a lymphatic vessel Chronic disease Disseminated disease rare

[a]Systemic disease = disease in deep organs or disseminated.
[b]Intermediate = mucous membrane or subcutaneous infections.

mycosis. A similar effect is seen in response to many of the disease-producing bacteria. The reasons for these differences in response to infection are considered in Chapter 10.

Some fungi regularly have the potential to infect and often cause disease, whereas most fungi seldom infect and rarely cause disease in normal people. However, if immune defenses are lowered for any reason, almost any organism may be able to invade and cause serious or fatal disease. Such infections are called opportunistic infections. Table 18–4 gives examples of some opportunistic mycoses.

Opportunistic, page 209

Candida albicans is found among the normal oral and gastrointestinal flora of humans and is thought to invade from these sites as an opportunist.

TABLE 18–3   CHARACTERISTICS OF SOME SUPERFICIAL MYCOSES[a]

Mycosis	Characteristics
Ringworm (tinea) of the scalp and hair	Most common in children Causative mold often transmitted from one child to another
Ringworm (tinea) of the smooth skin	Often transmitted from cats, dogs, or other animals to people
Athlete's foot	More common in adults than in children and occurs more often in men than in women Usually worse in hot, moist conditions
Tinea of the nails	Caused by the same fungi that cause athlete's foot, but occurs most often in 40- to 50-year-old women
Tinea versicolor	Caused by a special fungus (*Malassezia furfur*) that lives only in outer layers of the skin Lesions are different color from the skin (e.g., dark spots on white skin)

[a]Molds of three different genera cause most cases of superficial mycoses, with much overlap between genera causing the same diseases. The three genera are *Microsporum*, *Epidermophyton*, and *Trichophyton*, all of the Fungi Imperfecti (class Deuteromycetes).

TABLE 18–4   SOME COMMON OPPORTUNISTIC MYCOSES

Mycosis	Causative fungus	Factors predisposing to infection
Candidiasis (candidosis)	Candida species, especially C. albicans	Hormonal imbalances, e.g., pregnancy, oral contraceptive therapy, parathyroid abnormalities, diabetes, and others Immunosuppression caused by cancer or other diseases Suppression of normal bacterial flora by intensive antibiotic therapy Many other factors
Aspergillosis	Aspergillus species, especially A. flavus	Cancer and other immunosuppressive diseases or treatments
Cryptococcosis	Cryptococcus neoformans	As above
Phycomycosis	Species of Phycomycetes	As above

Other, more virulent disease-producing fungi, such as *H. capsulatum* and *C. immitis*, are soil molds whose airborne spores cause respiratory disease when they are inhaled. Note that some of the same fungi included in Table 18–1 are found in Table 18–4. A good example is *Candida albicans*. This yeast is of low virulence, yet very slight aberrations in the balance between the host and yeast in the normal flora can allow the organism to invade and cause disease; thus, it is an opportunist. Even physiological changes such as pregnancy can upset the balance. Some other conditions predisposing to Candida infections are discussed in Chapter 24. Other fungi in Table 18–4, for example, the common molds of the genus *Aspergillus*, infect only when host defenses are more seriously suppressed; therefore, these are not included in Table 18–1.

Because of the eukaryotic structure of fungal cells, the antibacterial antibiotics are useless in treating mycoses. However, several antifungal agents that are selectively toxic for fungi are available. An important antifungal antibiotic is amphotericin B, produced by species of the bacterial genus *Streptomyces*. This antibiotic binds to sterol-containing cytoplasmic membranes; therefore, it is also toxic for human cells. As a result of its dangerous side-effects, amphotericin B is used only for treating life-threatening or very serious mycoses.

Superficial Mycoses

The fungi that cause superficial mycoses are molds that are transmitted from animals to people or from person to person. Some of these molds live in the soil; all of them are able to degrade keratins, the major proteins of hair, skin, and nails, and to use keratins as nutrients. Many of the superficial mycoses are familiar to most people, since they occur worldwide and are very common. They include ringworm, athlete's foot, and other skin diseases known collectively by the name **tinea.** Table 18–2 includes the common names of some of these maladies, along with the names of the causative molds. The fungi grow well in warm, moist areas of the body, and the diseases are usually worse during hot, humid weather.

Superficial mycoses are not life-threatening diseases, since these fungi are not able to invade past the superficial layers of the body. Rather, they tend to be chronic* and difficult to cure. They are characterized by itching skin lesions, which can sometimes be disfiguring. Usually they are self-limiting* after a time, often months or years. They can be treated with antifungal medicines, often available without prescription, that are applied externally and frequently produce good results.

Severe cases of superficial mycoses can often be treated successfully with the antibiotic griseofulvin, produced by another fungus, *Penicillium griseofulvum.* This antibiotic is fungistatic; that is, it inhibits the growth of fungi but does not kill them. Therefore, the antibiotic must be taken orally over a period of weeks or months until the affected tissue containing the fungi is removed, until hair or nails grow out and are cut off, or until external layers of the skin are shed, removing the inhibited fungi.

Toxic and Allergic Effects of Fungi

Medical mycology is largely restricted to the study of pathogenic fungi because they are the infectious ones; however, it also properly includes a study of fungi that are toxic or that cause allergic reactions in people. Throughout history it has been known that some fungi produce compounds that profoundly affect the body. The hallucinogenic properties of certain mushrooms have long been used as a part of religious ceremonies in some cultures. The lethal effects of many mushrooms and toadstools have also been known for centuries. The poisonous effects of a rye smut called ergot were known during the Middle Ages, but only centuries later was the active chemical purified from the fungus to yield the drug ergot, used medicinally to stop uterine bleeding and for other purposes.

Allergic diseases can result from exposure to fungi to which an individual has become sensitized. For reasons discussed in Chapter 11, immediate allergic reactions of the hayfever or the asthma type can result from inhalation of the fungi, and sometimes severe, chronic allergic lung disease results from immune-complex or other forms of allergic reactions.

Some fungi produce toxins that appear to contribute to the development of cancer. The best-studied of these carcinogenic toxins, produced by species

---

*chronic—of long duration; prolonged; lingering

*self-limiting—do not continue to spread

of *Aspergillus*, are called **aflatoxins.** There is good evidence that ingestion of aflatoxins in moldy foods, such as grains and peanuts, is associated with the development of a type of liver cancer, or hepatoma. However, infection with hepatitis B virus is also associated with hepatoma. Much research is now directed toward learning what role the fungal toxins play in the development of these liver cancers.

PRACTICAL USES OF FUNGI

Some of the many practical uses of fungi were considered in the earlier sections of this chapter. A few others are worthy of special attention.

Fungi are the source of several very important antibiotics. They are the original source of penicillin, griseofulvin, and other antimicrobial agents.

Certain fungi form symbiotic relations with tree roots. These mycorrhizas* are essential for the growth of some trees. For example, the pine tree industry in Puerto Rico was about to perish before the proper fungi were introduced to form mycorrhizal relationships with the trees; now the industry is flourishing. Mycorrhizas also make it possible for useful trees to grow on marginal soils, and thus they assist in environmental restoration of mining operations.

Fungi can be used as biological control agents to replace some of the toxic chemicals that have been used in past years to control insect pests. They are vitally important as commercial sources of enzymes and other organic products.

The fungi have been quite useful tools for genetic and biochemical studies. *Neurospora crassa*, a common mold, became famous as a model system for modern genetic studies, and some of the yeasts provide model systems for studying biochemical reactions.

SUMMARY

Fungi are nonphotosynthetic eukaryotic organisms that are either saprophytes or parasites. They include molds, yeasts, mushrooms, rusts and smuts of plants, and many other species of varying morphology. Some fungi of medical importance occur in more than one form, usually as a mold in the soil and as a yeast or other tissue form in the human host, and thus they are known as dimorphic fungi.

The fungi can degrade and use many organic substances. They are important recyclers of organic wastes. They reproduce by either asexual or sexual reproductive spores.

Fungal diseases or mycoses of human hosts may be systemic or intermediate infections, such as histoplasmosis, coccidioidomycosis, and candidi-

---

*mycorrhizas—symbiotic associations between certain fungi and plant roots

asis, caused by fungi that are pathogenic or others that are of low virulence but are opportunistic. Other mycoses, caused by certain molds, are superficial, being limited to infections of the hair, skin, or nails, for example, ringworm and athlete's foot. Some fungi can cause disease by producing toxic or allergic effects. Examples are mushroom poisoning, some forms of asthma and other allergic lung diseases, and even certain cancers associated with fungal toxins called aflatoxins.

The fungi are a source of some important antibiotics and are useful participants in alcohol and food production, symbiotic relationships, and nontoxic forms of biological control.

SELF-QUIZ

1. The fungal mycelium is
   a. an important part of yeasts
   b. anaerobic
   c. enclosed by a mucopeptide-containing cell wall
   d. characteristic of molds
   e. important in sexual reproduction of fungi
2. A medically important dimorphic fungus is
   a. *Cryptococcus neoformans*
   b. *Histoplasma capsulatum*
   c. *Saccharomyces cerevisiae*
   d. *Malassezia furfur*
   e. *Aspergillus flavus*
3. Characteristics of fungi include all of the following *except* that
   a. most disease-producing species have capsules
   b. saprophytic or parasitic species occur
   c. they reproduce sexually or asexually, or both
   d. many species are plant pathogens
   e. they germinate from single-celled spores
4. *Histoplasma capsulatum* and *Coccidioides immitis* share the following characteristics:
   a. They are common soil molds in Ohio.
   b. Infections are uniformly fatal unless treated.
   c. Infections respond well to griseofulvin.
   d. They are dimorphic, pathogenic fungi.
   e. They infect via the oral route.
5. Superficial mycoses generally
   a. are transmitted via airborne spores
   b. affect the lungs
   c. are acute infections
   d. occur mainly in children
   e. are caused by molds

QUESTIONS FOR DISCUSSION

1. What properties of molds common in the environment, for example, *Rhizopus* bread mold, probably account for the fact that these molds are not pathogenic?
2. What kinds of conditions would be expected to permit common molds to act as opportunists?
3. Is it likely that vaccines will soon be available against disease-producing fungi? Why or why not? Should the Public Health Service be encouraging the development of fungal vaccines?

FURTHER READING

Ainsworth, G. C., and Sussman, A. S.: *The Fungi. An Advanced Treatise*, vols. I and II. New York: Academic Press, 1965.

Booth, C.: "Form and Function. II. Fungi." In Norris, J. R., and Richmond, M. H. (eds.): *Essays in Microbiology*. New York: John Wiley & Sons, 1978. A general treatment of the fungi. Concise.

Christensen, C. M.: *The Molds and Man: An Introduction to the Fungi*. Minneapolis: University of Minnesota Press, 1951.

Emmons, C. W., Binford, C. H., and Utz, J. P.: *Medical Mycology*, 3d ed. Philadelphia: Lea and Febiger, 1977. The third edition of a classic text in mycology.

Human Mycoses. Scope monograph. Kalamazoo, Michigan: The Upjohn Company, 1968. A well-illustrated manual of the major fungi responsible for human diseases.

McKenny, M., and Stuntz, D. E.: *The Savory Wild Mushroom*. Seattle: University of Washington Press, 1971. Scientifically accurate and esthetically pleasing, this well-illustrated book was written to enable the mushroom hunter to answer the question, "Is it good to eat?"

Myrvik, Q. N., Pearsall, N. N., and Weiser, R. S.: *Fundamentals of Medical Bacteriology and Mycology*. Philadelphia: Lea and Febiger, 1974. The last four chapters present an abbreviated overview of medical mycology.

Wilson, J. W., and Plunkett, O. A.: *The Fungous Diseases of Man*. Berkeley and Los Angeles: University of California Press, 1965. An excellent work concerned with human diseases caused by fungi.

Chapter 19

THE PROTOZOA

On the African savanna, a hunter surveyed the landscape looking for an antelope. Absentmindedly, he slapped a fly that lit on his shoulder. A drop of blood oozed from the small fly bite—an apparently insignificant event. In fact, however, the small tsetse fly has hindered humans from making inroads into many parts of Africa, thereby preserving much of the diverse animal life native to that region. The tsetse fly plays an important part in the life cycle of the trypanosome that causes African sleeping sickness. This trypanosome has been successful in keeping humans from inhabiting its environment because humans are unable to make antibodies for this protozoan. On the other hand, the ungulate population (antelopes, for example) is able to tolerate the protozoan and acts as a reservoir for the infection. The trypanosome cannot live in the more temperate zones of the world.

OBJECTIVES

To know
1. The morphology of the four basic groups of protozoa.
2. Some information about protozoan physiology.
3. How the protozoa reproduce both sexually and asexually.
4. The major pathogenic protozoa and the diseases they cause.

The protozoa are eukaryotic microorganisms; a few species are large enough to be seen with the unaided eye. In contrast to most fungi and algae, the other eukaryotic microbes, species of protozoa are characteristically

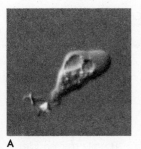
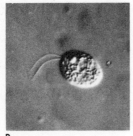
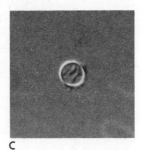

A   B   C

Figure 19–1
Polymorphism in a protozoan. This species of *Naegleria* may infect humans. In human tissues the organism exists in the form of (a) an ameba (10 to 11 μm in its widest diameter). After a few minutes in water the flagellate form (b) appears. Under adverse conditions a cyst (c) is formed. (Courtesy of M. Roth and F. Schoenknecht.)

motile, and with few exceptions they do not carry out photosynthesis. Species of protozoa often have complex life cycles and differ in form at different stages of the cycle; these species are called **polymorphic*** protozoa.

The ability to exist either in a vegetative form* (the **trophozoite***) or in a resting form (the **cyst***) is another characteristic of many protozoa (Fig. 19–1). Certain environmental conditions trigger the development of a protective cyst wall, within which the cytoplasm becomes dormant. Cysts provide a means for dispersal of the protozoa and survival under adverse conditions. When the cyst encounters a favorable environment, the trophozoite again emerges.

Some of the protozoa appear to be fairly complex organisms containing organelles that carry out specialized functions. However, as with other protists, each protozoan cell is a complete unit, capable of reproducing itself. A majority of protozoa are free-living and are widespread in nature, but many species are parasitic for plants or animals, including humans.

MORPHOLOGY

The protozoa are divided into four major groups, largely according to their morphology and modes of motility (Table 19–1), although there are other methods of classification in use. Some scientists consider these groups to be subphyla and classes, while others call them subdivisions and classes. In this discussion, the general term "class" will be used for these four major groups. The protozoa range from amebae of the class Sarcodina, to the ciliated Ciliata, to the flagellated organisms Mastigophora, and finally, to

*polymorphic—having different distinct forms at various stages of the life cycle

*vegetative form—the growing or feeding form, as opposed to the resting form

*trophozoite—the vegetative form of some protozoa

*cyst—dormant resting cell form characterized by a thickened cell wall

TABLE 19–1   SOME PROPERTIES OF THE MAJOR CLASSES OF PROTOZOA

Class	Movement	Reproduction	Other
Sarcodina	Pseudopods (some also have flagella)	Binary fission	Some species have hard coverings (tests) or internal skeletons
Mastigophora	Flagella	Longitudinal fission	
Ciliata	Multiple cilia	Transverse fission; also sexual reproduction involving micronuclei	Macronucleus and usually micronuclei
Sporozoa	Often nonmotile; may have flagella or creeping motility	Multiple fission; asexual reproduction in one host; some species reproduce sexually in a second host	

spore-forming organisms of the class Sporozoa, many of which have complex life cycles. Although the morphology varies considerably among the groups, each group has characteristic modes of moving about, at least during some stages of the life cycle.

All protozoa have the cell structure typical of eukaryotic cells. Membrane-bound lysosomes and other vesicles* are typically found in the cytoplasm. The lysosomes contain digestive enzymes; other vesicles are concerned with the uptake or excretion of materials, storage of food, or other functions. When cilia or flagella are present, they have the eukaryotic arrangement of two central fibrils surrounded by nine pairs of peripheral fibrils (Fig. 17–3). Many protozoa have **pellicles,*** outer coats exterior to the cytoplasmic membranes, and some have an internal skeleton composed of silicon compounds, which gives the organisms a characteristic shape.

Figure 17–3, page 359

Sarcodina

Amebae (class Sarcodina) use pseudopods for movement. These extensions of cytoplasm are thrust out from the rest of the cell, which then flows into the pseudopod, thus allowing the whole organism to move forward (Fig. 19–2). Since amebae lack an outer pellicle, the cytoplasmic membrane is able to move freely, giving the cell great plasticity.

Some other members of the class Sarcodina have hard coverings called **tests**, or solid skeletal structures within the cell made of deposits containing calcium, silicon, or strontium compounds. Even these rigid organisms carry

---

*vesicles—small membrane-enclosed sacs

*pellicles—thickened, elastic, nonrigid cell coverings

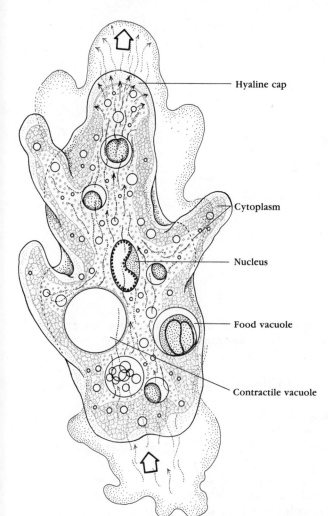

Hyaline cap

Cytoplasm

Nucleus

Food vacuole

Contractile vacuole

Figure 19–2
Ameboid movement in a
typical species of an ameba.
Arrows indicate direction of
movement.

out ameboid movement by extending pseudopods through or around the
solid structures, as indicated in Figure 19–3.

Ciliata

Paramecium species are examples of the class Ciliata (Fig. 19–4). The outer
pellicle is more apparent in paramecia than in some other protozoa. Extending through the membrane, but still covered by the pellicle, are numerous cilia. They originate from basal granules in the cytoplasm. These basal
granules are connected by fibers that are thought to help coordinate the
movement of the multiple cilia, although this has not been proved.

Ciliates commonly have a large nucleus (macronucleus) and one or more
small nuclei (micronuclei) (Fig. 19–4). The macronucleus is concerned with

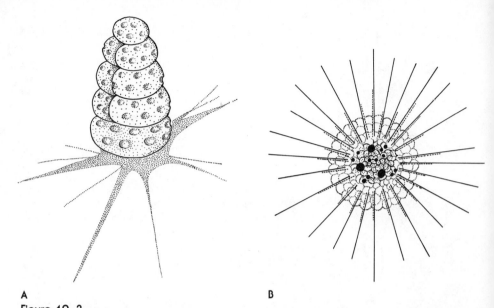

A
B

Figure 19–3
(a) Protozoan form with a multichambered shell, characteristic of some foraminifera. (b) A freshwater heliozoan. The marine radiolarians have forms similar to the heliozoan.

programming most cellular activities, and the micronuclei function in sexual reproduction.

Some ciliates, and also some flagellates, have organelles called **trichocysts** under the body surface within the cytoplasm (Fig. 19–5). There are different kinds of trichocysts, all of which can be "fired off" or ejected from the cell. Some of them contain toxic materials that can immobilize or kill prey as large as some of the smaller metazoans.* Other trichocysts act as anchors. After being ejected, the tip of the trichocyst adheres to a surface and remains attached to the protozoan cell by a long filament. This kind of anchoring mechanism could permit the organism to stay near a suitable food source without being washed away.

Mastigophora

The Mastigophora are usually ovoid or elongated cells with one or more flagella. These flagella resemble cilia in cross section but are longer and move in a different manner. They are attached to cytoplasmic bodies as the cilia are and sometimes to an undulating membrane on the cell surface (Fig. 19–6).

A well-known example of this class of protozoa is the species *Trypanosoma*

---

*metazoans—multicellular animals that show tissue differentiation

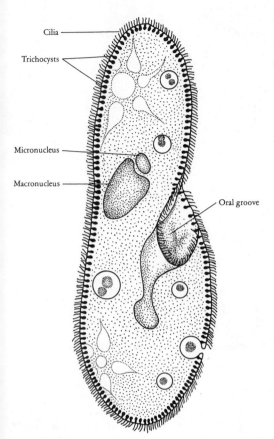

Cilia

Trichocysts

Micronucleus

Macronucleus

Oral groove

Figure 19–4
The morphology of *Paramecium*. The trichocysts are dartlike structures that can be ejected as a defense mechanism.

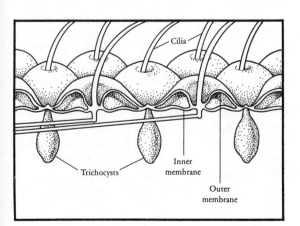

Cilia

Trichocysts

Inner membrane

Outer membrane

Figure 19–5
A cross section of the pellicle of a *Paramecium* as seen by electron microscopy. The many cilia emerge through the inner and outer membranes of the pellicle. Undischarged trichocysts extend below the pellicle.

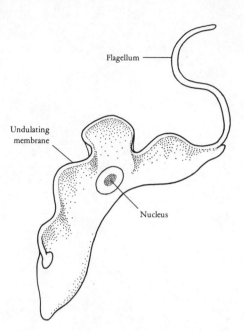

Flagellum

Undulating
membrane

Nucleus

Figure 19–6
A trypanosome as it appears in the human bloodstream. (Courtesy of A. Frazier.)

gambiense, which along with closely related species is the causative agent of African sleeping sickness.

Sporozoa

Members of the Sporozoa are parasitic, polymorphic protozoa that form spores at some stage of development. The parasites that cause malaria are examples of Sporozoa; the diagram of their life cycle in Chapter 28 (Fig. 28–2) shows some of the various forms of these parasites.

Figure 28–2, page 566

PHYSIOLOGY

Most protozoa require oxygen, although some protozoa live under anaerobic conditions in environments such as the human intestine or the rumen of animals. Aerobic protozoa have many mitochondria containing the enzymes necessary for aerobic metabolism and the production of ATP through the transfer of electrons and hydrogen atoms to oxygen.

Protozoans are adept at ingesting foods, and they can efficiently regulate their osmolarity* and excrete materials. Soluble substances can be either absorbed through the cell membrane or **pinocytized** (Fig. 19–7) by the formation of a tiny vacuole around the molecules, bringing them into the interior of the cell. Larger, particulate substances are ingested in various ways, depending upon the nature of the organism. The amebae and related pro-

---

*osmolarity—osmotic pressure within the cell

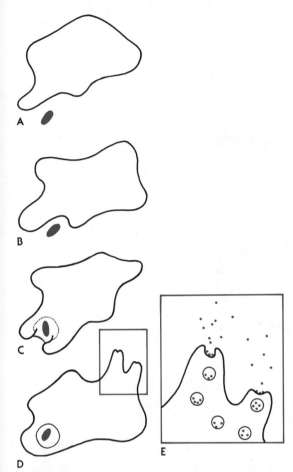

Figure 19–7
Phagocytosis and pinocytosis by an ameba. Phagocytosis (a through d) entails the engulfment and ingestion of particulate material. Pinocytosis (e), an enlargement of part of the cell shown in (d), is a similar ingestion of soluble molecules.

tozoa **phagocytize** particles by forming vacuoles similar to but larger than pinocytotic vesicles, which contain the nutrients (Fig. 19–7). Inside the amebae, these vesicles fuse with lysosomes that contain degradative enzymes. This mixture of the enzymes with food particles results in the degradation of the particles to readily utilized nutrients.

The ciliates, flagellates, and sporozoa usually have permanent mouthlike cavities called **cytostomes** that serve specifically for ingesting foods, and these species have developed interesting means of getting food into the cavity. They may beat their cilia or flagella in such a way as to produce a current that guides food particles toward the cytostome. Some protozoa produce netlike structures that trap food particles around the cytostome. Certain ciliates that feed on other kinds of ciliates even have feeding tentacles with harpoonlike organelles inside (Fig. 19–8). These harpoonlike organelles will react with specific prey organisms, but not with other organisms, when suitable target ciliates collide with the tentacles. The harpoonlike organelles are released and attach to the target cell, a portion of the pellicle of the target

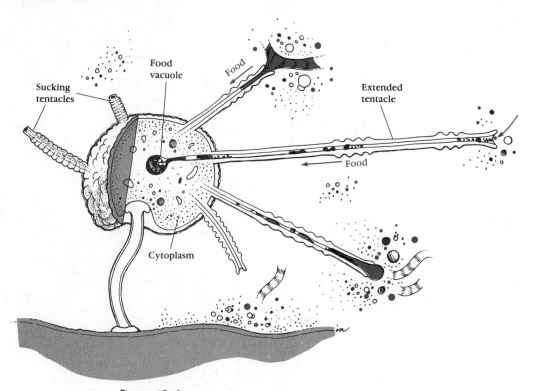

Figure 19—8
Harpoon organelles in ciliates. The organisms can use these harpoons for gathering food or attaching themselves to structures.

cell is digested, and the tentacles act as straws through which the nutrients of the hapless target cell are sucked into the attacking protozoan.

Contractile vacuoles of various protozoa differ in their numbers and locations, but they all release their contents outside the cell. Their primary function is to discharge water and other materials from the cell and to regulate the osmolarity of the cytoplasm. It is possible that many waste materials are excreted in this manner, but there is little evidence to support this concept. In a number of protozoa, wastes from nitrogen metabolism are known to be disposed of as diffusible ammonia, rather than excreted through contractile vacuoles.

REPRODUCTION

Both asexual (the cell simply divides or otherwise reproduces as haploid cells) and sexual (two cells unite, then reproduce as diploid cells) reproduction are common in the protozoa, and these modes of reproduction may alternate. The method of division of protozoan cells is another means, in addition to cell morphology, of classifying the organisms.

Binary fission occurs in many groups, but in the flagellates it usually occurs longitudinally, and in the ciliates it occurs transversely. Since some protozoa possess both flagella and cilia, their method of binary fission helps classify them into one group or the other. Of course, as in any biological system, there are examples that do not fit neatly into either category; for instance, some protozoa divide by binary fission slantwise across the cell!

While many protozoa divide by binary fission, which results in the production of two organisms from one, other protozoa reproduce asexually by **multiple fission,** in which one cell gives rise to multiple daughter cells simultaneously. This occurs in some of the flagellates, in *Entamoeba histolytica* (an ameba parasitic for humans), in sporozoa (such as malarial parasites), and others.

The alternation of sexual and asexual processes of reproduction is well exemplified by malarial parasites, *Plasmodium* species. Multiple fission of the asexual forms in the human host results in large numbers of parasites being released into the circulation at regular intervals to produce the characteristic cyclic symptoms of malaria. Some of the plasmodia develop into specialized sexual forms. The feeding mosquito takes these sexual forms into its intestine. The drop in temperature from 37°C in the human body to the cooler temperature of the cold-blooded mosquito triggers the maturation of the male sex cells and the initiation of sexual reproduction by the parasite in the mosquito host.

Sexual reproduction in paramecia has been studied extensively. These ciliates, which have both a macronucleus and micronuclei, undergo conjugation. Mating types, analogous to male and female, recognize each other, and the organisms fuse to exchange genetic information in a fairly complex manner (Fig. 19–9).

PROTOZOA PATHOGENIC FOR HUMANS

Protozoa representative of all the major classes can parasitize humans and cause disease. Some common examples of pathogenic protozoa are included in Table 19–2.

As indicated in the table, several protozoa can infect the human gastrointestinal tract and cause disease there. One of the most widespread of the intestinal protozoa is the flagellate *Giardia lamblia* (Fig. 19–10), which infects humans and other animals. Numerous epidemics of infections with *G. lamblia* (giardiasis) have occurred among travelers, especially to the Soviet Union and those who go to Southeast Asia, and in rural communities in the United States and Canada. The symptoms of this disease are usually chronic and mild and include diarrhea, nausea, indigestion, flatulence, fatigue, and weight loss. The infecting organism attaches to the lining of the small intestine by means of a "sucker." Even though it does not actually invade the tissues, it causes an inflammatory reaction and can interfere with the absorption of nutrients. Diagnosis of giardiasis is usually made by identifying the cysts or trophozoites of *G. lamblia* in the feces or in fluids of the small

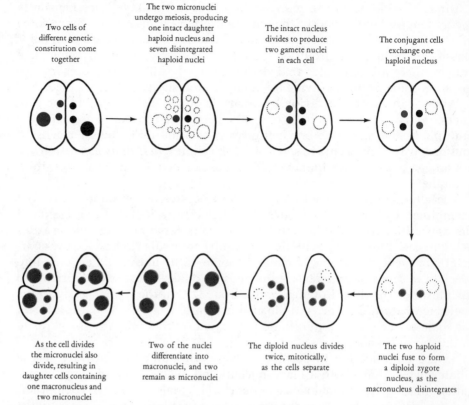

Two cells of different genetic constitution come together

The two micronuclei undergo meiosis, producing one intact daughter haploid nucleus and seven disintegrated haploid nuclei

The intact nucleus divides to produce two gamete nuclei in each cell

The conjugant cells exchange one haploid nucleus

As the cell divides the micronuclei also divide, resulting in daughter cells containing one macronucleus and two micronuclei

Two of the nuclei differentiate into macronuclei, and two remain as micronuclei

The diploid nucleus divides twice, mitotically, as the cells separate

The two haploid nuclei fuse to form a diploid zygote nucleus, as the macronucleus disintegrates

Figure 19–9

Reproduction in *Paramecium* species can occur by the process of conjugation. Cells of different genetic constitution are indicated by the color of the micronuclei. The daughter cells produced after conjugation contain a mixture of genes from each of the original conjugating cells.

intestine. Sometimes a biopsy* of the small intestine is necessary. Surface water contaminated by the feces of wild animals or humans is the most common source of giardiasis, but person-to-person spread has occurred in nurseries.

Another flagellate of even more widespread occurrence is *Trichomonas vaginalis,* but this member of the class Mastigophora infects the genital tract rather than the gastrointestinal tract. It most often causes a mild infection of the vagina.

Perhaps the most important of the protozoa producing intestinal infections in humans is *Entamoeba histolytica* (Fig. 19–11). This ameba, one of the causes of dysentery, is found worldwide, and it is especially prevalent in areas where there is poor sanitation. Encysted *E. histolytica* organisms are

*biopsy—the surgical removal of a small amount of tissue for microscopic examination

TABLE 19–2   SOME EXAMPLES OF PROTOZOA PATHOGENIC FOR HUMANS

Class, genus, and species	Disease produced
Mastigophora	
Giardia lamblia	Chronic or acute gastroenteritis
Trichomonas vaginalis	Vaginitis
Leishmania species	Leishmaniasis: Kala-azar; espundia; chiclero ulcer
Trypanosoma species	Trypanosomiasis: sleeping sickness; Chagas' disease
Sarcodina	
Entamoeba histolytica	Intestinal amebiasis; liver or lung or other abscess
Naegleria species	Meningoencephalitis
Sporozoa	
Plasmodium species	Malaria
Toxoplasma gondii	Toxoplasmosis; infection of the reticuloendothelial system, especially in the fetus or newborn
Pneumocystis carinii	Pneumocystosis (plasmacellular pneumonia)
Ciliata	
Balantidium coli	Diarrhea and dysentery

usually ingested in contaminated food or water, and once in the intestine they liberate trophozoites. Upon reaching the upper portion of the large bowel, these trophozoites begin feeding on mucus and cells lining the intestine. The amebae also produce several digestive enzymes that aid them in penetrating the lining cells into the intestinal wall. In fact, they may burrow into blood vessels and be carried to the liver or other body organs. Continued multiplication and tissue destruction in the intestine and in other body tissues result in the formation of abscesses. The irritant effect of the amebae

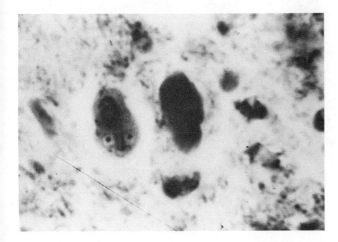

Figure 19–10
Giardia lamblia. A stained fecal specimen.

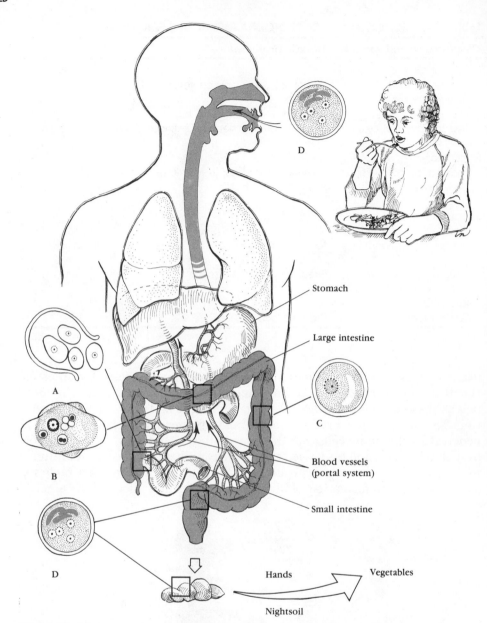

Figure 19–11

Life cycle of *Entamoeba histolytica*. Infectious quadrinucleate cyst enters the mouth on contaminated food and passes through the stomach to the lower small intestine. Four daughter protozoa (A) are released from each cyst and develop into the feeding form (B). Dehydration of intestinal contents in the lower portion of the large intestine stimulates progressive stages in cyst development (C, D). Mature cysts (D) are passed in the feces to contaminate soil, water, and hands. Trophozoites that burrow into blood vessels of the intestine can be carried by the portal system to the liver and cause abscesses.

on the intestinal lining cells results in increased movement of the bowel and production of fluid. Thus diarrhea—often tinged with blood—is a symptom of amebic dysentery. Direct examination of diarrheal fluid on a warm microscope slide may reveal the motile trophozoites of *E. histolytica*, which generally range from about 20 to 40 μm in diameter. Many people develop a chronic infection that may be asymptomatic and therefore may go unnoticed. The trophozoites of *E. histolytica* are not usually seen in examination of the feces of such patients, but cysts of the organism commonly are present. On passage through the large bowel, the cysts typically develop four nuclei, and in this form are infectious for the next host. The cysts of *E. histolytica* can be identified by direct microscopic examination of the feces of infected persons, or, if only small numbers are present, they can be concentrated by various laboratory methods from fecal material. This technique aids in their detection.

Entamoeba histolytica can be grown anaerobically in pure cultures or aerobically in fluid cultures containing bacteria. The organisms grow better when certain bacteria are added to the medium, and a mixture of added bacteria gives better amebic growth than does a single bacterial species. Indeed, in studies with some germ-free animals, *E. histolytica* has caused little or no injury to the intestinal lining unless bacteria were also introduced. In the treatment of people with amebic dysentery, antibiotics active against bacteria are often effective against the disease even though they have no direct activity against the amebae. This indicates that a synergistic* action of amebae and bacteria is required to produce amebic dysentery.

Among the Sporozoa, the most important pathogens are undoubtedly the malarial parasites. Two other species, *Toxoplasma gondii* and *Pneumonocystis carinii*, are important opportunists worthy of mention.

The Ciliata are of least importance as human pathogens. Only one species, *Balantidium coli*, is thought to cause gastrointestinal disease, and the pathogenicity of this species is questionable.

SUMMARY

Protozoa are eukaryotic microorganisms that are generally unicellular, motile, and nonphotosynthetic. They can often exist in several forms and thus are polymorphic. Most can live in a vegetative trophozoite state or in a resting form as cysts.

The four major groups of protozoa are Sarcodina (amebae and others), Ciliata (the ciliates), Mastigophora (the flagellates), and Sporozoa (spore-forming organisms). These classifications are based largely on a number of morphological characteristics and on the mode of division of the protozoa. The morphology is typical of eukaryotic cells and unique in a number of

*synergistic—cooperative

respects; for example, some protozoa possess trichocysts, which carry out various functions.

The metabolism of protozoa is often aerobic but may also be anaerobic. Specialized means of feeding are often observed. Contractile vacuoles for excretion regulate the osmolarity of the cytoplasm. Both sexual and asexual modes of reproduction occur in protozoa, often with an alternation of the modes in a given species.

Species of protozoa from all four major groups are pathogenic for humans. The various species can cause many different kinds of human diseases, some of them, including malaria, leishmaniasis, and trypanosomiasis, being extremely common in some parts of the world.

SELF-QUIZ

1. All of the following are true of trichocysts *except* they
   a. are found in some ciliates
   b. are found in most amebae
   c. can be ejected from the cell
   d. may be an anchoring mechanism
   e. may immobilize or kill prey
2. A flagellate that often infects the genital tract and is found in urine is
   a. *Giardia lamblia*
   b. *Entamoeba histolytica*
   c. *Trichomonas vaginalis*
   d. *Toxoplasma gondii*
   e. *Balantidium coli*
3. All of the following are true of *Entamoeba histolytica* except
   a. The trophozoite stage is infective.
   b. It forms both trophozoites and cysts.
   c. Cysts may contaminate food or water.
   d. Infection may be chronic and long-lasting.
   e. Trophozoites can phagocytize food.
4. *Toxoplasma gondii* is characterized by being
   a. an ameba
   b. an important saprophytic sporozoan
   c. an important opportunistic organism
   d. the only pathogenic ciliate
   e. an aerobic flagellate
5. Gastrointestinal disease is caused by
   a. *Giardia lamblia*
   b. *Trichomonas vaginalis*
   c. *Leishmania* species
   d. *Trypanosoma* species
   e. *Plasmodium* species

QUESTIONS FOR DISCUSSION

1. The soil contains large numbers of protozoa that are not pathogenic for humans. What properties of protozoa and of humans account for this?
2. What might be the outcome if protozoa lacked the ability to form cysts?
3. Protozoa often exist in a number of different developmental forms. How could this affect attempts to produce vaccines against them, for example, an antimalarial vaccine?

FURTHER READING

Chen, T. T., ed.: *Research in Protozoology*. Vol. I to IV. New York: Pergamon Press, 1967–1971.

Corliss, J.: "Systematics of the Phylum Protozoa." In Florkin, M., and Scheer, M. (eds.): *Chemical Zoology*, vol. 1. New York: Academic Press, 1967, pp. 1–20.

Curds, C. R., and Ogden, C. G.: "Form and Function—IV. Protozoa." In Norris, J. R., and Richmond, M. H. (eds.): *Essays in Microbiology*. Chichester: John Wiley and Sons, 1978, pp. 1–32.

Curtis, H.: *The Marvelous Animals—An Introduction to the Protozoa*. New York: Natural History Press, 1968.

Dogiel, V. A.: *General Protozoology*, rev. ed., Translated by G. I. Poljansky and E. M. Chejsin. London: Oxford University Press, 1965.

Garnham, P. C. C.: *Malaria Parasites and Other Haemosporidia*. Oxford: Blackwell Scientific Publishers, 1966.

Hall, R. P.: *Protozoa—The Simplest of All Animals*. New York: Holt, Rinehart and Winston, 1964. A small paperback containing clearly written, basic information.

Jahn, T. L., and Jahn, F. F.: *How to Know the Protozoa*. Dubuque, Iowa: William C. Brown, 1949.

Kidder, G. W., ed.: "Protozoa." In Florkin, M., and Scheer, B. T. (eds.): *Chemical Zoology*, vol. 1. New York: Academic Press, 1967.

Kudo, R. R.: *Protozoology*, 5th ed. Springfield, Ill.: Charles C Thomas, 1966.

Pitelka, D. R.: *Electron Microscopic Structure of Protozoa*. New York: Pergamon Press, 1963.

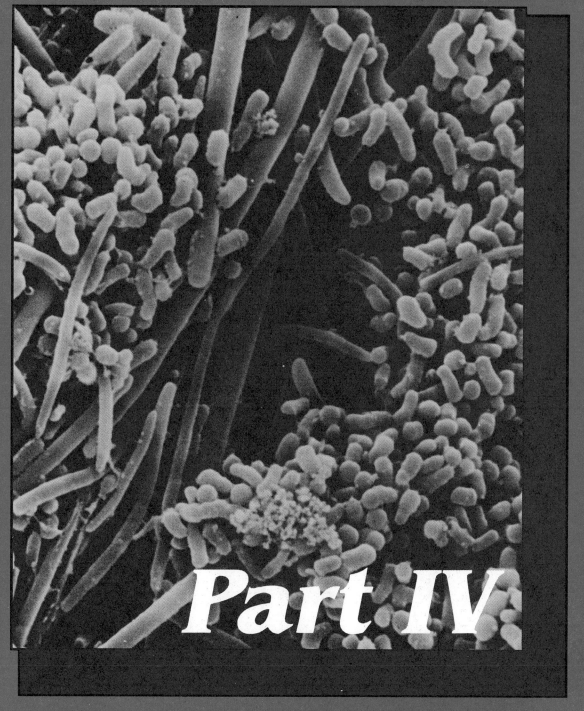

Part IV

INFECTIOUS DISEASES

Chapter 20

SKIN INFECTIONS

It had been a tough match. He had won, of course, but had not pinned his opponent. He had felt his abraded knees where the coarse texture of the wrestling mat had torn away at the outer layers of his skin, as it had done to so many contestants before him.

But why now, three and a half months later, did he think of that wrestling match as he stared at the warts just beginning to appear on his otherwise healthy knees?

OBJECTIVES

To know
1. The characteristics of skin that make it resistant to infection.
2. The kinds of microorganisms that normally inhabit the skin.
3. The principal bacterial causes of skin infection.
4. Three important viral diseases that involve the skin.
5. The fungi that often cause skin infections.

The major part of the body's contact with the outside world occurs at the surface of the skin. As long as it is intact, this tough, flexible outer covering is remarkably resistant to infection. But because of its exposed state, it is frequently subject to cuts, punctures, burns, or chemical injury, as well as to hypersensitivity reactions (allergies). As a result of such injuries, the skin and subcutaneous tissues frequently become infected. In addition, microorganisms and viruses carried by the bloodstream after entering the body through the respiratory or gastrointestinal system often infect the skin from within. Examples are the viruses that cause measles and chickenpox.

ANATOMY AND PHYSIOLOGY

The surface layer of the skin is composed of flat, scalelike material made up of cells containing **keratin,** a durable protein also found in hair and nails (Figs. 20–1 and 20–2). The more superficial cells of this layer are dead and continually peel off to be replaced by cells from a deeper layer. These in turn become flattened and die as keratin is formed within them. From the micro-

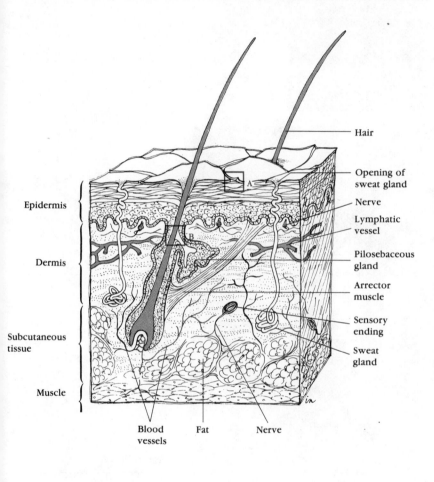

Hair

Opening of
sweat gland

Nerve

Lymphatic
vessel

Pilosebaceous
gland

Arrector
muscle

Sensory
ending

Sweat
gland

Epidermis

Dermis

Subcutaneous
tissue

Muscle

Blood
vessels

Fat

Nerve

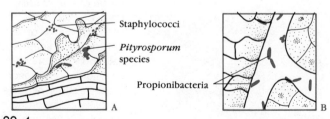

Staphylococci

Pityrosporum
species

Propionibacteria

Figure 20–1
Microscopic anatomy of the skin.

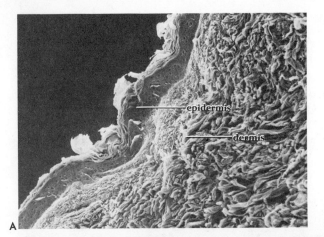

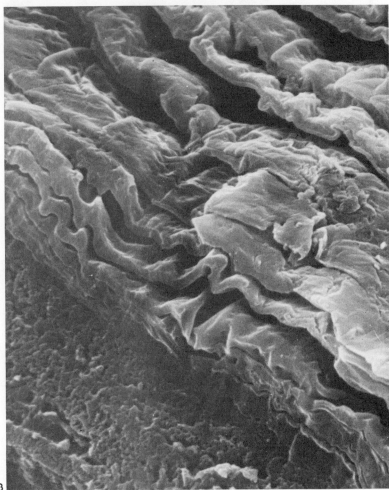

Figure 20–2
(a) Scanning electron micrograph of a cross section of human skin. The outermost layer of the epidermis consists of flat keratin containing scalelike cells that are no longer living. (b) High magnification of the cross section of superficial scalelike cells. (a and b courtesy of K. A. Holbrook.)

Figure continues on the opposite page

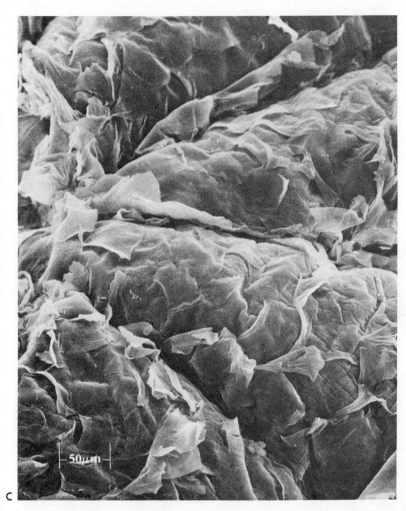

C

Figure 20–2 (Cont.)
(c) Scanning electron micrograph of the surface of the human skin. (Courtesy of K. A. Holbrook.)

bial point of view, the outer layer of the skin is far from being a smooth impenetrable surface. This superficial layer of the skin is the **epidermis.** Supporting the epidermis is a second, deeper layer of skin cells through which a large number of tiny nerves, blood vessels, and lymphatic vessels penetrate. This layer, called the **dermis,** blends in a very irregular fashion with the fat and other cells that make up the subcutaneous tissue.

Almost completely traversing the dermis and epidermis are the fine tubules of sweat glands and hair follicles (Fig. 20–3). Both structures are potential passageways through which microorganisms can pass below the skin to reach the deeper body tissues. Feeding into the sides of the hair follicles are tiny **pilosebaceous** glands that produce an oily secretion. This secretion normally flows up through the follicles and spreads out over the skin surface.

The secretions of the sweat and sebaceous glands are very important to the microbial population of the skin because they supply water, amino acids,

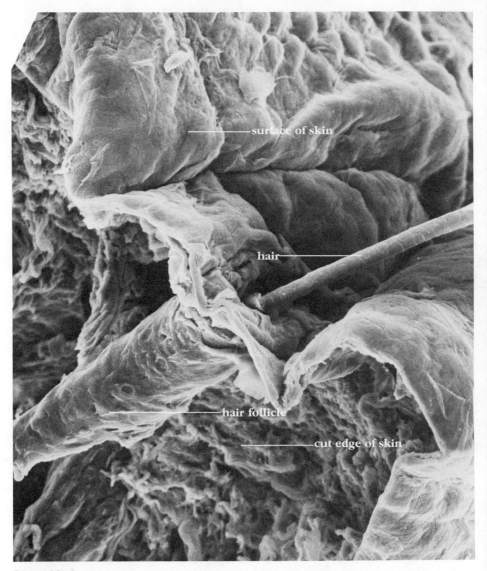

Figure 20–3
Scanning electron micrograph of a hair follicle. The upper right half of the figure is the outer surface of the skin. The lower left portion of the picture looks at the cut surface of the skin. (Courtesy of K. A. Holbrook.)

and lipids, which serve as nutrients for microbial growth. Breakdown of the lipids by the microbial residents of normal skin results in fatty acid by-products that inhibit the growth of many potential disease-producers. In fact, the normal skin surface is a rather unfriendly terrain for most pathogens, being too dry, too acidic, and too toxic for their survival.

TABLE 20–1 PRINCIPAL MEMBERS OF THE NORMAL SKIN FLORA

Name	Characteristics
Diphtheroids	Pleomorphic, nonmotile, Gram-positive rods of the genera *Corynebacterium* and *Propionibacterium*
Micrococci and staphylococci	Gram-positive cocci arranged in packets or clusters; micrococci are nonfermentative, and staphylococci are fermentative
Pityrosporum species	Small yeasts that require oily substances for growth

Factors that interfere with the normal functioning of the skin tend to make these antimicrobial properties useless. For example, an allergic condition that results in the oozing of plasma onto the surface of the skin neutralizes the antimicrobial action of the skin's fatty acids.

THE MICROBIAL INHABITANTS OF THE NORMAL SKIN

Many microorganisms live among the various components of the normal human skin (Table 20–1). For example, depending on the body location, amount of skin moisture, and other factors, counts of bacteria on the skin surface may range from only about a thousand organisms per square centimeter on the back to more than ten million on the scalp and in the axilla (armpit).

Diphtheroids

The principal organisms inhabiting the normal human skin fall into three main groups. The first group is the "diphtheroids," organisms resembling the diphtheria bacillus *Corynebacterium diphtheriae*. Diphtheroids are small Gram-positive pleomorphic ("many-shaped") rods of very low virulence. They are nonmotile, do not form spores, and are distinguished from *C. diphtheriae* by their different morphology, carbohydrate fermentation patterns, and nonproduction of toxin.

C. diphtheriae, page 300

The diphtheroid species found on the skin in the largest numbers is *Propionibacterium acnes*, which is present on virtually all humans that have been studied. Surprisingly, *P. acnes* is anaerobic; the primary site of its growth in the skin is not on the skin surface but within the hair follicles. Growth of *P. acnes* is enhanced by the oily secretion of the sebaceous glands, and the organisms are usually present in large numbers only in areas of the skin where these glands are especially well developed—the face, upper chest, and back. These are also the areas of the skin where acne develops, and the frequent association of *P. acnes* with acne undoubtedly inspired its name, even though most people who carry the organisms do not have acne. It was thought at one time that fatty acids released from oily skin secretions by the action of *P. acnes* enzymes contributed to the severity of acne. However, it has since been shown that medicines that completely suppress production of fatty acids fail to relieve acne. Although many patients with acne appear to be helped by antibiotic treatment, we do not understand why. It certainly is not due to the elimination of *P. acnes*, since this is rarely, if ever, accomplished by such treatment.

The main significance of diphtheroids is that they are commonly present in cultures of clinical material and must be distinguished from more virulent species; however, occasionally they produce abscesses* and other serious infections.

Micrococci and Staphylococci[1]

The second large group of microorganisms universally present on the normal skin is the Gram-positive cocci. Like the diphtheroids, these Gram-positive organisms are rarely pathogenic. On most parts of the body, staphylococci and micrococci are the most common skin bacteria able to grow aerobically. The principal species is *Staphylococcus epidermidis*.

The chief importance of the skin's micrococci and staphylococci is probably in preventing colonization by pathogens and in maintaining the balance among the skin's various flora. The two groups have been shown to produce antimicrobial substances highly active against *P. acnes* and other Gram-positive bacteria, and this may be a factor in maintaining a balance among the various species of microorganisms on the skin.

Yeasts

The third group of organisms inhabiting the normal skin consists of tiny yeasts. They are round, oval, or sometimes short, fat rods. A constriction in the middle of these organisms has given rise to the informal name "bottle bacillus." They grow on laboratory media containing fats such as olive oil and belong to the genus *Pityrosporum*. As with some of the diphtheroids, large numbers of these organisms are found in certain skin conditions, and attempts have been underway for years to define their role in causing dandruff and tinea versicolor. The latter is a common skin disease that causes few symptoms other than a patchy increase in pigment in light-skinned persons or a decrease in pigment in those with dark skin, as well as a slight scaliness.

SKIN DISEASES CAUSED BY BACTERIA

Furuncles and Carbuncles

S. aureus, p. 295

A pimple is sometimes the result of an infection of the skin. Presumably, such infection occurs when a virulent organism, such as *Staphylococcus au-*

---

[1]Staphylococci can be distinguished from micrococci by their ability to ferment carbohydrates; micrococci are nonfermentative.

*abscesses—localized collections of pus surrounded by inflamed tissue

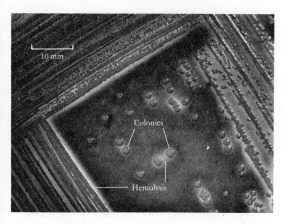

Figure 20–4
Staphylococcus aureus colonies growing on sheep blood agar medium. Note the zone of clearing around the colonies as a result of destruction of erythrocytes in the medium. (Courtesy of F. Schoenknecht.)

reus (Fig. 20–4), is rubbed into the opening of a pilosebaceous gland; the organisms then multiply and penetrate the wall of the neighboring hair follicle. An inflammatory response results, followed by the formation of pus and death of tissue. Pressure increases as a result of the breakdown of host materials. The infection then bulges outward through the superficial skin layers, with eventual rupture, discharge of pus, and healing of the lesion.

With pimples, this process occurs relatively close to the skin surface; there is not much swelling or effect on blood vessels and nerves to cause pain. If, however, the infection penetrates deeper than the dermis, a **boil** (**furuncle**) results. Even deeper skin infections are called **carbuncles.** They occur when the initial infection works its way down into the tissues beneath the skin. The pus then spreads out in all directions. Following the path of least resistance, it may penetrate neighboring structures and discharge pus at multiple points on the skin. The area involved can be extensive.

A carbuncle is the most dangerous of these skin infections because of the possibility that the infection will be carried by the blood vessels to areas such as the heart, brain, or bones. Although pimples may or may not be the result of infection, boils and carbuncles are almost always caused by the pyogenic (pus-causing) bacterium *Staphylococcus aureus* (Table 20–2). People who have a boil shed large numbers of *S. aureus* and should certainly not prepare food or work around sick people until the lesion is healed.

Staphylococcal infections are best treated by a penicillinase-resistant penicillin, but effective alternative medicines are also available for people who are allergic to penicillins.

Penicillinase, page 296

Staphylococcal Scalded Skin Syndrome (SSSS)

Staphylococcal scalded skin syndrome, or "Ritter's disease," is caused by an exotoxin produced by *Staphylococcus aureus* and appears to the eye, just as its name suggests, as scalded skin (color plate 33). It begins as a diffuse redness of the skin affecting 20 to 100 percent of the body. Other symptoms,

TABLE 20–2   CHARACTERISTICS OF *STAPHYLOCOCCUS AUREUS* AND *STREPTOCOCCUS PYOGENES*

Identification	Extracellular products	Pathogenic potential
S. aureus Gram-positive cocci arranged in clusters; cream-colored colonies; produce catalase and coagulase and most ferment mannitol; strain identification using bacteriophages	Hemolysins, leukocidin, hyaluronidase, deoxyribonuclease, staphylokinase, proteinase, lipase, penicillinase, and others	A prominent cause of boils, wound infections, abscesses, impetigo, and food poisoning
S. pyogenes Gram-positive cocci in chains or pairs; small beta-hemolytic colonies; catalase-negative; cell wall contains group A polysaccharide; strains distinguished by cell wall proteins (M and T antigens)	Hemolysins (streptolysins) O and S, streptokinase, DNase, hyaluronidase, and others	Causes impetigo, pharyngitis,* wound infections, puerperal fever; late complications of infection include glomerulonephritis and rheumatic fever

*pharyngitis—throat (pharynx) inflammation (-itis)

such as malaise,* irritability, and fever, are also present. The nose, mouth, and genitalia may be painful for one or more days before the typical features of the disease become apparent. Within 24 hours after the redness appears, the skin becomes wrinkled, and large blisters filled with clear fluid develop. The skin is tender to the touch and looks like sandpaper. The causative organism—*S. aureus*—is not present in the blister fluid but may be cultured from other places in the body, such as the blood, skin, nose, or sputum. The growing staphylococci produce a toxin that is absorbed by the body and carried to the skin, where it causes a cleavage (split) in the deep cellular layers of the epidermis. After about seven days of blistering and peeling, the skin develops a bronze appearance similar to that of a healing sunburn.

Plasmids, page 135

The genetic information for the toxin production resides in certain plasmids carried by only a few strains of staphylococci, and therefore the disease is rare. It can appear in any age group but is seen most frequently in babies and children under the age of five. Staphylococcal scalded skin syndrome usually appears in isolated cases, although small epidemics in nurseries have been reported. Rapid diagnosis is important to prevent death. Because the outer layers of skin are lost as in a severe burn, the mortality can range up to 40 percent, depending on age and whether the patient is weakened by another disease. People suspected of having the disease are placed in protective isolation,* and cultures of the eyes, nose, skin, and blood are ob-

*malaise—a vague feeling of being ill

*protective isolation—group of procedures used to help keep other pathogens that might be in the environment from infecting the damaged skin; usually involves a private room. The only persons allowed to enter are those who are wearing a mask and clean hospital gown. (See Appendix IX.)

tained. Initial drug therapy should include a bactericidal antistaphylococcal antibiotic such as methicillin. All dead skin and other tissue are removed to help prevent secondary infection with Gram-negative bacteria such as *Pseudomonas* species and fungi such as *Candida albicans* that can easily colonize such tissue.

Pseudomonas, page 311

C. albicans, page 380

Impetigo

Impetigo (color plate 34), another bacterial skin disease, is occasionally seen in infants and children, among whom it can spread epidemically. Unlike boil infections, the infection of impetigo is superficial, involving patches of epidermis just beneath the skin's dead scaly layer. The growth of the causative bacteria typically results in thin-walled blisters that break and are replaced by crusts and "weeping" of plasma through the skin. Usually no fever or pain accompanies impetigo, although lymph nodes near the involved areas often enlarge, indicating that bacterial products enter the lymphatic system.

Staphylococcus aureus, the same bacterium that causes boils, is the organism most commonly found in the blister fluid of people with impetigo. But *Streptococcus pyogenes*, another Gram-positive coccus (Table 20–2), is also an important cause of this disease. *Streptococcus pyogenes* strains (color plate 35) are frequently referred to as "group A streptococci" because their cell walls contain a substance called "polysaccharide A."

A number of extracellular products may contribute to the virulence of *S. pyogenes*. These include enzymes that break down protein, nucleic acid, and the hyaluronic acid of host tissues. As with staphylococci, it is doubtful that any one of these factors plays an essential role in streptococcal pathogenicity. However, the antigens of the *S. pyogenes* outer cell wall are known to be very important in enabling this organism to cause disease. Penicillin and certain other antibacterial medicines taken by mouth or injection or even applied locally* are highly effective in treating streptococcal impetigo.

Sometimes people convalescing from untreated infections caused by *S. pyogenes* suddenly experience fever and high blood pressure and excrete blood and protein in their urine. This potentially fatal sequel to streptococcal infection, called **acute glomerulonephritis,** is caused by inflammation of small tufts of tiny blood vessels (glomeruli) within the kidney (nephros). **Rheumatic fever,** another late complication of streptococcal infection, occurs much more commonly as a result of streptococcal "strep" throat rather than skin infections, whereas the reverse has generally been true of glomerulonephritis. In both of these complications, it is unusual to find *S. pyogenes* or other bacteria in the diseased tissues, since the pathological changes in the tissues represent a hypersensitivity reaction induced by the streptococci rather than a direct attack by these organisms. This form of glomerulonephritis is considered by some scientists to be an example of **autoimmune** disease, although the idea is controversial.

Autoimmune disease, page 231

---

*locally—directly on the affected area

SKIN DISEASES CAUSED BY VIRUSES

Several childhood diseases are caused by viruses that are carried to the skin by the blood. The principal features of these viral skin diseases are shown in Table 20–3. This group of diseases is usually diagnosed by clinical symptoms, but when the disease is not typical, tests can be performed for specific antibody, or the virus can be cultivated from body fluids in the laboratory. In some cases, the virus causes changes in certain cells of the body; these cells can be identified under the microscope. For example, in chickenpox and measles, clinical specimens show very large cells called giant cells. In some instances, these cells have one large nucleus; at other times there are many small nuclei. Virus-infected cells can also be detected in infected tissues by electron microscopy or the use of fluorescent antibody.

Antibody tests, page 246

Chickenpox (Varicella)

A few years ago, a 29-year-old woman complained to her physician of a painful rash that had appeared on one side of her chest. She had had chickenpox (color plate 36) years ago while in elementary school and had recovered without incident. The doctor examined the rash on her chest and noticed that it consisted of small red bumps, blisters, and scabs and that it seemed to follow the branches of one of the nerves of skin sensation. He informed her that she had "shingles," also known as **herpes zoster** (Fig. 20–5 and color plate 37), and that it represented a reactivation of the chickenpox infection she had had many years ago. This seemed somewhat unlikely to the patient, but her skepticism lessened appreciably when the first of her four children developed chickenpox two weeks later, followed by the other three in succession, and then their friends, until half of the younger children in the immediate community had become infected.

The varicella virus is transmitted when air contaminated with virus particles by a person with chickenpox is inhaled by another person. The virus multiplies in cells lining the second person's respiratory tract, invades the bloodstream, and from there ultimately reaches the cells of the epidermis. The infected skin cells initially enlarge and then break down, leaving a fluid-filled blister beneath the skin. This later becomes filled with pus, ruptures, and forms a crust or scab before finally healing. Most cases of chickenpox are mild, and recovery is uneventful. The varicella virus is, however, a threat to the babies of women who contract chickenpox near the time of delivery and to people whose immune mechanisms are impaired. In such instances, the virus causes extensive damage to internal organs, often resulting in death. In older children and adults, the virus sometimes causes severe pneumonia.

So far, the development of herpes zoster is not fully understood. However, it is quite clear that in many people who have chickenpox the varicella virus is not entirely eliminated from the body at the time of recovery. The surviving virus is thought to enter the cells of some nerves of sensation.

TABLE 20—3 VIRAL DISEASES INVOLVING THE SKIN

Disease and characteristic features	Infectious agent	Specimen needed for diagnosis	Incubation	Communi-cable period	Immunization	Treatment	Potential problems
Rubeola (measles): fever, rash, conjunctivitis, bronchitis, Koplik spots	Rubeola virus, a paramyxovirus	Serum or plasma collected on or before first day of rash	8–13 days (usually 10)	From 4 days before to 4 days after rash appears	Live virus vaccine given at 1 year of age or older	Symptomatic*: aspirin for high temperatures; adequate fluids	Damages respiratory mucosa; serious bacterial infections may result
Rubella (German measles or 3-day measles): fever, malaise, rash. Tender lymph glands at base of skull	Rubella virus, a togavirus	Throat and nasal secretions (amount of virus present is high during time of rash)	14–21 days (usually 18)	From 7 days before to 4 days after rash appears	Live virus vaccine given between 12 months and puberty. Not given to girls and women after puberty starts or when pregnant	Symptomatic: aspirin	Infections that develop during pregnancy commonly cause fetal damage
Varicella (chickenpox): small itchy bumps and blisters over skin and mucous membranes; fever	Varicella-zoster virus, a herpesvirus	Smear of new blister fluid	14–20 days (usually 13–17)	From 5 days before vesicles* appear to 6 days after last vesicles appear	No vaccine available, although ZIG (zoster immune globulin) may be given to highly susceptible persons such as those with leukemia	Symptomatic: rest until temperature is normal; keep skin clean; antihistamines by mouth and lotions applied to skin to relieve itching; mittens for very young children to prevent scratching	Secondary bacterial infections may occur; herpes zoster (shingles) may appear at a later time if virus becomes latent

*symptomatic treatment—treating symptoms brought on by the disease when the disease itself cannot be treated
*vesicles—small blisters

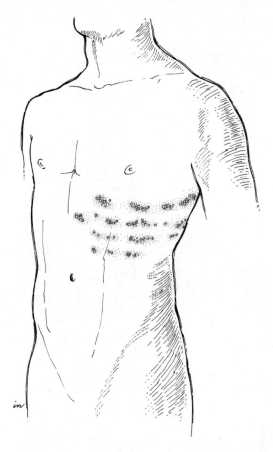

Figure 20–5
Herpes zoster. The rash follows the distribution of one of the nerves of sensation.

These nerves, responsible for transmission of sensations from the skin, are very complex and interesting structures. The nuclei of the cells in these nerves are located in small bodies called **ganglia** that lie near the spinal cord. Each cell has an extremely long extension that runs all the way from the region of the nucleus to the skin, as well as a shorter extension that runs to the spinal cord. A nerve is actually made up of a number of these long cell extensions all enclosed together in a sheath of other cells.

The first change occurring when herpes zoster begins is an inflammatory reaction in the ganglion of a nerve; this reaction is followed by the appearance of infectious virus in the skin supplied by the nerve cells originating from the ganglion, as well as by virus in the cerebrospinal fluid* in some cases. Even though high levels of antibody appear within a few days, skin blisters containing infectious virus are present for much longer periods in herpes zoster patients than in patients with chickenpox, where the appearance of antibody is slower.

---

*cerebrospinal fluid—the fluid that surrounds the brain and spinal cord

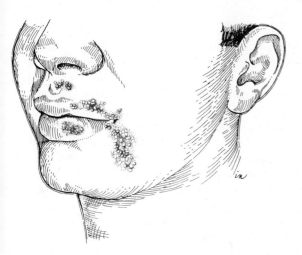

Figure 20—6
Herpes simplex or cold sore. Small blisters containing herpes simplex virus develop in the involved area.

The mechanism by which the varicella virus persists may represent an important adaptive solution to the problem of its survival in small isolated populations. It has been noticed, for example, that when a highly infectious virus such as that causing measles is introduced into such a population, it spreads quickly and infects most of the susceptible individuals, who then either become immune or die. The measles virus then disappears from the community. By contrast, when the chickenpox virus is introduced into such a community, it persists indefinitely, creating recurrent epidemics whenever sufficient numbers of susceptible children have been born. Cases of shingles are probably the cause of these recurring epidemics.

Significantly, the varicella virus is not a true pox virus but a member of the herpes group, which includes the virus of **herpes simplex,** the cause of fever blisters and cold sores (color plate 38 and Fig. 20—6).

Because chickenpox and shingles are rarely fatal for normal children, and because most people have been infected by the varicella virus by the time they reach adulthood, there has been little effort to develop control measures for these diseases. Indeed, chickenpox is so much more severe in adults than in children that it may be wise to promote childhood infection. Very susceptible people (such as patients with leukemia) can be protected by injecting them with gamma globulin from persons who have recovered from shingles. A live vaccine is being evaluated but is not available for general use.

Only about 150,000 cases of varicella are reported each year in the United States, but the incidence is much higher, many cases being so mild as to go unnoticed. Most cases occur in the winter months.

Measles (Rubeola)

Measles (rubeola) (color plate 39) develops in a manner similar to chickenpox, but the measles virus typically produces pronounced changes in the respiratory tract as well as the skin. The lesions in the mouth are called

H. influenzae, page 316

S. pneumoniae, page 299

Koplik spots and have a characteristic appearance (color plate 40). Serious consequences of measles are quite common and result from secondary infection of the ears and lungs by bacteria such as *Streptococcus pyogenes*, *Haemophilus influenzae*, and *Streptococcus pneumoniae*. Measles is now controlled effectively by using a living vaccine derived from measles virus considerably lessened in virulence (attenuated) by prolonged growth in the laboratory. In general, live measles vaccine should be given to children at about the age of 15 months. However, in the event of an outbreak of measles, the vaccine can be given as early as six months of age. At 15 months, these babies should be reimmunized. Some cases of measles are followed years after their occurrence by a very rare disease[1] marked by progressive degeneration of the brain and lasting months or years. Measles virus antigen can be demonstrated in the brains of such patients, and high levels of measles antibody are present in their blood. The incidence of measles has been steadily declining since the advent of the vaccine in 1963, and the number of reported cases per year in the United States is now about 13,500, compared with the 1950's when there were more than 30 times that number. Epidemics of measles still occur in inner cities and in other populations where the practice of routine immunization is often neglected. The goal of the elimination of measles from the United States should be reached within a few years, provided that intensive immunization programs are continued.

German Measles (Rubella)

German measles (rubella) (color plate 41) develops in a manner similar to rubeola, but its effects on both the skin and respiratory tract are mild, and it is less contagious than rubeola. For this reason, many people never develop rubella as children and are therefore susceptible to it as adults; rubella epidemics may thus occur among hospital personnel, college students, and military recruits. Unfortunately, during the stage when the virus is circulating in an infected pregnant woman's blood, her fetus is likely to contract the infection as well. In fact, if the woman contracts rubella during the first eight weeks of her pregnancy, there is at least a 90 percent chance that the fetus will become infected. During this stage of development, the tissues of the fetus are easily damaged by the virus, which stops tissue cell multiplication and produces chromosome abnormalities. Miscarriages often result from such infection, but more commonly the child lives and is born with one or more impairments, such as deafness, a defective heart, poor vision, or mental retardation. These defects, however, do not always follow infection of the fetus. The risk of serious defects among live-born children has ranged from only 30 to 50 percent when rubella was contracted during the first month of pregnancy; the risk declines progressively with infections occurring thereafter and is very low in late pregnancy.

---

[1]The disease is called subacute sclerosing panencephalitis (SSPE).

To control rubella and reduce the possibility of rubella infection in pregnant women, many states require that schoolchildren show proof of immunization. For example, in the state of Washington, 96% of the students in grades one through six have been immunized for both rubeola and rubella. A live vaccine is used, derived from the German measles virus through many subcultures of the virus in tissue cell cultures. It produces a milder disease than the "wild type"* rubella virus. It is not generally administered to girls and women in the childbearing years because they might be pregnant, and it is remotely possible that the vaccine virus, like the "wild type" virus, might damage the fetus. However, the lack of immunity in young adults, ranging from 12 to 50 percent in different groups, has sometimes made it necessary to vaccinate them, too. For example, a dietary worker with mild rubella unknowingly introduced the disease into a large hospital in the eastern United States and caused an outbreak of more than 40 cases among various employees. Because of this and similar epidemics, some hospitals allow only those proved to be immune to rubella to work with pregnant patients. Hospital staff members are required to have a blood test, and if it does not show rubella antibody, then they must receive the vaccine or work elsewhere.

Since widespread use of the vaccine in children began in 1969, the incidence of rubella in the general population of the United States has fallen by about 90 percent; only about 4000 cases per year are now reported.

Warts

A few viruses can infect the skin through abrasions and other injuries, in contrast to the viruses that enter the body through the respiratory tract. One of the most interesting examples is the group of viruses that cause warts. Warts are in fact small tumors called papillomas* that are caused by wart virus infection. These tumors do not become malignant, except in very rare instances with genital warts. Moreover, in more than half the cases they eventually disappear without any treatment. At least seven different papilloma viruses are known to cause human warts; they are all DNA viruses of about the same size and electron microscopic appearance. They have been very difficult to study because of difficulties in finding a suitable tissue cell culture to reproduce them consistently in vitro. Wart viruses can survive on inanimate objects such as wrestling mats, towels, and shower floors, and infection can be acquired from such contaminated objects. The virus infects the deeper cells of the epidermis and reproduces in the nuclei. The cells grow abnormally and produce the papilloma. The incubation period ranges from 2 to 18 months. Infectious virus is present in the wart and contaminates

Cell culture, Appendix II

---

*wild type—found under natural conditions, as opposed to altered by cultivation in the laboratory

*papillomas—benign tumors consisting of nipple-like protrusions of tissue covered with cells of the skin or mucous membrane

fingers or objects that pick or rub it. Like other tumors, warts can only be treated effectively by killing or removing all the abnormal cells. This can usually be accomplished by freezing the wart with liquid nitrogen, by cauterization* with an electric arc, or by surgical removal.

SKIN DISEASES CAUSED BY FUNGI

Earlier in this chapter, we mentioned the possible role of one of the components of the normal skin flora, yeast of the genus *Pityrosporum*, in causing a mild skin disease, tinea versicolor. Other fungi are responsible for more clear-cut and often more serious infections of the skin, although even in these cases the condition of the host's defenses against infection is often crucial. The yeast *Candida albicans* commonly lives harmlessly among the normal flora of the skin, yet in some people it invades the deep layers of the skin and subcutaneous tissues. In many people with candidal skin infections, no precise cause for the invasion can be determined; in others, such factors as increased skin moisture and defective immunity undoubtedly play a role. Similarly, a variety of mold-type fungi may invade the skin, hair, and nails, producing conditions with such colorful names as athlete's foot, jock itch, and ringworm, but such diseases occur only in a minority of the total number of people the organisms colonize. Most people colonized by the same organisms show no abnormality. These skin-invading molds belong to the genera *Trichophyton*, *Microsporum*, and *Epidermophyton* and are collectively called **dermatophytes.**[2] They are peculiar in that they are unable to grow in living tissues but instead multiply in inert structures containing keratin. Most of their unpleasant effects result from superinfection* or allergic reactions to the presence of these fungi, but the thickening of nails and the loss of hair they promote cause complaints as well. Some of these agents have major reservoirs* in soil or on the skins of animals, while others are known to attack only humans.

SUMMARY

Normal skin appears scaly and uneven when examined with a microscope. Oily secretions and sweat are produced by glandular elements in the skin and support the growth of a rich aerobic and anaerobic normal flora. The most important microbial groups on the skin are diphtheroids (including

[2]The skin diseases dermatophytes cause are called dermatomycoses.

*cauterization—tissue destruction

*superinfection (secondary infection)—invasion by another microorganism after the host has been weakened by the first one

*reservoirs—sources of a disease-producing organism

Propionibacterium acnes), micrococci and *Staphylococcus epidermidis*, and yeasts of the genus *Pityrosporum*.

Bacterial pathogens most frequently invade skin damaged by trauma or other factors. *Staphylococcus aureus* is one of the most important of these invaders, producing furuncles, carbuncles, staphylococcal scalded skin syndrome, and impetigo. *S. aureus* characteristically produces tissue destruction and abscesses in the skin and subcutaneous tissue, and if carried to other parts of the body by the blood, produces the same effects in new locations. *Streptococcus pyogenes* (group A beta-hemolytic streptococci) also causes skin infections such as impetigo. Superficial infections by certain strains of this bacterium sometimes lead to glomerulonephritis, an inflammatory condition of the kidneys, which develops during healing of the skin disease.

Several viral diseases involve the skin, although the mode of entry of the viruses into the body is usually through the respiratory or gastrointestinal tract. Chickenpox (varicella) is caused by the varicella-zoster virus, a member of the herpesvirus group, to which the herpes simplex (cold sore) virus also belongs. These two viruses can exist in tissues in a latent form, only to produce active disease years after their initial infection. Measles (rubeola) characteristically produces its most severe damage in the respiratory system, resulting in superinfection by bacteria. An unusual degenerative brain disease is now thought to be a late complication of measles virus infection. German measles (rubella) is a mild disease, medically important because of its ability to infect fetuses and produce birth defects. The wart viruses enter the skin at sites of injury and cause the formation of warts, which are small benign tumors of a type called papillomas.

Several fungi infect the skin. *Candida albicans* is one of the most invasive. Others of the genera *Trichophyton*, *Microsporum*, and *Epidermophyton* can attack only keratinized structures such as the outer layers of epidermis, the hair, and the nails. Inflammatory reactions of the skin may result from allergy to fungal products or superinfection by bacteria.

SELF-QUIZ

1. The diphtheroid species found on the skin in largest numbers is
   a. *Corynebacterium diphtheriae*
   b. *Propionibacterium acnes*
   c. tinea versicolor
   d. *Staphylococcus epidermidis*
2. One important feature of the micrococci and staphylococci that normally inhabit the skin is their ability to
   a. change lipids to fatty acids
   b. prevent colonization of the skin by pathogens
   c. maintain a constant temperature of the skin
   d. take the red dye in the Gram stain

3. A cause of impetigo is
   a. *Pityrosporum*
   b. *Corynebacterium diphtheriae*
   c. *Streptococcus pyogenes*
   d. *Candida albicans*
4. The varicella virus can persist for years unnoticed in a person's body and then cause a disease known as
   a. herpes zoster or shingles
   b. rubeola
   c. acute glomerulonephritis
   d. Ritter's disease
5. Immunization against rubeola (measles) should generally be given at the age of
   a. 1 day
   b. 3 months
   c. 6 months
   d. 15 months
   e. 5 years
6. Many birth defects can be prevented by immunizing against
   a. rubella
   b. chickenpox
   c. herpes zoster
   d. rubeola
   e. impetigo
7. Which of the following properties of normal skin enable it to resist microbial colonization?
   a. low pH (acid)
   b. dryness
   c. shedding of surface cells
   d. all of the above
   e. none of the above

QUESTIONS FOR DISCUSSION

1. Why is the incubation period for warts so long?
2. What is the difference between the way the wart virus causes disease and the way the varicella virus causes disease?
3. Why do *Streptococcus pyogenes* skin infections cause glomerulonephritis, whereas skin infections by *Staphylococcus aureus* do not cause this disease?

FURTHER READING

Beneke, E. S.: *Human Mycoses.* Kalamazoo, Michigan: Upjohn Company, 1970. Many good color photographs of fungi and skin diseases caused by fungi.

Cherry, J. D.: "The 'New' Epidemiology of Measles and Rubella," *Hospital Practice* (July 1980).

Selekman, J.: "Immunization: What's It All About?" *American Journal of Nursing,* *80*:1440–1445 (1980). A clear, comprehensive article containing a table listing the childhood diseases, their causative agents, incubation periods, main symptoms, and other pertinent information.

"Skin Rashes in Infants and Children." A programmed unit prepared by Stephen Cohen. *American Journal of Nursing, 78*:1–32 (1978). A self-instruction course that includes localized dermatological conditions and systemic conditions involving skin rashes. Good photographs.

Sykes, J., Kelly, A. P., King, M. L., and Kenney, J. A.: "Black Skin Problems." *American Journal of Nursing, 79*:1092–1094 (1979).

"Viral Infections of Athletes' Skin." *The Physician and Sportsmedicine* (November 1978). A recorded discussion among physicians concerning herpes simplex lesions and their spread. Test included.

Chapter 21

UPPER RESPIRATORY SYSTEM INFECTIONS

We were hiking in Wales. We saw, beside the sagging slate-roofed church, a line of weathered and lichened tombstones, tilted irreverently by the roots of an ancient tree. Each marker had the year 1838 etched upon it. Mother, father, five children—all dead within six weeks of each other. The cause was diphtheria.

OBJECTIVES

To know
1. What the upper respiratory system is.
2. The principal defenses of the upper respiratory system against pathogenic microorganisms and viruses.
3. Some of the microorganisms that normally inhabit the upper respiratory system and where they do and do not reside.
4. The cause and significance of colds and other viral diseases of the upper respiratory system.
5. The cause and significance of "strep throat" and diphtheria.

Diseases of the upper respiratory system are by far the most common human infectious diseases, far outweighing any others in cumulative misery and loss of productivity. The following sections illustrate the features of microbial interactions with the respiratory tract.

ANATOMY AND PHYSIOLOGY

A person normally breathes about 16 times per minute, inhaling about 500 ml of air with each breath, or more than 11,500 liters of air per day. This enormous volume of air, with its accompanying microbes, flows into the respiratory system as a result of the vacuum produced when chest and abdominal muscles enlarge the chest cavity during inspiration. The air enters the respiratory system at the nostrils (a common site of colonization by *Staphylococcus aureus*), flows into the nasal cavity, and curves downward through the throat. It enters the lower respiratory tract below the epiglottis, a muscular fold of tissue that closes off the lower respiratory tract during swallowing.[1] The nasal cavity is a chamber above the roof of the mouth, incompletely divided into right and left halves by a vertical wall extending almost back to the throat. Spongy masses of tissue bulge into each half of this chamber from its outside walls. These tissues are similar to the erectile tissues of the genitalia in that they expand and contract with alterations in blood flow controlled by nervous reflexes. With extreme enlargement, these spongy tissues contribute to "nasal congestion" or obstruction of the airways. Many stimuli can cause these changes, including variations in temperature and humidity, emotional factors, and the irritating effects of smoke and infections.

As shown in Figure 21–1, the eyes, ears, and air-filled chambers of the skull (sinuses and mastoid air cells) are all connected to the main respiratory passage of the nose and throat. Thus, the conjunctiva* of the eye and the membranes of the middle ear (the space just behind the eardrum) and sinuses, along with the passages connecting them to the nose and throat, are all part of the respiratory tract and are frequently involved in upper respiratory infections. Collections of lymphoid tissue (the tonsils and adenoids) are located where the mouth and nasal chamber join the throat. The tonsils and adenoids are important producers of immunity to infectious agents, but paradoxically they can also be the sites of certain infections and with enlargement can contribute to ear infections by interfering with normal drainage through the eustachian tubes. The lining cells of the nasal chamber, tubes to the eyes, eustachian tubes, sinuses, mastoid air cells, and middle ear all have tiny hairlike projections (cilia) along their exposed free border (Fig. 21–2). These cilia beat synchronously and continually propel a film of mucus (secreted by other cells lining the respiratory system) into the nose and throat. This ciliary action of the lining cells normally keeps the ears and sinuses free of microorganisms. Viral infection, tobacco, alcohol ingestion, and narcotics all interfere with the action of the cilia.

Lymphoid tissue, page 214

The primary functions of the upper respiratory tract are to regulate the temperature and humidity of inspired air and to remove from it or destroy microorganisms and other foreign material. When cold air enters the upper tract, nervous reflex mechanisms immediately increase the blood flow; with entry of warm air, these mechanisms decrease blood flow. Transfer of heat between the blood and the air usually adjusts the air temperature to within

---

[1]The epiglottis is the site of a serious infection of children by *Haemophilus influenzae*. The disease produced is called epiglottitis.

*conjunctiva—mucous membrane that lines the eyelids and eyes

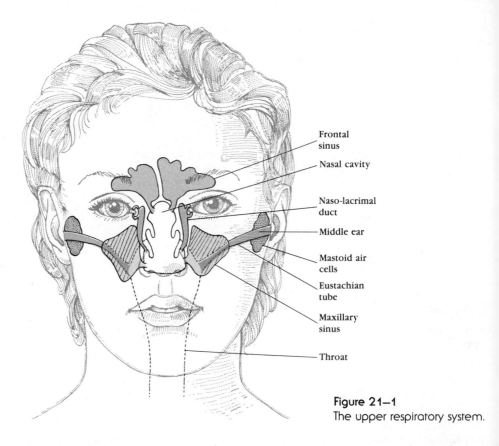

Frontal
sinus

Nasal cavity

Naso-lacrimal
duct

Middle ear

Mastoid air
cells

Eustachian
tube

Maxillary
sinus

Throat

Figure 21–1
The upper respiratory system.

two or three degrees of the normal body temperature by the time it reaches
the lung. This air also becomes saturated with water vapor, the nasal passage
having the ability to give up more than a quart of water per day to inspired
air. Expired air is cooled on passing out of the respiratory system, giving
back much of its water vapor to the system's mucous membranes. Trapping
of inspired particles by the upper respiratory system is also very efficient.
Turbulence produced by nasal hairs causes larger particles to impinge on
the system's mucous film and become trapped. Even small particles the size
of bacteria are often trapped, only 50 percent escaping to pass through the
upper airways. Infections of the upper respiratory tract interfere with these
normal functions and may thereby promote lower respiratory tract infection.

NORMAL FLORA OF THE UPPER RESPIRATORY SYSTEM

The principal types of bacteria inhabiting the upper respiratory system are
presented in Table 21–1. Surprisingly, even though the eyes are constantly
exposed to a multitude of microorganisms, about half of all normal healthy
people have no bacteria on the conjunctiva. Presumably this sparsity of
microorganisms results from the frequent mechanical washing of the eye

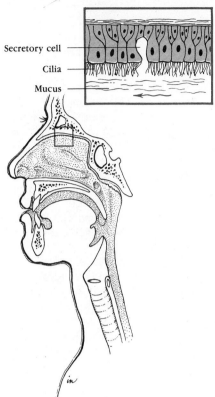

Secretory cell

Cilia

Mucus

Figure 21–2
Ciliated membrane lines the respiratory system.

with lysozyme-rich tears and from the eyelid's blinking reflex, which clears the eye surface. Thus, viruses and microorganisms impinging on the moist membrane covering the eye are almost immediately swept into the naso-pharynx.* Indeed, cold viruses can infect by this route. When organisms are recovered in cultures of the normal conjunctiva, they are usually scant in number and consist of species normally found on the skin.

The secretions of the nasal entrance usually contain large numbers of diphtheroids, micrococci and staphylococci, and smaller numbers of bacteria that normally inhabit the environment, such as *Bacillus* species. About a third of all normal humans also carry *S. aureus* in the nose. Further inside the nasal passages, the microbial population increasingly resembles that of the nasopharynx. The nasopharynx contains large numbers of micro-organisms, mostly alpha-hemolytic streptococci of the viridans group, non-hemolytic streptococci, *Branhamella catarrhalis*,[2] and diphtheroids. Anaerobic Gram-negative bacteria, including the genus *Bacteroides*, are also present in large numbers in the nasopharynx. In addition, potential pathogens such as

[2]Formerly called *Neisseria catarrhalis*.

*nasopharynx—the area behind the nasal chamber and above the throat

TABLE 21–1 NORMAL FLORA OF THE UPPER RESPIRATORY SYSTEM

Genus	Appearance	Comments
Micrococcus and Staphylococcus	Gram-positive cocci in clusters	Commonly includes the pathogen S. aureus
Corynebacterium	Pleomorphic* Gram-positive rods; nonmotile; non–spore-forming	Nonpathogens collectively referred to as "diphtheroids"
Branhamella	Gram-negative diplococci	Resembles Neisseria species
Haemophilus	Small pleomorphic Gram-negative rods	Commonly includes the pathogen H. influenzae
Bacteroides	Small pleomorphic Gram-negative rods	Strict anaerobes
Streptococcus	Gram-positive cocci in chains	Alpha (especially viridans streptococci), beta, and gamma types; commonly includes the pathogen S. pneumoniae

*pleomorphic—having many shapes

S. pneumoniae, H. influenzae, and N. meningitidis are commonly found in this area, especially during the cooler seasons of the year. Streptococcus pyogenes, Mycoplasma pneumoniae, Corynebacterium diphtheriae, Bordetella pertussis, and other virulent respiratory pathogens are generally found in only a small percentage of healthy humans.

VIRAL INFECTIONS OF THE UPPER RESPIRATORY SYSTEM

The Common Cold (Acute Afebrile Infectious Coryza)

The all-too-familiar symptoms of the common cold result from upper respiratory tract infections by any of approximately 120 known viruses and a few bacteria. Myxoviruses (influenza, parainfluenza, and respiratory syncytial viruses), adenoviruses, and enteroviruses (Coxsackie and ECHO viruses) have all been recovered from people with colds. However, the majority of common colds are caused by rhinoviruses ("rhino": nose, as in rhinoceros—"horny nose"), members of the picorna group ("pico" is Italian for "small" and "rna" is ribonucleic acid [RNA]; thus, small RNA viruses). The rhinoviruses were initially difficult to cultivate in the laboratory, since they failed to infect laboratory animals or tissue cell cultures. Researchers in England, however, discovered that if cell cultures of monkey or human origin were incubated at 33°C instead of at body temperature, and at a slightly acidic pH instead of at the alkaline pH of body tissues, positive cultures could be obtained in many cases. It is noteworthy that the lower temperature and pH of this method are conditions that normally exist in the upper respiratory tract. Rhinoviruses, however, are killed by strong acids, such as those found in the human stomach, and this characteristic is used to distinguish them from other picornaviruses.

Humans are the only significant source of cold-producing rhinoviruses, and close contact with an infected person appears to be necessary to transmit these agents. Once the viruses lodge on the respiratory-lining cells, infection of a few of these cells becomes established; the virus replicates intracellularly, is discharged, and infects adjacent cells. The injury stimulates nervous reflexes, which cause an increase in nasal secretions, sneezing, and swelling of the mucosa and nasal erectile tissue. This swelling causes the airway to become partially or completely obstructed. Later, an inflammatory reaction occurs, with dilation of blood vessels, oozing of plasma, and congregation of leukocytes in the infected area. Secretions from the area may then contain pus and blood. The infection is eventually halted by the inflammatory response, interferon release, and cellular and humoral immunity, but it can extend into the ears, sinuses, or lower respiratory tract before it is stopped.

Many people believe that colds occur more often when the body is cold, and scientific studies have been undertaken repeatedly to define the relationship between exposure to cold and the common cold. An important study, reported in 1933, showed that colds disappeared from the Arctic island of Spitzbergen during the long winter, when no ships came, and reappeared in Spitzbergen when ships arrived in late spring. This finding indicated that new sources of infectious virus were necessary to produce colds, regardless of the temperature of a region. Indeed, studies have shown that the incidence and severity of colds were the same when a rhinovirus was administered to two groups of nonimmune volunteers, one group exposed to chilling and the other not exposed to chilling. Finally, it has been shown that in semitropical areas a sharp increase in colds occurs with the onset of the rainy season, even though the mean temperature stays about the same. Thus, the influence of season on the incidence of colds and other respiratory diseases has been clearly demonstrated, but the mechanism for this phenomenon has not. It seems possible that the physiological state of the nasal chamber (including its temperature, pH, air velocity, and mucus flow) is influenced by climatic conditions and might be important in aiding the growth and dissemination of virus, in increasing the severity of the symptoms produced by infection, or in decreasing the number of viral particles necessary to produce illness. More important, however, is the fact that people tend to congregate indoors when the weather is bad, increasing the opportunities for infectious viruses to spread.

Specific control measures are not available for colds. The large number of different causative agents of the cold makes the development of vaccines impractical; and rhinoviruses, like other viruses, are unresponsive to antibiotics and other medications that control bacterial infections. Generally speaking, a person is most infectious to others during the first day or two of a cold, when symptoms are the worst, and at least in the case of adults, a person with severe symptoms is much more likely to transmit a cold than someone whose symptoms are mild. For the brief period of one to two days after onset, very high concentrations of virus are found in the nasal secretions and *on the hands* of infected people, with substantial amounts also present in the saliva. The nasal mucosa is highly susceptible to infection, and the disease is contracted by another person when the virus is inhaled, or when it is unwittingly rubbed into the eyes or nose by contaminated hands. Colds, however, are not highly contagious; less than half of nonimmune adults exposed in a family or dormitory setting contract colds. On the other hand, young children may transmit colds and other respiratory viruses much more effectively because of decreased care in dealing with their respiratory secretions. Washing the hands, even in plain water, readily removes rhinoviruses.

Adenovirus Infections

Children and young adults commonly experience an illness resembling a cold, but with high fever, very sore throat and severe cough, and swelling of the lymph nodes of the neck. A whitish-gray exudate may spread over the tonsils and throat. Conjunctivitis (eye inflammation) is commonly present, and occasionally pneumonia develops, or a lung infection resembling whooping cough. Recovery occurs spontaneously in one to three weeks. The cause of this illness in most cases is an adenovirus, of which more than 30 strains are infectious for humans.

Little is known about the pathogenesis of the disease caused by adenoviruses. The source of the infectious agent is humans; related viruses infect animals but not people. Once inside the cells of the host, the virus grows in the cellular nuclei. In severe infections, extensive cell destruction and inflammation occur. Different types of adenoviruses vary in the tissues they affect. For example, adenovirus types 4, 7, and 21 typically cause illness characterized by sore throat and enlarged lymph nodes, whereas type 8 is likely to cause extensive eye infection with few other symptoms. Adenoviruses 1 and 2 produce a mild throat infection in young children and then become latent in the tonsils and other lymphoid tissues, where they remain for years. Adenoviral throat infections may resemble infectious mononucleosis and "strep throat." Some human adenoviruses cause tumors when injected into baby hamsters but are not oncogenic* in humans.

There are no vaccines marketed for civilian use because adenovirus infections are ordinarily not serious enough to require widespread vaccination. Antibiotic treatment is of no value for treating adenovirus infections and sometimes does harm by suppressing some of the body's normal bacteria and allowing overgrowth of any resistant opportunists that might be present in the body.

---

*oncogenic—tumor-causing

BACTERIAL INFECTIONS OF THE UPPER RESPIRATORY SYSTEM

"Strep Throat"

The most important bacterial infection of the throat is caused by *Streptococcus pyogenes*, the group A beta-hemolytic streptococcus described previously as a cause of impetigo. Streptococcal infections of the throat often resemble adenoviral infections very closely but generally cause greater enlargement and tenderness of the lymph nodes in the neck. Conjunctivitis and pneumonia are not usually present. Although abscess formation and other local complications may prolong the illness, most patients with streptococcal sore throat recover spontaneously after about a week. In fact many infected people have only the symptoms of a mild cold. Some strains of *S. pyogenes* produce a toxin (erythrogenic* toxin) that is absorbed and carried by the bloodstream to the skin, resulting in a red rash. When this happens, the disease is called **scarlet fever.**

Streptococcal throat infections are spread both by inhalation into the respiratory tract and by ingestion of contaminated food. The incidence of sore throat caused by *S. pyogenes* varies greatly with the age of the person, the time of year, and the geographic location. Among students with sore throats at a large West Coast university, *S. pyogenes* was isolated from less than 5 percent; among some groups of military recruits, however, the incidence has been 25 percent.

Streptococcal throat infections, like skin infections, may lead to glomerulonephritis as a result of antibody reacting with streptococcal products. In addition, people with severe untreated *S. pyogenes* throat infections carry about a 2.5 percent risk of developing **acute rheumatic fever**. Rheumatic fever may develop in individuals with untreated *mild* infections, although the risk is very much lower than it is for those with severe infections. Mild infections do result in many cases of rheumatic fever, however, because such infections are relatively common and are much more likely to go untreated. For this reason, people with fever and sore throat should have a throat culture performed to rule out *S. pyogenes* before deciding that antibiotic treatment of their condition is unnecessary.

Rheumatic fever usually occurs about three weeks after the onset of a streptococcal sore throat and is characterized by inflammation of the joints, heart, skin, and other tissues. Heart failure and death may occur, but the process usually subsides with the help of medications. Unfortunately, the heart valves may be left permanently damaged by rheumatic fever, and if the afflicted person has subsequent *S. pyogenes* infections, rheumatic fever often recurs promptly, producing still more damage to the heart.

In contrast to acute glomerulonephritis, which is caused by only a few types of *S. pyogenes*, infection with any type of *S. pyogenes* may result in rheumatic fever. The disease is far more common following streptococcal throat infections than it is after infections of the skin or other body sites.

---

*erythrogenic—"erythro-" = red, as in "erythrocyte" (red cell); "-genic" = development, as in "genesis"

Indeed, in many cases, by the time rheumatic fever develops, *S. pyogenes* can no longer be cultured from the throat, and cultures of the blood, heart, and joint tissue are also usually negative. These facts suggest that rheumatic fever, while a consequence of the initial infection, is due to an immunological response that takes time to develop. The mechanism involved in this process is still a mystery. It is possible that a reaction between antibody and some streptococcal product is responsible for the ensuing tissue damage. Alternatively, antibody, increasing in response to some streptococcal antigen, may cross-react with the infected individual's tissues. Indeed, some streptococci have been shown to have a cytoplasmic membrane antigen in common with human heart muscle.

Adequate ventilation and avoidance of crowding help to control the spread of streptococcal infections. Currently, people suspected of having streptococcal sore throats have throat cultures performed, and if group A streptococcal infection is confirmed, they are treated with penicillin. This treatment eliminates the organisms in about 90 percent of these cases, and the risk of rheumatic fever then becomes infinitesimal. People recovering from rheumatic fever are usually advised to take penicillin continuously for at least five years, and sometimes for life, to prevent reinfection and the high risk of recurrent heart disease.

The incidence of rheumatic fever in the United States has been dropping steadily for many years, perhaps because of the widespread practice of administering penicillin for sore throats. In 1979 about 650 cases of rheumatic fever were reported, compared with about 9000 cases in 1960.

Diphtheria

A 62-year-old man was seen by a doctor because of a sore throat. The man had exudate* on his epiglottis and one of his tonsils, and the doctor prescribed an antibiotic without taking a specimen for culture. Ten days later, it was learned that the man had been exposed to diphtheria. Specimens were obtained and cultured for *C. diphtheriae*. These were negative. He was given diphtheria toxoid immunization, since he could not remember ever being immunized against diphtheria. Six weeks after his sore throat started, he experienced difficulty in swallowing, had double vision, and became very weak. Two days later, he was dead from diphtheria.

Diphtheria usually begins with a mild sore throat and slight fever, accompanied by a great deal of fatigue and malaise. Swelling of the neck is often dramatic. A whitish-gray membrane forms on the tonsils and throat or in the nasal cavity. Heart and kidney failure and paralysis may follow these symptoms.

The cause of diphtheria is *Corynebacterium diphtheriae* (Table 21–2), a nonmotile Gram-positive rod of variable shape that often stains metachro-

*exudate—pus

TABLE 21–2   CHARACTERISTICS OF *CORYNEBACTERIUM DIPHTHERIAE*

Identification	*Reservoir*	*Pathogenesis*
Pleomorphic, nonmotile Gram-positive rods that show metachromatic staining; grow on media containing tellurite as gray or black colonies; diphtheria exotoxin produced identified by specific antiserum	Humans with nose, throat, or skin colonization	Growth in the upper respiratory tract results in formation of a membrane that can cause respiratory obstruction; absorbed toxin damages heart, nervous system, and kidneys

matically.* These bacteria require special media and microbiological techniques for their recovery from throat exudate or other material. It is unlikely that they would be recovered by the usual methods for culturing S. *pyogenes* from sore throats. Their identification depends on staining appearance with a dye called methylene blue, on colonial appearance, on biochemical activity against carbohydrates and other substances, and on demonstration of the diphtheria toxin (color plate 42). Differences in the results of these procedures allow identification of different strains of C. *diphtheriae*, and this can aid in tracing epidemics. The ability of C. *diphtheriae* to produce toxin requires that the bacteria contain a certain bacteriophage, an example of phage conversion.

Humans, either carriers or those with active diphtheria, are the source of C. *diphtheriae*. The organisms are inhaled and establish infection in the upper respiratory system. They have very little invasive ability, but the powerful exotoxin they release is absorbed by the bloodstream. The toxin kills cells by interfering with protein synthesis. The gray-white membrane that forms in the throat of diphtheria victims is made up of clotted blood along with cells of the host mucous membrane and inflammatory cells* that have been killed by exotoxin from growing diphtheria bacteria. This membrane may come loose and obstruct the airways, causing the patient to suffocate. Absorption of the toxin by body cells results in damage to the heart, nerves, and kidneys. Even with modern treatment, about one of ten diphtheria patients dies of the disease; however, simple immunization procedures are extremely efficient in preventing the disease, and fewer than 200 cases per year are generally seen in the United States.

Effective treatment of diphtheria depends on giving antiserum against the

---

*metachromatic staining—situation in which binding of the dye to substances within the cell causes a change in the color of the dye, for example, from blue to red

*inflammatory cells—cells from blood and tissue that congregate in the area as part of the inflammatory response (see Chapter 9)

TABLE 21–3 DIPHTHERIA FACTS

Incubation period	Communicable period	Immunization	Treatment
2–6 days	About 2 weeks; two cultures from throat and two from nose should be taken not less than 24 hours apart. If they fail to show diphtheria bacilli, then the person can resume activities	DPT[a] vaccine given at age 2 months, 4 months, 6 months (optional), 18 months, 4–6 years TD[b] vaccine given at age 14–16 years and repeated every 10 years	Diphtheria antitoxin (test for sensitivity to horse serum first); penicillin and erythromycin given for 7–10 days

[a]DPT = diphtheria, pertussis (whooping cough), tetanus.
[b]TD = tetanus and diphtheria.

diphtheria toxin. This must be done as soon as the disease is suspected, since a delay of several days to obtain cultural confirmation may be fatal. The bacteria are sensitive to antibiotics such as erythromycin and penicillin, but antibiotic treatment only serves to stop transmission of the disease; it has no effect on the toxin that has been absorbed.

Toxoid, prepared by formalin* treatment of diphtheria toxin, is used for immunization against diphtheria. Antibodies produced in response to toxoid administration specifically neutralize the diphtheria toxin. Because serious damage in diphtheria results from toxin absorption rather than microbial invasion, control of diphtheria can be accomplished effectively by immunization with toxoid. The well-known childhood "shots" (DPT) consist of diphtheria and tetanus toxoids and pertussis vaccine, all three generally given together. Unfortunately, these immunizations are often neglected, particularly among socioeconomically disadvantaged groups, and serious epidemics of the diseases occur periodically. For example, 66 cases of diphtheria with three deaths occurred in San Antonio during the first eight months of 1970. Seventy-five percent of these people had had no previous immunization. An epidemic that occurred in the Pacific Northwest in 1972 and 1973 had associated skin ulcerations in many cases. People with diphtheritic ulcers represented an important reservoir in this epidemic because their ulcers were chronic, caused few symptoms, and discharged large numbers of C. diphtheriae organisms. Table 21–3 summarizes some important facts about diphtheria.

Earache (Otitis Media) and Sinus Infections (Sinusitis)

Infections of the middle ear result when infectious agents from the nasal passages and throat spread upward through the eustachian tube. This may occur simply by cell-to-cell spread of an infection, undoubtedly assisted at

---

*formalin—approximately 37% solution of formaldehyde and water

TABLE 21–4 PRINCIPAL BACTERIAL CAUSES OF SINUS INFECTIONS

S. pneumoniae	H. influenzae	Bacteroides *species*
Gram-positive diplococci, often encapsulated; bile soluble; alpha-hemolytic colonies	Gram-positive pleomorphic rods; facultative anaerobes; require X- and V- factors* for growth	Gram-negative pleomorphic rods; strict anaerobes

*X- and V-factors—X is a fraction of hemoglobin called hematin; V is NAD.   NAD, page 317

times by changes in airway pressure forcing infected secretions upward in the tube. Most people have experienced such sensations of pressure change while driving down a steep hill or landing in an airplane. The infection damages the ciliated cells, resulting in inflammation and build-up of pressure from fluid or pus collecting behind the eardrum. The throbbing ache of a middle ear infection is produced by pressure on nerves supplying the middle ear. Normal drainage through the eustachian tube is impaired by the poor ciliary action of its damaged mucous membrane and by inflammation, which tends to decrease the effective diameter of the tube.

The organisms that cause otitis media are those that infect the upper respiratory system. The majority of bacterial middle ear infections are caused by S. pneumoniae; many others are caused by H. influenzae. Streptococcus pyogenes may also cause severe otitis media but less commonly than the other two species. Occasional cases of otitis media are caused by a species of Mycoplasma. Cultures for bacteria are negative in about half the cases of otitis media, and such infections are presumed to be caused by respiratory viruses.

The respiratory viruses associated with the common cold frequently infect the sinuses. Bacterial sinus infection with S. pneumoniae, H. influenzae, or anaerobes such as Bacteroides species (Table 21–4) are also quite common. Bacterial infections of the ear may result in perforation of the eardrum. On rare occasions, infections of the ear and mastoid spaces may extend into the bone to involve nerves passing through the skull or to produce infection of the brain and its covering membranes (meningitis). Occasionally, bacterial infections of the sinuses spread in a similar fashion.

Scientific studies of the relationship between viral and bacterial upper respiratory diseases are still incomplete. However, upper respiratory viral infections are often followed closely by infections with S. pneumoniae, H. influenzae, S. pyogenes, N. meningitidis, and other bacterial respiratory pathogens.[3] Indeed, as mentioned in Chapter 20, one of the greatest hazards of measles (rubeola) is bacterial infections of the respiratory system, because the measles virus infects the mucous membranes just as it does the skin. Damage to the membrane caused by the virus predisposes one to bacterial invasion.

[3]These infections are often called "superinfections" or secondary infections.

Conjunctivitis (Pink Eye)

Symptoms of conjunctivitis include excessive tears and redness of the conjunctiva of one or both eyes followed by swelling of the lids, sensitivity to light, and sometimes a large amount of yellowish discharge. All or some of these symptoms generally last one to three weeks, depending on the cause. The most common infectious agents are *Haemophilus influenzae*[4] and pneumococci. Other bacteria, including *Moraxella lacunata, Chlamydia trachomatis, N. gonorrhoeae, S. aureus,* and *C. diphtheriae,* sometimes infect the conjunctiva, but the symptoms generally are different from those of pink eye.

Viral conjunctivitis is usually milder but may be long-lasting and produces little yellow exudate. The disease occurs throughout the country but is most prevalent in the warmer climates and is often associated with symptoms of "colds." Other people become contaminated through contact with conjunctival discharge from hands and clothing of the infected person. Children under five years of age are most often affected.

Bacterial conjunctivitis is effectively treated by eye drops or ointments containing an antibacterial medicine to which the infecting strain is sensitive.

SUMMARY

The middle ears, mastoids, sinuses, and nasal corners of the eyes all connect with the nasal passages and throat to compose the upper respiratory system. Inhaled organisms may establish infections in the lining cells of the nasal airways, and these infections may spread along common membranes to other parts of the upper respiratory system and to the lung. Rhinoviruses and adenoviruses, as well as the bacterial agents *Streptococcus pneumoniae, Haemophilus influenzae,* and *Streptococcus pyogenes,* are frequent causes of infections of the upper respiratory system. *S. pneumoniae* and *H. influenzae* frequently cause conjunctivitis.

Among the most serious complications of upper respiratory tract infections are structural damage to the eardrums and extension of infection to the bone of the skull and the nervous system. Viral infections of the upper respiratory system increase the risk of bacterial superinfections.

Diphtheria is a serious bacterial disease that can be completely controlled with adequate immunization. Streptococcal throat infections can be controlled with penicillin. Late complications of the latter disease are rheumatic fever, which can cause severe heart damage, and glomerulonephritis, a serious kidney disease.

---

[4]This species includes strains formerly known as *Haemophilus aegyptius* (Koch-Weeks bacillus).

SELF-QUIZ

1. Antibiotic treatment is not recommended for the common cold because
   a. it is too expensive
   b. antibiotics do not kill viruses
   c. a very large amount would have to be taken to be effective
   d. antibiotics should be saved for "big infections"
   e. all of the above
2. The chance of developing an effective vaccine against colds is small because
   a. rhinoviruses are not antigenic
   b. vaccine is not needed since colds are not a significant economic problem
   c. antibodies do not reach the respiratory tract surfaces
   d. hypersensitivity could be a problem
   e. there are too many different viruses that can cause colds
3. Recovery from colds involves
   a. interferon
   b. the inflammatory response
   c. T lymphocytes
   d. antibody
   e. all of the above
4. The possible complications of colds include
   a. ear infections
   b. sinus infections
   c. pneumonia
   d. all of the above
   e. none of the above
5. Transmission of cold viruses commonly occurs via
   a. air
   b. hands
   c. mosquitoes
   d. both a and b
   e. none of the above
6. Which of the following can interfere with the effectiveness of the cilia lining the upper respiratory system?
   a. narcotics
   b. viral infection
   c. tobacco
   d. alcohol
   e. all of the above
7. Rheumatic fever is a complication of infection with
   a. an adenovirus
   b. *Streptococcus pyogenes*
   c. *Haemophilus influenzae*
   d. *Streptococcus pneumoniae*
   e. a rhinovirus

8. Antibiotic treatment is given to people with diphtheria because
   a. it neutralizes the exotoxin of *Corynebacterium diphtheriae*
   b. it prevents the complications of diphtheria
   c. it interrupts the spread of the disease
   d. all of the above
   e. only a and b

QUESTIONS FOR DISCUSSION

1. Why do you think that some respiratory viruses, such as adenoviruses, typically cause fever, while others, such as cold viruses, usually do not?
2. If you were the manager of a business depending on a large number of employees, what policies would you set for minimizing the chances that the employees would catch colds and thus lose time from work?
3. Can you think of any steps in the pathogenesis of colds where the process might conceivably be stopped?
4. Immunization against diphtheria toxin appears to cause a decline in the number of people colonized with *C. diphtheriae*. Can you think of any reason why this decline might be expected to happen?

FURTHER READING

Brock, T. (ed.): *Milestones in Microbiology.* Englewood Cliffs, N.J.: Prentice-Hall, Inc., 1961. (Reprinted by the American Society for Microbiology in 1975). A collection of original papers written by microbiologists from the 1600's to the early 1900's. Two of the articles concern diphtheria.

D'Allesio, D. J., Peterson, J. A., Dick, C. R., and Dick, E. C.: "Transmission of Experimental Rhinovirus Colds in Volunteer Married Couples." *Journal of Infectious Diseases* 133:28–36 (1976).

Satir, P.: "How Cilia Move." *Scientific American* (October 1974) pp. 44–52. Good scanning electron micrographs of cilia with many interesting details.

Chapter 22

LOWER RESPIRATORY SYSTEM INFECTIONS

Several months after visiting her ill grandmother, a schoolgirl developed a persistent cough. Some time later, still coughing, she took a ten-hour school bus ride with 48 schoolmates. Unknown to anyone, her cough was a symptom of tuberculosis, which she had contracted from her grandmother. During that bus ride, everyone except the bus driver became infected with tuberculosis.

OBJECTIVES

To know
1. The basic anatomy of the lower respiratory system and how it relates to lung infections.
2. The characteristics of pneumonia caused by *Streptococcus pneumoniae* and *Klebsiella* species.
3. The nature of whooping cough, its cause, and its prevention.
4. The characteristics of tuberculosis that make it different from most other bacterial lung diseases.
5. The importance of influenza and how it can be controlled.
6. The nature, cause, and treatment of two fungus infections of the lung.

ANATOMY AND PHYSIOLOGY

The lower respiratory system consists of the windpipe (trachea) and its various branching divisions and subdivisions (bronchi and bronchioles) ending

439

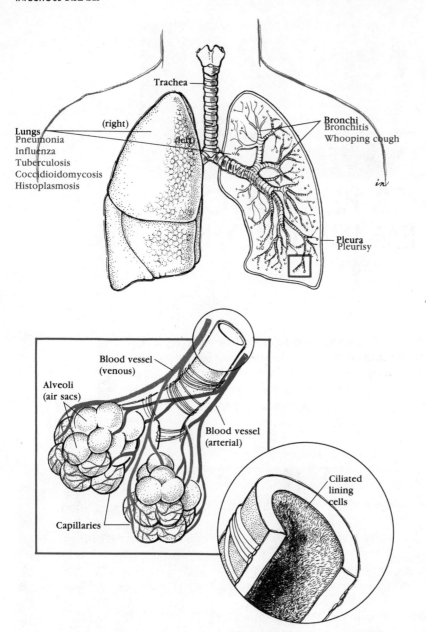

Lungs
Pneumonia
Influenza
Tuberculosis
Coccidioidomycosis
Histoplasmosis

Trachea

(right)

(left)

Bronchi
Bronchitis
Whooping cough

Pleura
Pleurisy

Blood vessel
(venous)

Alveoli
(air sacs)

Blood vessel
(arterial)

Capillaries

Ciliated
lining
cells

Figure 22–1
The lower respiratory system.

in the tiny, thin-walled air sacs (alveoli) that make up the lungs (Fig. 22–1). The lungs are surrounded by two membranes: One adheres to the lung, and the other adheres to the wall of the chest and diaphragm. These membranes (pleura) normally slide against each other as the lung expands and contracts. Thus there is a potential space between the two membranes where products of infection can accumulate and compress the lungs.

In contrast to their presence in portions of the upper respiratory system, microorganisms normally are absent from the entire lower tract. As with the upper tract, however, much of the lower respiratory system is lined with ciliated cells and a film of mucus (the "mucociliary escalator"). This film is constantly swept upward from the bronchioles and bronchi toward the throat at the rate of about an inch per minute under normal conditions, and its function is to trap and remove microorganisms and other foreign material. Extraneous factors such as tobacco smoke and chilling may decrease the ciliary action of the lower tract, and prolonged exposure to irritants leads to loss of cilia and flattening of the lining cells. The cough reflex, which also aids in expelling foreign materials, is activated by irritants and excessive secretions and can be depressed by external factors, such as alcohol and narcotics. Finally, phagocytic macrophages are numerous in the lung tissues and move readily into the alveoli and airways in response to foreign substances, such as microorganisms. These protective mechanisms are very efficient, especially against bacterial pathogens, and it is unusual to see bacterial lung infections except in people whose defenses are impaired.

Macrophage, page 236

PNEUMONIA

Pneumonia results from inflammation of the lung and is usually manifested by fever, cough, chest pain, and production of sputum. Pneumonia due to microbial infection may result when inhaled pathogenic organisms escape the defense mechanisms and lodge in the lower respiratory system. Perhaps more commonly, such microorganisms first establish themselves in the upper respiratory system, and a large inoculum is then accidentally carried into the lung with a ball of mucus during a deep breath or when the cough reflex is depressed. This is one way in which upper respiratory infection can be a prime factor in the development of lower tract infection. Serious pneumonias in babies and adults have also occurred when large numbers of organisms of low virulence are inhaled from air conditioning or hospital inhalation equipment containing contaminated water. Most pneumonias are bacterial or viral, but other microorganisms, chemicals, and allergies may also cause pneumonias.

Bacterial Pneumonias

Pneumococcal Pneumonia

One of the most common bacterial causes of lung infection is the Gram-positive, encapsulated diplococcus *Streptococcus pneumoniae* (pneumococcus)[1] (color plate 43). Pneumococci (Table 22–1) are parasites of humans, and both encapsulated and nonencapsulated variants of these organisms are

S. pneumoniae, page 299

---

[1]Formerly known as *Diplococcus pneumoniae*.

TABLE 22–1 CHARACTERISTICS OF BACTERIA INVOLVED IN LOWER RESPIRATORY SYSTEM INFECTIONS

Organism	Identification	Normal habitat	Pathogenic potential	Control
Streptococcus pneumoniae	Gram-positive cocci in pairs; colonies show collapsed centers and alpha hemolysis; bile-soluble	A common inhabitant of the upper respiratory tract of humans	Encapsulated strains are opportunistic pathogens; cause eye, sinus, and ear infections and pneumonia; infections sometimes spread to heart or brain; immunity results from antibody to the capsule, but there are many types of capsules	Immunization with capsular polysaccharides; treatment with antibiotics such as penicillin
Klebsiella pneumoniae	Encapsulated, nonmotile enterobacteria; lactose-fermenting; usually can use citrate as carbon source and give positive Voges-Proskauer* tests	Intestinal tract of humans and other animals; numerous environmental sources	Causes severe pneumonia in alcoholics and people with debilitating illnesses; can cause urinary and other infections	Maintenance of good health
Mycoplasma pneumoniae	Grows best under aerobic conditions; colonies become visible in 5 to 10 days; growth inhibited by specific antiserum	Humans; present in the respiratory tract for many weeks following infection	Common cause of pneumonia that is usually mild and self-limiting	No measures generally effective
Bordetella pertussis	Small, encapsulated, nonmotile Gram-negative rod; nonfermentative; requires special medium highly enriched with blood for growth	Respiratory tract of humans	Serious, long-lasting lung infections, especially in infants; causes violent coughing	Immunization of children using killed *B. pertussis*
Myco-bacterium tuberculosis	Acid-fast rods; aerobic; require special growth media; slow-growing; appear in culture as masses of bacterial cells in cordlike arrangement	Humans with tuberculosis	Highly infectious; initial infection is often asymptomatic and is confined to small area by host defenses, becoming inactive; infection commonly becomes active again later in life	Relatively resist-ant to disinfec-tants; disease detected by tuberculin tests or chest x-rays, treated with antimi-crobial drugs; prevention by BCG, a live vaccine

*Voges-Proskauer test—a method for detecting the production of acetylmethylcarbinol from glucose

commonly present among the upper respiratory flora in normal people. Pneumococcal pneumonia develops when encapsulated (virulent) pneumococci enter the alveoli of a susceptible host, multiply rapidly, and cause an inflammatory response. Serum and phagocytic cells pour into the air sacs of the lung, causing production of sputum and difficulty in breathing. This increase in fluid produces abnormal shadows on x-ray films of the chest in patients with pneumonia. The inflammation may involve nerve endings, causing pain, and when this pain comes from the pleura, the result is **pleurisy.** Pneumococci commonly enter the bloodstream from the inflamed lung and occasionally produce **septicemia** (infection of the bloodstream), **endocarditis** (infection of the heart valves), or **meningitis.** Infected people who do not have such complications usually develop sufficient specific antibodies within a week or two to permit trapping and destruction of the infecting organisms by lung phagocytes. Complete recovery usually results. Most pneumococcal strains cannot destroy lung tissue.

Opsonins, page 235

Treatment of most pneumoccocal infections is very effective if given early enough. Pneumococci are generally susceptible to penicillin, erythromycin, and tetracycline, but resistant strains are occasionally encountered. Strains resistant to several antibiotics were reported from South African hospitals in 1977, and there have been subsequent reports from other countries, including the United States. The mechanism that accounts for the rising incidence of antibiotic resistance in pneumococci is unknown. It has been ascribed partly to DNA transformation, but this is questionable.

As with several other common bacterial infections in which resistance to antimicrobial medicines has become a problem, renewed interest has been shown in developing preventive measures against pneumonia. Indeed, vaccines that stimulate the production of anticapsular antibodies to pneumococci have been prepared, giving immunity to pneumoccocal disease. More than 80 capsular types of pneumococcus are known, but about three fourths of all cases of pneumonia are due to one of only 12 types. To be fully effective, the vaccines against pneumonia must include pneumococci of each of the capsular types prevalent in a community. Immunization is especially important for certain high-risk patients, such as those suffering from chronic lung disease or alcoholism and those who lack a functioning spleen. In the United States alone, an estimated 20,000 deaths each year from pneumococcal disease could be prevented by immunization, and the benefits of immunization would be even greater in countries with less well-developed facilities for diagnosis and treatment of pneumococcal diseases.

Klebsiella Pneumonia

Among the less common but more serious causes of pneumonia are Gram-negative rods such as enterobacteria of the genus *Klebsiella* (Table 22–1) (color plate 44). The pathogenesis* of the pneumonia caused by these orga-

Klebsiella, page 316

---

*pathogenesis—the process by which disease develops

nisms is similar to that of pneumonia caused by pneumococci, except that permanent damage to the lung usually results and complications are more frequent. Most cases of klebsiella pneumonia occur in alcoholics, because alcohol interferes with lung defense mechanisms at several levels. These include impairment of ciliary action, cough reflex, and phagocytosis.

Klebsiellae persist better in the environment than do other enterobacteria, but they are killed by potent disinfectants such as phenolics and halogens. They are generally susceptible in vitro to certain antibiotics, including amikacin, tobramycin, and the cephalosporins, although the response of *Klebsiella* infections to therapy is often slow. *Klebsiella* strains tend to become resistant to antimicrobial medications very quickly because each population of the bacteria contains a few mutants highly resistant to the medicines usually employed. They also commonly contain R-factor plasmids, which confer resistance to several antibiotics, including aminoglycosides, tetracyclines, and cephalosporins. In some hospitals, *Klebsiella* strains are now the principal source of R-factors transferable to other pathogens. The virulence of klebsiellae is due partly to the antiphagocytic property of their capsules, and endotoxin may well play a role in the tissue destruction characteristic of the pneumonia they cause.

Laboratory diagnosis of this disease is complicated by the possible presence of *Klebsiella* strains in the oropharynx* of normal people, which may contaminate sputum specimens. Thus, merely finding colonies of klebsiellae in the sputum culture is not sufficient evidence on which to make the diagnosis of klebsiella pneumonia. Other evidence is needed, such as Gram stain of the sputum showing massive numbers of large, plump Gram-negative rods and pus cells.

Mycoplasma Pneumonia

Pneumonia caused by *Mycoplasma pneumoniae* (Table 22–1), frequently referred to as **primary atypical pneumonia,** may resemble pneumococcal pneumonia very closely. However, patients with mycoplasma pneumonia are usually not so acutely ill, and serious complications are not as common.

M. pneumoniae cells attach to specific receptors on cells lining the respiratory tract (Fig. 22–2), but the exact mechanism by which they produce pneumonia is unknown. One interesting aspect of the disease is the appearance of serum IgM antibodies called **cold agglutinins,** which cause clumping of human erythrocytes at low temperatures. The clumps disaggregate when the suspension is warmed. The reason for the appearance of these antibodies is that *M. pneumoniae* possesses an antigen similar to one of the human erythrocyte antigens, prompting an immunological cross-reaction. Tests for cold agglutinins are useful diagnostically, because approximately 80 percent

R-factors, page 135

M. pneumoniae, page 328

IgM, page 234

---

*oropharynx—the part of the throat just behind the mouth (as opposed to the nasopharynx, the part just behind the nasal cavity)

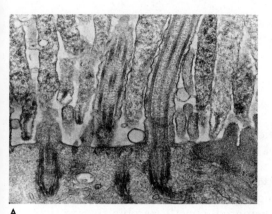

A

Figure 22–2
Electron micrograph (a) and diagram (b) showing attachment of *Mycoplasma pneumoniae* to respiratory membrane. Notice the distinctive appearance of the tips of the mycoplasmas adjacent to the host membrane. The tips probably represent a site on the microorganism that is specialized for attachment. (Courtesy of J. B. Baseman; from P. C. Hu, A. M. Collier, and J. B. Baseman, *J. Exp. Med.*, 145:1328, 1977.)

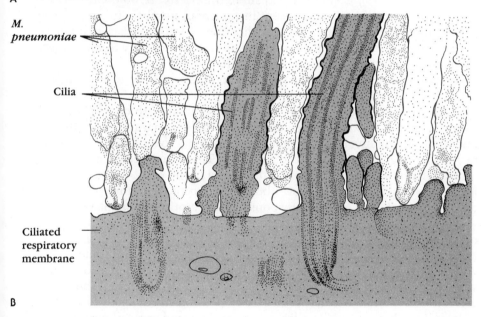

M. pneumoniae

Cilia

Ciliated respiratory membrane

B

of hospitalized people who have mycoplasma pneumonia have cold agglutinins. A specific diagnostic aid for this disease is a complement-fixation test using the patient's serum and mycoplasma antigen. The causative organisms can also be recovered in cultures, but because of their slow growth, diagnosis may take two weeks or more.

Complement-fixation, page 246

Not much is known about the survival of *M. pneumoniae* outside the host. *Mycoplasma pneumoniae* is presumed to spread from person to person by droplet infections. The organisms are not susceptible to penicillins because they lack a cell wall, but they are susceptible to several other antibiotics, such as erythromycin and the tetracyclines.

The diagnosis of mycoplasma pneumonia is complicated by the fact that, besides pneumococci, other bacteria and many viruses can produce pneumonia resembling mycoplasma pneumonia. Cultures, blood tests, sputum smears, and x-rays all help in identifying *M. pneumoniae* as the causative agent.

OTHER BACTERIAL INFECTIONS OF THE LUNG

Whooping Cough

Whooping cough is primarily a childhood disease and involves nasal congestion and a mild cough. After a week or two, these symptoms are replaced by spasms of violent coughing followed by a loud gasping noise as the patient draws a breath. Vomiting and convulsions frequently occur.

B. pertussis, page 313

The causative agent of whooping cough is the bacterium *Bordetella pertussis* (Table 22–1), a small, encapsulated, strictly aerobic Gram-negative rod. Early in the illness, it is present in enormous numbers in the respiratory secretions of the host. Specific identification of the organism is made using antiserum, and fluorescent antibodies can reliably identify *Bordetella pertussis* in smears of nasopharyngeal secretions.

Infection with *Bordetella pertussis* is confined entirely to the surfaces of the upper respiratory tract and tracheobronchial system,* on which the organisms grow in dense masses. The mucus becomes thick and sticky, and ciliary action is slowed. These factors result in the characteristically violent but relatively ineffective coughing of whooping cough. Some of the organisms disintegrate and release their toxin. This causes death of some of the

Lymphocyte, page 213

lining cells of the tract and stimulates a rise in the numbers of lymphocytes in the bloodstream of the host. About 10 percent of infants with pertussis die from the disease, and the response to antibiotic treatment is poor.

Bordetella pertussis organisms do not tolerate drying or sunlight and die quickly outside the host. They are spread entirely by inhalation of droplets from the respiratory system of an infected person, often an elderly person who has few symptoms except a mild cough. Intensive vaccination of infants with a killed vaccine can prevent the disease. This vaccine is about 70 percent effective, and its widespread administration to young children is probably responsible for the drop from 265,269 reported cases of pertussis in the United States in 1934 to only about 1500 per year at present. However, pertussis continues to be a major problem in poor areas of the Southeast and in big-city slums because of the lower percentage of infants who receive vaccine in these areas.

Tuberculosis

Tuberculosis is a chronic infectious disease of the lungs manifested by fever, weight loss, cough, and sputum production. The disease lasts months or years. In some patients, the infection spreads to the bones, meninges,* kid-

---

*tracheobronchial system—the windpipe, bronchial tubes, and their branches

*meninges—membranes covering the brain and spinal cord

neys, or other parts of the body. In rare instances, tuberculosis begins as an infection of the skin or gastrointestinal tract, rather than the lung (color plate 45).

Early in the twentieth century, the death rate from tuberculosis in the United States was about 200 per 100,000 of the population, and almost everyone in larger urban areas had contracted the infection by the time they reached adulthood. Although the initial infection was usually arrested by body defenses and went unnoticed, reactivation and serious disease during later life were very common. In the United States today, new cases of active tuberculosis (as opposed to asymptomatic infection) occur at the rate of only about 15 per 100,000 of the population per year, and most cases are permanently arrested by modern treatment. Schoolchildren have a very low infection rate, and fewer than 5 percent of young adults have ever been infected. The reasons for the dramatic decline in tuberculosis are incompletely understood, since the decline began before the availability of specific therapy and modern public health measures. Nevertheless, tuberculosis is far from being conquered, especially among the poor. In the United States, there are an estimated 16 million people who have been infected in the past and are subject to possible reactivation and the development of disease. Overcrowding and malnutrition are important in the reactivation and spread of tuberculosis. Worldwide, tuberculosis is still one of the leading causes of death from infectious diseases.

Tuberculosis is caused principally by *Mycobacterium tuberculosis* (Table 22–1), the "tubercle bacillus," a slender acid-fast rod-shaped bacterium (color plate 46). Unfortunately, the simple acid-fast staining procedure does not identify *M. tuberculosis* conclusively because other pathogenic and nonpathogenic mycobacteria are common. Nonpathogenic mycobacteria are present in the secretions around the human urinary opening, in ear wax, and in grain products and other foods. They are frequently also present in tap water and air and can thus contaminate stains and other reagents used in microbiological procedures. Moreover, acid-fast bacteria are difficult to remove from glassware, and washing and autoclaving often do not change their shape or acid-fast staining property. Thus, cultures must be performed in addition to acid-fast staining to identify *M. tuberculosis*. Since growth of *M. tuberculosis* is inhibited by substances present in most laboratory media, special media must be employed. Growth is slow, and cultural identification typically requires three to eight weeks.

Tubercle bacilli are inhaled in air that has been contaminated by a person with tuberculosis. The microorganisms lodge in the lung and produce an inflammatory response; they are then ingested by macrophages, where they survive destruction and may multiply, being carried to lymph nodes in the region and often to other parts of the body. As mentioned in Chapter 11, after about two weeks, a delayed-type hypersensitivity to the tubercle bacilli develops. An intense reaction then occurs at the sites where the bacilli have lodged. Macrophages collect around the bacilli, and some fuse together to form large multinucleated giant cells. Lymphocytes and macrophages then

M. tuberculosis, page 325

Delayed hypersensitivity, page 265

collect around these multinucleated cells and wall them off from the surrounding tissue. This localized collection of inflammatory cells is called a **granuloma** and is the characteristic response of the body to microorganisms and other foreign substances that resist digestion and removal. The granulomas of tuberculosis are called **tubercles.** The mycobacteria multiply little or not at all in the tubercles, but a few may remain alive for many years. In areas in which there are large collections of tubercles, the blood supply may be so poor that the tissues die. If this process involves a bronchus, the dead material may discharge into the airways, causing a large defect (cavity) in the lung and a spread of the organisms to other parts of the lung. Coughing and spitting transmit the organisms to other people as well. Indeed, a lung cavity may persist with prolonged shedding of bacteria into the bronchus.

Hypersensitivity to the tubercle bacilli is detected by injecting very small amounts of tuberculin, a sterile fluid obtained from a culture of *M. tuberculosis*, or, more commonly, by injecting a purified fraction of the fluid (purified protein derivative, PPD) into the skin. Purified protein derivative consists of bacterial antigens that have been separated from the medium's components in the culture fluid. Similar preparations are available to test for hypersensitivity to other species of pathogenic *Mycobacterium*. In people who are sensitive, redness and a firm swelling develop at the injection site, reaching a peak intensity after 48 to 72 hours. This test is called the tuberculin or **Mantoux**[2] **test** (Fig. 22–3). A strongly positive reaction to the test generally indicates the presence of living tubercle bacilli somewhere in the body of the person tested. In individuals who are tuberculin-positive, any impairment to their general health or treatments that suppress the immune response may result in renewed multiplication, enlarged tubercles, and actively progressing tuberculosis. This may happen years after the initial infection. A positive tuberculin test implies that the person is resistant to *reinfection* but because of the danger of reactivation is much more likely to develop tuberculosis than a tuberculin-negative individual. In the United States today, most cases of this disease occur in people who were infected with *M. tuberculosis* many years ago.

Control of tuberculosis is achieved by identifying unsuspected cases by using skin tests and x-rays. Individuals found to have active disease are then treated, thus interrupting the spread of *M. tuberculosis*. In addition, people found to have recent conversion of their Mantoux test from negative to positive are also treated, even when no evidence of disease exists. Treatment reduces the risk (estimated to be 12 to 20 percent) that these people will develop active contagious disease later in life. Milk is no longer a significant source of tubercular infection because of the widespread use of pasteurization and the surveillance of dairy cows to prevent the spread of *Mycobacterium bovis*, the principal cause of bovine* tuberculosis, which can also infect humans.

---

[2]Pronounced "Man-too."

*bovine—having to do with cows

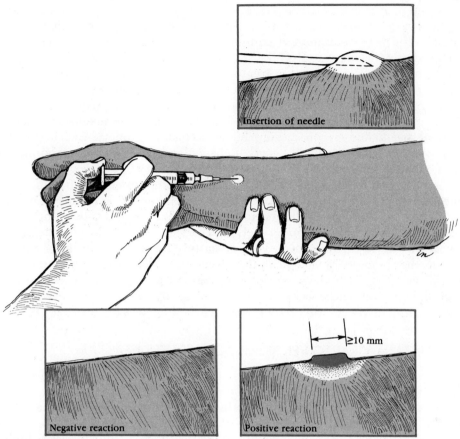

Insertion of needle

Negative reaction

Positive reaction

≥10 mm

Figure 22—3

Mantoux test for tuberculin hypersensitivity. A positive reaction requires 10 or more milli-
meters of induration (thickening of the skin). Redness of the skin alone is not a positive
reaction.

Antimicrobial medicines useful in the treatment of infections with *M. tu-*
berculosis include isoniazid (isonicotinic acid hydrazide, or INH), ethambutol, *para*-aminosalicylic acid (PASA), the aminoglycoside antibiotics (such
as streptomycin), and rifampin. However, other species of mycobacteria are
often highly resistant to these and other medicines. Even among sensitive
M. tuberculosis strains, drug-resistant mutants occur with high frequency.
Since mutants simultaneously resistant to more than one drug occur with
a very low frequency, two or more drugs are usually given together in treating tuberculosis. Because of the very slow generation times of most pathogenic mycobacteria and their resistance to destruction by body defenses,
drug treatment of tuberculosis must generally be continued for one or more
years to obtain permanent arrest of the disease. Even so, nonmultiplying
bacilli enclosed within old tubercles may not be killed by treatment.

Antitubercular medi-
cines, page 192

About 1 of every 14 strains of *M. tuberculosis* recovered from new untreated cases of tuberculosis is resistant to one or more antituberculous medicines. The first documented community outbreak of multiply resistant tuberculosis (resistant to INH, PASA, and streptomycin) occurred in Mississippi in late 1977. After being treated with a combination of INH, ethambutol, and rifampin, all but one person had a good response.

Vaccination against tuberculosis has been widely used in Scandinavia and many other parts of the world and is of proven value. The vaccinating agent, a living attenuated mycobacterium known as bacille Calmette-Guérin (BCG), is probably derived from *M. bovis*. Repeated subculture in the laboratory over many years resulted in selection of this strain of *M. bovis*, which has little virulence for humans but does result in some immunity. Bacille Calmette-Guérin vaccination is about 80 percent effective (the incidence of tuberculosis being one fifth that occurring in unvaccinated persons) and lasts at least several years. People receiving the vaccine also develop delayed hypersensitivity to tuberculin. Public health authorities have discouraged routine use of the BCG vaccine in the United States because it only prevents initial infection with *M. tuberculosis*. It does not prevent reactivation of infection acquired before BCG vaccination. Also, by causing a positive Mantoux test, BCG vaccination may prevent early diagnosis of tuberculosis.

INFLUENZA: A VIRAL LOWER RESPIRATORY TRACT DISEASE

Myxoviruses, Appendix VIII

Respiratory infection by myxoviruses of the influenza group is called **influenza.** It differs from bacterial pneumonia in its tendency to involve the bronchi and bronchioles and their supporting tissue to a greater extent than the alveoli. Even though the majority of people recover completely from influenza without any treatment, the number of deaths from this disease is high because so many people become infected. For example, in the period 1978 to 1980, more than 150,000 deaths were ascribed to influenza in the United States. Death may be caused by influenza itself, although it is more commonly the result of secondary invasion (superinfection) of the influenza-damaged lung by pathogenic bacteria. *Staphylococcus aureus, Haemophilus influenzae, Streptococcus pyogenes*, and *Streptococcus pneumoniae* are the chief bacterial offenders.

Influenza characteristically begins after an incubation period of one to two days with a sharp rise in temperature, muscle aches, and lack of energy. Headache, sore throat, nasal congestion, and cough follow quickly. Symptoms peak within a few days, and unless bacterial superinfection occurs, there is progressive improvement leading to recovery in several days. Some people continue to complain of fatigue for days or weeks after recovery. Bacterial superinfection typically is marked by an abrupt resurgence of fever and cough during the recovery phase of influenza.

The influenza virus enters the respiratory system by inhalation of air contaminated with virus from another infected person. Penetration of the respiratory mucus is presumably aided by neuraminidase, a viral enzyme at-

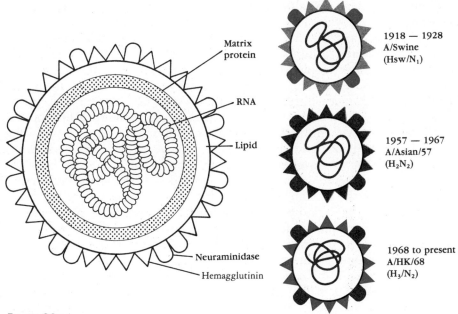

Figure 22–4
Stylized views of influenza A viruses.

tached to the viral envelope (outer coating of the virus). Infection becomes established in cells lining the air passages, the cells are destroyed, and large amounts of virus are then released to infect other cells. Death of the ciliated cells, inflammation, and leakage of plasma are prominent features of influenza.

The influenza viruses consist of a core composed of RNA and protein (nucleoprotein* core) surrounded by an envelope consisting of protein and lipid (lipoprotein* envelope). Two types of projections or spikes occur on the lipoprotein envelope, neuraminidase (N) and hemagglutinin (H). These spikes and the nucleoprotein are antigenic. Most human influenza viruses have one of two different kinds of nucleoprotein, A or B. The influenza A viruses are characterized by their H and N antigens. Five kinds of H antigens and two kinds of N antigens are found among the various known human influenza A viruses (Fig. 22–4). There are also different varieties of the influenza B viruses. Influenza viruses are designated by their nucleoprotein, the place and the year they were first recovered, and, in the case of influenza A viruses, their H and N antigens. For example, the predominant influenza virus in the United States in 1979–1980 was B/Singapore/79. There also were scattered cases caused by A/Bangkok/79(H3N2) and A/Brazil/78(H1N1).

---

*nucleoprotein—a macromolecule formed by complexing nucleic acid and protein

*lipoprotein—a macromolecule formed by complexing lipid and protein

For centuries, pandemics of influenza have swept the world at approximately ten-year intervals. The worst pandemic of modern times occurred in the fall of 1918, when the disease abruptly appeared in such diverse cities as Boston and Bombay, India, and then spread over much of the world. An estimated 500 million people were infected within a six-to-eight week period. At the peak of the epidemic, hundreds of people died each day in major cities. In the Seattle, Washington, area one of every 175 people between the ages of 20 and 30 died, and infants fared even worse. About one fifth of the fatalities occurred within the first four days of illness. The virus responsible for this major influenza epidemic was never recovered because the science of virology had not yet developed. Antibody tests performed years later on the survivors of the epidemics, however, showed that the virus had finally disappeared from the human population by 1929, but that a similar virus (swine influenza virus) had become established in pigs. The swine virus, perhaps a mutant of the epidemic human strain, has continued to be endemic* in pigs to the present time. The swine virus can infect humans but generally does not cause illness or spread from person to person. In January 1976, a new human influenza virus with an H antigen like that of swine influenza virus was identified at Fort Dix, New Jersey. The appearance of the new virus caused great alarm because, aside from a few military personnel who had been immunized against it, few people born after 1929 had antibody to swine influenza. Fortunately, spread of the virus was quite limited, and no major epidemic developed.

Table 22–2 lists some of the influenza A viruses that have been isolated during various epidemics. Generally speaking, the most serious epidemics occur when a strain with a new H antigen appears. The H antigen is associated with virulence, and a population that has not been exposed to a given H antigen will lack antibody to it and therefore will be highly susceptible. It is not known precisely where the genes for newly appearing H antigens come from, but there is strong experimental evidence that genes can be exchanged between human and animal viruses when cells are infected with both types of virus at the same time. This phenomenon may occur occasionally under natural conditions and give rise to major antigenic changes. However, another type of antigenic change, called **antigenic drift**, is seen between the major epidemics. This type of change is exemplified by A/Texas/77(H3N2) and A/Bangkok/79(H3N2), wherein there have been mutations that have altered the H antigen slightly. The antibody produced by people who have recovered from A/Texas/77(H3N2) is only partially effective against the mutant H3 antigen of A/Bangkok/79(H3N2). Thus, the newer Bangkok strain might be able to spread and cause a minor epidemic in a population previously exposed to the Texas strain.

Epidemics due to influenza B viruses generally produce fewer deaths than influenza A virus epidemics. As with other virus infections, a curious afflic-

*endemic—constantly present in a population

TABLE 22–2   SOME EPIDEMIC INFLUENZA A VIRUSES

Strain designation	Hemagglutinin	Neuraminidase
A/England/51	H1	N1
A/Singapore/57	H2	N2
A/England/64	H2	N2
A/Tokyo/67	H2	N2
A/Hong Kong/68	H3	N2
A/Texas/77	H3	N2
A/Bangkok/79	H3	N2
A/Brazil/78	H1	N1

tion known as **Reye's syndrome** sometimes occurs during influenza B infections (Fig. 22–5). The syndrome occurs predominantly in children between 5 and 14 years old and is characterized by liver and brain damage. The death rate, formerly 30 percent or more, has improved now that the syndrome is more widely known and therefore recognized and treated earlier. It is not known whether the influenza B viruses themselves are responsible or whether some other factor such as a medicine is the cause.

Prevention of influenza depends on the use of killed vaccines that are 80 to 90 percent effective in preventing disease when the vaccine is produced from the epidemic strain. Because of antigenic drift, vaccines produced from earlier strains having the same H type are less effective. Reactions from vaccination are usually minor with the newer purified vaccines and consist mostly of pain at the injection site and fever. However, a more serious reaction was observed with the swine influenza vaccine produced against the 1976 Fort Dix strain. This reaction, called the **Guillain-Barré syndrome,** occurred in about 1 of every 100,000 persons vaccinated. This syndrome is characterized by severe paralysis, and although most people recover completely, about 5 percent die of the paralysis. Use of the newer vaccines has not been associated with an increased risk of Guillain-Barré syndrome, and

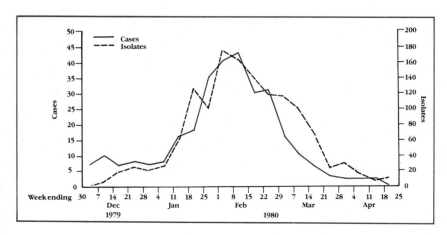

Figure 22–5
Reye's syndrome cases and influenza B isolates. (Adapted from *Morbid. Mortal. Wkly. Rep.*, United States Public Health Service, 29:27, 1980.)

the reason for its apparent association with swine influenza vaccine is not known.

It takes six to nine months after a new influenza strain first appears before adequate amounts of vaccine can be manufactured. Fortunately, amantadine, an antiviral medicine, can be employed to prevent influenza A disease when vaccine is not available. Amantadine is not effective in preventing disease caused by influenza B viruses.

Amantadine, page 200

FUNGAL INFECTIONS OF THE LUNG

Coccidioidomycosis

"Valley fever" and "desert rheumatism" are other names for coccidioidomycosis, which occurs commonly in certain hot, dry, dusty areas of the Americas. Fever, cough, chest pain, and loss of appetite and weight are common features of this disease. About one of ten people suffering the illness experiences hypersensitivity, manifested by a rash on the shins or on other parts of the body and pain in the joints. The majority of people afflicted with coccidioidomycosis recover spontaneously within a month and, in contrast to tuberculosis, have little risk of later reactivation of the disease. In a few people, the disease closely resembles tuberculosis. Death of lung tissue may lead to the formation of cavities, and, more rarely, the infection may spread throughout the body so that the skin, mucous membranes, brain, and internal organs become involved. About half of all those with disseminated coccidioidomycosis die unless proper treatment is given.

The causative agent of coccidioidomycosis, *Coccidioides immitis,* is a dimorphic* fungus living in the soil. It grows only in limited areas of the Western Hemisphere (Fig. 22–6). Infectious spores have unknowingly been transported to other areas (such as the southeastern United States), but there is no evidence that the fungus can establish itself in other climates. In the areas where *C. immitis* is endemic, infections occur only during the hot, dusty seasons when its spores are airborne; infections disappear during periods of rain. Virtually all individuals who live in these areas show evidence of having been infected. Indeed, people have become infected with *C. immitis* while traveling through contaminated areas.

The **tissue phase** of *Coccidioides immitis* is the form present in infected tissues and can be identified by microscopic examination of sputum or pus. It is a thick-walled sphere called a **spherule,** and in contrast to most other fungi growing in tissue, it never buds. The larger spheres of the fungus contain several hundred small cells called "endospores" (Fig. 22–7a). They have little resemblance to the endospores of bacteria, which are smaller and considerably more resistant to heat and disinfectants. The mold form of the organism grows readily on most laboratory media at room temperature and

---

*dimorphic—capable of existing as a mold under some conditions and as a yeast in others

Figure 22–6
Geographical distribution of
Coccidioides immitis.

is the form that grows in soil. On Sabouraud's medium (a specialized medium for growing fungi), the mold form of *C. immitis* usually grows in three to five days. Such cultures are extremely infectious, since most of the hyphae of the organism develop numerous very light barrel-shaped **arthrospores** (Fig. 22–7b), which separate easily from the hyphae and become airborne.

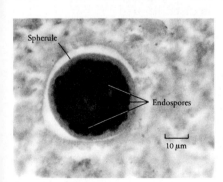
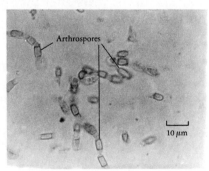

A B

Figure 22–7
Coccidioides immitis. (a) Tissue phase. Spherule containing endospores (stained preparation). (b) Mold phase. The barrel-shaped arthrospores are characteristic of this species.

Because other molds may resemble *C. immitis*, cultures may need to be converted to the spherule form to complete the identification. This conversion occurs when cultures are grown under conditions more nearly resembling those in body tissues.

Coccidioidomycosis is initiated by inhaling the arthrospores of *C. immitis* growing in the soil. Because these infectious arthrospores do not develop in humans or animals, transmission of *C. immitis* from animal to animal does not occur. After lodging in the lung, the arthrospores develop into spherules, which mature and discharge their endospores, each of which can then develop into another spherule. The pathogenesis of coccidioidomycosis is similar to that of tuberculosis.

The immune response of an individual can be measured by skin testing with **coccidioidin,** a liquid derived from the fluid portion of a *C. immitis* culture in a manner similar to that used for tuberculin. Most people show a positive coccidioidin skin test within three weeks after inhaling the fungal spores of *C. immitis*, and they retain the capacity to give a positive skin test for life if they remain in the area where the fungus occurs. Those who move away for several years and those whose disease spreads throughout their bodies often lose their skin reactivity to coccidioidin. Within the first month of illness, most people with coccidioidomycosis will develop precipitating antibodies, but these generally drop to low titers by the third month regardless of the progress of the disease. Complement-fixing antibodies to the organism arise after precipitating antibodies and often remain detectable for years, although the titer of these antibodies falls when recovery occurs. However, titers of complement-fixing antibodies continue to rise in disseminated infections. This laboratory finding means that treatment must be given to save the person's life.

Coccidioides immitis can be controlled by decreasing the amount of dust in the environment. Serious infections can be arrested by intravenous administration of the antibiotic amphotericin B. This treatment must be continued for months, and, even so, reactivation of the disease can sometimes occur months or years after therapy is discontinued.

Precipitating antibodies, page 243

Amphotericin, Appendix VI

Histoplasmosis

In many respects this disease is very similar to coccidioidomycosis, except for its distribution. Histoplasmosis occurs in tropical and temperate zones scattered around the world (Fig. 22–8). The causative organism of the disease is the dimorphic *Histoplasma capsulatum,* a fungus that prefers soils contaminated by bat or bird droppings (color plate 47). Caves and chicken coops are notorious sources of the infection. In its tissue phase, *Histoplasma capsulatum* is an oval yeast often growing within macrophage cells of the host (Fig. 22–9a). The mycelial (mold) form of the organism shows macroconidia characterized by numerous projecting knobs (Fig. 22–9b). Delayed hypersensitivity to the organism is detected by skin testing with **histoplasmin,** an antigen analogous to coccidioidin and tuberculin. Precipitating and comple-

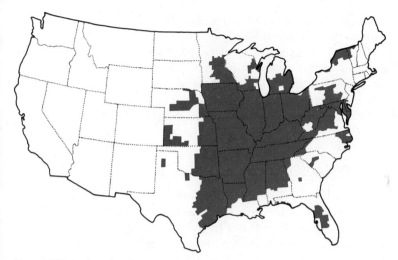

Figure 22–8
Geographical distribution of *Histoplasma capsulatum* in the United States, as revealed by high frequency of positive histoplasmin skin tests in humans. Human histoplasmosis has been reported from more than 40 countries other than the United States, including Argentina, Italy, Thailand, and South Africa.

ment-fixing antibodies develop in histoplasmosis as they do in coccidioidomycosis. These antibodies often cross-react with antigens of *Coccidioides* and *Blastomyces dermatitidis*, another pathogenic fungus. Skin tests and antibody determinations therefore help to establish the cause of a fungal infection only if antigens from all three fungi are tested at the same time. The antigen corresponding to the infecting fungus usually shows the largest skin test and greatest antibody response. Probably half of the reported cases of histoplas-

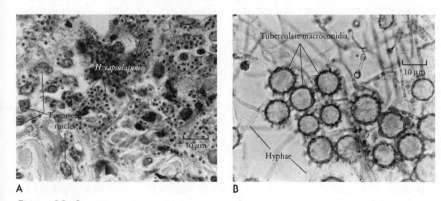

Figure 22–9
Histoplasma capsulatum. (a) Tissue phase. Numerous tiny yeast cells are present within the cytoplasm of macrophages. (b) Mold phase. The tuberculate macroconidia appear in cultures grown at 25°C.

mosis have occurred in the United States, primarily in the Mississippi River drainage area and in South Atlantic states such as Maryland and North Carolina; skin tests reveal that millions of people living in these areas have been infected. As with *Coccidioides*, only a tiny fraction of people infected with *Histoplasma* develop serious illness, and the antibiotic amphotericin B is helpful in arresting such serious infections. As with coccidioidomycosis, reactivation of the disease may occasionally occur after treatment is discontinued.

OTHER LUNG DISEASES CAUSED BY MICROORGANISMS AND VIRUSES

Besides the organisms already discussed, numerous other bacteria, viruses, and fungi produce infections of the lung. Other agents are unable to establish infection, but repeated exposure to them from inhalation may nevertheless result in serious damage from the inflammatory reactions they elicit. For example, repeated inhalation of hay dust containing certain thermophilic actinomycetes may produce an allergic reaction to the bacterial antigens. This reaction is characterized by severe inflammation and granuloma formation in the airways and alveoli. Spores of saprophytic fungi may be responsible for some cases of asthma. In such cases, there is little damage to lung tissue, but immediate hypersensitivity to the foreign antigen causes spasms of the airways and difficulty in breathing.

Legionnaires' disease, page 330

Legionnaires' disease, mentioned in an earlier chapter, is a recently discovered bacterial lung disease, with an estimated 26,000 to 71,000 cases per year occurring in the United States. The natural history of the causative organism, *Legionella pneumophila* (color plate 48), has not yet been fully clarified. However, the organism occurs in natural waters and apparently infects humans only in situations in which there is a high infecting dose or when there is some underlying lung impairment. The organisms can be made visible in lung tissue only by using special stains, and a medium containing unusual amounts of cysteine and iron is required for in vitro cultivation. The disease does not spread from person to person. The organisms are susceptible to erythromycin and rifampin. Sporadic and epidemic cases have now been identified in various parts of the United States and many other countries, usually by showing a rise in antibodies to *L. pneumophila* in the blood of the victims.

SUMMARY

Microorganisms enter the lung with inspired air or in infected material from the upper respiratory tract. They may be removed by bodily defense mechanisms, or they may produce a variety of diseases depending on the nature of the infecting organism. *Streptococcus pneumoniae*, *Klebsiella* species, and other bacteria can produce lung infections, as can viruses such as those causing influenza. Tuberculosis results primarily from infection with *Mycobacterium tuberculosis*. Chronic illnesses resembling tuberculosis may also result from infection by certain fungi, such as *Coccidioides immitis* and *Histoplasma capsulatum*. Severe lung disease can also result from inhaling saprophytic bacteria, as the consequence of hypersensitivity of lung tissues to these organisms' antigens. Legionnaires' disease is a lung infection caused by *Legionella pneumophila*. Vaccines can help prevent some lower respiratory infections, including pneumoccocal pneumonia, influenza, whooping cough, and tuberculosis.

SELF-QUIZ

1. The cold agglutinin test is used to diagnose which type of pneumonia?
   a. *Klebsiella*
   b. *Legionella*
   c. *Mycoplasma*
   d. *Streptococcus*
   e. all of the above
2. *Bordetella pertussis* organisms (the cause of whooping cough) are transmitted by
   a. the respiratory route
   b. contaminated needles
   c. insect bites
   d. contaminated food or water
   e. a and d only
3. The most likely source of tuberculosis in the United States today is
   a. contaminated milk
   b. patients known to have tuberculosis
   c. the unsuspected case
   d. laboratory accidents
   e. none of the above
4. A person has a positive skin test for tuberculosis if
   a. after 72 hours no reaction occurs after injection of the tuberculin under the skin
   b. there is an immediate reaction with swelling and redness
   c. there is an area of thickening of 10 mm or more in diameter reaching a peak intensity after 48 to 72 hours
   d. the Mantoux test is negative
   e. a complement-fixation test using patient's serum shows a titer of 1:10 with tuberculin antigen
5. The following is true of the drugs used to treat tuberculosis:
   a. most of the medicines used against other bacteria are effective against *M. tuberculosis*.
   b. Medicines used against *M. tuberculosis* are used for a shorter period of time than they would be used against other bacteria.

    c. Only a few of the medicines used against other bacteria are effective against tuberculosis.

    d. *M. tuberculosis* responds to antifungal agents only.

    e. Only one medicine is used at a time to avoid development of strains resistant to many antibiotics.

6. Most of the serious effects of influenza occur

    a. in schoolchildren

    b. as a result of bacterial superinfection (secondary invasion)

    c. in those who have received influenza vaccine

    d. in nonsmokers

    e. b and c only

7. Coccidioidomycosis

    a. has a worldwide distribution

    b. can be diagnosed by skin testing with coccidioidin

    c. is treated with penicillin

    d. is especially common in rainy climates

    e. a and c only

8. All of the following are useful in the diagnosis of bacterial pneumonia *except*

    a. blood culture

    b. sputum culture

    c. Gram stain of sputum

    d. x-ray of chest

    e. saliva culture

QUESTIONS FOR DISCUSSION

1. Do you think that cold viruses collaborate with pneumonia-producing bacteria in a "conspiracy" against humans? How much harder would it be for staphylococci, pneumococci, and other lower respiratory pathogens to get around if it were not for cold viruses?

2. What effect, if any, is the increasing life expectancy of the population having on the size of the tuberculosis reservoir?

3. You have a fever and a cough shortly after returning to New York City from a visit to the hot, dry central valley of California. Should your doctor warn the microbiology laboratory when sending your sputum for culture?

FURTHER READING

Devereux, P. D. and Goldstein, E.: "Legionnaires' Disease—Finding Answers to the Riddle." *American Journal of Nursing, 80*:81–85 (1980).

Fraser, D. W. and McDade, J. E.: "Legionellosis." *Scientific American* (October 1979).

Hoeprich, P. D.: "The Changing Face of Pneumonia." *Hospital Practice* (March 1979).

Jacobs, M. D., et al.: "Emergence of Multiply Resistant Pneumococci," *New England Journal of Medicine, 299*:735–740, (1978).

Kilbourne, E. D.: "National Immunization for Pandemic Influenza." *Hospital Practice* (June 1976).

Stuart-Harris, C.: "The Epidemiology and Prevention of Influenza." *American Scientist* (March–April 1981). A comprehensive article by one of the foremost authorities.

Chapter 23

ALIMENTARY SYSTEM INFECTIONS

> The face was sunken as if wasted by lingering consumption, perfectly angular, and rendered peculiarly ghastly by the complete removal of all the soft solids, in their places supplied by dark lead-colored lines. The hands and feet were bluish white, wrinkled as when long macerated in cold water; the eyes had fallen to the bottom of their orbes, and evinced a glaring vitality, but without mobility; and the surface of the body was cold.
>
> Description of cholera by army surgeon S.B. Smith in 1832

The alimentary tract, like the skin, is one of the body's boundaries with the environment and is continually exposed to microorganisms of the normal floral as well as to other organisms. It is, so to speak, our "inside outside," and consequently it is one of the major routes for invading germs.

OBJECTIVES

To know
1. The relationship between alimentary tract structure and function and infectious disease.
2. The pathogenesis of dental plaque and periodontal disease.
3. The characteristics of herpes simplex and of mumps.
4. The importance of the normal microbial flora of the alimentary tract.
5. The various types of viral hepatitis.
6. The important enterobacterial infections of the alimentary tract.
7. The common protozoan infections of the alimentary system.
8. The causes of food poisoning.

ANATOMY AND PHYSIOLOGY

The alimentary tract includes the mouth, esophagus, stomach, and small and large intestines. The system also includes some very important appendages: the salivary glands, the liver and gall bladder, and the pancreas. Fluids produced by all these appendages drain through ducts (tubes) into the alimentary tract, where they aid in digestion of food and carry out other functions. Figure 23–1 illustrates the relationships of the various parts of the alimentary system.

The Mouth and Salivary Glands

The teeth (Fig. 23–2) are made up largely of calcium phosphate but also contain some protein. The **enamel** of the teeth is especially dense, yet certain bacteria are able to penetrate it and produce cavities and tooth destruction. Both microscopic and large crevices in the surfaces of the teeth collect food particles and are sites for microbial colonization. The **gingival crevice** (space between the tooth and gum) is an ecological niche of great importance because both **gingivitis*** and **periodontal disease*** originate there. Saliva is secreted into the mouth from the various salivary glands at a rate of about 1500 ml per day. It serves to keep the mouth clean and lubricated, helps maintain a neutral pH, and, because it is saturated with calcium, tends to prevent the calcium phosphate of the teeth from dissolving in the acid produced by certain mouth bacteria. It has readily demonstrable inhibitory and killing powers against various groups of microorganisms, both members of the normal microbial flora and pathogens. Saliva also contains mucins* and IgA antibodies, both of which can coat bacteria and inhibit their ability to attach to and colonize the teeth and mouth tissues.

The Esophagus and Stomach

The esophagus connects the mouth and throat with the stomach, an elastic sac with a muscular wall. Some of the lining cells of the stomach produce hydrochloric acid and pepsinogen, a precursor of the protein-splitting enzyme **pepsin.** In response to the presence of food, the stomach begins a mixing action, bringing acid and enzymes into contact with the ingested material. Many types of microorganisms are destroyed by this action. As a result, the stomach lacks a normal resident flora. However, some microorganisms are very resistant to acid, and others may survive passage through the stomach when the acid is partially neutralized by food or antacid* medicines.

---

*gingivitis—inflamed gums

*periodontal disease—abnormality of the tissues surrounding the base of a tooth

*mucins—slimy sugar proteins like those found in mucus

*antacid—an agent that neutralizes acidity

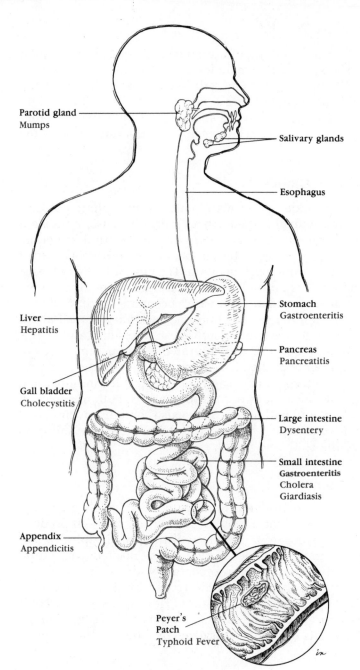

Parotid gland
Mumps

Salivary glands

Esophagus

Liver
Hepatitis

Stomach
Gastroenteritis

Pancreas
Pancreatitis

Gall bladder
Cholecystitis

Large intestine
Dysentery

Small intestine
Gastroenteritis
Cholera
Giardiasis

Appendix
Appendicitis

Peyer's
Patch
Typhoid Fever

Figure 23–1
The alimentary system.

The Small Intestine

A valve controls the passage of the well-mixed gastric contents from the stomach into the small intestine. The rate of passage slows down if there is considerable fat in the food or if it is very concentrated or dilute. The small intestine in an adult is about eight or nine feet long and has an enormous

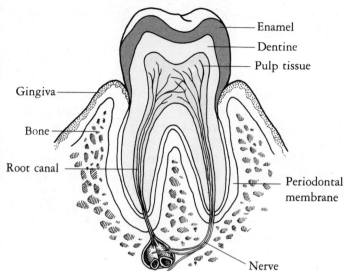

Enamel
Dentine
Pulp tissue

Gingiva

Bone

Root canal

Periodontal
membrane

Figure 23–2
A tooth and adjacent
structures.

Nerve

surface area available for absorption or secretion—about 30 square feet or
the area of a king-size bed. This large area results from the presence of many
small fingerlike projections (villi)—20 to 40 per square millimeter, each 0.5
to 1 mm long—along the inside surface of the intestine. Microbial diseases
may interfere markedly with the functions of the small intestine, such as
absorption and secretion, by affecting the cells lining the intestine. Failure
of absorption can cause vitamin deficiency and malnutrition, whereas too
much secretion is one of the important causes of diarrhea. Small patches of
lymphoid tissue spot the walls of the small intestine; they are sites of the
immune response to invading microbes, but paradoxically they may be selectively invaded by some pathogens, such as *S. typhi.*

The Liver and Gall Bladder

The liver has many functions, one of which is to remove the breakdown
products of hemoglobin from the bloodstream; these yellowish-green products are excreted with the bile, the fluid produced by the liver. Interference
with this excretory process, by infection or other factors, can produce **jaundice** (a yellow color of the skin and eyes). The gall bladder is simply a storage
area where bile is concentrated and stored and its acidity is neutralized.
Cholecystitis, infection of the gall bladder, is often a life-threatening disease,
but in other cases potentially pathogenic organisms actually grow in the gall
bladder for years without causing symptoms. The pathogens are discharged
into the intestine with the bile and are passed with feces. Since bile inhibits
the growth of many bacteria, it can be incorporated into media used to select
such pathogens. Bile is a detergent and thus can help to emulsify* fats in

---

*emulsify—to produce an emulsion, which is a mixture of two liquids not mutually soluble

the intestine; it aids in the absorption of fats and thus helps in the uptake of vitamins A, D, E, and K, which are fat-soluble.

The Pancreas

The pancreas also has multiple functions, including the production of insulin* and several digestive enzymes. About 2000 ml of pancreatic digestive juices and 500 ml of bile per day pour into the upper portion of the small intestine.

These fluids, as well as digestive juices of the small intestine itself, are alkaline and neutralize the stomach acid as it passes into the intestine. The pancreatic and intestinal fluids break down foods enzymatically into absorbable amino acids, fats, and simple sugars. These smaller molecules are then selectively taken up by the lining cells of the small intestine. Overproduction of small bowel fluids, induced by bacterial toxins, is a major cause of diarrhea.

The Large Intestine

Because of the great reabsorptive capacity of the small intestine, only 300 to 500 ml of gastrointestinal fluid normally reaches the large intestine per day. The large intestine absorbs water, electrolytes, vitamins, and amino acids from foods. After this further absorption, the semisolid feces remain. The feces represent an important source of opportunistic pathogens for other parts of the body, especially for the urinary tract and bloodstream. Infection of the large intestine may interfere with absorption and stimulate peristalsis,* thus causing diarrhea.

ORAL MICROBIOLOGY

It has been stated that if all the dentists in the United States were to work 24 hours a day filling cavities, as many new cavities would form each year as had been filled the previous year. Periodontal disease is no less significant, and the amount of money spent treating these two oral diseases clearly exceeds that spent for all other infectious diseases.

Bacteria of the Mouth

An enormous variety of bacteria inhabit the mouth, occurring on its mucous membranes, in tooth deposits, and in saliva. Two features of this oral bac-

---

*insulin—a hormone released into the blood from certain pancreatic cells; insulin deficiency results in diabetes

*peristalsis—the progressive wavelike contraction of intestinal walls that moves the contents in the proper direction

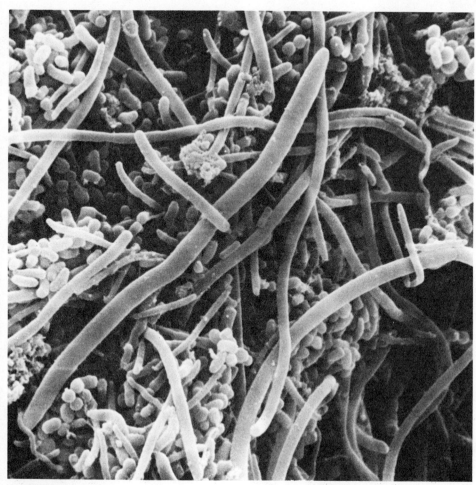

Figure 23—3
Scanning electron micrograph of dental plaque. (Courtesy of W. Fischlschweiger and D. C. Birdsell.)

terial population are especially noteworthy. First, of all the species of bacteria introduced into the mouth from the time of birth onward, relatively few are able to colonize and persist in the oral cavity. Second, the distribution of these bacterial species is far from uniform; certain members colonize one area, and others colonize another. Most prominent are species of the genus *Streptococcus*, Gram-positive chain-forming cocci that metabolize carbohydrates with the production of lactic acid. *Streptococcus salivarius* preferentially colonizes the upper part of the tongue; *Streptococcus sanguis* prefers the teeth; *Streptococcus mitis* likes the mucosa of the cheek. Collections of bacteria adhere to smooth surfaces and in the crevices of the teeth. These spongelike collections, called **dental plaque** (Fig. 23—3), are composed

Streptococci, page 297

TABLE 23–1 CHARACTERISTICS OF SOME OF THE BACTERIA FOUND IN DENTAL PLAQUE

Genus	Characteristics
Actinomyces	Gram-positive branching filaments; usually prefer or require anaerobic growth conditions; ferment sugars, producing acid end-products
Bacterionema	Gram-positive filaments with a bacillus attached to one end; facultative anaerobes; branching often occurs under aerobic conditions; produce several acid by-products during sugar fermentation
Leptotrichia	Long Gram-negative rods with pointed ends; anaerobic; mostly lactic acid produced during sugar fermentation
Nocardia	Gram-positive and sometimes acid-fast filaments; aerobic growth. Branching or fragmentation of filaments into bacilli commonly occurs at some stage of colonial growth
Rothia	Gram-positive; aerobic; mostly lactic acid produced during sugar fermentation; may show coccoid,* rod-shaped, or filamentous forms depending on growth conditions and age of the colony
Streptococcus	Gram-positive cocci in chains; usually prefer anaerobic growth conditions or are indifferent to oxygen; produce mostly lactic acid from sugars
Veillonella	Gram-negative anaerobic cocci that do not ferment carbohydrates; can use lactic acid as a nutrient

*coccoid—shaped like cocci

chiefly of streptococci intermixed with filamentous Gram-positive branching and nonbranching bacteria (*Actinomyces, Rothia, Nocardia, Bacterionema, Leptotrichia* genera) and anaerobic Gram-negative cocci of the genus *Veillonella* (Table 23–1). The gingival crevice characteristically is populated by strict anaerobes of the genera *Bacteroides* and *Fusobacterium* and by smaller numbers of spirochetes. Dental deposits of bacteria range up to 100 billion organisms per gram of plaque, and salivary organisms number up to one billion per ml of saliva.

Bacteroides, page 318

Ecological Considerations

Colonization of the mouth depends upon the ability of microorganisms to adhere to oral surfaces. Certain strains of streptococci and other bacteria have short surface filaments that attach specifically to certain host tissues, allowing microorganisms to resist the scrubbing action of food and the tongue and the flushing action of salivary flow. Ability to adhere to specific tissues accounts for the differences in location of different species of streptococci in the mouth. The host limits the extent of bacterial colonization of its mucous membranes by constantly shedding the superficial layers of cells and replacing them with new ones; the rate of this shedding correlates with the size of the microbial burden.* Furthermore, the antibodies and mucins

*burden—the number of bacteria stuck to the cells of the mucous membrane

in saliva cover up the bacterial attachment filaments and thus prevent or weaken adherence to the oral tissues.

Colonization of the mouth by strictly anaerobic flora requires the presence of teeth because only these structures can provide habitats of sufficiently low oxidation-reduction potential* to allow these organisms' growth. The blind pockets created by the gingival crevices and fissures in the teeth are one such habitat. Tooth surfaces also represent a nonshedding tissue where enormous bacterial concentrations are able to build up. The metabolic activity of the various organisms growing on teeth consumes oxygen and thus provides another area where some anaerobic organisms can grow.

Dental Caries

The presence of **dental caries** (tooth decay) in prehuman skeletal remains more than 500,000 years old demonstrates that this problem has been with us for a long time. In Europe, up through the medieval period, only 10 percent or less of human teeth had cavities, and they were mostly on parts of the teeth exposed by the receding gums of older people. However, in the seventeenth century a big increase in the incidence of carious teeth occurred, and the location of the cavities changed to include the sides and biting surfaces of the teeth, particularly of young people. This change in the frequency and location of the cavities is related to the introduction of two new substances into the European diet—refined flour and sugar (sucrose). We now know that although refined flour can produce dental caries, it is of minor importance compared with sucrose. Initially, only the royalty and other rich people had rampant tooth decay, since only they could afford the price of sugar. Soon, however, the development of ocean shipping and slave labor on huge sugar-cane planations brought sugar (and dental caries) within the financial reach of everyone. Sugar in the diet has spread progressively throughout the world since then. Today, the average American consumes about 90 pounds of sucrose per year, mostly in soft drinks, candy, ice cream, sweetened breakfast cereals, and baked goods. Six billion dollars per year are spent on repairing the effects of tooth decay in the United States.

Cause of Dental Caries

Dental caries is an infectious disease caused principally by *Streptococcus mutans*. These Gram-positive chain-forming cocci live only on the teeth and cannot colonize the mouth in the absence of teeth. They grow readily on a variety of laboratory media, generally preferring anaerobic conditions. The colonies are usually nonhemolytic on sheep blood agar, although many strains are beta- and some are alpha-hemolytic. They are distinguished from other streptococci by their ability to ferment the sugar alcohols mannitol

---

*low oxidation-reduction potential—condition in which there is little or no oxygen available to stop the growth of anaerobes

and sorbitol and to produce a particular kind of extracellular gluelike polysaccharide from sucrose. There are marked differences among various strains of *S. mutans,* and it is very likely that some of these strains will ultimately be given other species names.

Formation of Cariogenic* Dental Plaque

The first step in the formation of human dental plaque is the adherence of a cluster of *S. mutans* streptococci to the tooth **pellicle,** a thin film of proteinaceous material adsorbed from the saliva. The streptococci that in turn adsorb to this film have a specific affinity for the tooth pellicle and are not the predominant streptococci in the saliva. Their attachment is initially weak, but they soon produce their sticky extracellular polysaccharide and are thereby glued firmly to the tooth. Other organisms adhere to the developing plaque. Initially, these include aerobic organisms of the genera *Neisseria* and *Nocardia.* Growth of these organisms is accompanied by a reduction in the oxidation-reduction potential of the plaque, making it suitable for anaerobic species to colonize. Production of lactic and other acids by streptococci in the plaque provides a suitably acidic medium for lactobacilli. Thus these organisms can also colonize dental plaque but not other parts of the mouth. Plaque-colonizing anaerobes of the genus *Veillonella* also adhere to the polysaccharide and use some of the lactic acid produced from the metabolism of streptococci and lactobacilli. Filamentous organisms, too, colonize the plaque and by the end of a week make up most of its bulk.

Production of Dental Caries

Scientists have placed tiny electrodes into dental plaque and have studied what happens to its pH when sugar enters the mouth. The response is dramatic. The pH of the plaque drops from its normal value of about 7 to below 5 within minutes, more than a hundredfold increase in the acidity of the plaque, to a level at which the calcium phosphate of teeth dissolves. The duration of this acidic state depends on the duration of exposure of the teeth to sugars and on the concentration of the sugars. Thus, people who eat frequent snacks are more prone to dental caries than are those who take large meals less frequently. After food leaves the mouth, the pH of the plaque rises slowly to neutrality. This slow rise in pH is due to the ability of *S. mutans* to store a portion of its food as an intracellular starchlike polysaccharide, which is then metabolized to acid. Plaques thus act as tiny acid-soaked sponges closely applied to the teeth. Note that both *S. mutans* and a suitable sugar-rich diet are required to produce dental caries. *S. mutans* cannot colonize teeth and establish an acid-producing plaque in the absence of sucrose.

---

*cariogenic—causing dental caries

The process of tooth decay is slow. The outer portion of the tooth, called **enamel,** is very hard and only slowly dissolves in acid. However, acid diffusing into the tooth from overlying cariogenic plaque penetrates the enamel and attacks the more easily dissolved **dentin.** After a period of about one to two years, one sees a white or brown spot on the tooth under the plaque, representing an area of weakened enamel. Beneath this spot is the beginning of a small cavity where the dentin has dissolved. Eventually, the weakened enamel gives way, exposing the underlying cavity.

Prevention of Dental Caries

Studies undertaken years ago demonstrated that some populations had a far lower incidence of dental caries than other populations receiving the same diet. This led to the discovery that the water consumed by the people with the higher rates of caries lacked fluoride, whereas the water used by the more fortunate populations with low rates of dental caries contained fluoride. Trace amounts of fluoride are now known to be required for teeth to resist the acid of cariogenic plaques. Many communities now add small amounts of fluoride to their drinking water, which results in a 50 to 60 percent reduction in dental caries. An alternative is direct application of fluoride compounds to teeth, which can give a 20 to 40 percent reduction in dental caries.

Another method of dental caries control is restriction of sugar in the diet. Various studies have shown up to 90 percent reduction in caries by eliminating sweets from the diet.

A third preventive measure is mechanical removal of plaque by tooth brushing and use of dental floss. Careful cleaning of teeth once a day for five to ten minutes can reduce the incidence of dental caries by 50 percent.

Extensive research is underway to explore other methods of prevention. Vaccines against *S. mutans* show promise, as does the use of inhibitors of the enzymes involved in formation of plaque polysaccharide.

Dental Plaque and Periodontal Disease

Periodontal disease is a chronic inflammatory process involving the supporting tissues around the roots of the teeth (color plate 49). It usually develops very slowly, but it is the chief cause of loss of teeth from middle age onward. Formation of dental plaque is apparently necessary for the development of periodontal disease, but considerably less is known about the remaining details of how the disease progresses. Plaque and **calculus** (calcified plaque) in the region of the gingival crevice cause an inflammatory process which at first produces swelling, redness, and cellular exudate in the gum (gingiva). This gum inflammation is known as **gingivitis.** When gingivitis is chronic, the gingival crevice becomes widened and deepened, allowing a marked overgrowth of bacteria. An abscess then develops and penetrates through the base of the gingival crevice (color plate 50). The large population of microorganisms in the abscess produces the enzymes collagenase, protease, and hyaluronidase[1] and also produces endotoxins. These products penetrate into the tissues. The host's immune response to the bacterial products undoubtedly contributes to the inflammatory process. The membrane that attaches the root of the tooth to the bone becomes involved, and the calcium in the bone surrounding the tooth is gradually dissolved. Periodontal disease thus eventually leads to the loosening and loss of teeth, but it can be halted in its earlier stages by cleaning out the inflamed crevice and removing plaque and calculus. Careful tooth brushing is effective in preventing periodontal disease.

Severe acute cases of gingivitis are given the name **Vincent's angina** and are characterized by death of tissue (necrosis) at the edge of the teeth, pain, and fever. Associated with the presence of dead tissue is a marked overgrowth of fusiform and spirochetal bacteria (*Fusobacterium* and *Treponema* species) (color plate 51).

Oral Bacteria and Endocarditis

Endocarditis, page 548

Dental procedures can introduce oral bacteria into the bloodstream. Normally, these organisms are quickly removed by the body's mononuclear phagocyte system, but in individuals with abnormal heart valves they can colonize the valves and cause bacterial endocarditis (see Chapter 22). To prevent endocarditis, dentists treat such patients with an antibiotic to decrease the numbers of bacteria living around the teeth and to kill any bacteria that enter the blood as a result of the dental procedure. When an antibiotic such as penicillin is given, there is a transitory drop in the numbers of oral bacteria, followed in a few days by a return of the oral flora to normal levels even if the antibiotic treatment is continued. The rise in numbers of oral bacteria is due to growth of strains resistant to penicillin, and this is why, in attempting to prevent endocarditis, antibiotic medication is started only a few hours before a dental procedure, continued for two days afterward, and then promptly discontinued.

Herpes Simplex Virus

People with cold sores ("fever blisters," **herpes simplex**) and even those who are apparently healthy can excrete infectious herpes simplex virus in their saliva over long periods of time and may thus represent a hazard to nurses, dentists, and others who come in contact with their saliva. The herpes simplex virus causes not only cold sores but also ulcers of the mouth and eye and a painful infection called **paronychia,** or herpetic whitlow, in the area around the fingernail. This condition can be distinguished from bacterial

---

[1]Collagen is a tough fibrous protein found in skin, bone, cartilage, and ligaments; hyaluronic acid is a gelatinous material that helps tissue cells adhere together. The enzymes collagenase and hyaluronidase break down these substances. Protease breaks down protein.

paronychia by the absence of bacteria in cultured material and by the failure to respond to antibacterial medicines.

Mumps

Mumps is an illness characterized by fever and painful swelling beneath one or both ears. This condition arises from an infection of the parotid glands, which are salivary glands located between the angle of the jaw and the ear. The agent responsible for mumps is a paramyxovirus known as the mumps virus; it is transmitted by saliva, in which it is usually found one to two days before the onset of illness. In a few infected people, no illness occurs, and in others, the major effects of mumps appear in the central nervous system, heart, testes, ovaries, pancreas, or thyroid gland, with or without involvement of the parotid glands.

Paramyxovirus, Appendix VIII

Because other infectious agents may sometimes involve the parotid glands, not every patient with fever and painful swelling of the glands can be assumed to have mumps. Mumps virus infections are most commonly identified by testing the patient's serum for development of complement-fixing antibodies against known mumps virus antigen prepared in the laboratory. The mumps virus can also be isolated by inoculating saliva or urine from infected people into chicken embryos. A vaccine can give partial protection against mumps, but no antiviral medicine is available for its prevention or treatment.

GASTROINTESTINAL MICROBIOLOGY

The rich microbial flora of the mouth is absent in the fasting stomach, owing to the killing action of hydrochloric acid and other antimicrobial secretions. The small intestine is also relatively free of microorganisms. The predominant organisms of the small intestine are usually bacteria able to grow in air, including facultatively anaerobic Gram-negative rods and various streptococci. Lactobacilli are commonly found in small numbers, as are yeasts such as *Candida albicans*. The main reason for such low concentrations of microorganisms is thought to be the flushing action caused by the rapid passage of the various digestive juices through the small intestine.

In contrast to the scanty numbers of organisms in the stomach and small intestine, the colon (large intestine) contains very high microbial counts, usually 100 billion per gram of feces. The numbers of anaerobic organisms of the genera *Lactobacillus* and *Bacteroides* generally exceed the numbers of all other organisms by about one hundredfold. Facultatively anaerobic Gram-negative rods, particularly *Escherichia coli*, and members of the genera *Enterobacter*, *Klebsiella*, and *Proteus*, predominate among fecal microorganisms able to grow aerobically. In contrast to the anaerobes, these organisms are easy to cultivate, and some have been studied in exhaustive detail.

Enterobacteria, page 313

In the intestine, just as in the mouth, bacteria adhere to specific tissue receptors. This undoubtedly helps to explain the pattern of distribution of the normal intestinal flora as well as the sites of intestinal attack by pathogens.

Biochemical Activities of Intestinal Microbes

Because so many different kinds of bacteria live together in the alimentary tract, some of them depending on each other for growth, it is very difficult to predict which of their metabolic effects observed in the test tube are important to their animal host. It is clear, however, that some intestinal bacteria can use up vitamin C (ascorbic acid) and several B vitamins (choline, folic acid, cyanocobalamine, and thiamine). Under certain conditions, intestinal bacteria can therefore contribute to vitamin deficiency. They can also produce ammonia, form organic acids from carbohydrates, and degrade bile and digestive enzymes.

Some of the metabolic effects of the intestinal flora undoubtedly play a role in human disease. One example is an unusual condition called the "blind loop syndrome" in which, as the result of disease or surgery, a pocket of the small intestine becomes relatively isolated from the rest of the organ. In the pocket, the flow of intestinal juices is reduced, and large populations of the normal intestinal flora build up. Individuals affected by this condition may become severely anemic and lose weight because the large numbers of bacteria that accumulate in the loop use up certain vitamins and degrade bile. The condition may be corrected by repairing the structural abnormality or by giving antibiotics to reduce the numbers of organisms in the loop.

Although some intestinal microorganisms may perform potentially harmful metabolic actions, other common intestinal bacteria synthesize an excess of useful vitamins. These include niacin, thiamine, riboflavin, pyridoxine, cyanocobalamine (vitamin B_{12}), folic acid, pantothenic acid, biotin, and vitamin K. These biosynthetic reactions are of enormous importance for the nutrition of some animals. For humans they are of doubtful importance except when the diet is inadequate. Thus, when an individual is poorly nourished, oral antibiotics sometimes bring about vitamin deficiency by decreasing the numbers of intestinal microorganisms that produce vitamins.

A wide variety of microorganisms and viruses produce gastrointestinal disease. A few of the diseases caused by pathogenic microorganisms are discussed in the following sections.

Bacterial Diseases

Diarrhea Caused by Campylobacter fetus

Campylobacter fetus, only recently recognized as an enteric pathogen, accounts for 5 to 10 percent of acute diarrhea and leads all other known bacterial causes in the United States. Like *Vibrio cholerae*, discussed in the following section, *C. fetus* is a Gram-negative curved rod-shaped bacterium that requires special media and laboratory techniques for recovery in culture. Infected animals and contaminated water supplies and foods have been sources of the infection. Most patients recover without treatment, but antibiotic treatment may be required for some if the bacteria invade the blood and tissues outside the intestine.

Cholera

Cholera is a very severe form of diarrhea caused by *Vibrio cholerae*. There have been seven pandemics of cholera since the early 1800's. The most recent one started in Indonesia in 1961 and spread to Southeast and South Asia, the Middle East, parts of Europe and Africa, and groups of Pacific islands. The current pandemic continues to spread slowly but has not yet reached the Western Hemisphere. Interestingly, no known cases of cholera occurred in the United States from 1911 to 1973. However, in 1973 and again in 1978, cases were discovered along the Gulf Coast. The infections were traced to coastal marsh crabs that had been eaten, but bacteriophage typing showed that these cases were due to a strain of *V. cholerae* different from the pandemic strain. It is not known how long this strain has been in the United States or how it got here.

Vibrio, page 316

Cholera is characterized by profuse outpouring of fluid from the intestine over a period of a few hours. The loss often amounts to 15 percent of the body weight. So much of the body's water and electrolytes* are lost that the blood becomes thickened and reduced in volume, producing insufficient blood flow to keep vital organs, such as the kidney, working properly. Many people die unless this lost fluid can be replaced promptly.

Much research has been done to determine how *Vibrio cholerae* produces cholera. Unlike *Salmonella* and *Shigella* infections, there is no visible damage to the lining cells of the intestine in cholera. The pathogenesis of the disease depends on a heat-labile* enterotoxin that causes excessive secretion of water and electrolytes by the cells of the small intestine.[2] The colon is not affected, but it cannot absorb the huge volumes of fluid that rush through it.

Treatment of cholera depends on the rapid replacement of salt and water before irreversible damage to vital organs can occur. The discovery that the absorptive ability of the small intestine is not impaired even in the presence of cholera toxin has simplified therapy enormously. The small intestine's absorptive mechanism requires the presence of glucose, so this sugar along with the proper amounts of sodium and potassium salts is simply dissolved

---

[2]The toxin binds to specific receptors on the surface of the cell, causing an increase in the conversion of ATP to cyclic adenosine monophosphate (cAMP). Accumulation of cAMP in the cell causes increased excretion of salt and water.

*electrolytes—acids, bases, and salts that are dissolved in body fluids

*heat-labile—destroyed by heating; for example, 100°C for 20 minutes

in water and taken by mouth. Thus the difficulties and expense of giving intravenous therapy are avoided.

Between epidemics of cholera, the *V. cholerae* organisms persist in the intestines of their human carriers, who have been shown to excrete them for periods as long as six years. Control of cholera is aided by administering a vaccine of killed *V. cholerae;* this stimulates bactericidal and toxin-neutralizing antibodies in its recipients but gives only partial protection and lasts only a few months. Well-nourished people living under sanitary conditions appear to be quite resistant to cholera. Indeed, over the decade ending in 1970, only six cholera cases and no deaths were reported among the millions of Americans who traveled in areas of the world in which cholera existed. The requirement of cholera vaccination for travelers entering the United States has been dropped.

Gastroenteritis (Diarrhea and Vomiting) Caused by E. coli

Escherichia coli is an almost universal member of the normal intestinal flora of humans (and a number of other animals) and was therefore long ignored as a possible cause of gastrointestinal disease. It was not until the early 1960's that *E. coli* became generally recognized as a cause of life-threatening epidemic gastroenteritis in infants. More recently, *E. coli* has been recognized as a cause of gastroenteritis in adults, notably as the agent responsible for many cases of travelers' diarrhea (also called "Delhi belly," "Montezuma's revenge," and the like). Understanding the importance of *E. coli* in gastrointestinal disease has depended on the knowledge that there are in fact hundreds of distinct strains of the bacterium and that only those possessing certain virulence factors cause gastrointestinal disease. Table 23–2 gives some of the factors now known to be responsible for the virulence of *E. coli*, but there is strong evidence that not all factors have been identified.

Some of these virulence factors, such as enterotoxin production and ability to adhere to the small intestine, depend on the presence of plasmids. These plasmids can be transferred to other *E. coli* organisms by conjugation, thereby rendering the recipient strain virulent also. Gastroenteritis-producing strains of *E. coli* often have more than one type of virulence plasmid. It is interesting that the heat-labile toxin of *E. coli* is very closely related to the cholera toxin produced by *Vibrio cholerae*. This suggests a common ancestry for the genes responsible for the two toxins.

Treatment of *E. coli* gastroenteritis consists of replacing the fluid loss resulting from vomiting and diarrhea. In addition, infants may require oral treatment with antibiotics such as gentamicin or polymyxin for a few days. Travelers' diarrhea can be prevented by taking an antibiotic such as tetracycline if the *E. coli* strains in the area visited are sensitive to the antibiotic. However, the widespread use of an antibiotic to prevent diarrhea promotes the appearance of resistant strains by fostering the spread of R-plasmids.

R-plasmids, page 135

TABLE 23–2 VIRULENCE FACTORS OF STRAINS OF *E. COLI* THAT CAUSE
GASTROINTESTINAL DISEASE[a]

Virulence factor	Characteristics	Laboratory detection
Labile toxin (LT)	Exotoxin, molecular weight 10^6; antigenic; inactivated at 60°C; acts by causing increase in cyclic AMP in intestinal cells, thereby causing hypersecretion of intestinal fluids; acts like cholera toxin	Visible change in certain tissue cultures; causes fluid accumulation in small intestine of rabbits
Stable toxin (ST)	Exotoxin, molecular weight 10^3; not antigenic; resists boiling for 30 minutes; does not increase cyclic AMP	Fluid accumulation in small intestine of mice and rabbits
Invasive property	Permits invasion of intestinal cells similar to that with *Shigella* species	Invasion of eyes of guinea pigs or certain tissue cell cultures
Adherence property	Surface material complementary to intestinal cell receptors	Specific antiserum; electron microscopy

[a]*E. coli* strains that cause diarrhea may have any, all, or none of these factors.

Dysentery Caused by Shigella *Species*

Another frequent source of intestinal infection is bacteria of the genus *Shigella* (Table 23–3). One of the most severe and extensive *Shigella* epidemics of modern times began in Guatemala early in 1969 with antibiotic-resistant *Shigella dysenteriae* striking almost simultaneously at widely separated villages. Characteristically, the people infected had fever, diarrhea, and vomiting; pus and blood appeared in the feces 12 to 72 hours after onset of the disease. The condition subsided without treatment in five to seven days in most instances. The causative organism of this epidemic was not immediately determined because the epidemic strain would not grow on many of the usual laboratory media, and therefore most attempts to culture the organism failed to yield *Shigella* until special media were devised. State public health laboratories in the United States began to isolate the epidemic agent from travelers returning from Guatemala during the same year, but only rarely did the agent spread from these people to others in the United States.

TABLE 23–3   ENTERIC PATHOGENS OF THE GENERA *SHIGELLA* AND *SALMONELLA*

	Characteristics	*Isolated from*	*Pathogenic potential*
Salmonella	Lactose-nonfermenting enterobacteria; usually motile	Cold- and warm-blooded animals, including people	Gastroenteritis; sometimes invasion of the bloodstream; *S. typhi* causes typhoid fever
Shigella	Lactose-nonfermenting enterobacteria; nonmotile	Humans	Gastroenteritis; dysentery

Shigella dysenteriae has remained a rare cause of shigellosis* in the United States, generally constituting less than 1 percent of the total *Shigella* isolates.

    Shigella organisms enter the mouth on materials contaminated directly or indirectly by feces; their source, however, is essentially always human. Spread of *Shigella* species in day care centers has been a common problem, and waterborne epidemics are not rare. The organisms invade and multiply in the cells of the intestinal lining. Death of these cells results in intense inflammation and formation of small intestinal abscesses and ulcerations. Severe dysentery is the result. Children infected with some *Shigella* strains often have headache, a stiff neck, and convulsions. Adults commonly have painful joints for weeks or months following recovery. Because shigellae are not commonly found in the general blood circulation during dysentery, the absorption of a bacterial toxin may explain the extraintestinal symptoms of this disease. Production of one or more exotoxins has been demonstrated for *S. sonnei* and *S. flexneri*, as well as the rare but more virulent *S. dysenteriae*.

    Control of the spread of shigellae is almost entirely accomplished by sanitary measures and surveillance of food handlers and water supplies. Most cases do not require treatment, but antimicrobial medicines such as ampicillin or trimethoprim sulfa are useful in severe cases in shortening the time during which shigellae are excreted. Two specimens of feces, collected at least 48 hours after stopping antimicrobial medicines, must show negative cultures for shigellae before a person is allowed to return to a day care center or to a food-handling job.

Gastrointestinal Diseases Caused by Salmonella Species

Salmonella enteritidis, a Gram-negative facultatively anaerobic rod (Table 23–3), is a prominent cause of gastroenteritis, infecting more than a million people in the United States every year. These organisms generally enter the

---

*shigellosis—disease caused by species of *Shigella*

gastrointestinal tract with food, although contaminated water, fingers, and other objects may also be sources. The food products most commonly contaminated with *Salmonella* are eggs or egg products and poultry. Epidemics have also resulted from contaminated brewers' yeast, protein supplements and dry milk and even from a medication used to help diagnose intestinal disease. The reason for such a large variety of substances possibly being contaminated with *Salmonella* is not hard to explain: These organisms infect a wide variety of animals—from cows and chickens to pet turtles—and it is primarily the strains of *Salmonella enteritidis* that originate in animals that produce diarrhea in people.

Virulent salmonellae invade the lining cells of the colon and lower small intestine, but unlike shigellae they commonly pass through to the underlying tissue. An inflammatory response results and causes an increase in fluid secretion and a decrease in fluid absorption by the intestine. Exotoxins are produced by some strains, but their role in pathogenesis is not yet fully understood. A few strains, mostly of *Salmonella cholerae-suis* tend to invade the bloodstream and tissues, producing abscesses, fever, and shock, sometimes with little or no diarrhea.

The causative organisms of salmonellosis* are usually present in the feces in large numbers. Because many species of *Salmonella* are relatively resistant to toxic chemicals, selective media containing such substances as bismuth sulfite or brilliant green dye can be used in culture media to suppress the normal fecal flora and allow the *Salmonella* to grow.

Control measures for *Salmonella* have concentrated on establishing reporting systems for salmonellosis, tracing of sources by careful identification of individual strains, and routine sampling of animal products for contamination. Adequate cooking is very important, especially of frozen fowl, since heat penetration to the center of the carcass may be inadequate even when the outside appears "done." Unfortunately, the incidence of reported cases of salmonellosis has continued to rise, and the organisms have also shown increasing plasmid-mediated resistance to antimicrobial medicines. This is partly due to selection of resistant strains of *Salmonella* by the widespread use of antibiotics such as tetracycline in animal feeds. Fortunately, most cases due to *S. enteritidis* do not require treatment with antimicrobial medicines.

Typhoid Fever

Typhoid fever is caused by *Salmonella typhi* and is characterized by fever that increases over a three-day period and by severe headache and abdominal pain. In some cases, there is rupture of the intestine and shock from loss of blood. If treatment is not given, about one in five affected people dies of the disease.

*salmonellosis—disease caused by species of *Salmonella*

Salmonella typhi infects only humans. The organisms usually enter the gastrointestinal tract in contaminated food or water, as do other salmonellae, and like them readily penetrate the intestinal lining. Phagocytic cells appear in the gut wall and ingest the invaders, but the typhoid bacilli are not destroyed and multiply inside these cells. The infected cells are carried to other parts of the body by the bloodstream. *Salmonella typhi* also localizes in the collection of lymphoid cells (Peyer's patches; Figure 23–1) of the intestinal wall, and destruction of this tissue can lead to intestinal rupture and hemorrhage.

Typhoid organisms can be recovered from the blood or bone marrow of an infected person, and later from the feces, by using selective and enrichment media. Over a one- to two-week period, a rise in antibody titer against the organisms can often be demonstrated in the blood of an infected person and may be helpful in identifying the cause of the illness.

This disease is maintained in nature by human carriers, people who appear perfectly well but who excrete as many as ten billion typhoid bacilli per gram of their feces. Because normally far fewer of the organisms than this are required to infect, it is easy to see how dangerous typhoid carriers can be. One of the most notorious carriers was Typhoid Mary, a young Irish cook living in New York State in the early 1900's. She is known to have been responsible for at least 53 cases of typhoid fever transmitted during a 15-year period. At that time, about 350,000 cases of typhoid occurred in the United States each year. Today, with improved sanitation and public health surveillance measures, the reported incidence of typhoid fever in the United States is about 400 cases per year.

A vaccine against *S. typhi* and related organisms is widely used by the military and in countries where typhoid is prevalent. Although this vaccine is effective in reducing the incidence and severity of the disease, it does not offer complete protection. Typhoid carriers can be detected by isolating *S. typhi* from their feces and by serological tests for antibodies against the Vi (capsular) antigen of *S. typhi* in their sera. These carriers are then treated with antimicrobial medicines such as ampicillin, or trimethoprim combined with a sulfa (drug). In some carriers, the organisms reside in the gall bladder, and surgical removal of the gall bladder may be necessary to rid them of infection.

Protozoan Infections

Several species of protozoa produce intestinal infections and are important in causing human disease. All the major protozoan groups (see Chapter 19) are represented, for example, *Giardia lamblia* (flagellate), *Balantidium coli* (ciliate), *Isospora belli* (sporozoan), and *Entamoeba histolytica* (ameba)(color plate 52).

FOOD POISONING

Microbial growth in food can cause illness chiefly by two different mechanisms: (1) the contaminating microorganisms may infect the person who ingests the food or (2) products of microbial growth in the food may be poisonous. Gastroenteritis, dysentery, and typhoid fever caused by bacteria of the genera *Salmonella* and *Shigella* may originate from the first mechanism, whereas a wide variety of common microbial species (including members of the bacterial genera *Clostridium, Bacillus,* and *Staphylococcus*) may be responsible for illnesses originating by the second mechanism. Although all illnesses coming from microbial growth in ingested food are popularly considered "food poisoning," the following discussion deals with those illnesses caused by microbial products rather than those resulting from direct infection.

Staphylococcal Food Poisoning

One of the most common forms of food poisoning is due to *Staphylococcus aureus,* the same species of bacterium responsible for boils and other infections. The illness ensues a few hours after ingesting contaminated food, and is manifested by nausea, vomiting, and diarrhea. Recovery follows within another few hours. This demoralizing but rarely fatal type of food poisoning results from an enterotoxin produced when certain strains of *S. aureus* grow in a suitable food, usually one high in carbohydrates. The toxin may be produced quickly by staphylococci growing at room temperature, often within only one or two hours. Moreover, the toxin is relatively heat-stable, so subsequent cooking may kill the staphylococci but leave toxic activity. Several antigenically distinct varieties of enterotoxins are known to be produced by the various food-poisoning strains of *S. aureus* and can be identified in extracts of contaminated foods by using specific antisera.

S. aureus, page 295

Food Poisoning by *Clostridium perfringens* and *Bacillus cereus*

Clostridium perfringens, which produces a heat-labile enterotoxin during its process of sporulation, is another common cause of food poisoning. It produces mainly diarrhea and abdominal cramps, although half the victims also have nausea and vomiting. One outbreak in California was traced to bean-filled burritos, which illustrates that foods other than meat and poultry contain the amino acids necessary to support growth of *Clostridium perfringens*. The pathogenesis of the food poisoning caused by *Bacillus cereus* has also been ascribed to an exotoxin.

Clostridium, page 303

Botulism

Botulism, a microbial poisoning that is characterized by paralysis, is one of the most feared diseases. Its cause is the Gram-positive, rod-shaped, strictly anaerobic bacterium *Clostridium botulinum*, which is widely distributed in soils around the world. The name *botulinum* comes from the Latin word for sausage and was chosen because some of the earliest recognized cases of botulism occurred in people who had eaten contaminated sausage. However, many other foods, including vegetables, fruit, meat, seafood, and cheese, have been sources of this food poisoning.

Like other clostridia, *C. botulinum* produces endospores that may be unusually resistant to heat and can thus persist in foods despite cooking and canning processes. These spores can later germinate, and growth of the bacterium may result in the release of a powerful exotoxin into the food. The exotoxin production is the result of lysogenic conversion. When someone eats the contaminated food, the toxin is absorbed into the bloodstream, where it may continue to circulate for as long as three weeks. This circulating toxin is carried to the various nerves of the body, where it acts by blocking the transmission of nerve signals to the muscles, thus producing paralysis. Twelve to 36 hours (sometimes longer) after a person has eaten toxin-containing food, blurred or double vision gives the first indication of this paralysis. All muscles may be affected, but respiratory paralysis is the most common cause of death. Despite treatment, about one fourth of the victims of botulism die.

Botulism has no effect on sensation or on most functions of the brain, but the toxin causing this disease is one of the most powerful poisons known. A few milligrams would be sufficient to kill the entire population of New York City. Indeed, botulism has resulted from eating a single contaminated string bean or even from licking a finger contaminated with toxin. Fortunately, the toxin is moderately heat-labile, and even high concentrations are completely inactivated by boiling contaminated food for 15 minutes.

In the five years preceding 1922, there were 83 outbreaks of botulism in the United States, many traceable to commercially canned foods. The work of the distinguished microbiologist Karl F. Meyer and his colleagues showed that canning methods then in use were inadequate to kill the heat-resistant spores of *C. botulinum*. Strict controls were then placed on commercial canners to ensure adequate sterilizing methods. Since then, outbreaks caused by commercially canned foods have been infrequent. The occasional lapses in proper canning technique point up the potential for disaster when defective processing occurs in large food producing and distributing firms.

Most cases of botulism from inadequately processed canned food have been caused by organisms of the type A or B strains. Awareness of major outbreaks of type E botulism occurred in the early 1960's, primarily in Japan and the United States. Like other types of *C. botulinum*, type E strains are soil organisms, but they are more readily taken up by fish and sea mammals. Type E strains also differ in that they can produce toxin at lower temperatures and under greater exposure to aerobic conditions. For these reasons, type E botulism is generally associated with ingestion of contaminated seafood.

Botulism is treated by administering the appropriate antitoxin. This neutralizes toxin circulating in the bloodstream, but recovery of the nerves is slow, requiring weeks or months. Control of the disease depends on proper sterilization and sealing of food at the time of canning and on adequate heating prior to serving. One cannot rely on a "spoiled" taste or appearance to detect contamination, because such changes may not always be present.

Wound Botulism and Infant Botulism

Proper food handling cannot completely eliminate the danger of botulism, because *C. botulinum* can occasionally colonize wounds or the intestine and produce a mild form of the disease. In 1976, for example, four previously healthy infants were hospitalized in parts of California because of paralysis caused by botulism. *Clostridium botulinum* organisms and toxin were present in their feces, but the toxin was not detectable in their blood. Although one infant required respiratory support and another required tube feeding, all recovered promptly without receiving antitoxin treatment. Recovery probably resulted from their own antibody production. The source of these infants' infections was undoubtedly *C. botulinum* spores, which are ubiquitous in dust and in various foods such as honey. *Clostridium botulinum* organisms and toxin persisted in the infants' feces for a time after recovery but were gradually replaced by competing normal flora.

Poisoning Caused by Fungi

In earlier ages, ergot poisoning ("St. Anthony's Fire") followed the ingestion of grain contaminated with certain fungi. People afflicted with this condition had agonizingly painful convulsions, and some developed gangrene of the hands and feet because of spasms of the arteries. Epidemics of ergot poisoning still occur occasionally, but now effective medicines are available to counteract the ergot.

Proper standards of agricultural practice and public health surveillance of food grains now make it unlikely that ergot-producing fungi will contaminate food and produce ergot poisoning. However, another kind of fungal toxin is of great current interest. Called *aflatoxin*, this fungal poison is produced by common molds of the genus *Aspergillus* (color plate 53). Aflatoxin can cause acute poisoning in many species of animals, affecting chiefly the young. A few milligrams, for example, may cause severe liver damage and death in a dog within 72 hours. Much interest in aflatoxin arose with the observation that tiny traces of this poison in the food of certain animals caused tumors. In fact, aflatoxin is one of the most potent tumor-inducing substances known for trout. Aflatoxins have been demonstrated in many human foods, and high levels of an aflatoxin metabolite may appear in cow's milk. Aflatoxins produced by various strains of molds are not always the same, and a family of related compounds is now also known. The possible role of these substances in human disease is still under study.

HEPATITIS

Hepatitis, or inflammation of the liver, is occasionally caused by allergic reactions to antimicrobial and other medicines, certain toxic chemicals such as anesthetic gases and cleaning fluids, and various microorganisms and viruses. Typically, the symptoms of hepatitis are loss of appetite and vigor, fever, and jaundice. Two viruses, the hepatitis A virus (HAV) and the hepa-

TABLE 23—4 HEPATITIS VIRUSES

Characteristics	*Pathogenic potential*	*Epidemiology*
Small viruses, relatively resistant to heat and disinfectants; HBV contains double-stranded DNA; HAV contains RNA	Preferentially infect the liver, often producing jaundice and sometimes liver failure	Source: human cases and carriers; incubation period usually 15 to 50 days for HAV, 43 to 180 days for HBV; transmission usually fecal-oral for HAV, blood or blood products for HBV—as little as 0.0001 ml of blood may be infective

titis B virus (HBV), account for most cases of hepatitis, although there are other viral causes. The principal characteristics of both viruses appear in Table 23–4. Both are notable in their resistance to disinfectants and lack of satisfactory growth in vitro, but they are immunologically distinct and differ in their mode of spread and incubation period.

Hepatitis A Virus Disease

Hepatitis A virus disease (formerly called infectious hepatitis) spreads in epidemic fashion, principally through the fecal contamination of hands, food, or water. A prominent cause in recent years has been ingestion of raw shellfish, since these animals concentrate the hepatitis A virus from polluted sea water. Some cases are severe, requiring many weeks of bed rest for recovery. However, most cases are mild and self-limited, and many are asymptomatic. The infection is sufficiently widespread in the population that many people have hepatitis A virus antibody in their blood. For example, in New York City, 20 to 80 percent of the residents have HAV antibody, the higher percentages occurring in the lower socioeconomic groups. Since the antibody to HAV is so common, gamma globulin obtained from the donated blood of many people is pooled and administered to susceptible exposed people to provide passive immunity to the disease.

Hepatitis B Virus Disease

Hepatitis B virus disease (formerly called serum hepatitis) is spread principally by blood and blood products. If a minute amount of blood from an infected person is injected into the bloodstream or rubbed into minor wounds, the recipient is very likely to get hepatitis B. A small amount of plasma given experimentally by mouth has also produced the disease. Hepatitis caused by HBV tends to be more severe than that caused by HAV, producing liver failure and a mortality ranging from 1 to 10 percent of hospitalized cases. However, many people with HBV infection show few or no symptoms. Indeed, for every patient with jaundice there are at least three without. However, the incidence of infection with HBV has been increasing progressively over the past decade, and an estimated 150,000 cases per year occur in the United States. Symptomatic and asymptomatic individuals may unknowingly become carriers of HBV. The number of carriers in the United States is estimated to be about 900,000. Many hepatitis B virus infections result from needles shared by drug abusers. Unsterile tattooing and ear-piercing instruments, shared toothbrushes, razors, or towels, and similar objects can also transmit HBV infections. Blood transfusions and improperly sterilized dental instruments were a common source of HBV infection in the past, but with present precautionary practices, the risk from these sources is very small, estimated at 1 case in 30,000 or more transfusions. Hepatitis virus B antigen has been demonstrated in as many as 5 of every 1000 prospective blood donors. Although present methods of blood testing fail to detect the virus in a small percentage of donors, tests now under development should be capable of identifying essentially all HBV carriers.

Carriers, page 606

After an individual has become infected with HBV, virus is usually not detectable in the bloodstream for a month or longer, and signs of liver injury take an additional week or more to develop. The virus does not cause cell destruction, and damage to the liver is probably due in part to a cell-mediated immune response against virus-induced cellular antigens. In all but 5 to 10 percent of patients, the hepatitis B virus disappears from the blood during the latter part of the illness. Persisting viremia* can follow both symptomatic and asymptomatic cases, and the virus may continue to circulate in the blood for months or years. These carriers are of major importance in the epidemiology of hepatitis B. Besides being present in the blood, the hepatitis B viral antigen is often present in large amounts in saliva, breast milk, and other body fluids, but these fluids are of little or no epidemiological importance. Five percent or more of pregnant women who are carriers transmit the disease to their babies at the time of delivery, and more than two thirds of the women with hepatitis late in pregnancy or soon after delivery do so. The infected babies may die of liver failure, but most survive, usually to become long-term carriers.

There is no specific treatment for hepatitis, but prospects for its control have greatly improved. Carriers of HBV can now be identified by blood tests for hepatitis B antigens and antibodies and can be instructed in measures that prevent transmission of the disease. Likewise, health workers and others at high risk for contracting hepatitis B can undertake rational precautions in handling blood, wear protective clothing, wash their hands often, and keep their hands out of their mouths. Special precautions can also be taken during pregnancy. Finally, prospects are good for active and passive immunization to assist in the control of hepatitis B. A promising vaccine is now being evaluated in human subjects.

---

*viremia—presence of viruses in the blood

Non-A, Non-B Hepatitis

With the development of reliable methods to detect HAV and HBV, it became clear that one or more other hepatitis viruses exist. Viral hepatitis not due to HAV or HBV is called "non-A, non-B hepatitis," pending the characterization of the other viruses. More than 90 percent of hepatitis cases following blood transfusion are now of this type. As with HBV, chronic blood carriers exist and account for transmission via contaminated needles and blood transfusions.

SUMMARY

One of the major routes of microbial and viral invasion of the human body is the alimentary tract; resulting diseases may involve any part of the body, including the alimentary system itself. Members of the normal alimentary flora with little invasive ability can produce dental caries or the blind loop syndrome. Certain viral infections have specific tissue affinities, such as the parotid glands in mumps and the liver in hepatitis. The genus *Salmonella* is responsible for a wide spectrum of human diseases, ranging from gastroenteritis (generally from species colonizing other animals) to the more serious typhoid fever from *Salmonella typhi*, a strain infecting only humans. *Shigella* species, also facultatively anaerobic Gram-negative rods, are a frequent cause of dysentery. The toxins of *Vibrio cholerae* and *Escherichia coli* greatly increase intestinal fluid excretion, producing watery diarrhea and dehydration.

Microbial food poisonings result from eating foods contaminated with certain toxic microorganisms or with the byproducts of their growth in food. Bacteria of the genera *Bacillus, Clostridium,* and *Staphylococcus* are commonly responsible for poisoning manifested by nausea, vomiting, and diarrhea. In the case of *Staphylococcus aureus*, a heat-stable enterotoxin is responsible. Fungal toxins include the aflatoxins of *Aspergillus* species, which can produce liver damage and tumors in some animals and therefore cause concern when present in human food, although their significance to human health is unknown. One of the most feared types of poisonings is botulism, an often fatal paralysis caused by the exotoxins of *Clostridium botulinum*.

SELF-QUIZ

1. The pathogenesis of dental caries involves
   a. plaque formation
   b. sucrose
   c. *Streptococcus mutans*
   d. lactic acid
   e. all of the above

2. If sugar were eliminated from the diet
   a. the incidence of dental caries would drop dramatically
   b. there would be no change in incidence of dental caries unless *S. mutans* was eliminated
   c. the incidence of dental caries would probably rise
   d. dentists would go bankrupt because there would be no other dental disease to treat
   e. *S. mutans* would not mind since it can attach to the tongue
3. A person who has a boil and prepares food
   a. risks introducing food-poisoning strains of staphylococci into the food
   b. need not cause concern if the boil is under clothing where it cannot be seen
   c. should not worry because staphylococci that cause skin infections do not cause food poisoning
   e. none of the above
4. Potential sources of contamination of foods with *Salmonella* species include
   a. hands of a person preparing food who has forgotten careful hand-washing
   b. the slaughterhouse where food animals are prepared for marketing
   c. tabletops or other surfaces contaminated by uncooked food animals
   d. all of the above
   e. none of the above
5. Ways to help avoid botulism are
   a. to cook home-canned food just before eating it
   b. to taste food before eating
   c. to reject any food that appears spoiled
   d. a, b, and c
   e. a and c
6. Mumps is a viral disease that
   a. is transmitted by saliva
   b. can involve the central nervous system and thyroid gland
   c. characteristically involves the parotid salivary glands
   d. all of the above
   e. none of the above
7. Viral hepatitis acquired from blood transfusions is usually caused by
   a. hepatitis A virus
   b. hepatitis B virus
   c. non-A, non-B viruses
   d. a and b
   e. none of the above
8. Enterotoxigenic *E. coli* organisms produce an illness that most closely resembles illness due to
   a. *Vibrio cholerae*
   b. *Salmonella typhi*
   c. *Giardia lamblia*

d. *Entamoeba histolytica*
e. *Shigella dysenteriae*

QUESTIONS FOR DISCUSSION

1. Are you impressed by the large numbers of distinctly different strains within the species *Escherichia coli* and *Salmonella enteritidis?* What determines these differences within species? How many more distinct strains are likely to be detected in these and other species? Would it be better to identify specific properties (such as virulence factors) than to identify species?
2. Is the ability to produce enterotoxin of any use to the producing bacterium?

FURTHER READING

Bauer, D.: "Preventing the Spread of Hepatitis B in Dialysis Units." *American Journal of Nursing 80*:260–261 (1980).

Blake, P. A., et al.: "Cholera—A Possible Endemic Focus in the United States." *New England Journal of Medicine 302*:305–309 (1980).

Krogstad, D. J., et al.: "Amebiasis." *New England Journal of Medicine 298*:262–265 (1978).

Mandel, I. D.: "Dental Caries." *American Scientist 67*:680–687 (1979).

Rosenberg, C. E.: *The Cholera Years: The U.S. in 1832, 1849, and 1866.* Chicago: University of Chicago Press, 1962. Using newspapers, diaries, periodicals, and medical sources of the time, the author has written a most interesting account of this disease.

Roueche, B.: *Annals of Epidemiology.* Boston: Little, Brown, 1967. Popular accounts of episodes of botulism, typhoid fever, and *Salmonella* gastroenteritis.

Chapter 24

GENITOURINARY INFECTIONS

A 30-year-old man was seen by a physician in late November 1979 because of symptoms of acute urethritis, pain on urination, and discharge of pus from the urethra. The diagnosis of gonorrhea was made by a stained smear of the urethral pus showing Gram-negative intracellular diplococci. The presence of *N. gonorrhoeae* was confirmed by culture. The patient was treated with a single large dose of ampicillin given by mouth, but symptoms persisted, and in mid-December he was re-treated, this time with 2.4 million units of penicillin injected into each buttock. This treatment also failed, and in mid-January he was again re-treated, this time with large doses of tetracycline. When there was no response after a week, the patient was cultured again and the infecting strain of *N. gonorrhoeae* was finally sent to a reference laboratory to be tested for antibiotic sensitivity. The results showed that the patient's strain of *N. gonorrhoeae* produced penicillinase, indicating the presence of a resistance plasmid. Immediately the public health department began the task of identifying the man's sexual partners, interviewing them, and learning the names of their other sexual contacts. Cultures from all these people and all other gonococcal cultures coming into the public health laboratory were tested for penicillinase-producing *N. gonorrhoeae*. By the end of April, 28 cases of gonorrhea due to the resistant strain were identified. Unfortunately, because of the long lapse of time before the case was investigated, the source could not be traced.

The first case of gonorrhea due to a penicillinase-producing, R plasmid—containing gonococcus was reported in March 1976. The disease was contracted in southeast Asia. By April 1980, a total of 1022 cases had been reported in the United States. As in the above account, these plasmids often confer resistance to other antibiotics besides the penicillins, such as tetracycline.

OBJECTIVES

To know
1. The anatomical and physiological setting for genitourinary infections.
2. The distribution of normal flora in the genitourinary system and the influence of host hormones.
3. The causes and significance of urinary tract infections.
4. The causes and characteristics of genital bacterial infections produced by *Neisseria gonorrhoeae, Treponema pallidum, Chlamydia trachomatis, Staphylococcus aureus,* and *Streptococcus pyogenes.*
5. The cause and significance of nonbacterial genital infections such as yeast vulvo-vaginitis, genital herpes simplex, and trichomonal vaginitis.

The human body is efficiently constructed to prevent microbial invasion, a fact particularly evident in studying the anatomy of the genitourinary tract. Under certain circumstances, however, the genitourinary tract can be invaded by a variety of organisms from the normal flora, which act as opportunists, as well as by other pathogenic species.

ANATOMY AND PHYSIOLOGY

The urinary tract consists of the kidneys (in the upper urinary tract), the bladder (in the lower urinary tract), and accessory structures. The kidneys act as a specialized filtering system to clean the blood of many waste materials, selectively reabsorbing other substances that can be reused. Figure 24–1 shows that each kidney is drained by a tube called the **ureter,** which connects it with the urinary bladder. The bladder empties through the **urethra.** The downward flow of urine helps clean the system by flushing out microorganisms that may have invaded it before they have a chance to multiply and infect.

Many kidney infections can be very difficult to eradicate, are often chronic, and can lead to marked destruction of the kidneys. Death follows kidney failure promptly unless the patient is lucky enough either to be able to use one of the all-too-scarce artificial kidneys or to receive a kidney transplant.

The urinary tract is protected from infection by a number of mechanisms. Normal urine contains antimicrobial substances such as organic acids and small quantities of antibodies. During urinary tract infections, larger quantities of specific antibodies can be found in the urine. There is also evidence that antibody-forming lymphoid cells infiltrate* the infected kidneys or bladder and form protective antibodies locally at the site at which they are needed. In addition, during infection there is an inflammatory response in

Lymphoid cells, page 214

---

*infiltrate—pass into

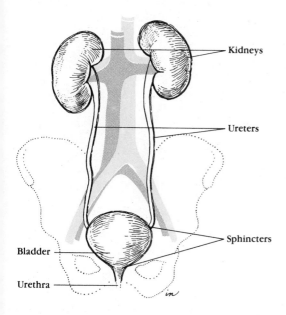

Figure 24–1
The anatomy of the urinary tract. Urine flows from the kidneys, down the ureters, and into the bladder, which empties through the urethra. Sphincter muscles help to prevent any contamination from ascending.

which phagocytes are of the utmost importance in engulfing and destroying the invading microorganisms.

The anatomy of the female and male genital tracts is shown in Figure 24–2. When an ovum is expelled from the ovary, it is swept into the adjacent fallopian tube, and then into the uterus, by the action of ciliated cells (Fig. 24–2a). Because the vagina is the portion of the female genital tract connecting the female genital system with the environment, it is the most frequent site of infection of this system.

Inflammatory response, page 216

NORMAL FLORA OF THE GENITOURINARY TRACT

Normally, the urine and the urinary tract above the entrance to the bladder are essentially free of microorganisms; the urethra, however, has a normal resident flora. Species of *Lactobacillus*, *Staphylococcus* (coagulase-negative), *Corynebacterium*, *Haemophilus*, *Streptococcus* (alpha-hemolytic or enterococci), and *Bacteroides* are common inhabitants of the normal urethra.

The normal flora of the genital tract of women is influenced by the action of estrogen hormones on the lining (epithelial) cells of the vaginal mucosa. When estrogens are present, glycogen* is deposited in these epithelial cells. The glycogen is then converted to lactic acid by lactobacilli, resulting in an acidic pH. Thus, the normal flora and resistance to infection of the female genital tract vary considerably with a woman's hormonal status. For several weeks after birth, the vagina of a newborn girl remains under the influence of maternal hormones and has an acidic pH. During this time, lactobacilli

---

*glycogen—a polysaccharide composed of glucose molecules linked together in a characteristic manner

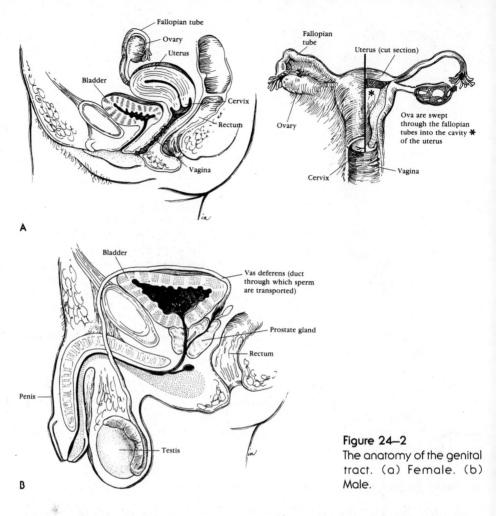

Figure 24–2
The anatomy of the genital tract. (a) Female. (b) Male.

are predominant in its normal flora. As the influence of the maternal hormones wanes, however, the pH of the infant vagina increases to neutrality and remains neutral until puberty. During this childhood period, the normal flora of the vaginal tract consists of a variety of cocci and rod-shaped organisms, including species of *Streptococcus* and many other genera. At puberty, lactobacilli again become predominant, although smaller numbers of yeasts and other bacterial species are also present. After the menopause, the tract returns to a neutral pH and to the mixed flora typical of childhood.

URINARY TRACT INFECTIONS

A number of factors can predispose the urinary tract to infection. Any situation that leads to stasis* of the urine increases the chance of such infection.

*stasis—stagnation due to fluid staying in one place instead of flowing naturally

After anesthesia and major surgery, for example, the reflex ability to void urine may be inhibited for a time. In this circumstance, urine accumulates and distends the very elastic bladder. Even a few bacteria that manage to evade the urinary tract defenses and enter the bladder can multiply during stasis, causing infection. Urine provides abundant nutrients for many bacteria. Paraplegics* also commonly suffer urinary tract difficulties. Because they lack nervous control, paraplegics are unable to void normally and require a tube (catheter) to carry urine from the bladder to an outside container. This connection to the "outside" makes it easier for pathogens to reach the bladder and cause urinary tract infections.

Because the vagina is adjacent to the urethra in women (Fig. 24–2a), the pressure associated with coitus* may cause the introduction of organisms into the urinary tract. Many women experience their first episode of bladder infection following their first sexual intercourse.

Unless such infection of the lower urinary tract is overcome, either by successful treatment or by natural defense mechanisms, it can ascend through the ureters to involve the kidneys. Thus, most kidney infections develop after bladder infection. Bacteria carried by the bloodstream can also infect the kidneys, but this occurs less often than ascending urinary tract infections.

Infections of the urinary tract occur far more frequently in women than in men because the female urethra is short (about 1.5 inches, compared with 8 inches in the male) and is adjacent to the genital and intestinal tracts. More than 90 percent of urinary tract infections are caused by bacterial species that are part of the normal intestinal flora and that consequently are readily available to invade the urinary tract when its natural defenses are interrupted (Fig. 24–3). Serological studies of infecting organisms have shown that they are usually derived from the patient's own intestinal bacteria. In these studies, for example, *Escherichia coli* isolated from the urine during urinary tract infections was almost always of the same *serotype** as *E. coli* from the intestine of the patient.

Escherichia coli accounts for the majority of initial urinary tract infections, but other Gram-negative rods of different genera, such as *Proteus* and *Enterobacter*, are also often the cause (Fig. 24–3). *Pseudomonas aeruginosa*, an aerobic Gram-negative rod, is a particularly troublesome urinary tract pathogen because it is difficult to treat successfully. *Streptococcus faecalis* (enterococcus), found in the normal bowel flora, is the most frequently isolated Gram-positive organism responsible for urinary tract infections.

P. aeruginosa, page 311

The causative organisms of urinary tract infections can usually be recovered easily and in large numbers from the urine of infected patients. In doing this, careful collection of voided urine specimens is necessary, so that

---

*paraplegics—individuals with paralysis of the lower half of the body

*coitus—sexual intercourse

*of the same serotype—having the same antigens, as detected by using known antisera

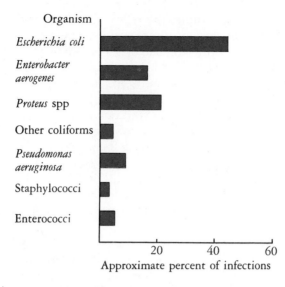

Organism

Escherichia coli

Enterobacter aerogenes

Proteus spp

Other coliforms

Pseudomonas aeruginosa

Staphylococci

Enterococci

20 40 60

Approximate percent of infections

Figure 24–3
Common causes of urinary tract infections. The percentages vary considerably from one study to another. Often infections are caused by more than one kind of organism.

external contamination from the patient's genitalia or intestinal tract is minimized. This is more difficult with women than with men. It is best accomplished by collecting a sample during midstream after carefully cleaning the external genitalia. Normal urine collected in this way usually contains less than 10,000 bacteria per milliliter, as determined by colony counts. Concentrations of bacteria greater than 100,000 per milliliter indicate infection. Bacteria multiply exponentially*; therefore, cultures for colony counts must be prepared promptly after the urine is collected in order to obtain meaningful counts of urinary bacteria (Fig. 24–4).

Urine specimens are cultured on media especially suitable for the growth of enterobacteria such as *E. coli* and on blood agar, which will also support the growth of enterococci and some other Gram-positive pathogens.

The causative organisms of diseases involving areas of the body other than the urinary tract may also be found in the urine. Thus, in typhoid fever, *Salmonella typhi* organisms become disseminated throughout the body and are excreted in the urine about one to two weeks after infection, sometimes establishing a chronic urinary tract infection. In like manner, slender spirochetes of the genus *Leptospira* (the cause of leptospirosis) may become established in the urine. Contamination of water with urine from infected humans or animals is the major means of transmission of leptospirosis. The *Leptospira* organisms may then continue to be excreted in the urine for many months after recovery from the illness. Leptospirosis is a common infection of animals and is one of three diseases against which dogs are commonly immunized.

Leptospira species, page 325

*exponentially—the numbers increase by a certain fixed multiple during each time interval; for example, rising from 200, to 400, to 800, to 1600, to 3200, and so on, every 15 minutes

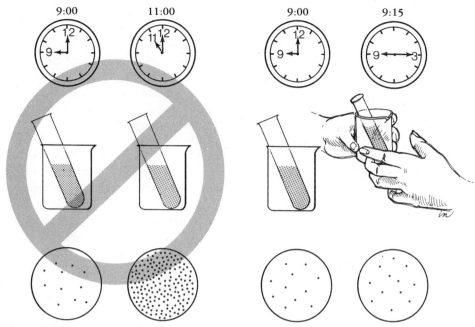

Figure 24—4

Importance of promptly transporting urine specimens to the laboratory. Urine is a growth medium, and delay in laboratory testing for infection can produce erroneous results.

BACTERIAL INFECTIONS OF THE GENITAL TRACT

Puerperal Fever

Historically, group A beta-hemolytic streptococci (*S. pyogenes*) have been a major cause of puerperal fever (childbirth fever). They are transmitted by direct contact at childbirth with contaminated instruments or hands. In addition to *S. pyogenes*, the causative bacteria of puerperal fever may be some of the anaerobic species (clostridia, bacteroides, and anaerobic streptococci) that infect wounds and find a fertile area for growth in tissues injured by birth trauma.

The history of puerperal fever illustrates the general human tendency to resist change. For centuries it had been an accepted fact that many mothers developed fever and died following childbirth. About the middle of the nineteenth century in the hospitals of Vienna (the major medical center of the world at that time), about one of every eight women died of puerperal fever following childbirth. A Hungarian physician named Semmelweis, who had come to Vienna to practice, studied this abhorrent situation and concluded that the disease was transmitted by the attending physicians or their instruments. He instituted a program of simple disinfection of all instruments used during delivery and was able to cut the maternal death rate at childbirth from about 12 percent to less than 1 percent. Instead of accepting these findings and the new technique of disinfection that Semmelweis used, his colleagues in Vienna refused to face the fact that they had been responsible, even unknowingly, for the death of so many patients. The work of Semmelweis was attacked, and he was forced to leave Vienna and return to his native Hungary. There he was again able to use disinfection techniques and to achieve a remarkable reduction in the number of deaths from puerperal fever. Dr. Oliver Wendell Holmes, the famous American physician, independently made the same finding at about the same time.

Over the years, with increasing improvement in (and acceptance of) disinfection and sterilization techniques, puerperal fever has become rare in countries with adequate medical facilities. Even today, however, outbreaks occasionally occur, and it may be very difficult to identify the source. Not long ago, several cases of puerperal fever occurred in the United States despite careful precautions and the use of proper antiseptic techniques. In one report, it was found that a medical attendant was an anal carrier of *S. pyogenes*, which was disseminated by the movement of his clothing.

Toxic Shock Syndrome

S. aureus, page 295

Growth of certain strains of *Staphylococcus aureus* within the vagina leads to fever, diarrhea, muscle aches, low blood pressure, and a sunburn-like rash. This collection of symptoms is called the **toxic shock syndrome**, and it occurs most frequently among women who use intravaginal tampons during their menstrual periods, especially when the tampon is left in place for extended periods of time. The symptoms of the disease probably result from absorption of one or more staphylococcal toxins, a situation analogous to staphylococcal scalded skin syndrome. About one woman in 12 reported as having the disease during 1980 died. Widespread dissemination of knowledge about toxic shock syndrome beginning in 1980 should promote earlier diagnosis, prompt treatment, and a lower mortality. Toxic shock syndrome can occur in both men and women as a result of infections elsewhere in the body, such as skin, bone, or lung, but such cases have been rare. Even among menstruating women, the incidence of toxic shock syndrome has been estimated to be 3 per 100,000 women.

Venereal Diseases

Gonorrhea

Although there are a number of venereal diseases, gonorrhea is probably the most prevalent. In the United States, the incidence of gonorrhea is the highest of any *reportable* bacterial disease. An increase from 400,000 cases in 1967 to more than a million in 1980 is an alarming development (Fig. 24–5). The causative organism, *Neisseria gonorrhoeae* (gonococcus), is a Gram-negative diplococcus, identical in appearance to the closely related *N. meningitidis*

N. gonorrhoeae, page 308

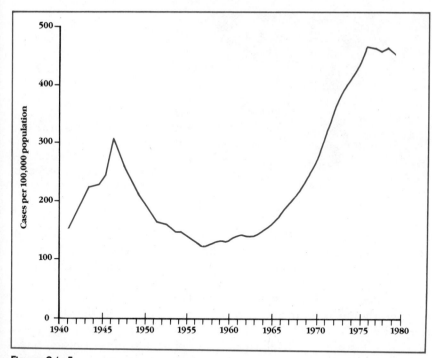

Figure 24–5
Reported civilian cases of gonorrhea in the United States, 1941–1979. (Adapted from *Morbid. Mortal. Wkly. Rep.*, United States Public Health Service, 28:33, 1979.)

(the cause of epidemic meningitis). The gonococci are flattened on adjacent sides and are frequently found within certain leukocytes in pus, as indicated in Figure 24–6. Gonococci are parasites of humans only; they prefer to live on the mucous membranes of their host. Most strains are susceptible to cold and drying and hence do not survive well outside the host. For this reason, gonorrhea is transmitted primarily by direct contact, and since the bacteria live primarily in the genital tract, this contact is almost always venereal. There are certain exceptions to this general rule. One form of very serious gonococcal disease, for instance, is an eye infection, **ophthalmia neonatorum,** that can be transmitted at birth from mother to infant during the infant's passage through an infected birth canal. This form of gonococcal infection has been controlled in the United States by laws requiring the use of silver nitrate or an antibiotic placed directly into the eyes of all newborn infants (Fig. 24–7). Some mothers who feel certain that they do not have gonorrhea have challenged these laws because they do not want their babies to have the unpleasant experience of receiving irritating eye drops. The long-standing and often asymptomatic nature of gonorrhea in women and some men makes omission of silver nitrate prophylaxis a risky business.

Gonococci infect the genitourinary tract, eye, and pharynx areas where a certain kind of epithelial cell—the columnar epithelial cell—is found. Path-

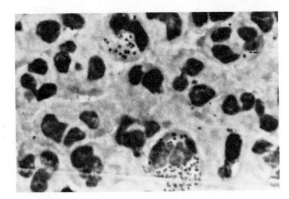

Figure 24–6
Neisseria gonorrhoeae in Gram-stained material from a patient with gonorrhea. The diplococci are abundant within polymorphonuclear leukocytes in the pus.

ogenic gonococci possess pili that attach selectively to columnar epithelial cells; thus, there is a positive correlation between the presence of pili and the ability of gonococci to adhere to mucous membranes and establish infection.

Gonorrhea differs in men and women, at least in part because the location of columnar epithelial cells in the two sexes differs. In men the disease is characterized by inflammation of the urethra, with pain during urination and a thick, pus-containing discharge that becomes apparent a few days after infection. Usually the disease in men is self-limiting and clears of its own accord, but the infection may extend itself in the genital tract to involve the prostate gland and testis, producing complications. As an example, extensive inflammatory reaction to the infection can lead to the formation of fibrous tissue* that partially obstructs the urethra, predisposing the man to future urinary tract infection. If fibrous tissue occludes the tubes that carry the sperm, or if tissue in the testes is destroyed by inflammation, sterility may result.

Gonorrhea follows a different course in women. Gonococci thrive in the cervix and fallopian tubes (Fig. 24–2b) as well as in other areas of the female genital tract. The early stages of infection, involving the cervix, commonly are mild, and the victim is usually unaware of the disease. Women are thus especially likely to become carriers of gonorrhea. If the disease is not limited at this stage, infection frequently progresses upward through the uterus into the fallopian tubes. Fibrous tissue formed in response to the infection may block normal passage of the ova through the tubes, and gonorrhea is thus one cause of sterility in women. Obstruction of a fallopian tube can also lead to a dangerous complication—ectopic pregnancy—in which the ovum is fertilized and develops in the fallopian tube or in the abdominal cavity outside the uterus. This commonly leads to life-threatening internal hemorrhage.

Another common complication of gonorrhea, in both women and men, is gonococcal arthritis caused by the growth of gonococci within the joint

---

*fibrous tissue—scar tissue

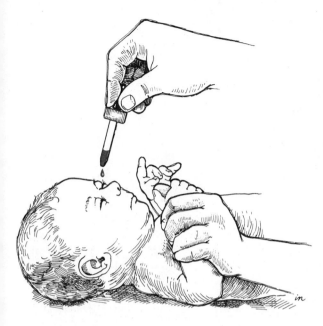

Figure 24–7
Instillation of silver nitrate solution into the eyes of a newborn to prevent ophthalmia neonatorum.

spaces. Any joint may be affected, especially the larger ones. It has been shown that some strains of gonococci have a greater tendency than others to disseminate to the joints or other areas away from the original site of the infection. The areas of the body affected by gonorrhea discussed in the preceding paragraphs are illustrated in Figure 24–8.

In both men and women, gonococci can be recovered from purulent* discharge or from the secretions collected from infected areas. Gram-stained smears often reveal the characteristic Gram-negative diplococci within the cells of pus, but in order to identify the organisms as *N. gonorrhoeae* with certainty it is necessary to culture them and to carry out certain biochemical and serological tests.

Gonococci are usually susceptible to penicillin, but they have gradually decreased in susceptibility over the years. Thus, 20 years ago 300,000 units of penicillin were usually sufficient to cure a case of gonorrhea, whereas today a dose of 4.8 million units is often necessary, and even this dose is not always successful. In 1976 totally penicillin-resistant penicillinase-producing strains of gonococci were isolated from patients in several areas of the world. This penicillin resistance has been shown to result from the acquisition of a plasmid. Plasmid-containing strains highly resistant to antimicrobial medicines have been introduced into the United States and are spreading rapidly.

The widespread use of oral contraceptives ("the pill") has contributed to the increased incidence of gonorrhea and other venereal diseases in a variety of ways, both social and medical. Several studies have shown that, on the

Biochemical tests,
page 82

---

*purulent—pus-containing

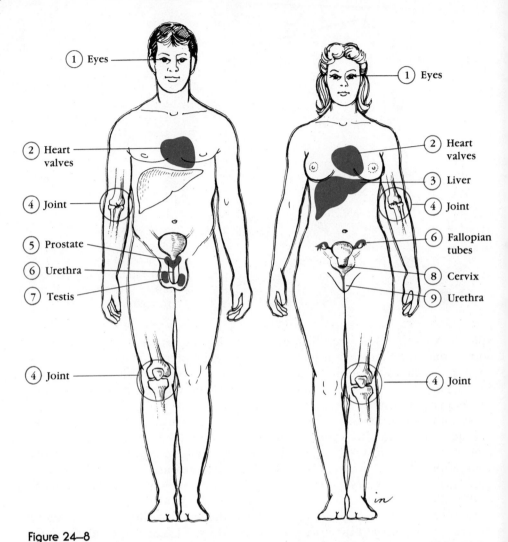

Figure 24—8
Complications of gonorrhea. (1) Eyes of both adults and children are susceptible to the gonococcus; serious infections leading to loss of the eye are likely in the newborn. (2) Organisms carried by the bloodstream affect the heart valves and cause damage. (3) The outer covering of the liver can become involved when gonococci enter the abdominal cavity and infect fallopian tubes. (4) Gonococcal arthritis results from organisms carried by the blood. (5) Prostatic gonococcal abscesses may be difficult to eliminate. (6) Infection of the fallopian tubes results in scarring, which can lead to sterility or ectopic pregnancy. (7) Scarring of testicular tubules can cause sterility. (8) The cervix is the usual site of primary infection in women. (9) Urethral scarring from gonococcal infection can predispose to urinary tract infections by other organisms.

average, women taking oral contraceptives begin having intercourse at an earlier age and have more sex partners than was the case before "the pill" became available. Then too, mechanical contraceptives, such as diaphragms and condoms, which impose a physical barrier to infection, are not used as frequently today as they were previously. In addition, long-term use of some oral contraceptives leads to a migration of susceptible epithelial cells from upper areas of the cervix into more exposed areas of the outer cervix. Oral contraceptives also tend to increase both the pH and the moisture content of the vagina. All these factors favor infection, not only with gonococci but with several other agents that cause venereal diseases as well. Women taking oral contraceptives are thus especially vulnerable to gonorrheal infection and are also more liable to develop serious complications if they contract the disease. Furthermore, such complications usually become apparent early in these women, often within a few days after infection.

Several other factors contribute to the increasing incidence of gonorrhea. Carriers of gonococci, both male and female, can unknowingly transmit these bacteria over months or even years. Identification of carriers is critical in controlling gonorrhea. Another factor contributing to the increased incidence of gonorrhea is the lack of lasting immunity following recovery from the disease. The individual who has recovered from gonorrhea is susceptible to reinfection. Studies of the immunology of gonorrhea are incomplete, and it is uncertain whether repeated infection results from brief immunity or from a multiplicity of immunologically distinct strains of the gonococcus. A great effort, so far unsuccessful, is being made to develop a vaccine that will provide effective protection against gonococci.

Syphilis

Syphilis, another venereal disease of major importance, is caused by a spirochete, *Treponema pallidum*. The organisms (treponemes) of syphilis are motile, extremely thin, tightly coiled spirals. They cannot be seen on Gram-stained smears because they are so thin; thus it is necessary to use special methods to make them visible. These treponemes cannot be grown in test tube cultures and under natural conditions are strictly parasites of humans. Like the gonococcus, *T. pallidum* is very sensitive to destruction by drying and temperature change, and for this reason it is transmitted almost exclusively by sexual or oral contact.

Syphilis occurs in three stages. As a rule, in **primary syphilis,** treponemes transmitted by sexual contact grow and multiply in localized areas of the genitalia, spreading from there to the lymph nodes and bloodstream. About three weeks after infection, an ulcer called a **hard chancre** appears at the site of infection. This represents a cellular response to the bacterial invasion, and examination of a drop of fluid expressed from the chancre reveals that it is teeming with infectious *T. pallidum*. Whether treatment is given or not, the chancre disappears within four to six weeks, and the patient may mistakenly believe that the disease is cured. However, about two to ten weeks

T. pallidum, page 323

later the manifestations of **secondary syphilis** may appear. These usually include running nose and watery eyes, aches and pains, sore throat, and a rash that involves the palms and soles. By this time, the spirochetes have spread throughout the body, and infectious lesions occur on the skin and mucous membranes in various locations, especially in the mouth; syphilis can be transmitted by kissing during this stage. The spirochetes also readily cross the placenta after about ten weeks' gestation and can be passed from a pregnant woman to her fetus if the woman has secondary syphilis at this stage of pregnancy. The secondary stage lasts for weeks to months, sometimes as long as a year, and gradually subsides. About 50 percent of untreated cases never progress pass the secondary stage; however, after a latent period of from 5 to 20 years, or even longer, some cases continue to **tertiary syphilis.** The primary and secondary stages pass unnoticed in many cases.

Tertiary syphilis represents a hypersensitivity reaction to small numbers of *T. pallidum* that grow and persist in the tissues. The treponemes may be found in almost any part of the body, and the symptoms of tertiary syphilis depend on where the hypersensitivity reactions occur. If they occur in the skin, bones, or other areas not vital to existence, the disease is not life-threatening. However, if they occur within the walls of a major blood vessel, the vessel may become weakened and even rupture, and death soon follows. Hypersensitivity reactions in the eyes cause blindness; in the central nervous system, the reactions can cause insanity, characteristic paralysis, and other manifestations.

Although some argue that syphilis was transported to Europe from the New World by Columbus's men, others find convincing evidence, including biblical references, that syphilis existed in the Old World for eons before Columbus. History and literature record that kings, queens, statesmen, and heroes degenerated into madness, mental incompetence, or other serious debility as a result of tertiary syphilis; Henry VIII, Ivan the Terrible, Catherine the Great, Arthur Schopenhauer, Heinrich Heine, Benito Mussolini, Emil von Behring, and Lord Randolph Churchill were just a few. It is difficult to comprehend the emotional as well as the physical anguish caused by this disease over so many centuries before the discovery of antibiotics. One can only speculate on the role that syphilis and other venereal diseases might have had in the development of moral codes controlling sexuality in various ancient cultures.

Primary or secondary syphilis can be diagnosed by finding the motile spirochetes in fresh material from lesions on the skin or the mucous membranes of the genitalia or the mouth. Dark-field illumination permits the thin spirals to be seen, and dried specimens can be examined by using fluorescent antibody techniques. In tertiary syphilis the patient is not infectious, and the internal lesions contain so few organisms that they cannot be detected by these means.

During all stages of syphilis, serological tests are helpful in diagnosis. Although *T. pallidum* cannot be grown in culture and thus is not readily available as a source of antigen, it was accidentally found that antibodies formed during syphilis will react with certain substances extracted from beef heart. This chance discovery allowed the development of tests for serum antibodies against *T. pallidum*, using the beef heart substance as "antigen." At first, complement-fixation reactions such as the Wassermann test were widely used, but today most laboratories use a form of precipitation reaction called the **flocculation test.** Examples of this are the Kahn test and the VDRL (Venereal Disease Research Laboratory) test. In these tests, a fluffy "antigen"-antibody precipitate, or flocculate, is formed. Such flocculation tests are useful for preliminary testing of large numbers of people, but a positive test does not always indicate that an individual has syphilis, since false positive reactions occur fairly frequently in people with certain other diseases. Other methods must be employed to confirm the suspected diagnosis, using *T. pallidum* itself as an antigen. The fluorescent treponemal antibody test (FTA-ABS) is especially valuable in this respect. In this indirect immunofluorescence test, the patient's serum is allowed to interact under carefully controlled conditions with killed *T. pallidum* organisms that have been grown in the testes of rabbits. If specific antitreponemal antibodies are present in the serum, they combine with the treponemes; excess serum proteins can then be washed from the slide. To detect the reaction, one adds fluorescent dye-tagged antibodies against human gamma globulin. If specific antitreponemal gamma globulin antibodies have combined with them, the treponemes will fluoresce when examined under ultraviolet light with the fluorescence microscope.

Sometimes it is difficult to establish a diagnosis of syphilis. If a chancre fails to develop, the primary stage of the disease may go unnoticed, or the symptoms that are observed may be confused with symptoms of another disease. Syphilis has been called the "the great imitator" because it resembles or imitates many other diseases as it progresses. Furthermore, congenital syphilis may go unrecognized if the disease is not apparent in the mother. It is possible for pregnant women who have passed through the infectious stages of secondary syphilis and have no symptoms of the disease to transmit treponemes across the placenta to the fetus. Babies born with congenital syphilis may show characteristic deformities such as changes in the teeth and facial bones (color plates 54 and 55).

Actively growing *Treponema pallidum* organisms are always susceptible to low doses of penicillin, so penicillin is the treatment of choice except for people who are allergic to penicillin. Treatment must be continued for a longer period during tertiary than during primary or secondary syphilis, probably because many of the organisms in tertiary syphilis are not actively multiplying. There is no lasting immunity to syphilis after either spontaneous recovery or treatment early in the disease, nor have effective methods of immunization been developed.

Theoretically, syphilis can be controlled by the use of effective antibiotics and an intensive program for finding and treating the people who have had contact with infected individuals. However, increases in the disease occur whenever control measures are relaxed. For example, the number of cases of syphilis in the United States dropped precipitously each year following the introduction of penicillin, as shown in Figure 24–9; however, an increase

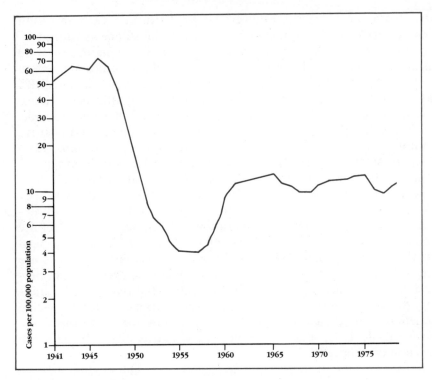

Figure 24–9
Reported civilian cases of primary and secondary syphilis in the United States, 1941–1979.
(Adapted from *Morbid. Mortal. Wkly. Rep.*, United States Public Health Service, 28:79,
1979.)

occurred in 1962. When more stringent precautionary measures were taken,
the rate decreased again. The general trend in the United States seems to be
downward, but even with public health precautionary measures and the use
of penicillin, we have not yet returned to the low incidence of 1958.

Several factors explain the lower incidence of syphilis compared with
gonorrhea in the United States. Case workers carefully trace all reported
cases of both syphilis and gonorrhea; they attempt to reach and treat all
people who have had sexual contact with these patients. This approach usually allows such individuals to be treated prophylactically during the long
(one to three weeks) incubation period before syphilis develops. However,
the incubation period of gonorrhea is so brief (one to three days) that the
disease has usually developed in the person's sexual contacts before they can
be traced and treated. Frequently, a number of other people will also have
been infected. A second important factor is that serological tests will identify
subjects with unsuspected syphilis, but reliable serological tests for gonorrhea are not yet available for screening large numbers of people in order to
detect both male and female carriers of gonococci. These carriers, often
women who have never had symptoms of gonorrhea, are the major reservoir

of gonococci, and there is no easy way to identify them. It is necessary to culture secretions from the genital tract and to identify *N. gonorrhoeae* from the cultures—a laborious and expensive procedure. Thus, it appears that the incidence of gonorrhea will continue to exceed the incidence of syphilis.

Homosexual or bisexual men present a problem in the control of syphilis, because about a third of the male cases of syphilis in the United States occur in this population. In fact, in urban areas such as Los Angeles, Seattle, and New York, more than half of all reported syphilis cases involve homosexual men. Eradication of the disease in this population has been particularly difficult, and it is probable that cases among homosexual men will continue to contribute significantly to the incidence of syphilis.

Chlamydial Infections

Although gonorrhea and syphilis are probably the best-known venereal diseases, other venereal diseases caused by bacteria occur with equal or greater frequency in the United States. For example, 80 to 90 percent of male college students with symptoms suggesting gonorrhea actually have other venereal infections. Twenty-five to 40 percent of such students are infected with *Chlamydia trachomatis*,[1] a common cause of venereal disease that mimics *N. gonorrhoeae* infections in several ways, including production of urethritis, epididymitis, and salpingitis.* Asymptomatic carriers are fairly common, as shown by one study of college women, which revealed that almost 5 percent had positive genital cultures in the absence of any symptoms. Nonvenereal transmission of this agent also occurs, for example, in nonchlorinated swimming pools. Moreover, newborn babies of infected mothers may experience eye infections or pneumonia from infections contracted during passage through the birth canal.

C. trachomatis, page 329

Chlamydia trachomatis differs importantly from *N. gonorrhoeae* in that it is an obligate intracellular bacterium and cannot be cultivated on cell-free laboratory media. Infections caused by *C. trachomatis* can be treated effectively with tetracycline and erythromycin. but not by the penicillins.

Other Venereal Diseases

A common cause of vaginal infections (vaginitis) is the Gram-negative rod *Gardnerella vaginalis*,[2] which is frequently acquired through sexual intercourse. An increased malodorous discharge is the most prominent symptom. The organism can be cultivated on several commonly used laboratory media.

---

[1]Other strains of *C. trachomatis* cause different types of illnesses. For example, some cause **trachoma**, a serious eye disease in economically deprived areas of the world, and **lymphogranuloma venereum**, a venereal disease that causes enlarged, painful, pus-filled groin lymph nodes and scarring in the genital area.

[2]Formerly known as *Haemophilus vaginalis*

*urethritis, epididymitis, salpingitis—inflammation of the urethra, epididymis (collection of tubules in the testes), and fallopian tubes, respectively

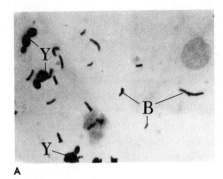
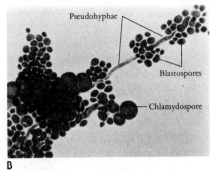

A   B

Figure 24–10

Candida albicans. (a) The morphology of *C. albicans* as it appears in the genitourinary tract. The budding yeasts (Y) are large (6 μm) and readily distinguished from bacteria (B). An epithelial cell from the genitourinary tract is seen in the upper right-hand corner. (b) *C. albicans* grown in culture. (Courtesy of S. Eng.)

Interestingly, infections with this bacterium respond poorly to antibiotics but well to metronidazole, a medication used principally against protozoa.

Several other bacterial venereal diseases are known, including **chancroid** (caused by *Haemophilus ducreyi*), which is characterized by a genital sore somewhat resembling the chancre of syphilis. All the bacterial venereal infections are effectively treated by antimicrobial medications, but in many cases the patient's sexual partner is a carrier and must be treated at the same time as the patient to prevent reinfection and recurrent disease.

NONBACTERIAL INFECTIONS OF THE GENITOURINARY TRACT

Yeast Vulvovaginitis

A common fungal cause of genitourinary disease in women is *Candida albicans* (Fig. 24–10). This yeast is part of the normal flora of the mucous membranes in about 35 to 40 percent of the female population. It is usually nonpathogenic. The interaction between large numbers of bacteria and small numbers of these fungi results in an ecological balance between the groups. When this balance is upset, disease may occur, usually as a vulvovaginitis.*
This happens most often after intensive antibacterial treatment, particularly with antibiotics such as ampicillin and tetracycline, which suppress the growth of much of the normal bacterial flora of the mucous membranes, allowing fungi to grow. The large numbers of *C. albicans* found in such a situation can cause persistent and severe infections of the mucous membranes of the genitourinary and alimentary tracts. Other predisposing factors to *Candida* infection include pregnancy, hormone therapy with oral contraceptives, and diabetes. Treatment of *C. albicans* infections with medicines such as nystatin or clotrimazole is usually effective.

---

*vulvovaginitis—inflammation of both the vulva and the vagina

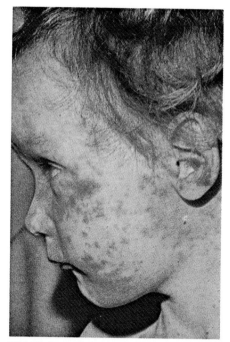

COLOR PLATE 41 Rubella (German measles). (Courtesy of Korting, G. W.: *Hautkrankheiten bei Kindern und Jugendlichen.* Copyright 1969 and 1972 by F. K. Schattauer Verlag GmbH, Stuttgart.)

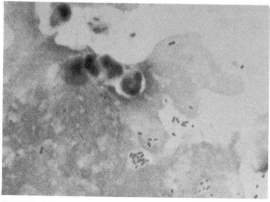

COLOR PLATE 44 Sputum from a patient with klebsiella pneumonia (Gram stain). (Courtesy of C. E. Roberts.)

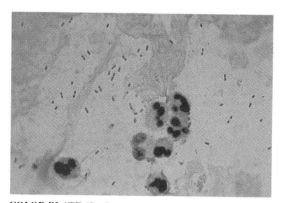

COLOR PLATE 43 Sputum from a patient with pneumococcal pneumonia (Gram stain). The small, dark dots occurring in pairs are the pneumococci. (Courtesy of C. E. Roberts.)

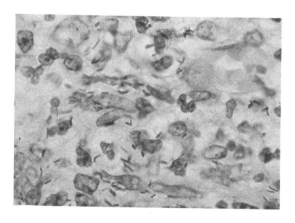

COLOR PLATE 45 (Right) Tuberculous infection of the small intestine (acid-fast stain), now a rare disease because of milk pasteurization and tuberculin testing of dairy cows to detect infected animals. The red rod-shaped bodies are the tubercle bacilli. The larger blue bodies are the nuclei of the cells of the intestine. (Courtesy of C. E. Roberts.)

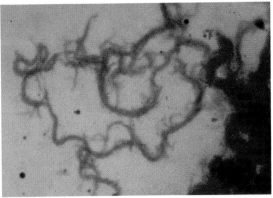

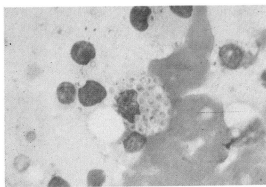

COLOR PLATE 46 Cording of *Mycobacterium tuberculosis* (acid-fast stain). Each cord consists of masses of tubercle bacilli adhering side by side. This is characteristic of growth of virulent strains in moist media. (Courtesy of C. E. Roberts.)

COLOR PLATE 47 Spleen of a mouse infected with *Histoplasma capsulatum*. The fungi, seen as small, pale blue ovals, completely fill the cytoplasm of the large cell in the center of the photograph. (Courtesy of C. E. Roberts.)

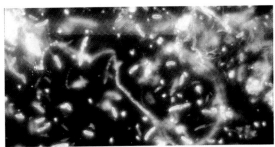

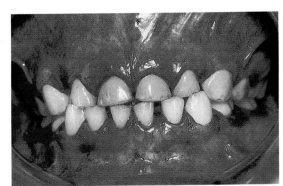

COLOR PLATE 48 Dark-field examination of *Legionella pneumophila*. (Courtesy of S. Eng and F. Schoenknecht.)

COLOR PLATE 49 Patient with early periodontal disease. (Courtesy of S. Eng and F. Schoenknecht.)

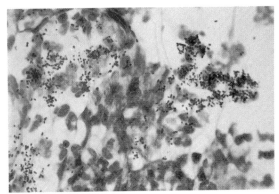

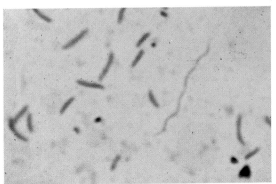

COLOR PLATE 50 Smear from a tooth abscess. Note the masses of tiny Gram-positive bacteria and the red material that represents nuclei of pus cells. (Courtesy of S. Eng and F. Schoenknecht.)

COLOR PLATE 51 Gram-stain smear from an acute periodontal infection. Note the single wave-shaped spirochetal organism and the multiple rod-shaped fusiform bacteria. (Courtesy of C. E. Roberts.)

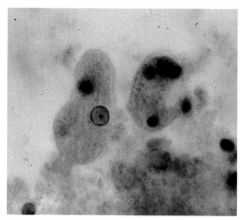

COLOR PLATE 52 Stained smear of trophozoites of *Entamoeba histolytica* in feces. Note the typical appearance of the spherical nucleus in the pear-shaped ameba near the center of the field. (Courtesy of S. Eng and F. Schoenknecht.)

COLOR PLATE 53 Culture of *Aspergillus fumigatus* recovered from a leukemic patient dying of aspergillosis. This common fungus is usually harmless to humans, but it can cause fatal disease in people with leukemia and other conditions of impaired immunity. (Courtesy of F. Schoenknecht.)

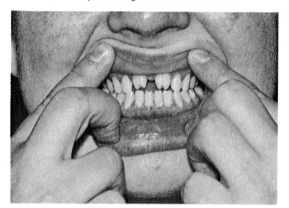

COLOR PLATE 54 Hutchinson's teeth, a late manifestation of congenital syphilis.

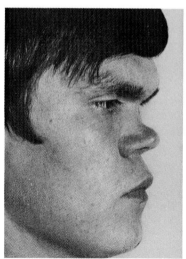

COLOR PLATE 55 Saddle-nose deformity caused by congenital syphilis.

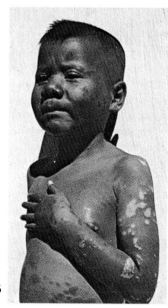

COLOR PLATE 56
Leprosy.

Note: Color plates 54, 55, and 56 are courtesy of Korting, G. W.: *Hautkrankheiten bei Kindern und Jugendlichen.* Copyright 1969 and 1972 by F. K. Schattauer Verlag GmbH, Stuttgart.

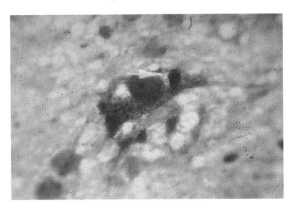

COLOR PLATE 57 Smear of brain tissue from a rabid dog (Seller's stain). The arrow points to one of several red Negri bodies within the triangle-shaped motor nerve cell. (Courtesy of C. E. Roberts.)

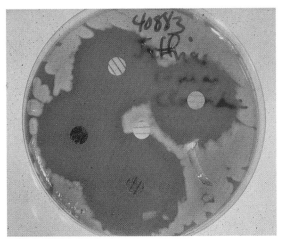

COLOR PLATE 58 Antibiotic susceptibility of *Clostridium perfringens* on sheep blood agar medium. Note the zones of inhibited growth and hemolysis about the antibiotic-containing discs. (Courtesy of C. E. Roberts.)

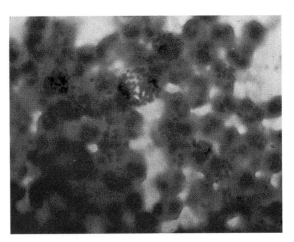

COLOR PLATE 59 Staphylococcal abscess (Gram stain). Staphylococci appear as masses of dark dots within and without the red pus cells. (Courtesy of C. E. Roberts.)

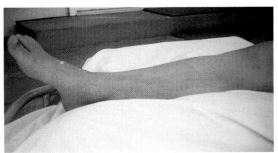

COLOR PLATE 60 Lymphangitis. Note the red streak extending up the leg from the infected toe. (Courtesy of C. Thelen and J. Roberts.)

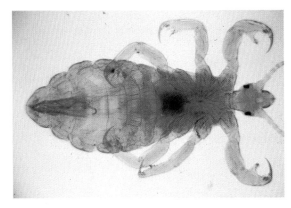

COLOR PLATE 61 *Pediculus humanus* (body louse). (×125) (Courtesy of S. Eng and F. Schoenknecht.)

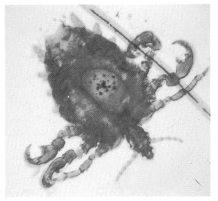

COLOR PLATE 62 *Phthirus pubis* (crab louse). (×125) (Courtesy of S. Eng and F. Schoenknecht.)

Candida albicans is an oval, budding yeast (Fig. 24–10a). When grown at 37°C on a rich medium, it forms smooth, creamy colonies. However, under less favorable conditions, such as a less nutritious medium at room temperature, pseudohyphae are formed along with the budding yeast cells. These pseudohyphae consist of very elongated budding cells in chains, resembling hyphae in a mycelium. Frequently, round, thickened spores (chlamydospores) form at the ends of the pseudohyphae of *C. albicans* (Fig. 24–10b). These morphological characteristics help to distinguish *Candida* from other yeasts. Carbohydrate fermentation patterns may also be used for identification of *C. albicans*. In addition, this organism produces germ tubes in human serum (Fig. 24–11).

C. albicans, page 378

Vaginitis Caused by *Trichomonas vaginalis*

Trichomonas vaginalis, a protozoan commonly present in the normal flora of the genitalia, may sometimes invade the genitourinary tract and cause disease. This organism frequently causes a mild infection of the vagina in women and less often an infection of the prostate in men or of the bladder in either sex. These flagellate, unicellular protozoa are easily identified by microscopic examination of the urine or infected exudate (Fig. 24–12). Infections usually respond well to metronidazole treatment.

Genital Herpes Simplex

Genital herpes simplex is now the third most common venereal disease, after gonorrhea and nongonococcal urethritis. The symptoms begin 2 to 20 days (usually about six days) after exposure, with genital itching, burning, and often severe pain. Clusters of small red bumps appear on the genitalia, which then become blisters surrounded by redness. The blisters break in three to five days, leaving an ulcerated area. The ulcers slowly become dry and crusted, then heal without a scar. Symptoms are the worst in the first one to two weeks and disappear in three weeks. They recur in an irregular pattern; about 75 percent of patients will have at least one recurrence within a year, and the average is about four recurrences a year. The recurrent symptoms are not as severe as those of the first episode.

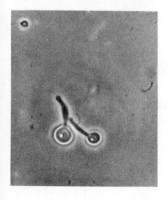

Figure 24–11
Candida albicans forms germ tubes when incubated in normal human serum or saliva. (Courtesy of L. H. Kimura.)

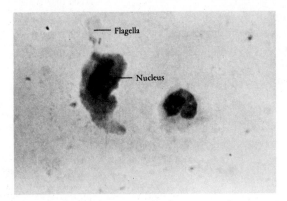

Figure 24–12
Flagellate of the genus *Trichomonas* in Gram-stained material from the human genitourinary tract. The protozoan is on the left and a leukocyte is on the right. Identification of *Trichomonas* sp. is easier in a fresh specimen than in a stained smear because the living organisms have a characteristic motility. (Courtesy of S. Eng.)

Labels on figure: Flagella, Nucleus

Herpes virus,
Appendix VIII

Herpes simplex,
page 472

The cause of the disease is herpes simplex virus type 2, a medium-sized DNA virus closely resembling the herpes simplex virus type 1 that causes cold sores and other problems. During the primary infection and for as long as one month thereafter, the virus is found in body secretions. During recurrences, the virus is usually present for less than a week. The virus may be present in the absence of symptoms.

Pathogenesis of the disease is poorly understood, but the viral chromosome is known to exist within nerve cells during times when there are no symptoms. The virus is not in its complete infectious form and may be analogous to a temperate bacteriophage in that, under proper circumstances, the viral chromosome can replicate and form complete infectious viruses.

At present, there is no generally available cure or effective treatment for genital herpes, although promising medications such as acyclovir are being evaluated experimentally. The disease is important not only because of the symptoms it produces but because, if the primary infection in a pregnant woman occurs at the time of delivery, the baby acquires the infection and often dies. There is also great concern that genital herpes may contribute to cancer of the uterine cervix, since women with genital herpes have more than five times the risk of precancerous changes of the cervix, compared with women who do not have herpes.[3] Women with recurrent genital herpes should have a Papanicolaou smear every 6 to 12 months to be sure of detecting cervical cancer in its early (curable) stage. Both infected men and women should alter their sexual activity to avoid spreading the virus.

SUMMARY

Urinary tract infections, much more common in women than in men, are usually caused by members of the normal intestinal flora, especially *Escherichia coli* and similar Gram-negative rod-shaped bacteria. Infection be-

---

[3]The association of herpes simplex type 2 infection with cervical cancer does not necessarily mean that the virus causes the cancer.

comes established when the body's normal defenses fail. As a rule, infections occur first in the bladder and, if not limited there, reach the kidneys by ascending through the ureters.

Historically important, but uncommon in the United States today, is puerperal fever (childbirth fever), often caused by *Streptococcus pyogenes*. Toxic shock syndrome occurs primarily in menstruating women who use intravaginal tampons. The cause is a toxin-producing *Staphylococcus aureus* growing in the vagina.

Among the extremely common bacterial infections of the genital tract are various venereal diseases. Gonorrhea results from infection with *Neisseria gonorrhoeae*, and syphilis is caused by the spirochete *Treponema pallidum*. *Chlamydia trachomatis* infections mimic gonorrhea and are very common, as are vaginal infections with *Gardnerella vaginalis*. Chancroid, caused by *Haemophilus ducreyi*, causes a genital ulcer that resembles the chancre of syphilis. Nonbacterial venereal infections include those caused by a fungus (*Candida albicans*) and a protozoan (*Trichomonas vaginalis*). Genital herpes simplex is the third most common venereal disease and carries with it an increased risk of cancer of the uterine cervix. Genital herpes simplex is commonly recurrent, and no cure is generally available at present.

SELF-QUIZ

1. The Gram-negative organism responsible for most acute bladder infections of young women is
   a. *Streptococcus faecalis*
   b. *Pseudomonas aeruginosa*
   c. *Escherichia coli*
   d. *Candida albicans*
   e. *Trichomonas vaginalis*
2. *Neisseria gonorrhoeae*
   a. is usually transmitted by toilet seats
   b. is highly resistant to cold and drying
   c. cannot be cultivated on cell-free media
   d. is a Gram-positive diplococcus
   e. sites of host colonization are determined by pili
3. Of help in diagnosing syphilis is
   a. the Gram stain
   b. bacteriological culture media
   c. dark-field microscopy
   d. all of the above
   e. none of the above
4. In diagnosing syphilis, the most significant test would be
   a. two blood samples, taken four weeks apart, the second one showing an increased amount of antibody to syphilis bacteria
   b. dark-field examination showing diplococci
   c. a single blood sample that gives a positive VDRL test

d. a stained smear of a genital ulcer showing Gram-negative rods

e. fluorescent anti-rabbit IgG attached to syphilis bacterium obtained from rabbits

5. Urine collected for culture should be transmitted promptly to the laboratory because

a. urinary antibodies will kill the causative bacteria of urinary infections

b. the bacteria causing a urinary infection may multiply rapidly in urine at room temperature

c. a few *E. coli* from among the normal flora bacteria can contaminate the urine specimen and may grow rapidly in number in standing urine

d. the doctor needs the culture results right away in order to begin treatment

e. all of the above

6. Characteristics of genital herpes include

a. caused by a virus

b. incurable

c. recurrent disease is common

d. painful genital ulcers

e. all of the above

7. Men and women with genital herpes must be concerned about

a. possible threat to a newborn baby

b. transmission to another person in the absence of symptoms

c. an increased risk of cancer in women

d. all of the above

e. none of the above

8. In between attacks of genital herpes, the causative agent

a. exists within nerve cells

b. is easily obtained by culture of nerve tissue

c. can be found in genital secretions by Gram staining

d. is never present in genital secretions

e. all of the above

QUESTIONS FOR DISCUSSION

1. If through the spread of R factors, gonorrhea joins genital herpes in being an untreatable disease, do you see a future for artificial insemination?

2. What patterns of daily living could influence the chance of developing a urinary tract infection?

FURTHER READING

Buckley, R. M. Jr., McGuckin, M., and MacGregor, R. R.: "Urine Bacterial Counts After Sexual Intercourse." *New England Journal of Medicine* 298:321–324 (1978).

Eschenbach, D. A., Buchanan, T. M., Pollock, H. M., et al.: "Polymicrobial Etiology of Pelvic Inflammatory Disease." *New England Journal of Medicine 293*:166–171 (1975).

"Follow-up on Toxic Shock Syndrome." *Morbidity and Mortality Weekly Report 29*:441–445 (1980).

Holmes, K. K., and Stamm, W. E.: "Chlamydial Genital Infections: A Growing Problem." *Hospital Practice* (October 1979), pp. 105–117.

Lukacs, J., and Corey, L.: "Genital Herpes Simplex Virus Infection: An Overview." *Nurse Practitioner* (May-June 1978), pp. 7–10.

Lum, B., Lortz, R., and Barnett, E.: "Reappraising Newborn Eye Care." *American Journal of Nursing 80*:1602–1603 (1980). A question-and-answer format regarding silver nitrate and the eyes of newborns. Recent research by the authors is included.

Pheifer, T. A., Forsyth, P. S., Durfee, M. A., et al.: "Nonspecific Vaginitis: Role of *Haemophilus Vaginalis* and Treatment with Metronidazole." *New England Journal of Medicine 298*:1429–1434 (1978).

Rapp, F.: "Herpesviruses, Venereal Disease, and Cancer." *American Scientist 66*:670–674 (1978).

Rosebury, T.: *Microbes and Morals.* New York: Viking Press, 1971.

Turck, M.: "Urinary Tract Infections." *Hospital Practice* (January 1980), pp. 49–58.

Wroblewski, S. S.: "Toxic Shock Syndrome." *American Journal of Nursing 81*:82–85 (January 1981). A good overall look at this disease, with emphasis on observing and documenting symptoms as they appear.

Chapter 25

NERVOUS SYSTEM INFECTIONS

When too little has been done for such a wound [bite of a mad animal], it usually gives rise to a fear of water which the Greeks call hydrophobia. . . . In these cases there is little hope for the sufferer. But still there is just one remedy, to throw the patient unawares into a water tank which he has not seen beforehand. If he cannot swim, let him sink under and drink, then lift him out; if he can swim, push him under at intervals so that he drinks his fill of water even against his will; for so his thirst and dread of water are removed at the same time. Yet this procedure incurs a further danger that a spasm of sinews, provoked by the cold water, may carry off a weakened body. Lest this should happen, he must be taken straight from the tank and plunged into a bath of hot oil.[1]

Celsus (25 B.C.–A.D. 50)
De Medicina, V. 27 (tr. by W.G. Spencer)

Today's treatment of rabies is less drastic but unfortunately no more successful than in the days of Celsus.

Infections of the nervous system are relatively uncommon but are apt to be serious because they strike at our ability to move, to feel, and to think. Nervous system infections may be caused by bacteria, viruses, fungi, and even some protozoa. Fungal and protozoan nervous system diseases are mentioned elsewhere in the text and are not discussed in this chapter.

See *Cryptococcus*, page 379, and *Naegleria*, page 387

---

[1]Strauss, M.: *Familiar Medical Quotations*. Boston: Little, Brown & Co., 1968.

OBJECTIVES

To know
1. The relationship of the structure and function of the nervous system to its susceptibility to infection.
2. The causes, pathogenesis, and other features of bacterial meningitis, including epidemiology and prevention.
3. The nature of leprosy and the role of immune mechanisms in its pathogenesis.
4. The pathogenesis and prevention of poliomyelitis and rabies.

ANATOMY AND PHYSIOLOGY

The brain and spinal cord make up the central nervous system. Both are enclosed by bone—the brain by the skull and the spinal cord by the spinal column (Fig. 25–1). The network of nerves throughout the body is connected with the central nervous system by large bundles of nerve fibers that penetrate the protective bony covering at intervals. These larger nerves can therefore be damaged if there are infections of the bones at sites of nerve penetration. Two kinds of nerves are involved in the body's nerve network: **motor nerves,** which cause different parts of the body to act, and **sensory nerves,** which transmit sensations, such as heat, pain, light, and sound. All nerves are made up of special cells with very long, thin extensions that transmit electrical impulses. These long cellular extensions are actually the fibers composing the nerve bundles. They can sometimes regenerate if severed or damaged, but they cannot be repaired if the nerve cell of which they are part is itself killed, as may occur, for example, in poliomyelitis.

Deep inside the brain are several cavities filled with a clear fluid called **cerebrospinal fluid.** This fluid is continually produced in these cavities, flowing from them through small openings to spread over the surface of the brain and spinal cord and out along the nerve trunks that penetrate the bone. The cerebrospinal fluid then flows into lymph* channels and eventually enters the bloodstream. In people with suspected infections of the central nervous system, a needle can be inserted safely between the vertebrae composing the spinal column in the small of the back, and a sample of cerebrospinal fluid can be withdrawn from the spinal canal for examination (Fig. 25–1). In this manner, the causative agent of a central nervous system infection can often be identified microscopically and cultivated on laboratory media. An effective treatment against the pathogen can then be administered.

Two membranes called **meninges** cover the surface of the brain and spinal cord; the cerebrospinal fluid flows between them. Inflammation of these membranes is called **meningitis.** Because blood vessels and nerves pass

---

*lymph—a body fluid that bathes the cells; similar to plasma

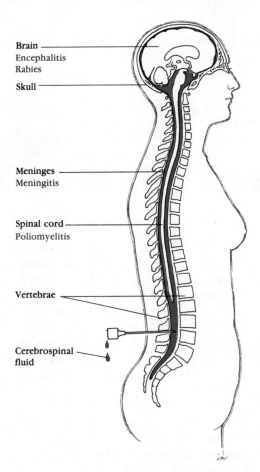

Brain
Encephalitis
Rabies

Skull

Meninges
Meningitis

Spinal cord
Poliomyelitis

Vertebrae

Cerebrospinal
fluid

Figure 25–1
The central nervous system.

through both the meninges and the film of cerebrospinal fluid between them, meningitis may involve these vessels and nerves and cause their malfunction.

The central nervous system lies in a well-protected environment that prevents infectious agents from getting to it readily. When they do, it seems that their entry must almost have been accidental, because most microorganisms that infect the nervous system infect other parts of the body much more frequently. Pathogenic microorganisms and viruses reach the brain and spinal cord by the following routes:

1. The bloodstream. This is the chief source of central nervous system infections, although it is very difficult for infectious agents to cross from the bloodstream to the brain. This barrier to the passage of harmful agents is often called the **blood-brain barrier.** Although its exact nature is unknown, this barrier is probably made up of a thin layer of cells, some of which are phagocytic. The blood-brain barrier is effective in preventing pathogens from entering nervous tissue in all but a few individuals afflicted with bloodstream infections. The reasons for these few

failures of the blood-brain barrier are not known, but they probably relate to the concentration of organisms in the bloodstream and to the length of time they circulate. Veins in the face may connect directly to those on the brain surface, and upper facial infections may therefore sometimes spread to the brain.

2. **Entrance by means of the nerves.** Some pathogenic agents penetrate the central nervous system by traveling up the nerves. It is uncertain whether these pathogens move inside the nerve fibers themselves or through passages between other kinds of cells that surround and protect the nerve fibers. In some instances, the pathogen grows in these surrounding cells.

3. **Direct extensions of infections elsewhere.** To reach the central nervous system, microorganisms must penetrate not only the bone that surrounds it but also tough outer membranes that surround the meninges and bone. Infections in bone surrounding the central nervous system may on rare occasions erode inward to reach the brain or spinal cord. Skull fractures may produce nonhealing injuries that predispose a person to infection of the central nervous system. Infections may also extend to the central nervous system from the respiratory sinuses,* mastoids* or, more commonly, the middle ear.

SOME BACTERIAL INFECTIONS OF THE NERVOUS SYSTEM

Meningococcal Meningitis

Meningococcal meningitis is often called epidemic meningitis because it can spread in epidemic fashion; however, it typically appears at widely separated locations and occurs throughout the year (Fig. 25–2). In a typical case the first symptoms are those of a mild cold, followed by the sudden onset of a severe throbbing headache and fever and marked pain and stiffness of the neck and back. Purplish spots may appear on the skin (petechial rash). Epidemic meningitis is frightening because the infected person may develop shock and die within 24 hours of having felt and appeared completely healthy. Usually, though, the illness lasts longer and is not fatal.

The causative organism of meningococcal meningitis, *Neisseria meningitidis* (the meningococcus) (Table 25–1), can usually be found on smears made from the centrifuged sediment of cerebrospinal fluid or in scrapings of the purple skin spots characteristic of the disease. Meningococci are Gram-negative cocci occurring in pairs (diplococci); each member of the diplococci has a flattened side where it faces its neighbor. Most of the organisms seen in cerebrospinal fluid occur within leukocytes, which enter the fluid in response to the infection. *N. meningitidis* organisms grow readily on media enriched with blood, especially if incubated in an atmosphere with an increased car-

N. meningitidis, page 310

Enriched media, page 72

---

*sinuses—air-filled cavities within the facial bones

*mastoids—air-filled cavities within the temporal bone, behind the ear

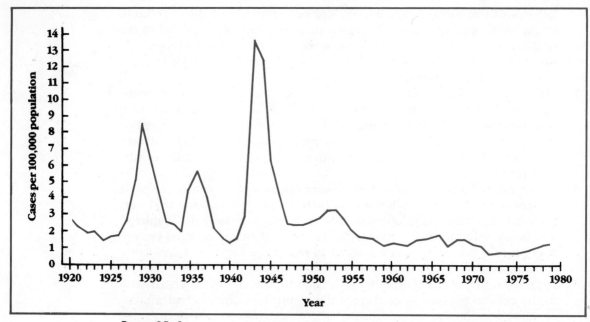

Figure 25–2
Meningococcal infection rate in the United States, 1920–1979. (Adapted from *Morbid. Mortal. Wkly. Rep., Annual Summary 1979*, United States Public Health Service, 28:53, 1979.)

bon dioxide concentration. Some meningococcal strains are killed quickly by moderate chilling, and this may account for some failures to recover the organism in cultures of material from patients. For this reason it is important to transmit specimens promptly to the laboratory and to culture them immediately.

N. meningitidis is strictly a parasite of humans, and infection is transmitted by exposure to a person carrying this species. The majority of meningococcal infections are in locations other then the central nervous system and pass unnoticed, being detected only by the chance discovery of a heavy growth of the causative organism in a throat culture from an infected person.

TABLE 25–1 CHARACTERISTICS OF *NEISSERIA MENINGITIDIS*

Identification	*Normal habitat*	*Pathogenic potential*
Gram-negative encapsulated diplococci with characteristic pattern of sugar fermentations; eight groups (A, B, C, etc.) identified by specific antisera	Upper respiratory tract of humans	Can cause serious infections marked by bloodstream invasion and meningitis

If a selective medium is used to suppress the growth of other nasopharyngeal flora, meningococci can be recovered from about 15 percent of normal people.

Over the years, the frequency of meningitis caused by different strains of *N. meningitidis* has varied markedly. For example, in 1945 strains of group A[1] were responsible for most cases; in 1970 group C organisms were dominant; while by 1978 most strains isolated from meningococcal meningitis patients were group B. Factors that may have played a role in these changing frequencies include (1) relative immunity of the population to type A strains, developed during previous epidemics; (2) the widespread use of sulfa drugs, which resulted in the selection of rare sulfa-resistant strains known to be present among types B and C even before the introduction of sulfa drugs; (3) crowding of susceptible populations, as in military barracks, which allowed rapid spread of virulent strains; and (4) transfer of virulence genes among different strains of meningococcus, possibly by bacterial transformation.

Bacterial transformation, page 136

The mechanisms by which meningococci produce cerebrospinal meningitis have long been studied. The organisms probably infect people who inhale droplets from another person, almost always a carrier, or one with an unnoticed infection. The meningococci establish an infection in the upper respiratory tract and then enter the bloodstream. The blood carries the organisms to the meninges and spinal fluid, where they multiply faster than they can be engulfed and destroyed by the leukocytes. The inflammatory response, with its formation of pus and clots, may obstruct the normal outflow of cerebrospinal fluid, squeezing the brain flat against the skull by buildup of internal pressure. Infection of brain tissue, and later scar formation, may damage motor nerves and produce paralysis. Moreover, *N. meningitidis* bacteria circulating in the bloodstream injure the blood vessels, so that normal blood pressure cannot be maintained and shock* results. The smaller blood vessels of the skin are also affected by circulating meningococci, which cause the small skin hemorrhages (the purple spots) characteristic of the disease.

Blood vessel damage by *N. meningitidis* probably results from endotoxin activity. Not only does the organism's endotoxin cause shock, but it also produces fever and increased sensitivity to any further exposure to endotoxin. This increased sensitivity, called the **Shwartzman phenomenon,** can be demonstrated in animals following injections of meningococci, and it may well be responsible for the damage to the small blood vessels seen in humans. The tendency of meningococci to autolyse (rupture spontaneously) may explain the release of their endotoxin.

---

[1]Different groups of meningococci are identified using antisera and are designated by capital letters: A, B, C, W, Y, etc. Numerous strains can be distinguished within these groups by differences in antibiotic or bacteriocin susceptibility, or by serological techniques.

*shock—a state in which there is insufficient blood pressure to supply adequate blood flow to meet the oxygen needs of vital body tissues

Type-specific opsonins* protect against meningococcal disease. In fact, vaccines composed of purified capsular polysaccharide have been prepared against *N. meningitidis* types A and C. These vaccines are effective in controlling meningitis and are being used to immunize certain high-risk human populations. Since epidemic meningococcal strains probably have high virulence, people intimately exposed to cases of meningococcal disease are given prophylactic treatment with the antibiotic rifampin. Patients with meningococcal meningitis are isolated for 24 hours after starting treatment to help prevent spread of the disease to other people. Meningococcal meningitis can usually be cured by penicillin or chloramphenicol. The mortality is about 15 percent.

Rifampin, page 198

Other Bacterial Causes of Meningitis

In addition to *N. meningitidis,* many other bacteria may cause infection of the meninges, and in most cases, just as with *N. meningitidis*, these organisms are commonly carried by healthy people, are transmitted by inhalation, and only rarely produce meningitis. Organisms in this category include the respiratory pathogens *Streptococcus pneumoniae* and *Haemophilus influenzae.* Although they are common inhabitants of the respiratory tract, these organisms seldom cause meningitis in newborn babies, presumably because antibody against them is transferred across the placenta from the mother. Instead, Gram-negative rods, such as *Escherichia coli* from the mother's intestinal tract or even *Flavobacterium meningosepticum* from the nursery water faucet, are more likely to cause meningitis in the newborn. The current view is that the newborn infant lacks protection against Gram-negative rods such as these because the antibodies effective against such organisms are of the IgM type and are unable to pass across the placenta. Other causes of meningitis in newborn infants are beta-hemolytic streptococci of serological group B and a Gram-positive rod, *Listeria monocytogenes*. These two species of bacteria are carried among the vaginal flora of some women and can act as opportunists in immunologically immature infants.

IgM, page 234

Leprosy (Hansen's Disease)

Although now a relatively minor problem, leprosy was once common in Europe and America (color plate 56). In 1868 G. A. Hansen, a Norwegian, demonstrated the causative bacterium (*Mycobacterium leprae*) in the tissues of leprosy patients. This is said to have been the first time a bacterium was causally linked to human disease. Yet, despite many attempts, *M. leprae* has still not been grown in test tube culture.* Like tuberculosis, leprosy began to recede in Europe and America for unknown reasons, and today it is chiefly

---

*opsonins—antibodies that aid phagocytosis

*in test tube culture—in cell-free media or tissue cells grown in laboratory containers

a disease of tropical and more economically underdeveloped countries, with an estimated worldwide incidence of between 10 and 20 million cases.

Because attempts to grow *M. leprae* in vitro have been so unsuccessful, many attempts have been made to infect laboratory animals with it. A tremendous advance was made in 1960 when it was demonstrated that the organisms would multiply in the footpads of mice. In addition, in the early 1970's, the nine-banded armadillo prevalent in Texas and Louisiana was shown to be susceptible. Wild armadillos infected with a bacterium indistinguishable from the human leprosy bacillus have been found in Louisiana, but there is no evidence that these animals are the source of human leprosy.

From studies of infected mice it has been possible to acquire some interesting information about *Mycobacterium leprae*. One of the more striking findings has been that these bacteria have a long generation time, estimated to be about 12 days. This finding helps to account for the very long incubation period of human leprosy, probably a minimum of two years and often ten years or longer. In addition, growth of *M. leprae* in mice has permitted studies aimed at finding new medicines for the treatment of this disease.

Mycobacterium species, page 325

The earliest detectable finding in human infection with *M. leprae* is the invasion of the small nerves of the skin. Indeed, *M. leprae* is the only known human pathogen that preferentially attacks the peripheral nerves. The bacterium also grows within macrophages and the cells that line blood vessels. The course of the infection depends on the immune response of the host. In most cases, cell-mediated immunity and delayed hypersensitivity develop against the invading bacteria. Immune macrophages limit the growth of *M. leprae,* and the bacteria therefore do not become numerous, but nevertheless nerve damage may progress and lead to disabling deformity. In most instances, however, the disease is arrested after a few years, and thereafter the nerve damage, although permanent, does not progress. This limited type of leprosy is called the **tuberculoid** type. People wth tuberculoid leprosy rarely, if ever, transmit the disease to others.

In some infected people, however, cellular immunity and delayed hypersensitivity to *M. leprae* either fail to develop or are lost, and unrestricted growth (rather then destruction) of *M. leprae* occurs in macrophages. This relatively uncommon form of leprosy is referred to as the **lepromatous** type. In addition to the nerves, it involves most tissues of the body. The tissues and mucous membranes then swarm with the leprosy bacteria, which can be transmitted to others by intimate direct contact. A similar disease can be produced in armadillos inoculated with the bacilli and also in mice whose capability for cellular immunity has been destroyed by thymectomy* and radiation. During lepromatous leprosy, there may be a general impairment of delayed hypersensitivity responses, not only to leprosy bacilli but also to many other antigens. On the other hand, a return toward normal immune function occurs when the disease is controlled by medication.

Cell-mediated immunity, page 265

---

*thymectomy—surgical removal of the thymus gland

Leprosy is one of the least contagious of all infectious diseases, and the morbid days when lepers were forced to carry a bell or horn to warn others of their presence are fortunately past. Today, people with leprosy can expect a normal life span, provided ill-founded fears and social pressures do not cause them to avoid early medical consultation.

The disease is diagnosed by microscopic inspection of scrapings or biopsies of infected tissues that have been stained to detect acid-fast bacteria. *Mycobacterium leprae* produces a highly characteristic enzyme that may also prove to be useful in diagnosis. The disease can usually be arrested by long-term treatment with sulfone drugs. Research is underway on vaccines to prevent leprosy.

SOME VIRAL DISEASES OF THE NERVOUS SYSTEM

Poliomyelitis (Infantile Paralysis)

A person having poliomyelitis usually first suffers the symptoms of meningitis: headache, fever, stiff neck, and nausea. In addition, pain and spasm of some muscles generally occur, later followed by paralysis and finally by a relative shrinking of muscle and failure of normal bone development in the affected area. In the more severe cases, the muscles controlling respiration are involved, and the victim requires an artificial respirator (a machine to pump air in and out of the lungs). Some recovery of function is the rule if the person survives the acute stage of the illness. The nerves of sensation (touch, pain, temperature) are not affected.

Historically, poliomyelitis has had a greater impact in the more economically advanced countries. The disease does occur in poorer, more crowded nations, but epidemics of paralytic poliomyelitis generally appear to affect only small groups of geographically or culturally isolated people in these countries; among the remaining population, the causative viruses are widespread, and very few people escape childhood without infection. The adults in these countries therefore have antibodies in their bloodstreams, and these cross the placenta during pregnancy and enter the fetal circulation. Newborn infants in these nations thus are partially protected against bloodstream infection and nervous system invasion by poliomyelitis virus for as long as their mother's antibodies persist in their bodies—usually about two or three months. During this time, because of exposure to the poliovirus through crowding and unsanitary conditions, infants are likely to develop mild infections of the throat and intestine and thereby achieve lifelong immunity. Even if exposure is delayed beyond two or three months, it is likely to occur at an early age, when the disease is less likely to be severe.

In contrast, in such areas as suburban American communities in the days before the vaccine was developed, the poliomyelitis virus sometimes could not spread fast enough to sustain itself because of efficient sanitation. If its reintroduction from another community was delayed sufficiently, some people of all ages, including older children and adults, would then lack antibody and would be susceptible to infection. When the virus was finally reintroduced, a high incidence of paralysis resulted.

Poliovirus, the causative agent of poliomyelitis, can be recovered from the throat and feces early in the development of the illness, later from the blood, and finally, in some patients, from the cerebrospinal fluid. Excretion of the virus (an enterovirus) in the feces continues for weeks or months but is transitory in the other sites. When introduced into appropriate tissue cell cultures (usually consisting of monkey kidney cells), the virus produces a cytopathic* effect (plaques), which can readily be seen with the unaided eye. Poliomyelitis is generally caused by one of three types of poliovirus, which are distinguished by using antisera.

Enteroviruses, Appendix VIII

Like other enteroviruses, polioviruses are small RNA-containing viruses that are quite stable under natural conditions but are inactivated by pasteurization and proper chlorination of drinking water. The viruses can infect a cell under natural conditions only if the surface of that cell possesses specific receptors to which the virus can attach. This helps explain the remarkable selectivity of the polioviruses for motor nerve cells of the brain and spinal cord, whereas they spare most other kinds of nerve cells. Destruction of the infected cell occurs upon release of the mature virus.

Until the late 1950's, poliomyelitis was a terrifying threat, especially in economically advanced nations. Fear of the disease was widespread, and hundreds of thousands of people in the United States contributed money to the National Foundation for Infantile Paralysis (March of Dimes) to help victims and to support research for the solution of this problem. Former President Franklin D. Roosevelt, himself a polio victim, helped dramatize the need for this support.

One of the most important developments toward overcoming polio was the demonstration in 1949, by Dr. John Enders and his colleagues at Harvard University, that polioviruses grow readily in cell cultures, producing clear-cut cytopathic effects. By 1954, Dr. Jonas Salk had perfected a formalin-inactivated virus vaccine that was widely employed, and it soon lowered the incidence of paralytic poliomyelitis dramatically. Meanwhile, cell culture techniques were used to select mutant poliovirus strains that had little ability to attack nervous tissue but could nevertheless stimulate antibody formation against the wild-type, virulent virus. Living orally administered vaccines consisting of such attenuated viruses were then quickly developed and proved more effective than the killed-virus vaccines. A living vaccine prepared by Dr. Albert Sabin was chosen for use. As a result of these historic research developments, generally fewer than a dozen cases a year of paralytic poliomyelitis are reported in the United States, as opposed to the many thousands reported in earlier years (Fig. 25–3).

Oral polio vaccine should be given at ages 2 months, 4 months, 6 months, and 18 months and between the ages of 4 and 6 years. Ironically, some of

*cytopathic—cell-damaging

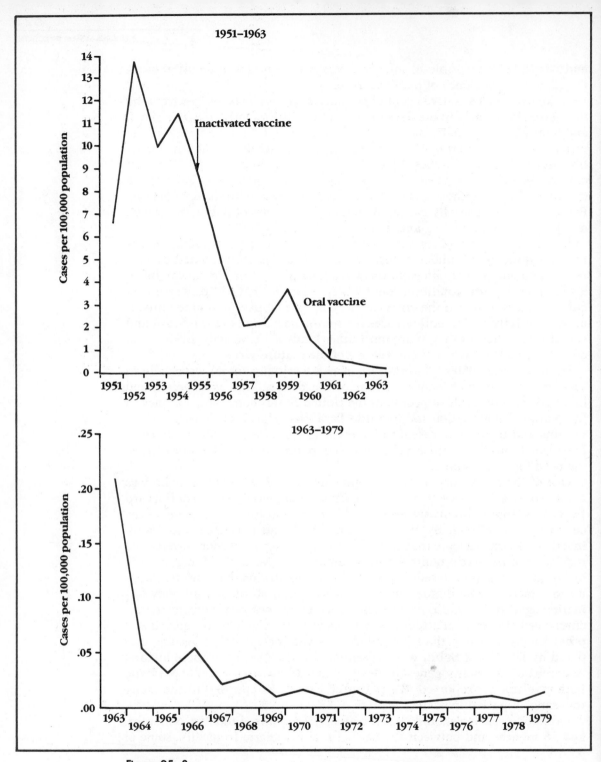

Figure 25–3
Number of cases of paralytic poliomyelitis in the United States per 100,000 population between the years 1951 and 1979. (Adapted from *Morbid. Mortal. Wkly. Rep., Annual Summary 1979*, United States Public Health Service, 28:64, 1979.)

TABLE 25–2 RECOMMENDATIONS FOR THE USE OF INACTIVATED POLIO VACCINE (IPV, SALK)

1. People over 18 who have never been immunized or have a questionable immunization. Three injections 6–8 weeks apart are given, followed by a "booster" of oral polio vaccine one year later.[a]

Other groups of all ages who should receive IPV:

2. People who are suspected of having an immune deficiency disease.
3. People with altered immune states, such as those occurring in leukemia, lymphoma, and advanced malignancy.
4. People who have lowered resistance resulting from prolonged treatment with corticosteroids, alkylating agents, or x-rays.
5. Household members of people in groups 2, 3, and 4.

For travel to a country where polio is endemic:

Adults who are not immunized against polio should be given IPV as in no. 1 above. However, if there is insufficient time to receive the IPV series, TOPV[b] should be given.

[a]Such immunization should be carried out before younger members of the family are immunized with oral vaccine, since the live (oral) vaccine readily spreads to older family members.
[b]TOPV—trivalent oral poliomyelitis vaccine

the more recent cases of polio have been caused by the live polio vaccine, which can produce paralysis in a rare recipient. This has led to resurgence of the use of the Salk vaccine (inactivated virus vaccine) in place of the live oral vaccine. Recommendations for the use of inactivated polio vaccine (IPV, Salk) are given in Table 25–2.

Rabies

Rabies is one of the most feared of all human diseases because its terrifying symptoms almost inevitably end with death. It is an acute infectious disease involving the central nervous system and is usually transmitted to warm-blooded animals by the saliva of animals infected with the rabies virus. In humans, the disease usually develops after an incubation period of 30 to 60 days, although the extreme range may be from ten days to more than a year. A prominent feature of the disease is spasm of the muscles of the mouth and throat at the sight (or even the thought) of water, and thus the popular name "hydrophobia" arose.

Identification of the causative virus of rabies is usually made most quickly by preparing smears from the conjunctiva, the fatty tissue of the neck, or the brain of a person or animal that has died from the disease. These smears are stained with a combination of dyes called **Seller's stain** or with fluorescent antirabies antibody and are inspected for nerve cells containing characteristic viral inclusions called **Negri bodies** (color plate 57). The rabies virus can be cultivated in animals such as laboratory mice or in cell culture. The virus is large, "bullet"-shaped, and enveloped and contains single-stranded RNA. It buds from the surface of infected cells.

Although rabies virus can infect by both the respiratory and oral routes under experimental conditions, the principal mode of transmission of rabies to humans is the passage of infected saliva through the skin barrier by way of a bite or other break in the skin. Once through the skin, the virus travels slowly, by an unknown mechanism, along the nerves leading to the central nervous system, where it multiplies in some of the motor nerve cells without causing cell destruction. The slow passage of the virus via the nerves is thought to account for the very long incubation periods often observed. The mode of spread of the rabies virus through the nerves and central nervous system is still poorly understood, and the explanation of how the virus moves into salivary glands and other tissues (such as the conjunctiva and fat) is obscure. The presence of the virus in the eye is of some practical significance, since cases may sometimes be diagnosed by staining smears made from the conjunctiva. Moreover, at least four cases have occurred in patients who received corneal transplants from donors who probably died of undiagnosed rabies.

Rabies is widespread in wild animals; about 6500 cases were reported in this country in 1980. This represents an enormous reservoir from which infection can be transmitted to humans and domestic animals. In the United States, skunks constitute the chief reservoir host; bats and raccoons are also important.

The majority of human cases of rabies in the United States have been caused by dog bites, and this is true in most areas of the world. The dog population in the United States is about 25 million, and an estimated one million people are bitten each year. Fortunately, since World War II the incidence of canine rabies (and human rabies) has dropped dramatically, and now almost 85 percent of reported rabies infections are in wild animals. Unfortunately, the data on the natural history of rabies in wild animals are incomplete. In dogs, anywhere from 10 to 100 percent will die of rabies following injection with wild rabies viruses from different sources. About three quarters of those that develop rabies excrete rabies virus in their saliva, and about one third of these begin excreting it one to three days before they get sick. Some dogs become irritable and hyperactive with the onset of rabies, produce excessive saliva, and attack people, animals, and inanimate objects. Perhaps more common is the "dumb" form of rabies, in which an infected dog simply stops eating, becomes inactive, and suffers paralysis of throat and leg muscles. Therefore, when someone is bitten by an unvaccinated, apparently healthy dog, the animal should be confined for ten days to see if rabies appears.[2] Table 25–3 is a guide for the prevention of rabies. The reported number of rabies cases in humans now generally ranges from zero to four per year in the United States.

---

[2]The bites of hamsters, guinea pigs, gerbils, squirrels, chipmunks, rats, mice, and rabbits almost never need antirabies treatment.

TABLE 25—3 A GUIDE FOR RABIES PREVENTION

Animal	Condition of animal at time of attack	Treatment of exposed person
Dog and cat	Healthy and available for 10 days of observation	None unless animal develops rabies[a]
	Rabid or suspected rabid	RIG and HDCV
	Unknown (escaped)	Consult public health officials. If treatment is needed, RIG and HDCV are given.
Skunk, bat, fox, coyote, raccoon, bobcat, and others	Regard as rabid unless proved negative by laboratory tests. The animal should be killed and tested as soon as possible without being held for observation.	RIG and HDCV

[a]In which case treatment should begin with rabies immune globulin (RIG) and human diploid cell rabies vaccine (HDCV). The sick animal should be killed immediately and tested.
From *Morbidity and Mortality Weekly Report,* United States Public Health Service, 29:23, 1980.

The dramatic decline in the incidence of rabies in people is largely the result of immunizing dogs and cats against rabies infection. In effect, this creates a partial barrier to the spread of the rabies virus from wild animal reservoirs to humans. Because pets vary in their tolerance of rabies vaccines depending on their age and species, several kinds of vaccines are available. In recent years, these have consisted of both attenuated and inactivated rabies viruses.

Attenuated vaccines, page 279

The risk of rabies developing in people bitten by dogs having rabies virus in their saliva may be as high as 30 percent, and in cases of bites by other wild animals the risk may be higher still. Louis Pasteur discovered that this risk can be lowered considerably by administering rabies vaccine as soon as possible after exposure to the virus. Presumably, the effectiveness of this measure is based on the accessibility of the rabies virus and its susceptibility to inactivation by immune mechanisms at some time during its long journey to the motor nerve cells of the brain. By giving large doses of rabies vaccine immediately after a bite, in most cases sufficient immunity can be developed in time to prevent infectious virus from reaching the central nervous system. In fact, this technique can reduce the risk of rabies by about 85 to 90 percent.

Pasteur's vaccine consisted of rabies virus grown in the brains of laboratory animals and then inactivated. Unfortunately, however, there is a risk of central nervous system disease from the vaccine itself. This arises because the vaccine virus cultivated in animal brain tissue contains an antigen com-

TABLE 25–4   PREVENTIVE REGIMEN FOR PEOPLE WITH SIGNIFICANT EXPOSURE TO
SUSPECTED RABID ANIMALS

Rabies vaccine	Route of administration	Doses
HDCV (human diploid cell rabies vaccine)	Intramuscular	Give one dose on the day of the bite and on days 3, 7, 14, and 28.[a]
DEV[b] (duck embryo vaccine)	Subcutaneous	21 consecutive daily doses followed by 1 dose 10 days later and 1 dose 10 days after that. OR 2 doses on each of the first 7 days, followed by 7 daily doses, 1 dose 10 days later, and 1 dose 10 days after that.[a]

[a]The patient's serum should be tested at this point to see if antibody to rabies has developed. Further immunization is required if the antibody response is inadequate. All patients receive rabies immune globulin (RIG) at the time of exposure.
[b]In the United States DEV has been supplanted by HDCV.
From *Morbidity and Mortality Weekly Report,* United States Public Health Service, 29:23, 1980.

mon to both the animal and the human brain. Those receiving such a rabies vaccine may thus develop an immune response to antigens of their own brain, producing **allergic encephalitis** in as many as one of every 2000 instances. This risk drops to only about 1 in 25,000 with rabies vaccine cultivated in duck embryos instead of in animal brains. The rate of serious nervous system disease with the newest vaccine, an inactivated rabies virus grown in human cell cultures, is even lower. The human cell culture vaccine can safely be used to immunize veterinarians and people who live in areas of the world where the rabies incidence is high before they are bitten by a rabid animal and to protect laboratory personnel who work with the rabies virus.

In the United States each year about 30,000 people receive treatment to prevent rabies after being exposed to suspected rabid animals. Immediate, thorough washing of the wound with soap and water is an important first step. The patient then receives a series of five injections of human diploid cell vaccine (HDCV) intramuscularly. If this vaccine is not available, then duck embryo vaccine (DEV) is given. The schedule of doses is given in Table 25–4. These people are further protected from rabies by injecting them with rabies **antibody.** Formerly, the source of this antibody was horses that had been immunized by injection with rabies vaccine. However, since horse antiserum is a foreign protein, many human recipients quickly formed antibodies against the horse antiserum. This often resulted in **serum sickness** or even **anaphylactic shock.** Today, the antiserum used is obtained from human

Anaphylaxis, page 262

volunteers who have been immunized against rabies, and the dangers of using horse serum therefore are avoided. As of 1980, 77 patients bitten by proven rabid animals were treated with human antirabies antiserum and human cell culture rabies vaccine, and none developed rabies.

SUMMARY

In comparison with other tissues, the brain and spinal cord are relatively protected from infections, and central nervous system involvement generally occurs in only a fraction of the total infections produced by a microbial agent. In many cases, examination of cerebrospinal fluid helps to recover and identify the infecting microorganism.

The respiratory pathogens *Neisseria meningitidis, Haemophilus influenzae,* and *Streptococcus pneumoniae* are the most frequent causes of bacterial meningitis, although in newborn babies the cause is more likely to be *Escherichia coli,* group B streptococci, or environmental bacteria.

Mycobacterium leprae, the cause of leprosy, is the only bacterial pathogen that preferentially attacks peripheral nerves. Leprosy has a prolonged incubation period, probably because of the slow growth of *M. leprae.* The disease exists in two main forms, tuberculoid and lepromatous; the latter is associated with defective delayed hypersensitivity.

Poliomyelitis and rabies are viral diseases of the nervous system. The former is generally caused by one of three types of poliovirus that attack the motor nerve cells of the brain and spinal cord. Dramatic control of poliomyelitis has been achieved by the development of vaccines. Rabies is not a common disease in humans in the United States, but a huge reservoir of rabies virus exists in wild animals. The main source of human infection is the domestic dog. Human rabies control rests largely on vaccination of dogs against rabies virus. The long incubation period of rabies allows time for active and passive immunization before disease can develop in an exposed individual. Human cell vaccine and human antirabies antibody are employed for this purpose and markedly decrease the chance of developing rabies. Routine vaccination of humans is avoided because of a small danger of allergic encephalitis. However, vaccination is justified for people with a high risk of exposure to the virus.

SELF-QUIZ

1. The "blood-brain barrier" is
   a. important in preventing microorganisms from entering nervous tissue from the blood
   b. a bony structure that surrounds the brain and spinal cord

   c. composed of the membranes between which the cerebrospinal fluid flows

   d. designed to keep the blood out of the brain

   e. composed of bundles of nerve fibers

2. Cerebrospinal fluid specimens from patients with meningitis should be taken to the laboratory promptly and cultured without delay. This is especially important in the case of suspected *Neisseria meningitidis* infections because

   a. the organism multiplies rapidly at room temperature, giving a distorted cultural result

   b. *N. meningitidis* is often killed by chilling

   c. the specimen can become contaminated if it sits around

   d. Gram-positive cocci autolyse easily

   e. being extracellular, the diplococci are unprotected from antibody in cerebrospinal fluid.

3. The heightened sensitivity to meningococcal endotoxin that results from previous exposure to the endotoxin is called

   a. opsonin

   b. shock

   c. Shwartzman phenomenon

   d. inflammatory response

   e. Hansen's disease

4. The key factor that determines whether a person develops tuberculoid or lepromatous leprosy is

   a. intracellular growth

   b. nerve invasion

   c. pathogenicity for armadillos

   d. ability to cultivate in vitro

   e. the immune status of the host

5. The type of specimen from which poliovirus can be recovered for many weeks after the start of poliomyelitis is

   a. cerebrospinal fluid

   b. urine

   c. blood

   d. feces

   e. sputum

6. The reason inactivated virus vaccine may be preferable to live vaccine for immunizing some people against poliomyelitis is

   a. some adults get poliomyelitis from the live vaccine

   b. inactivated vaccine produces better immunity than live vaccine

   c. attenuation enhances virulence

   d. enteroviruses are killed by stomach acid

   e. live vaccines may cause throat or intestinal infections

7. Rabies can sometimes be diagnosed during the illness by staining smears made from the

   a. sputum

b. saliva
c. nasal secretions
d. conjunctiva
e. cerebrospinal fluid
8. The finding of Negri bodies is diagnostic for
a. poliomyelitis
b. meningococcal meningitis
c. *Flavobacterium meningosepticum*
d. Hansen's disease
e. rabies

QUESTIONS FOR DISCUSSION

1. If all the polio vaccine factories were destroyed by an earthquake and only one million doses of inactivated polio vaccine were available, which population group would have the highest priority for receiving the vaccine? Why?
2. How can rabies virus continue to exist when it is generally fatal to its host?

FURTHER READING

"Human to Human Transmission of Rabies via Corneal Transplant—Thailand." *Morbidity and Mortality Weekly Report 30*:473 (1981).
Martin, B.: *Miracle at Carville.* New York: Doubleday, 1950. A famous novel concerning leprosy.
Paul, J. R.: *A History of Poliomyelitis.* New Haven, Conn.: Yale University Press, 1971. A fascinating account of the conquest of this ancient, terrible disease.
Roueche, B.: *The Incurable Wound.* In *Annals of Epidemiology* series. Boston: Little, Brown, 1967. A story about rabies.

Chapter 26

WOUND INFECTIONS

A 26-year-old man injured the knuckle of the long finger of his right hand during a tavern brawl when he punched an assailant in the mouth. Because of his inebriated condition, a night spent in jail, and the insignificant early appearance of his knuckle wound, the man did not seek medical help until more than 36 hours later. At that time his entire hand was massively swollen, red, and tender. Furthermore, the swelling was spreading to his arm. The surgeon cut open the infected tissues, allowing the discharge of pus. He removed the damaged tissue and washed the wound with sterile fluid. Smears and cultures of the infected material showed several varieties of aerobic and anaerobic mouth bacteria, including species of *Bacteroides* and *Streptococcus*. The patient was given antibiotics to combat the infection, but the wound did not heal well. Several weeks later x-rays revealed that infection had spread to the bone at the base of the finger. To cure the infection, the finger had to be amputated.

OBJECTIVES

To know
1. The factors that control the development and nature of wound infections.
2. The principal causes of surgical wound infections.
3. The role of *Pseudomonas aeruginosa* in burn infections.
4. The causes and principal characteristics of tetanus and gas gangrene.
5. The importance of wound infections resulting from animal and human bites.
6. The characteristics of sporotrichosis, its origin, and causative organism.

Whether a wound is caused accidentally, by trauma, or intentionally, such as by surgery, the exposed tissue is extremely susceptible to microbial invasion and, if infected, pus often forms in the injured area, resulting in a

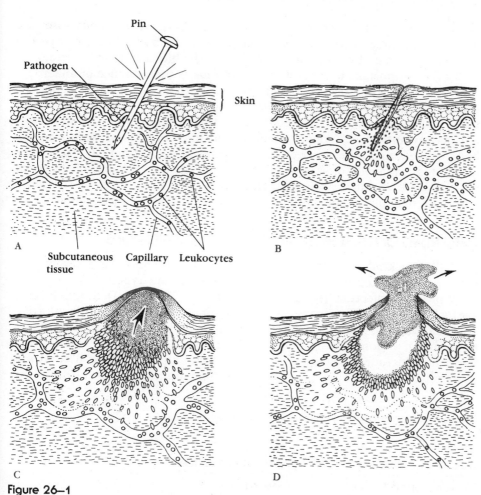

Figure 26–1

Abscess formation. (a) A pathogenic microorganism is deposited in the subcutaneous tissue. (b) In response to tissue injury, blood vessels dilate and leukocytes migrate from them to the area of developing infection. (c) Pus formation is the result of the breakdown of cells, leukocytes, and bacteria. (d) Build up of pressure results in rupturing of the abscess wall and discharge of its contents.

wound abscess. An abscess (Fig. 26–1) is a localized collection of pus (including leukocytes, components of tissue breakdown, and any infecting organisms that may be present) and is devoid of blood vessels. A surrounding area of inflammation tends to isolate the abscess from normal tissue. Consequently, abscess formation helps to localize an infection and prevent its spread. However, microorganisms in abscesses often are not affected by antimicrobial medicines. Generally, this is so because many of the microorganisms cease multiplying in the abscesses, and active multiplication is generally required for antibiotics to be effective. In addition, the chemical nature of pus interferes with the action of some antibiotics, and diffusion of antibiotics of low therapeutic ratio into abscesses may sometimes be inadequate because of the lack of blood vessels. Microorganisms in abscesses are a potential source of infection of other parts of the body if they escape the surrounding area of inflammation and enter the blood or lymph vessels.

Another important feature of many wounds is the presence of anaerobic conditions, allowing the growth of anaerobic pathogens. Such conditions are especially likely in dirty wounds, wounds with crushed tissue, and puncture wounds. Puncture wounds caused by nails, thorns, splinters, and other sharp objects can introduce foreign material and microorganisms deep into the body. Bullets and other projectiles that enter at high speed can carry contaminated fragments of skin or cloth into the tissues. Because of the force with which they enter, projectiles cause relatively small breaks in the skin, and these may close quickly, masking areas of extensive tissue damage.

ORGANISMS OFTEN RESPONSIBLE FOR WOUND INFECTIONS

Virtually any kind of pathogenic or opportunistic organism can cause wound infections under appropriate conditions, but certain types of wounds are more prone to infection by particular organisms (Table 26–1). Staphylococci are by far the most frequent cause of wound infections. Aerobic or facultative bacteria that also commonly infect wounds include streptococci, enterobacteria, and pseudomonads.* Other microorganisms that infect wounds less frequently include a variety of fungi and some animal pathogens that are transmitted by animal bites or scratches. Techniques used to culture these

---

*pseudomonads—*Pseudomonas aeruginosa* and bacteria closely related to it

TABLE 26–1 IMPORTANT CAUSES OF WOUND INFECTIONS

Type of wound	Causative agent(s)
Wound from accidental trauma	Staphylococcus aureus Streptococcus pyogenes Clostridium species Sporothrix schenckii
Surgical wound	S. aureus S. pyogenes Pseudomonas aeruginosa Enterobacteria
Burn	P. aeruginosa S. aureus S. pyogenes Enterobacteria
Animal-inflicted wound	Pasteurella multocida

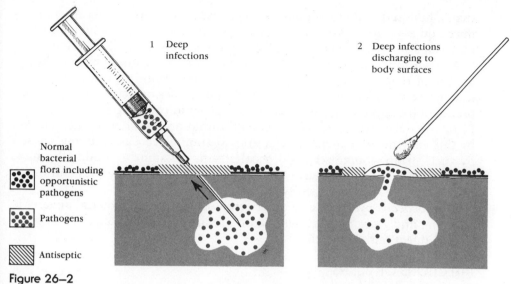

1 Deep
  infections

2 Deep infections
  discharging to
  body surfaces

Normal
bacterial
flora including
opportunistic
pathogens

Pathogens

Antiseptic

Figure 26–2
Techniques used to culture wounds. In the case of 2 one could easily mistake opportunists from the normal flora for the pathogen responsible for the infection, especially when the latter grows more slowly.

wounds are shown in Figure 26–2. Depending on the type of wound, it may be difficult to distinguish the cause of the infection from members of the normal flora that are growing in the wound.

Anaerobic microorganisms are responsible for about 10 percent of all wound infections. It is therefore necessary to use special culture media and anaerobic conditions, along with the usual aerobic techniques, when examining material from wounds. A major reason why anaerobes infect wounds is that these organisms are included among the normal human flora and may be the dominant organisms numerically, so they are readily available to invade wounded areas. Species of *Clostridium*, *Bacteroides*, *Peptostreptococcus*, and many others are anaerobic opportunistic pathogens.

INFECTIONS OF SURGICAL WOUNDS

Even when the most careful precautions are taken, a surgical wound may become infected. *Staphylococcus aureus* is the most common cause of surgical wound infections as well as infections of wounds from accidental trauma. The presence of sutures* in surgical wounds is an important predisposing factor to *S. aureus* infection. Experiments have shown that hundreds of thousands of *S. aureus* organisms can be injected beneath the skin without causing serious damage in a normal individual, but that less

S. aureus, page 295

---

*sutures—special "thread" the surgeon uses in stitching parts of the body together

than a hundred of these organisms can cause abscess formation when introduced on a suture. Tissue reactions to sutures and to other kinds of foreign material strongly favor the establishment of staphylococcal infection.

As mentioned in Chapter 11, strains of *S. aureus* commonly produce a variety of tissue-injuring extracellular products. Many strains of *S. aureus* also produce penicillinase, an enzyme that degrades penicillin; penicillinase production depends on the presence of a plasmid carried by the bacterium. Penicillin derivatives (such as methicillin) that are resistant to degradation by this enzyme are generally used to treat staphylococcal infections caused by such strains. Some staphylococcal strains resistant to methicillin have emerged, but so far they have been sensitive to other antibiotics and therefore can be treated successfully.

S. pyogenes, page 297

Streptococcus pyogenes and other streptococci, enterobacteria, anaerobic bacteria, and *Pseudomonas aeruginosa* are also common causes of surgical wound infections (Table 26–2).

INFECTIONS OF BURNS

Burned areas, with their damaged skin, offer ideal sites for infection by bacteria of the environment or of the normal flora. Almost any opportunistic pathogen can infect burns, but one of the most common and hardest to treat

TABLE 26–2   PRINCIPAL CAUSES OF SURGICAL
WOUND INFECTIONS

Cause	Percentage of infections[a]
Staphylococci	20.2
S. aureus	15.4
S. epidermidis	4.8
Streptococci	15.7
Group D	10.3
Group B	1.6
Group A	0.7
Other groups	3.1
Enterobacteria	31.9
Escherichia coli	15.2
Proteus species	7.2
Klebsiella species	5.4
Enterobacter species	4.1
Pseudomonas aeruginosa	4.6
Anaerobic bacteria	9.6
Bacteroides fragilis	2.9
Other anaerobes	6.7

[a]Most of the remaining cases either were not cultured or were cultured but no growth was obtained.

Source: *Morbidity and Mortality Weekly Report* 29:27 (January 25, 1980). Data based on approximately 41,000 surgical wound infections reported during 1975–1978.

is the Gram-negative rod *Pseudomonas aeruginosa*, which can actually color the burned tissues with its blue-green and fluorescent pigments. It is especially dreaded because of its resistance to a wide variety of antibiotics. In fact, infections with pseudomonads are a major cause of death in burn patients; however, recently introduced immunological methods appear to be effective in combating these infections. Although more experience is needed to demonstrate the usefulness of these methods in treating humans, they appear to have substantially reduced the high death rate from infected burns in some treatment centers. One method involves immunization of badly burned patients with killed polyvalent *P. aeruginosa* vaccine* within the first few days after injury and at intervals during the first two weeks thereafter. Antibodies against the bacteria are produced, usually before the wounds have time to become infected, and the antibodies give some protection against *Pseudomonas* invasion. Another approach is to administer gamma globulin from volunteers who have been immunized with the polyvalent vaccine, and such globulin can be given along with active immunization of the patient.

P. aeruginosa, page 311

Pseudomonas aeruginosa (color plate 22) is commonly present in the hospital environment, growing on plants, in water of condensation in certain medical equipment, and even in dilute disinfectant solutions such as hexachlorophene and quaternary ammonium compounds. About 10 percent of humans carry the bacterium in their intestinal tracts. Generally, *Pseudomonas aeruginosa* is readily killed by most phenolic disinfectants and by drying. However, *P. aeruginosa* in pus and burned tissue can survive both disinfectants and long periods of drying. High-concentration mafenide cream (a sulfa drug) applied to burns inhibits the growth of these bacteria, and invasion of the bloodstream or deep tissues can be treated effectively by intravenous administration of certain aminoglycosides (gentamicin, tobramycin) and penicillins (carbenicillin, ticarcillin) given concurrently in high doses.

ANAEROBIC WOUND INFECTIONS CAUSED BY CLOSTRIDIA

Tetanus (Lockjaw)

One of the most dangerous diseases resulting from anaerobic infection of wounds is tetanus, caused solely by the action of a powerful exotoxin produced by vegetative cells of *Clostridium tetani*. Even though this disease is easily prevented by immunization with tetanus toxoid, hundreds of unvaccinated people in the United States contract the disease each year, and many die from it. During the period 1970–1979, 1025 cases of tetanus were reported to the Center for Disease Control. The mortality was about 45 percent. Age is an important factor, probably because of inadequate immunity in many older people. In 1979, for example, the incidence of tetanus in the age group

C. tetani, page 303

---

*polyvalent vaccine—vaccine containing a number of different—in this case *Pseudomonas*—strains

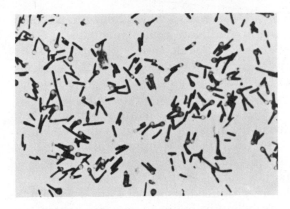

Figure 26–3
The terminal endospores of *Clostridium tetani*. (Courtesy of Turtox/Cambosco; Macmillan Science Co., Inc.)

60 years and older was three times the next highest incidence, which was in the age group from birth to four years. About 97 percent of the people who developed tetanus in recent years had never been immunized with the toxoid. *Adequate immunization would have prevented their disease.*

Tetanus occurs much more frequently on a global basis than it does in the United States. In many parts of the world, it is a common cause of death in newborn babies as a result of cutting the umbilical cord with unsterilized instruments contaminated with *C. tetani*.

Clostridium tetani organisms are widely distributed in soil. They are also commonly found in the gastrointestinal tract of humans and other animals; hence, manured soil is apt to be especially rich in the spore-forming *C. tetani* (Fig. 26–3). The organisms rarely invade past the wound area, and the widespread symptoms of tetanus result from the action of the exotoxin that diffuses from the infected wound. The isolation of *C. tetani* from wounds is not definite proof that a person has tetanus; conversely, failure to find the organism does not eliminate the possibility of tetanus. The reasons for these observations are that tetanus endospores (spores) may contaminate wounds that are not sufficiently anaerobic to allow germination and toxin production, or the person may simply be immune. Only vegetative cells, not endospores, synthesize toxin. Spores may remain dormant in the tissues for a long time and may germinate and produce toxin after a wound has healed. In a recent study in the United States, it was found that *C. tetani* could not be isolated from 7 percent of the tetanus cases studied, probably because of the small numbers of organisms present or because of technical difficulties in achieving the strict anaerobic conditions required to culture this species.

The wounds that lead to tetanus may be very small and seemingly insignificant; however, the toxin is potent in extremely small amounts. It acts on nerve cells by a mechanism that is not fully understood. The toxin is a protein that diffuses away from the infected wound and may enter the circulation. It can also travel along the nerves to the spinal cord. In the central nervous system, it interferes with control of certain reflex activities, resulting in violent spasmodic contractions of muscles that counteract each other.

These contractions can be triggered by any minor stimulus, such as sound or light. The muscles of the jaw are commonly involved, hence the origin of the common name "lockjaw."

The incubation time of tetanus may be as short as four days after infection or as long as weeks if endospores, which must first germinate, are present in the tissues. Tetanus begins with restlessness, irritability, stiffness of the neck, contraction of the muscles of the jaw, and sometimes convulsions, particularly in children. As more muscles tense, the pain grows more severe, just like that of a severe leg cramp; breathing becomes labored; and after a period of almost unbearable pain, the patient often dies of lung problems such as pneumonia or from aspiration of regurgitated stomach contents into the lung.

Tetanus is treated by the administration of specific tetanus antitoxin. Gamma globulin from humans immunized with tetanus toxoid is the treatment of choice. This antitoxin does not cause severe hypersensitivity reactions in humans and is much more effective than the horse antitoxin formerly used. The antitoxin, however, cannot neutralize the exotoxin that is already bound to nervous tissue. This may explain why tetanus antitoxin often has only limited effectiveness. In addition to antitoxin treatment, a wound must be thoroughly cleaned, and all dead tissue and foreign material that would provide anaerobic conditions must be removed. Penicillin can kill any actively multiplying clostridia it reaches and help to prevent the formation of more exotoxin. It would of course not affect endospores.

Prophylactic immunization with tetanus toxoid is by far the best weapon against tetanus. Three injections of the toxoid are given to stimulate development of antibody-producing cells. Immunization is usually begun during the first year of life. For infants and young children, the tetanus toxoid is usually given in combination with diphtheria toxoid and pertussis vaccine. The three together are commonly known as DPT. A "booster" dose is given after a year and again when the child enters school. Once immunity has been established by this regimen, "booster" doses of tetanus toxoid given at about ten-year intervals will maintain an adequate level of protection. More frequent doses of toxoid are not recommended because of the danger of an allergic reaction following intensive immunization. This sort of hypersensitivity reaction can result when antigen (the toxoid) is introduced into an individual who already has large quantities of antibodies against the antigen.

Gas Gangrene

Bacteria of the gas gangrene–producing group of clostridia can also cause serious wound infections. This group of gas-forming organisms includes *C. perfringens* (formerly known as *C. welchii*) and several other species. At present, gas gangrene is quite unusual except on the battlefield, where wounds cannot be treated promptly; however, it occurs occasionally as a result of surgery, unskilled induced abortion, and accidents. Impaired oxy-

C. perfringens, page 303

genation of tissue, for instance, when blood vessels are damaged by arteriosclerosis* or diabetes, is a predisposing factor.

The natural habitat of *C. perfringens* includes both the soil and the human intestine, and contamination of clothing, skin, and wounds is very common. Only rarely does wound contamination result in gas gangrene, however; according to one series of reports in the medical literature, only 1.76 percent of 187,936 major open wounds of violence resulted in gas gangrene. The principal factors that foster the development of gas gangrene are the presence of large amounts of dirt in the wound and long delays before careful cleaning and removal of dirt and dead tissue. *Clostridium perfringens* is unable to infect healthy tissue but grows readily in dead and poorly oxygenated tissue, releasing a powerful exotoxin called **alpha toxin.** This toxin damages adjacent normal tissue, which is then invaded. The pressure of a flammable mixture of gases, produced as a by-product of clostridial metabolism, pushes the organisms into normal tissue. This process is fostered by several tissue-attacking enzymes also produced by the pathogen. The extension of this process into muscle tissue constitutes gas gangrene and is marked by severe pain, modest amounts of gas and thin brownish fluid seeping from the wound, and blackening of the overlying skin. Shock and death commonly follow.

Diagnosis depends on the patient's symptoms, supported by a stained smear of the brownish wound fluid showing bits of partly digested muscle and plump Gram-positive rods without spores. Cultures are of little value, since gas-forming and non–gas-forming infections of wounds and subcutaneous tissues by *C. perfringens* are common but do not constitute gas gangrene. However, cultures help in distinguishing clostridial from other rarer causes of gas gangrene.

Treatment depends primarily on surgical removal of all infected tissues. In addition, **hyperbaric oxygen treatment*** inhibits growth of the clostridia and release of alpha toxin and also improves oxygenation of injured tissues. Antibiotics such as penicillin also stop bacterial growth and toxin production but do not dependably diffuse into large areas of dead tissue (color plate 58). Antiserum against the alpha toxin has also generally been administered to patients and may be of some value in protecting erythrocytes from any alpha toxin that diffuses into the bloodstream. However, its use has not convincingly decreased mortality from gas gangrene, and the antitoxin is no longer produced commercially in the United States.

NONCLOSTRIDIAL ANAEROBIC INFECTIONS OF WOUNDS

As a result of using better methods for isolating and identifying anaerobic bacteria, it has become apparent that many wound infections are caused by

---

*arteriosclerosis—hardening of the arteries

*hyperbaric oxygen treatment—regimen in which the patient is placed in a special chamber and breathes pure oxygen under three times normal air pressure

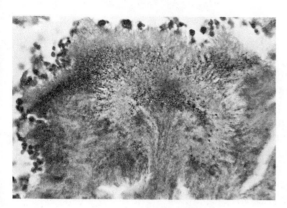

Figure 26–4
Actinomyces israelii. These fungus-like bacteria are part of the normal flora of the oral cavity of humans, but under unusual circumstances they can invade and cause an infection. This specimen is part of a colony of *A. israelii* in pus from an infected patient. The colony appears as a yellow mass; therefore it is called a "sulfur granule." Filaments of bacteria can be seen radiating from the edge of the colony. (Courtesy of Turtox/Cambosco; Macmillan Science Co., Inc.)

anaerobes other than clostridia. Although much remains to be learned about these anaerobic infections, they are caused principally by certain organisms of the normal flora, namely, Gram-negative rods of the genera *Bacteroides* and *Fusobacterium,* and by Gram-positive cocci, principally streptococci of the genus *Peptostreptococcus.*

In addition to these organisms, certain filamentous bacteria can cause wound infections. Prominent among these is *Actinomyces israelii,* a Gram-positive anaerobic branching bacterium that is an occasional member of the normal flora of the alimentary and upper respiratory tracts. Within pus from infected areas, the actinomycete forms microcolonies known as "sulfur granules," so named because of their yellow color. When examined microscopically, a "sulfur granule" is seen to consist of filaments of the bacteria arranged in radial fashion around the edge of each granule (Fig. 26–4). As might be expected, wounds caused by dental procedures sometimes become infected with *Actinomyces israelii* or other anaerobic bacteria that are part of the normal flora of the mouth. Actinomycetes can also occasionally invade the lung or gastrointestinal tract in the absence of any obvious injury.

A. israelii, page 327

BITE WOUNDS

Both normal flora and pathogens of animals can be transmitted to humans by animal bites. The viral disease rabies is one of the most feared infections transmitted in this manner, but less serious bacterial infections also occur. For example, bite wounds from cats, dogs, and some other mammals are frequently contaminated with *Pasteurella multocida.* This small Gram-negative rod is similar in appearance to the causative agent of the plague, *Yersinia pestis,* and also resembles the bacteria that cause tularemia, *Francisella tularensis. Pasteurella multocida* is facultatively anaerobic and fastidious but grows readily on enriched media. An inhabitant of the normal flora of the mouth and upper respiratory tract of its natural animal hosts, it can cause severe infection when host defenses are suppressed or stressed (as by a viral infection). Different strains of *P. multocida* show differences in virulence and host preference. Human infections with *P. multocida* following animal bites can be severe, with marked redness, swelling, and pus formation; however, they usually respond to penicillin treatment and soaking of the wound, which promotes drainage of the infected site.

Wounds from cat scratches and bites occasionally lead to a disease known as **"cat-scratch fever."** The disease affects the lymph nodes near the area of injury. Often the nodes become swollen and pus-filled and must be lanced.* Typically, the infection is mild and localized, lasting from a few weeks to a few months. The causative agents have not been definitely identified, but chlamydia are probably responsible for at least some cases. The disease affects humans only, and cats that transmit it give no evidence of illness.

Chlamydia, page 329

Rat bites can also lead to infections with bacteria carried in the oral cavity of rats, notably *Spirillum minus* and *Streptobacillus moniliformis*.

Wounds resulting from one person biting another are not uncommon (Table 26–3), and they have the potential both for transmitting certain infectious diseases and for causing severe infections in the area of the wound. Among the diseases reported to be transmitted by human bites are syphilis, tuberculosis, and hepatitis B. However, much more frequent are the severe wound infections resulting from opportunistic bacteria that normally live in the mouth and nose. Similar infections of wounds can occur from objects such as forks or toothpicks that have been in someone's mouth.

The most important factor in the development of severe infections of human bite wounds is delay in starting treatment. The crushing nature of the tissue injury produces anaerobic conditions in the wound and allows growth of anaerobic streptococci, fusiforms, spirochetes, and *Bacteroides* species. These anaerobic species are particularly prevalent in the mouths of individuals with poor dental hygiene, and if such individuals must be punched, it should be elsewhere than in the mouth. As one would expect because of the proximity of the nose and mouth, *Staphylococcus aureus* is also a common cause of these infections (color plate 59).

---

*lanced—cut open with a surgical knife

TABLE 26–3 BITE WOUNDS REPORTED TO THE NEW YORK CITY HEALTH DEPARTMENT, 1977

Animal	Number
Dog	22,076
Cat	1,152
Human	892
Rodent	548
Pet rabbit	40
Lion	3
Anteater	1

Source: Marr, J. S., et al.: *Public Health Reports* 94:514 (1979).

Treatment consists of opening the wound widely, with a scalpel if necessary, washing it thoroughly, and removing dirt and dead tissue. An antibiotic active against the infecting organisms is also an important part of therapy.

FUNGAL INFECTIONS OF WOUNDS

For two months a necrotic, ulcerated area on the chin of a two-year-old boy puzzled his doctors, until finally a culture was taken of material from deep inside the wound. This culture showed the cause of the boy's chin ulcer to be the fungus *Sporothrix schenckii*.[1] Frequently his great-grandmother, an enthusiastic gardener, took care of the child, and she reported that similar persistent skin ulcers would occur on her arms after working on her rosebushes. Since this fungus is common on vegetation, it was likely that her great-grandson had been scratched by a rose thorn and thus had contracted the fungus. The disease is seen mainly in the Mississippi and Missouri river valleys of the United States, but it has also been reported from many other areas of the world.

Sporothrix schenckii also grows well on timbers in mines where both the temperature and humidity are consistently high. Over the 20-year period from about 1925 to 1945, more than 3000 cases of *S. schenckii* wound infections were reported in South African mine workers. Typically, wounds caused by splinters from mine timbers became ulcerated, often without causing pain. After about a week, nodules developed in a chain along the lymphatic vessel draining the wound, and these nodules then developed into ulcers. The lymph node draining the area usually became enlarged, and the lesions persisted for long periods of time without making the patients very ill.

This disease rarely becomes generalized but usually follows the chronic, persistent course previously described. The fungus may also spread in the skin, causing scaly, flat plaques or wartlike lesions, but inoculation by a puncture wound more often produces lymphatic involvement.

Sporothrix schenckii is a dimorphic fungus with a worldwide distribution. In nature or in cultures incubated at room temperature, it forms a fluffy mold mycelium, tan or brown in color. In vivo and in vitro at 37°C, it grows as a budding yeast. In the mold form, the hyphae of *S. schenckii* are much thinner than those of most fungi, and spores are formed along the hairlike hyphae. The spores are oval or rounded and occur in clusters on branches at right angles to the hyphae, producing a flowerlike appearance (Fig. 26–5). The yeast phase in cultures grown at 37°C and in tissues is characterized by elongated cigar-shaped cells with one to three buds at both ends of the cell.

Other fungi can invade and infect wounds, but none produces the characteristic, chronic disease that results from *S. schenckii*. Treatment with potassium iodide is usually successful.

---

[1] Formerly known as *Sporotrichum schenckii*.

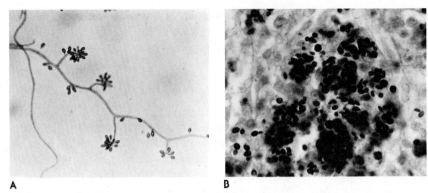

A   B

Figure 26–5
Sporothrix schenckii, a dimorphic fungus. (a) The mold phase exhibits slender hyphae with elongated ovoid conidia in a flowerlike arrangement. (b) The yeast phase from infected tissue. Some budding cells can be seen.

SUMMARY

Wounds damage normal host defenses and therefore predispose the host to infection. *Staphylococcus aureus* is the most frequent cause of wound infections. Other aerobic and facultative microorganisms often responsible for wound infection include streptococci, enterobacteria, and pseudomonads. Wounding often leads to the development of anaerobic conditions in the tissues and subsequent invasion by anaerobic bacteria such as clostridia, anaerobic streptococci, bacteroides, fusobacteria, and actinomycetes.

Surgical wounds commonly become infected with penicillinase-producing *S. aureus* strains that are therefore resistant to penicillin. The presence of sutures in a wound predisposes a patient to staphylococcal infection. Streptococci, including *Streptococcus pyogenes,* and various enteric bacterial species are less common causes of surgical wound infections.

Burns are particularly susceptible to infections. *Pseudomonas aeruginosa* is one of the most prevalent agents in burn infections and a frequent cause of death in burn patients because of its resistance to antibiotics.

Tetanus is an important disease resulting from anaerobic infections of wounds with *Clostridium tetani.* The disease is caused by the potent tetanus nerve toxin; it is therefore treated with tetanus antitoxin and prevented by immunization with tetanus toxoid. Gas gangrene is also produced by members of the genus *Clostridium* and is marked by progressive destruction of muscle tissue. *Clostridium perfringens* is the chief offender.

Animal and human bites and scratches may cause serious disease by transmitting microorganisms from animals to humans and from humans to humans. Among the infectious agents transmitted in this manner are rabies virus, *Pasteurella multocida, Staphylococcus aureus,* anaerobic streptococci, fusiforms, spirochetes, and *Bacteroides* species.

Various fungi can infect wounds. One of the most readily recognized fungus infections is caused by *Sporothrix schenckii*, which invades small wounds and causes a characteristic lymph node involvement and skin ulceration.

SELF-QUIZ

1. The bacterial species that is the most frequent cause of wound infections is
   a. *Streptococcus pyogenes*
   b. *Escherichia coli*
   c. *Sporothrix schenckii*
   d. *Staphylococcus aureus*
   e. *Bacteroides melaninogenicus*
2. The Gram-negative organism that produces blue-green pigment and is prevalent in burn infections is
   a. *Pseudomonas aeruginosa*
   b. *Staphylococcus aureus*
   c. *Streptococcus pyogenes*
   d. *Clostridium perfringens*
   e. *Pasteurella multocida*
3. A helpful means of preventing tetanus is
   a. administration of penicillin
   b. administration of antitoxin
   c. booster injections of toxoid
   d. removal of foreign material and dead tissue from wounds
   e. all of the above
4. A fungus that grows on vegetation and causes ulcerated areas when introduced into a wound is
   a. *Streptococcus pyogenes*
   b. *Sporothrix schenckii*
   c. *Pseudomonas aeruginosa*
   d. *Staphylococcus aureus*
   e. *Proteus* species
5. Diagnostic of gas gangrene is
   a. gas in the tissue
   b. wound culture positive for *Clostridium perfringens*
   c. stained smear of wound drainage showing plump, non–spore-forming Gram-positive rods
   d. all of the above
   e. none of the above

QUESTIONS FOR DISCUSSION

1. In the event of a national disaster with many wounded people and no doctors or medicines, what could you do to minimize the prevalence of tetanus and gas gangrene?

2. In treating burns involving large areas of denuded skin, some authorities advocate leaving the burns exposed to air if at all possible. Does this make sense?

FURTHER READING

Altemeier, W. A., and Fullen, W. D.: "Prevention and Treatment of Gas Gangrene." *Journal of the American Medical Association 217*:806 (1971).
Duncan, C. L.: "Role of Clostridial Toxins in Pathogenesis." *In* Schlessinger, D. (ed.): *Microbiology—1975.* Washington, D.C., American Society for Microbiology, 1975, pp. 283–291. A brief summary of information concerning the nature and mode of action of the clostridial toxins, including (among others) the toxins that cause tetanus and gas gangrene.
Fisher, M. W.: "Polyvalent Vaccine and Human Globulin for Controlling *Pseudomonas aeruginosa* Infections." *In* Schlessinger, D. (ed.): *Microbiology—1975.* Washington, D.C., American Society for Microbiology, 1975, p. 416.
Mann, R. J., Hoffeld, T. A., and Farmer, C. B.: "Human Bites of the Hand: Twenty Years of Experience." *The Journal of Hand Surgery,* 2:97–104 (1977).
Marr, J. S., Beck, M. A., and Lugo, J. A.: "An Epidemiologic Study of the Human Bite." *Public Health Reports,* 94:514–521 (1979).

Chapter 27

BLOOD AND LYMPHATIC INFECTIONS

Thirty-two patients undergoing cardiac catheterization* experienced fever, shaking chills, and a drop in blood pressure. These symptoms strongly suggested the presence of endotoxin in their blood, but blood cultures showed no bacteria. Further investigation revealed that the catheters had been thoroughly washed, rinsed in distilled water, and sterilized after each use. A sample of the distilled water used to rinse the catheters was found to contain large numbers of the Gram-negative bacterium *Acinetobacter calcoaceticus.* This organism was the source of the endotoxin, contaminating the catheters when they were rinsed before sterilization. Since even distilled water becomes contaminated with Gram-negative organisms when it is exposed to the air, it is difficult to obtain water free of endotoxin. Species of *Pseudomonas* and *Acinetobacter* as well as a variety of other Gram-negative bacteria are widely found in samples of water.

OBJECTIVES

To know
1. The basic features of the blood and lymphatic circulations and their relationship to the development of infectious diseases.
2. The mechanisms by which bacteremia can produce endocarditis and septicemia (blood poisoning).
3. The cause and pathogenesis of a viral infection of the heart muscle.
4. The important characteristics of tularemia and brucellosis.
5. The cause and significance of toxoplasmosis.
6. The characteristics of infectious mononucleosis.

---

*cardiac catheterization—the process of inserting a long plastic tube (catheter) into the heart via one of the blood vessels of the leg; the purpose is usually to measure pressures in the heart

ANATOMY AND PHYSIOLOGY

Most people today are familiar with the general structure and function of the body's blood vascular* system. The heart, a muscular pump enclosed in a sac called the **pericardium,** supplies the force that moves the blood. As shown schematically in Figure 27–1, the heart is divided into a right and a left side, separated by a wall of tissue through which blood normally does not pass after birth. The right and left sides of the heart are each divided into two chambers, one that receives blood (atrium) and another that discharges it (ventricle). Although not common, infections of the heart valves, muscle, and pericardium may be disastrous because of their effect on the vital functions of the total circulatory system. Blood from the right ventricle flows through the lungs and into the atrium on the left side of the heart. This blood then passes into the left ventricle and is pumped through the aorta to the arteries and capillaries that supply almost all the tissues of the body.

A system of veins collects the blood from the capillaries in the tissues and carries it back to the right atrium of the heart. Because pressure in the veins is low, one-way valves help keep the blood flowing in the right direction. Thus, as the blood flows around the circuit, it alternately passes through the lungs and through the tissue capillaries. Also during each circuit, a portion of the blood passes through other organs, such as the spleen, liver, and lymph nodes, all of which, like the lung, contain fixed phagocytic cells of the mononuclear phagocyte system. Foreign material, such as microorganisms, may be removed by these phagocytes as the blood passes through such tissues.

Mononuclear phagocyte system, page 213

The system of lymphatic vessels begins in tissues as tiny tubes that differ from blood capillaries in having closed distal ends and in being somewhat larger. Inside these lymphatic vessels is **lymph,** an almost colorless fluid. Lymph originates from plasma that has made its way through the blood capillaries to become the **interstitial fluid.** It bathes the tissue cells and then enters the lymphatics. Unlike the blood capillaries, the lymphatics readily take up foreign material such as invading microbes or their products. The tiny lymphatic capillaries join progressively larger lymphatic vessels. Many one-way valves in the lymphatic vessels keep the flow of lymph moving away from the lymphatic capillaries. Both contraction of the vessel walls and compression by the movements of muscles force the lymph fluid along.

At many points in the system, lymphatic vessels drain into small, roughly bean-shaped bodies called **lymph nodes.** These nodes are so constructed that foreign materials such as bacteria are trapped in them; the nodes also contain cells that are involved in phagocytosis and antibody production. Lymph flows out of the nodes through vessels that eventually unite into one or more large tubes that discharge the lymph into a large vein such as the vena cava* and thus back into the main blood circulation.

An infection of a hand or a foot is sometimes made apparent by the spread

---

*vascular—referring to vessels

*vena cava—largest vein in the body; it conducts blood into the right side of the heart

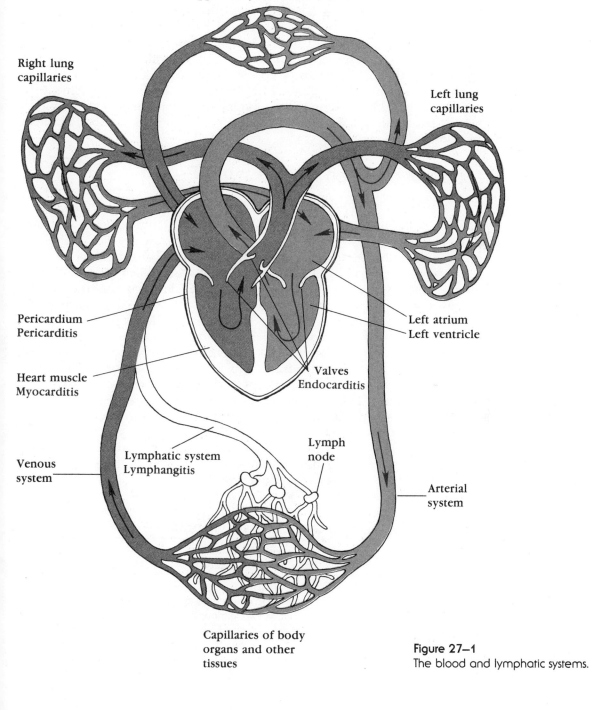

Capillaries of head,
upper body and other tissues

Right lung
capillaries

Left lung
capillaries

Pericardium
Pericarditis

Left atrium
Left ventricle

Heart muscle
Myocarditis

Valves
Endocarditis

Lymphatic system
Lymphangitis

Lymph
node

Venous
system

Arterial
system

Capillaries of body
organs and other
tissues

Figure 27–1
The blood and lymphatic systems.

of a red streak up the limb from the infection site (color plate 60). This streak represents the course of lymphatic vessels that have become inflamed in response to the infectious agent. This condition is called **lymphangitis.** It may stop abruptly at a swollen and tender lymph node and later continue to yet another lymph node. This arrest of an inflammation demonstrates the ability (even though sometimes temporary) of the lymph nodes to clear the lymph of an inflammatory agent. As indicated earlier, the blood and lymph carry such antimicrobial agents as leukocytes, antibodies, complement, lysozyme, beta lysin, and interferon. Lymph and blood may clot in regional vessels as a result of inflammation arising from infection or antibody-antigen reactions.

Host defenses, page 211

DISEASES OF THE BLOOD VASCULAR SYSTEM

Subacute Bacterial Endocarditis (SBE)

Table 27–1 outlines the important features of subacute bacterial endocarditis. This infection of the inner lining of the heart is usually localized to one of the heart valves. It commonly occurs in hearts that are abnormal as a result of rheumatic fever or some other disease or a birth defect. Afflicted people typically become ill very gradually; fever develops, and they slowly lose energy and vigor over a period of weeks or months. The causative organisms of endocarditis are usually shed from the infected valve into the circulation and can often be identified by culturing samples of blood drawn from an arm vein. They are almost always organisms frequently present among the normal flora of the infected person, most commonly oral alpha-hemolytic streptococci or other streptococci of low virulence or *Staphylococcus epidermidis*. This is a good example of how pathogenicity depends on both host factors and the virulence of the microorganism. Even bacteria of low virulence can cause serious, even fatal, infections.

Rheumatic fever, page 431

As discussed in Chapter 23, a few organisms of the normal flora frequently gain entrance to the bloodstream during dental procedures, brushing of teeth, or trauma. These organisms normally are eliminated quickly by body

TABLE 27–1   FEATURES OF SUBACUTE BACTERIAL ENDOCARDITIS

Predisposing host factors	Usual microbial causes	Pathogenesis
Structural abnormality of heart, such as birth defect or deformed valve from rheumatic fever or other diseases	Normal flora: *Staphylococcus epidermidis*, viridans streptococci, and other bacteria of low virulence	Turbulent blood flow; clot formation; colonization of clot; enlargement of infected clot; release of clot fragments; plugging of vessels; damage due to immune complexes

defense mechanisms. In an abnormal heart, however, turbulent blood flow fosters the formation of a thin blood clot that in turn traps such organisms in areas where phagocytes have difficulty functioning. The high levels of antibodies often present in bacterial endocarditis are of little value and may even aid the progress of the infection by clumping the causative bacteria and depositing them on the clot. The organisms multiply extensively, and more clot may be progressively deposited around them, gradually building up a fragile mass. Bacteria continually wash into the circulation, and pieces of infected clot may break off and, if large enough, may block important blood vessels and lead to tissue death or to the weakening and ballooning out of larger vessels.[1] Circulating immune complexes may lodge in the kidney and produce a form of glomerulonephritis. Even though the organisms normally have little invasive ability, great masses of them growing in the heart are sometimes able to burrow into heart tissue to produce abscesses or damage valve tissue, resulting in a leaky valve.

In 5 to 15 percent of cases of infective endocarditis, culturing the blood from an arm vein fails to yield the causative agent. This occurrence is especially likely when the infection is on the right side of the heart, and blood from the infected site must pass through both the lung and the tissue capillaries (with their phagocytic cells) before reaching the arm vein. In other instances, an organism (e.g., *Veillonella*) may be too fastidious to grow in the usual bacteriological media. In the case of subacute bacterial endocarditis caused by the Q fever rickettsiae, a medium containing living cells is required.

Infections of the heart can also be caused by more virulent bacteria such as *Staphylococcus aureus* or *Streptococcus pneumoniae*. These organisms can produce a rapidly progressing illness called **acute bacterial endocarditis.** Such organisms are much more likely to invade the heart and its valves and permanently destroy tissue than those organisms causing the more common subacute bacterial endocarditis. Acute bacterial endocarditis is often seen as a complication of drug addiction, where unsterile needles are used to inject substances intravenously. It is also an occasional complication of pneumonia and gonorrhea.

Septicemia

Septicemia (blood poisoning) is an illness caused by microorganisms or by their products circulating in the bloodstream. (Bacteria sometimes enter the bloodstream without causing illness; this is referred to as *bacteremia*.) The symptoms of septicemia are fever and a drop in blood pressure, probably due at first to impaired strength of the heart and the walls of blood vessels. In severe septicemia, the blood vessel damage may be irreversible. If the drop in pressure is marked and prolonged, there will be insufficient flow of

---

[1]This change in a blood vessel is called an aneurysm.

blood to supply adequate oxygen to the organs and other tissues, and shock*
results. Septicemia almost always originates from an infection somewhere
else in the body (e.g., boil, kidney infection, abscess, pneumonia) that has
become uncontrolled. Thus, septicemia is generally caused by one of the
usual bacterial pathogens. In addition, alterations in normal body defenses
as the result of medical treatment (e.g., surgery, catheters, and drugs that
interfere with the immune response) have in recent years resulted in septicemia from microorganisms that normally have little invasive ability. These
infections commonly originate from environmental sources as well as the
normal flora of the body.

Gram-negative rod-shaped bacteria are most often responsible for septicemia, although Gram-positive bacteria and fungi can also be causative
agents. Septicemia caused by Gram-negative bacteria tends to be more serious than that caused by Gram-positive bacteria. Shock is common, and
only about half of all people afflicted with this kind of infection survive.
Cultures of the blood in these cases usually reveal facultatively anaerobic
Gram-negative rods such as *Escherichia coli*, *Enterobacter aerogenes*, *Serratia
marcescens*, and *Proteus mirabilis*. Among the aerobic Gram-negative rod-
shaped bacteria encountered in septicemia are organisms commonly found
in nature, such as *Pseudomonas aeruginosa*. This organism has extraordinary
biochemical capabilities, is very resistant to antibiotics, and can grow under
a variety of conditions unfavorable to the usual bacterial pathogens.

Not all the Gram-negative organisms causing septicemia will grow in the
presence of air. As mentioned in Chapter 23, anaerobic Gram-negative rods
of the genus *Bacteroides* make up a major percentage of the normal flora of
the large intestine (as well as the upper respiratory tract), and these frequently cause septicemia. This group of organisms was poorly classified, and
many strains that were isolated from cultures of abscesses and blood failed
to fit into existing species. Today, better procedures and equipment have
evolved for studying the obligate anaerobes. **Gas chromatography**[2] is an
especially useful tool for identifying this group of bacteria. This technique
readily determines the characteristic metabolic products of the bacteria.

The manner in which Gram-negative bacteria produce septicemia and
shock is still not completely understood, but the probable sequence of events
is shown in Table 27–2. Gram-negative bacteria contain endotoxin as a part
of their outer cell walls. When injected into the bloodstream of laboratory
animals in milligram doses, the endotoxin produces fever and shock, often
leading to death. Despite these effects of endotoxins defined under laboratory
conditions, there is as yet no agreement about the extent to which endotoxins

---

[2]In this technique, volatile (gaseous) microbial products from a culture are passed through a device that
separates them from each other in a sequence determined by their chemical properties. The time of appearance of each substance is automatically recorded and is used to identify the substance by comparison with
the times of appearance of known substances.

*shock—a condition having numerous causes, characterized by inadequate blood flow to body tissues re-
sulting in an insufficient supply of oxygen for normal function

TABLE 27–2   PROBABLE SEQUENCE OF EVENTS IN SEPTICEMIA CAUSED BY GRAM-NEGATIVE ROD-SHAPED BACTERIA

1. Bacteria enter the bloodstream from an infected area such as the urinary tract.

2. Endotoxin is released from the bacteria.

3. Endotoxin damage to blood vessels causes impaired ability of the blood vessel walls to contract. This impairment becomes irreversible if prolonged.

4. Dilation of blood vessels results in a drop in blood pressure.

5. Insufficient pressure and maldistribution of blood result in an inadequate supply of oxygen to vital organs.

6. Death results in many cases, even though the infection is controlled, because of damage to the circulatory system from endotoxin.

are released in infections nor about the exact role they play in specific diseases caused by Gram-negative bacteria. Nevertheless, it is very likely that endotoxins cause the principal damage to the host in many septicemias, especially those resulting from relatively avirulent bacteria of the normal flora and the environment. Other bacterial products, including protein exotoxins, probably also play a role in some cases of septicemia, such as those due to *Pseudomonas* species.

Myocarditis

Myocarditis, or inflammatory disease of the heart muscle, may be caused by bacteria, fungi, or protozoa but is most commonly the result of a viral infection. When the inflammation primarily involves the sac around the heart (pericardium), it is called **pericarditis.** Involvement of the heart muscle may produce only pain in the chest and abnormalities of the electrical impulses of the heart, but if the infection is extensive, the heart is unable to contract effectively, and heart failure may result. This problem is most likely to occur in infants but can arise at any age.

The virus or microorganism causing myocarditis can sometimes be identified from cultures of the fluid accumulating in the pericardial sac or from cultures of feces and throat secretions, plus demonstration of a rise in neutralizing antibody* during the course of the illness. In most of the instances in which an agent has been recovered in myocarditis, it has been a Coxsackie virus, a small RNA-containing virus of the enterovirus group of picornaviruses.

Picornaviruses, Appendix VIII

The virus responsible for myocarditis presumably reaches the heart by way of the bloodstream or lymphatic system after infecting the respiratory

---

*neutralizing antibody—immunoglobulin produced in response to infection which reacts specifically with the virus and makes it noninfectious

or gastrointestinal tract. However, it is not known why only a few infected people manifest myocarditis during any given Coxsackie virus epidemic. Once in the heart, the virus replicates in the muscle cells, causing extensive damage, with some cells dying and being replaced by scar tissue. In some patients dying of myocarditis, viral antigens can still be detected in the heart muscle many months after the illness starts. These antigens can be demonstrated by using fluorescent antibody against Coxsackie virus. Mysteriously, the virus usually cannot be recovered in cultures even though large amounts of viral antigen are demonstrable. This suggests the possibility of a persistent infection with a defective form of the virus and progressive heart damage from an immune response to the infected muscle cells.

Coxsackie viruses are able to infect and produce disease in laboratory mice, but susceptibility of these mice to the virus is markedly age-dependent. In fact, only newborn mice are affected consistently. Moreover, Coxsackie viruses differ from each other in the types of mouse tissue cells they will attack. Of the two different serological groups A and B, group A viruses characteristically damage the muscles, whereas group B viruses attack the central nervous system, liver, pancreas, and fatty tissues. Different Coxsackie viruses also show differing tissue affinities in humans. Most people with viral myocarditis have group B Coxsackie virus infections.

Because only a small fraction of people infected with Coxsackie viruses experience serious illness and very few of these die, there has not been much effort to establish control measures against these organisms. No vaccines are available, and as with other enteroviruses, sanitary practices are unlikely to limit spread of these agents, although they should help decrease the infecting dose of virus received by its victims and thereby decrease the likelihood of severe disease. Excretion of the virus in the feces of an infected person may continue for weeks, even after mild, unnoticed infections. The presence of healthy disseminators in a community undoubtedly aids in the spread of Coxsackie viruses.

DISEASES INVOLVING THE LYMPH NODES AND SPLEEN

Enlargement of the lymph nodes and spleen is a prominent feature of certain infectious diseases in which the mononuclear phagocytes are involved. This enlargement may result from infection of the mononuclear phagocyte cells and other cells involved in body defense and immunity. Both inflammation and multiplication of cellular elements play a role.

Tularemia

Not long ago in Vermont an outbreak of disease characterized by fever and enlargement of lymph nodes occurred. Altogether 72 cases were identified, and all occurred among people who had been in contact with fresh muskrat pelts or skinned muskrats. These people were diagnosed as having tularemia, a disease caused by a bacterium commonly found in wild animals and arthropods. Tularemia was often fatal in the days before antibiotics were available for treatment.

The causative microorganism of tularemia, *Francisella tularensis*,[3] is a pleomorphic, nonmotile, aerobic Gram-negative rod that derives its name from Tulare County, California, where it was first studied. The organism can be cultured from blood and other materials taken from infected humans and animals. Unfortunately, other bacteria from the environment or the normal flora, if present in the inoculum, may sometimes obscure the growth of *F. tularensis*. If this occurs, a portion of the material can be injected into a laboratory animal such as a guinea pig or white mouse. The defense mechanisms of the animal quickly destroy the contaminating flora, while the pathogen invades the animal's tissues. Fluorescent antibody or other serological techniques can then be used to identify *F. tularensis* from the infected tissues.

F. tularensis, page 312

Tularemia occurs in many areas of the Northern Hemisphere, including all the states of the United States except Hawaii. In the eastern United States, human infections usually occur in the winter months, as a result of skinning rabbits (thus the common name **rabbit fever**). In the West, infections result primarily from the bites of ticks and deer flies and thus usually occur during the summer. *Francisella tularensis* may seem to infect through unbroken skin, but it probably enters the body through small cuts or scratches that may not be visible or through the mucous membranes of the eye or mouth. It can also occasionally be acquired by inhalation, producing pneumonia. Typically, it causes an ulcer at the site of entry and enlargement of the lymph nodes in the surrounding area of the body; these nodes may become filled with pus. The organisms spread to other parts of the body via the lymphatic and blood vessels. *F. tularensis* is of interest because (like *M. tuberculosis*) it is ingested by but readily grows within phagocytic cells. This may explain why tularemia persists in some people despite the high titers of antibody in their blood. The mechanisms of cell-mediated immunity are probably responsible for ridding the host of this and other organisms that tend to persist intracellularly. Both delayed-type hypersensitivity and serum antibodies quickly arise during infection, and demonstration of their presence can be used to help identify tularemia when positive cultures of *F. tularensis* cannot be obtained.

Cell-mediated immunity, page 246

About 150 to 200 cases of tularemia are now reported each year in the United States. This is one sixth the number seen 30 years ago. Most cases respond well to treatment with streptomycin or tetracycline.

Brucellosis

Brucellosis (undulant fever; Bang's disease) is contracted primarily from diseased cattle, pigs, and goats. In animals, brucellosis is typically a chronic

---

[3]Formerly known as *Pasteurella tularensis*.

B. abortus, page 312

infection involving the mammary glands and the uterus, thereby contaminating milk and causing abortions. A disease characterized by fever, body aches, weight loss, and enlargement of lymph nodes and spleen develops in humans and characteristically subsides without treatment but may then recur one or more times during a period of weeks or months.

The Gram-negative rod-shaped bacterium *Brucella abortus* is one of the causative species of this disease and is most commonly transmitted to humans by cattle. However, five additional *Brucella* species can infect humans. These organisms usually originate from animals other than cattle. In the United States, 200 to 250 cases are reported each year, and there are 10 to 20 unreported cases for each one that is reported. Sixty percent of the cases of brucellosis occur in workers in the meat-packing industry; less than 10 percent arise from ingestion of raw milk or other unpasteurized dairy products. Occasional cases are acquired from household dogs or from eating raw reindeer bone marrow. Worldwide, brucellosis is a major problem in animals used for food, causing yearly losses of many millions of dollars.

As with tularemia, the organisms responsible for brucellosis penetrate mucous membranes or breaks in the skin and are disseminated via the lymphatic and blood vessels to the heart, kidneys, and other parts of the body. Like *F. tularensis*, *Brucella* species are resistant to phagocytic killing and can grow intracellularly.

Antibiotic treatment with tetracycline, ampicillin, or certain other antibiotics is usually effective. Without treatment, the disease is debilitating but rarely fatal; 80 percent of the patients recover within a year. The most important control measures against brucellosis are pasteurization of dairy products and inspection of domestic animals for evidence of the disease.

Infectious Mononucleosis: A Viral Disease

Infectious mononucleosis ("mono") is a disease familiar to many students because of its high incidence among people between the ages of 15 and 24 years. The term **mononucleosis** refers to the fact that people infected with this condition have an increased number of mononuclear leukocytes in their bloodstreams.[4] The most dramatic symptoms of the disease are fever, sore throat, and enlargement of the lymph nodes and spleen. Occasionally, complications such as myocarditis, meningitis, hepatitis, or paralysis develop. One very interesting aspect of infectious mononucleosis is the development of an antibody that reacts in the laboratory with an antigen present on the surface of erythrocytes taken from sheep or horses. Such antibodies, which react with the cells of other species, are called **heterophile antibodies.** In most laboratories, testing for the heterophile antibody is the only practical way to separate patients with infectious mononucleosis from those with other conditions closely resembling it.

---

[4]In this instance, "mononuclear" refers only to the lack of a segmented nucleus; the cells are lymphocytes.

In the late 1960's, scientists reported some interesting findings relating to infectious mononucleosis. These were reports of studies of a peculiar virus seen in cell cultures prepared from patients with Burkitt's lymphoma, a malignant tumor of the lymphatic tissues commonly found in children in parts of Africa. The virus was shown to be a member of the herpes group of viruses but not closely related to any known virus. The virus is now commonly known as the Epstein-Barr (or EB) virus after its discoverers. With EB virus grown in tissue cell cultures as an antigen, it was possible to detect and measure anti-EB virus antibody in the bloodstreams of various people. The highest titers of this antibody were found in people with the type of malignant lymphatic tumor that had originally yielded the virus, but many others were also found to have the antibody. Of special interest was the finding that people who contracted infectious mononucleosis consistently lacked the antibody to EB virus before their disease started but developed high titers of this antibody during their illness. The antibody has been found to persist for years after infection. Moreover, a virus apparently identical or closely related to the EB virus could be cultured from the lymphocytes of people infected with mononucleosis in a manner similar to its culture from the malignant tumor. People with infectious mononucleosis do not develop the lymphoma. Although EB virus probably plays an essential role in causing the African tumor, other unknown factors are of crucial importance in causing the malignancy.

Herpes viruses, Appendix VIII

Present evidence indicates that the EB-like virus of infectious mononucleosis is widespread and infects most individuals in economically disadvantaged groups at an early age without producing significant illness. In such populations, infectious mononucleosis is quite rare. On the other hand, more than 50 percent of the students entering some United States colleges have shown no antibody to EB virus and presumably have escaped past infection with the agent. Infectious mononucleosis occurs almost exclusively in such antibody-negative subjects. Epidemiological evidence indicates that kissing may be an important mode of transmission of infectious mononucleosis in young adults. The virus is not highly contagious, and it rarely spreads even within households.

Toxoplasmosis: A Protozoan Disease

Toxoplasmosis is a disease that may closely resemble infectious mononucleosis, with fever and enlarged lymph nodes and spleen. However, it gives a negative test for heterophile antibodies. The causative agent is a small, crescent-shaped protozoan, *Toxoplasma gondii* (Fig. 27–2). This microorganism can infect a broad range of hosts, including both birds and mammals, and disastrous epidemics among commercially valuable animals occasionally occur.

T. gondii, page 399

Actual illness in humans as a result of this organism is unusual, although infection is common. Three main disease patterns are seen: (1) Infection of

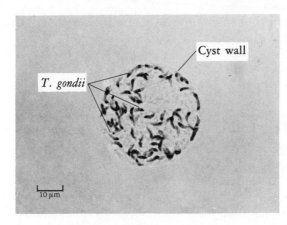

T. gondii

Cyst wall

10 μm

Figure 27–2
Toxoplasma gondii cyst. (Courtesy of J. Remington.)

the fetus that results from the asymptomatic* infection of the expectant mother. Blood tests indicate that about one of every 1000 babies is infected by this route. If the fetus is infected during the last three months of pregnancy, the baby is often born mentally defective or shows damage to the retina of its eyes. (2) Infections of adults who are immunologically compromised by cancer or other serious underlying diseases. Such cases may be marked by myocarditis, brain damage, and involvement of most other body tissues. (3) A mild illness that resembles infectious mononucleosis and that is usually seen in young adults. Tests for *Toxoplasma* antibodies in the blood of students at the University of California and at Harvard University showed that about one of five had been infected with *Toxoplasma*. Infection is thought to occur mainly from eating improperly cooked meat, and an outbreak of five cases once occurred among a group of New York medical students who had eaten inadequately cooked hamburger. Their illness resembled infectious mononucleosis, and they had lymph node enlargement lasting about three months before they recovered completely.

The life cycle of *T. gondii* was long a mystery, but it is now known that the organism is a coccidium (a protozoan parasite) of cats, infecting the cells lining the cat intestine. The coccidia are sporozoa, as are the protozoa that cause malaria. In one American city, about half the cats tested had antibodies against *Toxoplasma*, indicating past infection, since they showed no signs of illness. Acutely infected cats usually have diarrhea. The protozoa are excreted by the sick cat for only about two weeks. After excretion, the protozoa change into an infectious form after two to four days and remain infectious for about one year thereafter. Other animals become infected when they ingest food or water contaminated by cat feces. The organisms then enter the tissues of the new host—perhaps a sheep or cow—where they develop into a cystic form containing many infectious progeny. These live for an indefinite period and can be transmitted when another animal, such as a human, ingests the inadequately cooked meat of the host. However, the com-

---

*asymptomatic—without apparent symptoms

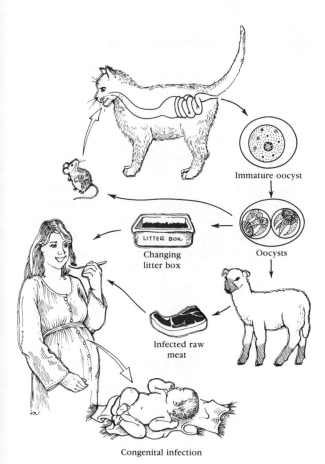

Immature oocyst

Oocysts

LITTER BOX

Changing
litter box

Infected raw
meat

Congenital infection

Figure 27–3
Epidemiology of toxoplas-
mosis. Both infected meat
and cyst forms from cat feces
can infect humans.

plete cycle of growth of *T. gondii*, as shown in Figure 27–3, can develop only within the intestinal lining cells of a cat or a closely related animal species. It has been established that people who are often in close contact with cats have a higher *Toxoplasma* infection rate than those who are not.

Toxoplasma gondii is of general interest because of its wide host range. The organisms are infectious for almost all cells of warm-blooded animals except nonnucleated erythrocytes.[5] *Toxoplasma gondii* was also one of the first infectious agents shown to produce general enhancement of resistance to intracellular infections by other species of microorganisms. The practical significance of this observation is as yet undefined, but in view of the widespread and harmless nature of most human *Toxoplasma* infections, it is interesting to speculate on the possible role of this coccidium of cats in preventing or increasing resistance to other human diseases. Nevertheless, expectant mothers would do well to avoid cat litter and rare meat.

---

[5]The erythrocytes of some species, such as fowl, contain nuclei, whereas other species, including humans, lack nuclei in their erythrocytes.

SUMMARY

There are two networks of vessels in the body, the blood vascular system and the lymphatic system. As blood is pumped by the heart around the blood vascular circuit, any microorganisms or viruses present are likely to be removed by the mononuclear phagocytic cells of the spleen, lung, and other organs. Unlike the blood vessels, the lymphatic vessels begin as blind-ended capillaries, the walls of which can readily be crossed by infectious agents. Lymph is usually cleared of contaminants by mononuclear phagocytes in the lymph nodes. Failure of body defense mechanisms may result in endocarditis, myocarditis, pericarditis, and lymphangitis. Microorganisms circulating in the bloodstream may release toxins and produce septicemia. Tularemia, brucellosis, and infectious mononucleosis are diseases in which the lymph nodes and spleen are often prominently involved. Toxoplasmosis is an unusual manifestation of the widely spread infections caused by *Toxoplasma gondii,* a protozoan parasite of cats.

SELF-QUIZ

1. Microorganisms and other foreign material are normally removed from the blood
   a. by the mononuclear phagocyte system
   b. through the left atrium
   c. by the heart valves
   d. by tetracycline treatment
   e. none of the above
2. The alpha-hemolytic streptococci that commonly cause subacute bacterial endocarditis (SBE) usually originate in the
   a. liver
   b. lymph nodes
   c. mouth
   d. spleen
   e. heart
3. The Gram-negative rod-shaped bacteria *Escherichia coli* and *Serratia marcescens* are likely causes of
   a. subacute bacterial endocarditis
   b. septicemia
   c. tularemia
   d. brucellosis
   e. myocarditis
4. A man diagnosed as having a disease caused by *Francisella tularensis* most probably contracted the infection by involvement in what activity?
   a. drinking raw milk
   b. eating undercooked meat
   c. raising cats

   d. skinning a wild animal
   e. injecting heroin intravenously
5. Brucellosis can be contracted by drinking milk contaminated with *Brucella abortus*, but most of the cases of brucellosis occur
   a. through working in the meat-packing industry
   b. in microbiology laboratories where safety measures are inadequate
   c. in small children with pet cats
   d. by contact with immunologically compromised patients
   e. through kissing (people)
6. Coxsackie myocarditis is
   a. effectively treated with ampicillin
   b. a frequent complication of drug addiction
   c. often marked by prolonged fecal excretion of the causative agent
   d. unlikely to affect the heart
   e. least common in infants
7. Infectious mononucleosis
   a. is a tumor occurring in African children
   b. invariably results from EB virus infection
   c. mostly occurs through contact with sheep and horses
   d. is distinguished from other diseases with enlarged lymph nodes and spleen by the absence of a sore throat
   e. characteristically is accompanied by the appearance of heterophile antibodies
8. Toxoplasmosis
   a. may be acquired by the fetus during asymptomatic infections of the mother
   b. can show involvement of the heart and brain of immunologically compromised adults
   c. is often a prolonged illness of young adults marked by enlargement of the lymph nodes and spleen
   d. all of the above
   e. none of the above

QUESTIONS FOR DISCUSSION

1. How long could *Toxoplasma gondii* continue to exist if all cats were abruptly exterminated?
2. If cellular immunity is responsible for eliminating EB virus–infected lymphocytes during recovery from infectious mononucleosis, what happens to the patient's ability to fight other infections?
3. Which would be more likely to cause lung abscesses, endocarditis on the right side of the heart or on the left? Why?
4. Can you think of any reasons why, among all the thousands of people with Coxsackie infections, so very few develop myocarditis? Could a viral infection such as this one, which slowly destroys the heart, destroy some other organ such as the pancreas?

FURTHER READING

Bayer, A. S., et al.: "Circulating Immune Complexes in Infective Endocarditis." *New England Journal of Medicine 295*:1500 (1976).

Duma, R. J.: "Thomas Latta, What Have We Done?—The Hazards of Intravenous Therapy." *New England Journal of Medicine 294*:1178 (1976).

Parrillo, J. E., et al.: "Endocarditis Due to Resistant Viridans Streptococci During Oral Penicillin Prophylaxis." *New England Journal of Medicine 300*:296 (1979).

Teutsch, S. M., et al.: "Pneumonic Tularemia on Martha's Vineyard." *New England Journal of Medicine 301*:826 (1979).

Chapter 28

ARTHROPODS AND PARASITIC WORMS

Ring a ring o'roses
A pocket full of posies
Atishoo! Atishoo!
We all fall down.

This innocent-sounding nursery rhyme is believed to have originated during a fourteenth-century epidemic of bubonic plague in Europe. A "ring o' roses" refers to skin changes, and the sneezing ("atishoo") to respiratory symptoms. The "posies" were the herbs popularly worn with the hope of preventing contagion. Many who "fell down" in real life never got up again. Such was the severity of the plague, a disease transmitted by fleas. Fortunately, plague is seen much less often these days.

OBJECTIVES

To know
1. The principal features of malaria, yellow fever, and arthropod-borne encephalitis, particularly the causative agents, transmission, diagnosis, and prevention.
2. The cause of bubonic plague, and its transmission, diagnosis, and control.
3. The cause of typhus and Rocky Mountain spotted fever; their transmission, diagnosis, treatment, and prevention; and the importance of mites in human disease.
4. The natural history of certain diseases caused by nematodes, particularly how they are diagnosed and controlled.
5. The natural history of schistosomiasis, especially how it relates to incidence of the disease and its control.
6. The natural history of tapeworms and how infestation of humans is detected and controlled.

The main focus of this book is on unicellular microorganisms (bacteria, protozoa, algae, fungi) and subcellular agents (viruses). However, some of the small multicellular parasitic animals are studied by using microscopic and immunological principles similar to those that have been presented in earlier chapters. Moreover, many of these small parasitic animals are themselves infected by microorganisms and viruses that can be pathogenic for humans and the animals and plants on which we depend. The present chapter gives some examples of the interactions among these various biological agents, particularly those that relate to human health.

Almost all of these small multicellular parasites are members of three broad animal groups (phyla)—Platyhelminthes, Nematoda, and Arthropoda—that include both parasitic and nonparasitic members. The first two of these phyla are composed of primitive wormlike creatures. Only a few instances are known in which the wormlike parasites transmit pathogenic microorganisms or viruses to mammals. In contrast, the Arthropoda are highly advanced on the evolutionary scale, and many of the parasitic arthropods carry infectious agents to their hosts. Thus, in some instances the main importance of the multicellular parasite is as a reservoir,* transmitter, and injector of an infectious agent, while in other instances the main factor is the parasite's ability to invade and damage the host. In general, parasitic arthropods are an example of the former, and parasitic worms are an example of the latter. The extent of the problem is described in the following paragraph.

The parasitic diseases are the "cancers" of developing nations. They bring death, suffering and long-term disability to many of the world's people, but to residents of the United States they are distant problems that rarely rage into epidemics and rarely cross international borders. They are the forgotten problems of forgotten people.[1]

ARTHROPODS AND ARTHROPOD-BORNE INFECTIONS

Examples of some important arthropods, the agents they transmit, and the resulting diseases are shown in Table 28–1.

Mosquitoes

In a medical context, the mosquito (Fig. 28–1) is one of the most important insects. The mouth parts of the female mosquito consist of sharp stylets, which are forced through the host's skin to the subcutaneous capillaries.[2] One of these needlelike stylets is hollow, and salivary secretions of the mos-

---

[1]Schultz, M.G.: New England Journal of Medicine 297:1259 (1977).

[2]Only the female mosquito has mouth parts adapted to piercing animal skin; the male mosquito feeds instead on plant juices. Ovarian development and egg production in mosquitoes are markedly ehanced by a blood meal, indicating that blood is the source of important hormones or nutrients for this insect and contributes to its reproductive functions.

*reservoir—source of a disease-producing organism

TABLE 28–1   EXAMPLES OF ARTHROPODS THAT TRANSMIT INFECTIOUS AGENTS

Arthropod	Infectious agent	Disease and characteristic features	Laboratory diagnosis	Control
Mosquito (*Anopheles* species)	*Plasmodium* species	Malaria: chills, bouts of recurring fever	Stained smear of peripheral blood; measurement of antibody against plasmodia	Screening blood smears for carriers; treating infected persons; mosquito control by insecticides and habitat control
Mosquito (*Culex* species)	Arbovirus	Equine encephalitis: chills, fever, nausea, malaise, brain inflammation	Measurement of antibody against virus	Mosquito control; vaccine for horses
Mosquito (*Aedes aegypti*)	Arbovirus	Yellow fever: fever, aches, vomiting, jaundice*	Measurement of antibody against virus	Mosquito control; vaccine
Flea (*Xenopsylla cheopis*)	*Yersinia pestis*	Plague: chills, fever, headache, confusion, enlarged lymph nodes, skin hemorrhage*	Cultures of blood, sputum, lymph nodes	Rodent control; insecticides for flea control; vaccine; quarantine*
Louse (*Pediculus humanus*)	*Rickettsia prowazekii*	Typhus: fever, hemorrhage, rash, confusion	Weil-Felix test; complement-fixation test using rickettsial antigen	Insecticide for lice control; vaccine; cleanliness and avoidance of overcrowding
Tick (*Dermacentor* species)	*Rickettsia rickettsii*	Rocky Mountain spotted fever: fever, hemorrhagic rash, confusion	Weil-Felix test and other tests for antibody	Prompt removal of ticks; vaccine

*jaundice—yellowing of skin caused by bile pigments
*hemorrhage—bleeding
*quarantine—any isolation or restriction on travel imposed to keep contagious diseases from spreading

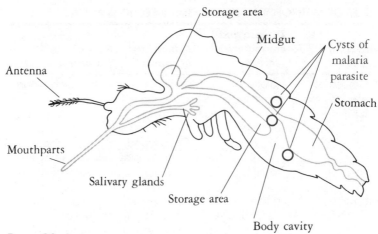

Figure 28–1
Internal anatomy of the mosquito.

quito are pumped through it. These secretions help keep the victim's blood from clotting as it is drawn into a tube formed by the other mouth parts of the insect, and the secretions may also act on the victim's blood vessels to increase blood flow. The injected mosquito saliva contains antigens to which some individuals develop hypersensitivity, and it may also contain agents that can cause serious infectious diseases.

The muscles of the throat of the mosquito suck blood into its digestive system and into large storage areas. The latter allow the mosquito to take in as much as twice its body weight in blood, giving it a relatively good chance of picking up microorganisms such as any malarial parasites circulating within the host's capillaries. Other parts of the mosquito's digestive tract are the stomach, in the wall of which the malaria parasite develops, and the midgut, through which the parasite penetrates the mosquito's body cavity to infect the insect's salivary glands.

Most of the medically important mosquitoes are included in the genera *Aedes, Culex,* and *Anopheles.* Precise identification of species and subspecies of these genera is extremely important and depends largely on microscopic examination of antennae, wings, claws, mating apparatus, and other features. Correct identification is often essential to control medically important vectors* because different species of mosquitoes differ greatly in their breeding areas, time of feeding, and choice of host.

Although mosquitoes may act as vectors for certain parasitic worms and for bacterial agents such as *Francisella tularensis,* the protozoan and viral agents transmitted by mosquitoes are numerically far more important. A few of these mosquito-borne infections that are important to humans are considered in the following sections.

*vectors—animal carriers of an infectious agent

Malaria

Malaria, an ancient scourge, was suspected of being mosquito-borne as long ago as the second decade of the eighteenth century. This suspicion was dramatically confirmed in the late 1800's when mosquito control measures in Havana led to a drop in the incidence of malaria from about 9 in 10 people to 1 in 50. In the early 1900's, effective mosquito control reduced the incidence of malaria and yellow fever and allowed completion of the Panama Canal.

Human malaria is caused by protozoa of the genus *Plasmodium*. Four species are involved—*P. vivax, P. falciparum, P. malariae*, and *P. ovale*—which differ in microscopic appearance and, in some instances, life cycle, type of disease produced, severity, and treatment. In recent years, the majority of patients diagnosed in the United States have been infected with *P. vivax*, but up to 30 percent have had the more dangerous *P. falciparum*. As will be mentioned in Chapter 29, there are important racial differences influencing susceptibility to malaria.

Plasmodium species, page 395

Racial susceptibility, page 607

The most characteristic feature of human malaria is the occurrence of bouts of fever at one- to three-day intervals, with periods of relative well-being in between. The protozoan parasite of the infection, one of the four species of *Plasmodium*, grows and divides in the erythrocytes of the host (Fig. 28–2). After a time, the offspring of this division rupture the erythrocytes in which they are growing and are released into the plasma. The victim's intermittent fevers result from this periodic release of the plasmodia. Most of the young plasmodia then enter new erythrocytes and multiply; the cycle is repeated. Some of them, however, develop into **gametocytes,** specialized sexual forms different from the other circulating plasmodia in both their appearance and susceptibility to antimalarial medicines. These sexual forms cannot develop further in the human host and are not important in causing the symptoms of malaria. They are, however, infectious for certain species of *Anopheles* mosquitoes and are thus ultimately responsible for the transmission of malaria from one person to another.

Infection of a mosquito begins when it ingests the gametocytes of the malarial parasite. Shortly after entering the intestine of the mosquito, stimulated by the drop in temperature, the male gametocyte produces tiny whip-like bodies that unite with the female gametocyte in much the same way as the sperm and ovum unite in higher animals. The resulting zygote enters the wall of the midgut of the mosquito and forms a cyst, which enlarges as the zygote undergoes meiosis and division into numerous asexual organisms. The cyst then ruptures into the body cavity of the mosquito, and the released parasites find their way to the salivary glands from which they may be injected into a new human host. Malarial parasites are then carried by the human bloodstream to the liver, where they infect liver cells. In these cells, each parasite enlarges and subdivides, producing thousands of daughter cells. These are released into the bloodstream and establish the cycle involving the erythrocytes. In some forms of malaria, the infected erythrocytes

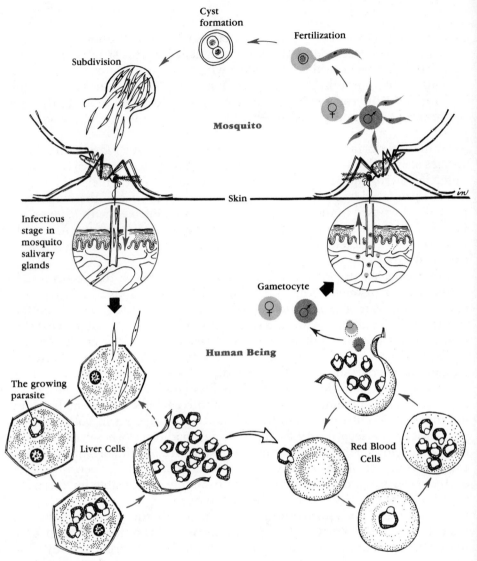

Figure 28–2
Life cycle of a malaria parasite.

can adhere together and block capillaries in the brain and other parts of the body (Fig. 28–3). Some species of *Plasmodium* parasites released from the liver cells may also infect new liver cells, thus establishing a cycle of infection in the liver. A sustained liver infection may result and persist despite effective treatment of the bloodstream infection. Malaria may recur later, when treatment is discontinued.

Malaria can also be transmitted through the placenta to the fetus, result-

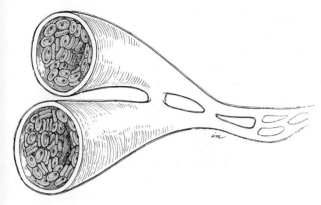

Figure 28–3
Plugged capillaries in *Plasmodium falciparum* malaria. The infected red blood cells have become sticky and adhere to each other, blocking the capillaries and depriving the tissues of oxygen.

ing in a spontaneous abortion* or a child born with active malaria. Such a child was born to Kampuchean[3] parents who arrived in the United States two months before his birth in 1979. After two days of fever and vomiting, the child was admitted to the hospital, where blood smears revealed *Plasmodium vivax* (color plate 5). Treatment was effective, and he was then sent home. Many people (like this child's mother) recover from an acute attack of malaria and live with few or no symptoms despite the presence of parasites in their blood. Often these parasites can be detected by examining blood smears. Although there were no symptoms of malaria during the mother's pregnancy, nevertheless sufficient organisms were present in her blood to infect the unborn child.

The ancient Turks had an ingenious treatment for malaria. The infected individual was made to run for two miles while being chased by a person with a horsewhip. There are no scientific data that support this treatment, and to the best of our knowledge, it has been entirely replaced by modern chemotherapy*. Medicines are also effective preventives, and in most areas of the world malaria can be avoided by taking chloroquine once weekly for two weeks before and during the stay and for six weeks after leaving the area. Two weeks of treatment with primaquine phosphate is also required after leaving the area to rid the body of any parasites that might reside in the liver, since these parasites are not eliminated by chloroquine.

Chloroquine, Appendix VI

Malaria is best controlled by using insecticides, by eliminating breeding areas of the principal mosquito vectors, and by giving treatment to infected patients. Patients are identified by preparing stained smears of their blood and examining these microscopically for the malarial parasites. Intensive efforts, often with assistance from the World Health Organization, have markedly reduced or eliminated mosquito-borne malaria from many coun-

---

[3]Natives of the country formerly called Cambodia.

*spontaneous abortion—commonly known as a miscarriage

*chemotherapy—treatment with medicines, in this case, quinine and chloroquine

tries. Eradication of human malaria is theoretically possible but difficult to accomplish. Difficulties in control arise partly because there are strains of mosquitoes resistant to insecticides and strains of *Plasmodium* resistant to chemotherapeutic agents. In addition, control efforts in some areas have been relaxed too soon. Sri Lanka,[4] for example, formerly was subject to serious malarial epidemics. In 1935 malaria caused an estimated 80,000 deaths. With intensive eradication efforts, the incidence of malaria in Sri Lanka had dropped to only 17 cases in 1963; control measures were then largely discontinued, and surveillance efforts were relaxed. Subsequently, the incidence of the disease promptly rose once again, so that in the month of February 1968 more than 42,000 cases of malaria were diagnosed in Sri Lanka.

Malaria once was common in the United States, but it became very rare with the establishment of control measures. A dramatic reappearance of malaria in the United States occurred with the return of military personnel from Southeast Asia (more than 3000 cases in 1970, with at least one case in every state), but most of these cases were acquired outside the country. In Alabama, however, a minor epidemic of malaria among teenagers was traced to veterans of the Vietnam War who had malaria and who had attended outdoor movies. The local mosquitoes had feasted on the veterans and on other members of the audience, and ten days later malaria developed in teenagers who had attended the movies. A national low of only 237 cases was again reached in 1973, but the incidence has tended to rise each year since then. This steady increase in the number of malaria cases reflects both the amount of international travel to areas of endemic* malaria and the resurgence of this disease worldwide. More United States citizens are traveling to malarious areas these days, and more people are immigrating to the United States from these areas. In some instances, transmission of the disease has been the result of blood transfusions. Transmission can also occur when drug addicts share syringes. This type of transmission causes only erythrocyte infections, not liver infections, and is therefore easier to cure.

Yellow Fever

Arboviruses, Appendix
VIII

Yellow fever is caused by a mosquito-borne arbovirus. This disease is characterized by fever, aches and vomiting (often associated with bleeding from the nose and gastrointestinal tract), and jaundice. The mortality from yellow fever often reaches 40 percent. The reservoir of the disease is now mainly infected mosquitoes and primates* living in tropical jungles of Central and South America and Africa. No cases of yellow fever have been acquired in the United States since 1911, although the disease was once common here.

---

[4]Formerly known as Ceylon.

*endemic—continually existing in a population

*primates—group of animals that includes monkeys, apes, and humans

In late 1978 there was an outbreak of yellow fever in Trinidad. To control the epidemic, health officials immunized 75 percent of the population, and by March 1979 the epidemic was over. Control of yellow fever is achieved by control of its principal mosquito vector, *Aedes aegypti*, and by using a live attenuated vaccine to immunize people who might become exposed, such as those traveling to areas of Africa or Central and South America where the disease is endemic.

Arthropod-borne Encephalitis

In addition to carrying the arbovirus responsible for yellow fever, mosquitoes may also bear viruses that cause *encephalitis*, an inflammation of the brain. Human and domestic-animal encephalitis caused by arboviruses occurs in endemic or epidemic fashion every year in the United States. Usually about 250 cases occurring in more than 20 states are reported yearly. During epidemics, however, the prevalence can be much higher. Among the causative viruses are the equine encephalitis viruses, which normally infect horses but can also cause death, mental disorder, or motor nerve impairment in humans.

The encephalitis viruses can pass from one mosquito to another through common exposure to another host, normally one of the birds living in a swamp. The infected mosquito transmits the virus to the bird, in which the virus multiplies and enters the blood. A second mosquito bites the bird, taking in virus particles with the blood meal. The virus multiplies in the cells of the mosquito, and viral particles are discharged in the salivary juices. The viruses remain in the infected mosquito for the life of the mosquito, and some species of virus can also be transmitted to the next generation of mosquitoes through the eggs (transovarial passage). Every now and then, when the host mosquitoes are unusually plentiful and the number of infected birds is large, the encephalitis virus may be spread to other areas and to less well-adapted hosts. Mosquitoes may then transmit the virus to domestic fowl, horses, or people. The transfer of virus from mosquito to bird to mosquito to bird, and so on, is often referred to as a "cycle," although it is more like a never-ending chain with repeating links. Through evolutionary selection, the mosquitoes and birds involved in the cycle are often affected very little by the presence of the viral infection. On the other hand, the spread of the virus to other hosts is usually a dead end because of the high mortality the viral infection causes among the new hosts. Furthermore, different species of mosquitoes usually bite these new hosts, and these mosquitoes are less able to sustain the infection and transmit the causative virus. It is likely, however, that on rare occasions mutants among infecting encephalitis viruses may multiply and persist in their new vector and cause less serious disease in the new host, so that a new cycle develops.

Control of equine encephalitis is achieved chiefly through mosquito control, although vaccines given to animals may be helpful in some instances. In some states, vaccination of horses is required by law.

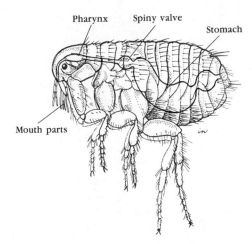

Figure 28–4
Anatomy of a flea.

Fleas

Fleas are wingless insects that depend on their powerful hind legs to jump quickly from place to place. The anatomy of a flea is diagramed in Figure 28–4. Points of importance in the identification of flea species include the spines or "combs" about the head and first thorax segment of this insect, the muscular pharynx (for sucking blood from the host), and the long esophagus, followed by the spiny valve composed of rows of teethlike cells. Fleas, like most other ectoparasites,* are themselves more of a nuisance than a health hazard. However, one species of flea is of major importance because of its susceptibility to infection by the bacterium that causes plague.

Bubonic Plague

Y. pestis, page 316

The bacterium *Yersinia pestis*,[5] which causes plague, exists in nature by infecting rats and other rodents and the fleas they carry. The flea contracts the plague bacillus by feeding on an infected rodent. So many of the bacilli may be present in the blood of an infected rat that a flea can ingest 25,000 of these microorganisms in the course of a single feeding! The organisms pass into the stomach of the flea and multiply there, producing sticky masses of bacteria, which can cause obstruction by adhering to the spiny valve. When obstruction occurs, blood sucked into the flea's esophagus cannot pass into its stomach but is regurgitated* into the puncture wound in the host's skin, carrying the plague bacilli with it. Moreover, the hungry flea tries again and again to feed and may infect several hosts with *Y. pestis*. Infected fleas may

---

[5]Formerly called *Pasteurella pestis*.

*ectoparasites—parasites that attack the skin of the host

*regurgitated—returned to the mouth from the stomach

live for a long time, because the obstruction in the spiny valve often cures itself, but the infection persists and *Y. pestis* is then excreted with the flea feces. This fecal material may also infect humans and other animal hosts when rubbed or scratched into flea bites.

At times, the mortality among rodents infected with *Y. pestis* may be so high that the flea vectors of the disease find their supply of natural hosts diminishing very rapidly. In this case, they will begin to feed on humans, who thereby become infected with *Y. pestis* and develop bubonic plague. Carried by tissue fluids to the lymph nodes near the site of a flea bite, the bacteria cause enlargement of the nodes (an enlarged node is called a "bubo," thus "bubonic" plague). Large numbers of the bacilli are then released into the bloodstream and distributed to all areas of the human body. In some cases, bleeding beneath the skin is a prominent feature of bubonic plague; the darkened blood and the dusky appearance of dying patients may have suggested the name "Black Death" for the European epidemic of the fourteenth century.

Yersinia pestis bacilli in the blood are carried to the lungs, where they may produce a form of the disease that spreads easily from person to person via the respiratory route without the mediation of fleas. The bacilli enter the bronchial secretions; with coughing, they are sprayed into the air and can thus be inhaled by others. The disease resulting from inhalation rather than flea bite is called "pneumonic plague" and accounts for rapid epidemic spread. In 1975, a 14-year-old boy acquired plague in New Mexico and then traveled to California, where he died from plague pneumonitis.* Fortunately, the boy's coughing did not transmit the disease to others. Although cases of bubonic plague are seen in the United States every year, there has been no spread from person to person reported since 1924.

In both bubonic and pneumonic plague, the causative bacteria can be seen in infected material and readily grown on laboratory media. *Y. pestis* is a Gram-negative enterobacterium that stains more intensely at each end, looking somewhat like a safety pin.

In 1968, 90 cases of plague with 36 deaths occurred in villages of central Java. Indonesian and American health officers worked together to impose strict quarantines to prevent person-to-person spread of the disease. Plague vaccine was given to people living close by who were threatened by spread of the epidemic. Rats were trapped, and their ectoparasites were carefully identified microscopically. The most frequent parasite found on them was the rat flea, *Xenopsylla cheopis*. Villages were heavily sprayed with DDT to reduce the numbers of fleas. The epidemic was quickly brought under control, and cases ceased to appear within about two months. Plague in humans is an uncommon disease in the United States today, although *Y. pestis* has been identified in the wild rodent population of at least 15 states and thus still remains a threat.

---

*pneumonitis—inflammation of the lung

In many cases of human plague in the United States there has been no obvious contact with wild rodents, and it is probable that domestic pets and other animals may transport infected rodent fleas to human residences. A fatal case reported from New Mexico in 1981 was contracted by skinning a bobcat. In addition, wild rodents may be attracted to the vicinity of houses by garbage or other food.

Lice

Like fleas, lice are small, wingless insects that prey on warm-blooded animals by piercing their skin and sucking blood. However, the legs and claws of lice are adapted for holding onto body hair and clothing, rather than for jumping from place to place. Like fleas and mosquitoes, lice are susceptible to microbial infections that can be transmitted to people.

Pediculus humanus, the most notorious of the lice, is 2 to 3 mm long, with a characteristically small head and thorax and a large abdomen (color plate 61). This louse has a membranelike lip with tiny teeth that anchor it firmly to the skin of the host. Within the floor of the mouth is a piercing apparatus somewhat similar to that of fleas and mosquitoes. *Pediculus humanus* has only one host—humans—but easily spreads from one person to another by direct contact or by contact with personal items, especially in situations of crowding and poor sanitation. A similar organism, *Phthirus pubis* ("crab louse") (color plate 62), is not uncommon among young adults; it is most frequently transmitted during sexual intercourse. It is not known to be a vector of infectious disease, but it does cause considerable discomfort and embarrassment.

Epidemic Typhus

Typhoid, page 479

Epidemic typhus (not to be confused with typhoid, caused by *Salmonella typhi*) is caused by *Rickettsia prowazekii*, a tiny obligate intracellular bacterium transmitted by *P. humanus*. This louse-borne disease tends to occur under crowded and unsanitary living conditions, such as might occur in areas of water shortage or during wartime. Napoleon, during his invasion of Russia, lost an estimated 150,000 men to typhus. In fact, typhus has been responsible for many more wartime deaths than have battles. Fever develops abruptly in patients with typhus, followed in a few days by rash and confusion (typhus means "hazy"). Damage to small blood vessels then becomes prominent, resulting in gangrene and hemorrhages beneath the skin. Involvement of the heart, brain, and kidneys may contribute to the illness, and mortality from typhus can range from about 10 to 40 percent.

Lice feeding on typhus-infected individuals ingest the infectious rickettsiae. The microorganisms are then taken up and multiply within the cells that line the gut of the louse. The infected cells die and liberate the rickettsiae, which then pass through the louse intestine and are excreted in the feces. Scratching the itchy louse bites causes a new victim to rub some of the contaminated louse excreta into the bite wound, thereby causing inoculation with *R. prowazekii*. Infection may at times also occur by inhalation of dried louse feces or by accidentally rubbing them into the eyes.

Like other rickettsiae, *R. prowazekii* does not grow on cell-free media but only in susceptible living cells. The organisms can be recovered from human patients or lice by inoculating infected blood or other material into laboratory animals, embryonated eggs,* or tissue cell cultures. However, it is easier and safer for laboratory workers to look for a rise in the titer of antibody to typhus rickettsiae in a patient's blood. This is done by making use of a dramatic example of antigenic sharing by members of two greatly different genera of bacteria, *Rickettsia* and *Proteus*. Certain strains of *Proteus vulgaris* and *P. mirabilis* have cell wall antigens that react with the antibodies formed against *R. prowazekii* and some other rickettsiae. The resulting antibody-antigen reaction produces agglutination of the *Proteus*. This procedure, employing strains of *Proteus* to diagnose rickettsial disease, is known as the Weil-Felix test.

Rickettsia species, page 328

Epidemic typhus can be readily controlled by using insecticides to kill lice. For all practical purposes, transmission of the disease from person to person can result only from *P. humanus* infestation, and humans are the only other reservoir for *R. prowazekii*. Unfortunately, complete eradication of typhus is complicated by the fact that people who recover from the disease may continue to have living rickettsiae in their body tissues for many years without apparent ill effects. Under conditions of stress, fading immunity, and other unknown factors, these organisms may multiply and again cause typhus. If *P. humanus* is present, the disease may be transmitted to others, producing a new wave of epidemic typhus.

Rickettsia prowazekii probably persists in tissues through a delicate balance between defensive factors of the host and attributes of the microbe that protect it against these factors. However, the exact mechanisms involved are unknown. Except for occasional cases of this recurrent type of disease, louse-borne typhus disappeared from North America years ago. It is still a problem in poor areas of South and Central America, Europe, and Asia. About 50 cases of a milder disease, **murine typhus,** are seen in the United States each year. This disease is caused by a different rickettsia and is transmitted by rodent fleas.

Ticks

Ticks are not insects but **arachnids.** Arachnids differ from insects in their lack of wings and antennae and in the fact that their thorax and abdomen are fused. Furthermore, adult arachnids have four pairs of legs compared with the insect's three pairs. *Dermacentor andersoni* (Fig. 28–5), a common American tick, has a broad host range—more than 25 animals. Most of the

---

*embryonated eggs—chicken eggs containing a developing embryo

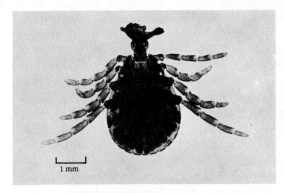

Figure 28–5
The tick, *Dermacentor andersoni.*
The mouth parts are imbedded in
a piece of the host's skin that was
torn away when the tick was removed. Note that four pairs of legs
are present as opposed to the three
pairs that insects have.

ticks that attack humans are intermittent parasites, attaching from time to time to feed on human blood, then dropping off to rest, grow, or deposit eggs. Their bite is usually completely undetected, and one discovers the tick by chance, firmly attached to the skin and becoming swollen with ingested blood.

The attachment of ticks to the skin is very secure. In fact, if one tries to pull a tick off, the mouth parts may separate from the tick rather than tear from the skin of the host. How the tick attaches itself without causing pain to its host is unclear. Some ticks produce a toxin powerful enough to cause paralysis. Paralyzed humans and animals usually recover rapidly following removal of the tick.

Rocky Mountain Spotted Fever

A four-year-old boy was brought to the emergency department of a Cape Cod hospital because of fever, sore throat, and a rash that looked like the rash of measles. Unlike many other children who come to emergency departments with similar symptoms and prove to have mild viral illnesses, this boy failed to improve over the next few days. Indeed, after a week of illness, high fever abruptly developed, and he became confused. Hemorrhages appeared in his skin, his eyes became puffy, his pulse rose to 200 beats per minute, and his blood pressure dropped to dangerously low levels. Blood persistently oozed from the places where blood samples had been taken. Negative Gram stains and cultures for bacterial pathogens, plus the discovery that the boy had removed an attached tick from his body four days before his illness, heightened suspicion that he had Rocky Mountain spotted fever. Unfortunately, despite heroic attempts to treat this disease, it was diagnosed too late, and the boy died. Some blood that had been obtained before his death was sent to an infectious disease research laboratory where it was injected into embryonated chicken eggs. After several days, *Rickettsia rickettsii* could be demonstrated growing in the embryo cells. Tests for antibody to the organisms in the boy's blood were negative because he died before sufficient antibody could develop. The boy's dog, which was sick at the same time, survived and

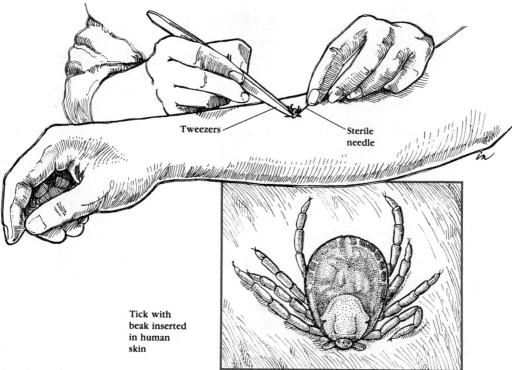

Tweezers

Sterile
needle

Tick with
beak inserted
in human
skin

Figure 28–6
Procedure for removing a tick. The tick is pried out, with care taken not to separate the beak from the rest of the body.

was shown to develop large amounts of antibody to Rocky Mountain spotted fever. The boy was unlucky; only one or two cases of spotted fever occur each year on Cape Cod, and most patients respond to treatment with antibiotics such as tetracycline and chloramphenicol.

Rocky Mountain spotted fever is transmitted to people by the bite of a tick infected with *Rickettsia rickettsii*. These ticks may remain infective for their lifetime of 18 months or more. Rickettsia can pass from generation to generation of ticks through tick eggs (transovarial passage), demonstrating the ability to coexist with their arthropod host for long periods without harming the host.

Contrary to its name, Rocky Mountain spotted fever is common in different parts of the country, including the South Atlantic states. The dog tick *Dermacentor variabilis* and the wood tick *Dermacentor andersoni* are the chief vectors.

If the ticks are removed within 4 hours of attachment, this disease usually does not result. When a person is traveling in tick-infested country, it is important to check frequently for ticks on the body and to remove them promptly (Fig. 28–6).

When the diagnosis of Rocky Mountain spotted fever is made quickly and

TABLE 28–2   OTHER HUMAN PATHOGENS THAT INFECT TICKS

Infectious agent	Disease and characteristic features[a]	Laboratory diagnosis
Coxiella burnetii	Q fever: mild pneumonia	Measurement of complement-fixing antibody against rickettsia
An arbovirus	Colorado tick fever: fever without rash, aching muscles, headache	Measurement of antibody against virus; viral isolation
Francisella tularensis	Tularemia: fever, enlarged lymph nodes	Measurement of antibody against the bacteria; recovery of organism from lesion or blood is possible but hazardous to laboratory personnel
Borrelia species	Relapsing fever: fever recurring one to three times, headache, malaise	Stained smear of blood

[a]All except Colorado tick fever are treated with antibiotics.

treatment is started, death is uncommon, but the mortality is about 20 percent when diagnosis and treatment are delayed. Black people and people over 40 are affected most severely by the disease. There has been a general tendency for the incidence to increase over the past decade. In 1980, 1132 cases were reported in the United States. North Carolina reported more cases than any other state.

Other Tick-borne Diseases (Table 28–2)

In addition to carrying *R. rickettsii, Dermacentor andersoni* may also harbor the pathogenic rickettsia *Coxiella burnetii*, the cause of **Q fever,** usually contracted by people through inhalation of infected tick feces or by drinking milk from an infected cow. An epidemic that occurred in 1979 at a University of California research center was traced to infected sheep. The disease can be fatal but is more likely to be a mild pneumonia. Another tick-borne agent of medical importance is the arbovirus of **Colorado tick fever.** This virus is endemic in wild rodents of the Rocky Mountain region of the United States. In humans it typically produces an uncomfortable illness with head and muscle aches and fever without rash; recovery occurs without treatment. Finally, ticks may be infected with the bacterium *Francisella tularensis*, the Gram-negative rod that causes **tularemia.** Like *R. rickettsii, F. tularensis* can pass transovarially. Members of another genus of ticks transmit species of *Borrelia*, the spirochetes responsible for **relapsing fever.**

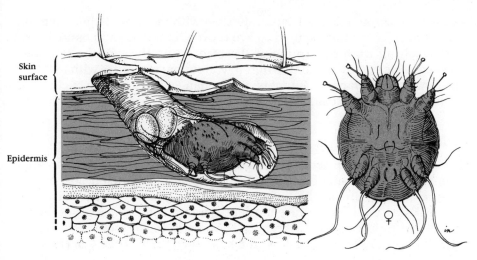

Skin surface

Epidermis

Figure 28–7
Sarcoptes scabiei (scabies mite) burrowing under the outermost layer of skin.

Mites

Mites, like ticks, are arachnids. They are generally tiny, are fast-moving, and live on the outer surfaces of animals and plants. *Demodex folliculorum* lives in the oil-producing glands of the skin, usually without producing symptoms. Other species of mites are of greater medical significance.

Scabies

Scabies is characterized by an itchy rash most prevalent in areas of moist skin between the fingers, under breasts, and in the genital area. Except in children and infants, the rash usually does not occur above the neck. The incidence has been increasing over the last one to two decades, and some speak of an existing pandemic* of scabies. Scabies is easily transmitted by personal contact, and the disease is commonly acquired during sexual intercourse. The cause of scabies is the mite *Sarcoptes scabiei* (color plate 63), which measures less than half a millimeter in length. The female burrows into the outer layers of skin and feeds and lays eggs over its lifetime of about one month (Fig. 28–7). Hypersensitivity to the mites develops and is largely responsible for the itchy rash. The diagnosis is often difficult to make unless the mites can be detected, since scabies can mimic other skin diseases. The mites are best detected by placing a drop of mineral oil over the burrow, scraping out the mite with a scalpel blade, and transferring the scrapings and oil to a microscope slide. The characteristic microscopic appearance is

*pandemic—a worldwide epidemic

diagnostic. Treatment is easily accomplished with medication applied to the skin, although care must be used to avoid overtreatment and possible toxic effects resulting from absorption of the medication through the skin. *Sarcoptes scabiei* is not known to transmit infectious agents. Mites with identical appearance cause **mange** in animals. Some mange mites can infect children, but the rash produced usually heals spontaneously.

Other Diseases

Other animal mites can act as vectors of infectious agents. In North America the most important mite-borne disease is probably **rickettsial pox,** caused by a species of rickettsiae that infects mice. The disease is characteristically mild and often does not require antibiotic treatment. Epidemics occur periodically in Eastern cities, especially New York. Much more serious rickettsial diseases are transmitted by mites in other parts of the world.

NEMATODA (ROUNDWORMS)

The roundworms comprise a very large group of organisms, most of which are harmless or beneficial to humans. Several others are parasitic and, although frequently unnoticed, can produce mere annoyance or serious disease. Some parasitic forms are small enough to be carried by insects such as mosquitoes. In general, diagnosis of infestation* depends on microscopic identification of the worms or their eggs (ova) or on blood tests for antibody to the worms.

The following paragraphs give a few examples of human parasites. In Table 28–3, additional features of these wormlike parasites are described. Included in this table are the worms *Ascaris lumbricoides* (Fig. 28–8 and color plate 64), which infects about one of every four people in the world; *Trichuris trichiura* (color plate 65), which infects about 350 million people; and *Wuchereria bancrofti* and *Brugia malayi*, both of which cause **filariasis,** affecting 300 million people.

Hookworm Disease

Hookworms infest 400 million people and are found in warm areas such as the southern part of the United States, Europe, Southeast Asia, and the Far East. The worm is about 10 mm long (Fig. 28–9 and color plate 66) and lives in the small intestine of humans, attaching by means of small hooks or plates located about its mouth. It feeds by sucking the blood of the host. The life cycle of the hookworm is shown in Figure 28–10. More than 1000 worms may be found in a single person, and the constant loss of blood frequently

---

*infestation—word often used in place of "infection" when talking about multicellular parasites living on or in the human body

TABLE 28–3 NEMATODA (ROUNDWORMS)

Disease and causative agent	Entry to body	Laboratory diagnosis	Disease charac-teristics[a]	Environmental control	Drug control
Hookworm disease *Necator americanus* and *Ancylostoma duodenale*	Larvae penetrate bare feet	Microscopic examination of stool for eggs of parasite	Mild symptoms progressing to anemia, weakness, fatigue, and physical and/or mental retardation in children	Sanitary disposal of human feces; wearing shoes	Mebendazole for treatment of *N. americanus*
Ascariasis *Ascaris lumbricoides*	Ingestion of eggs of parasite along with contaminated food or water	Microscopic examination of stool for eggs of parasite	Absent or mild symptoms progressing to live worms vomited or passed in stool	Sanitary disposal of human feces	Piperazine citrate
Enterobiasis (pinworm) *Enterobius vermicularis*	Hand to mouth; inhalation (in heavily contaminated areas); contaminated food or water	Microscopic examination of Scotch tape swab of anal region to identify eggs; female worms found in anal region	Itching of anal region; restlessness, irritability and nervousness due to poor sleep; scratching may lead to secondary infection	Frequent handwashing and showers; daily change of under-clothing and bed sheets; examination of other family members and drug treatment if infected	Pyrantel pamoate
Trichinosis *Trichinella spiralis*	Eating raw or undercooked meat, usually pork	Biopsy of muscle showing organism; blood tests for antibodies	Mild fever progressing to swelling of upper eyelids, muscle soreness and high fever	Federal regulations to assure "adequate processing of pork products"; cooking pork adequately (should reach 66°C)	Thiabendazole; in severe cases, corticosteroids are also used

Table continues on the following page

TABLE 28–3   NEMATODA (ROUNDWORMS) (*Cont.*)

Disease and causative agent	Entry to body	Laboratory diagnosis	Disease characteristics[a]	Environmental control	Drug control
Trichuriasis (whipworm) *Trichuris trichiura*	Ingestion of eggs of parasite along with contaminated food or water	Microscopic examination of stool for eggs of parasite	No symptoms but may progress to abdominal pain, bloody stools, diarrhea, and weight loss	Sanitary disposal of human feces; frequent handwashing	Mebendazole
Strongyloidiasis (threadworm) *Strongyloides stercoralis*	Larvae penetrate bare feet	Larvae in fresh stool	Skin rash at site of penetration; cough; abdominal pains; weight loss	Sanitary disposal of human feces; wearing shoes	Thiabendazole
Filariasis *Wuchereria bancrofti; Brugia malayi*	Bite of infected mosquito	Larvae in blood smear	Fever; swelling of lymph glands, genitals, and extremities	Eliminate breeding places of mosquito	Diethyl-carbamazine

[a]Symptoms may be mild or severe depending on intensity of infection.

produces anemia. Children who acquire hookworms may be undernourished already, and the anemia resulting from hookworm may cause weakness, fatigue, and physical and probably mental retardation. The lifespan of the worm is about six years, but medicine can reduce or eliminate the infestation.

The female worm releases her eggs—perhaps 25 million in her lifetime—and they are discharged with human feces. Thus, hookworm eggs can be found on the ground along with human feces in places where toilets are not common. The eggs hatch, releasing tiny larvae* that develop in the soil and become infectious. These larvae infect by penetration through the skin, generally of the feet. They then enter the bloodstream and are carried with the blood to the heart and lungs. In the lungs, they push through the blood vessel walls into the alveoli* and thence gain entrance to the gastrointestinal system when lower respiratory tract secretions are swallowed. Once in the small intestine, they attach and mature to complete their life cycle.

Diagnosis of hookworm can be made by microscopic examination of fecal material for the eggs (ova) (Fig. 28–11). The four-celled ovum identifies the

---

*larvae (singular: larva)—immature stage in the development of worms, insects, and certain other creatures

*alveoli—air cells of the lung

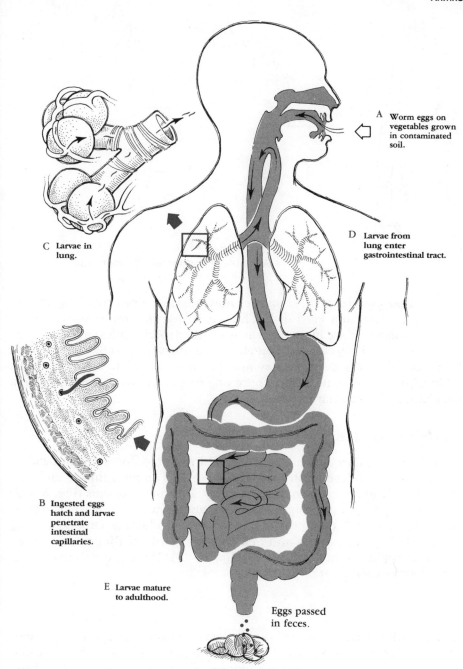

A Worm eggs on
vegetables grown
in contaminated
soil.

C Larvae in
lung.

D Larvae from
lung enter
gastrointestinal tract.

B Ingested eggs
hatch and larvae
penetrate
intestinal
capillaries.

E Larvae mature
to adulthood.

Eggs passed
in feces.

Figure 28–8
Life cycle of *Ascaris lumbricoides,* a roundworm averaging 25 cm in length.

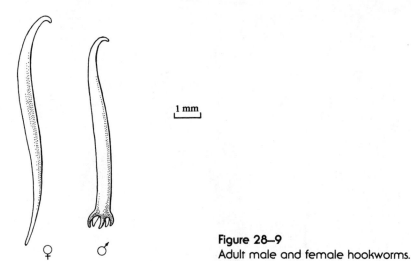

1 mm

Figure 28–9
Adult male and female hookworms.

♀ ♂

worm as a hookworm, and the species may be determined by allowing it to develop and hatch in a Petri dish. The species *Necator americanus* and *Ancylostoma duodenale* are human parasites. Other species infest animals such as dogs and cats. They can penetrate human skin, causing inflammation and itching, but soon die because they are unable to complete their life cycle. Children playing in sandboxes where dog or cat feces are present have a good chance of becoming infected.

Pinworm Disease (Enterobiasis)

Enterobius vermicularis, the pinworm of children, is cosmopolitan and widely known. Infestation develops from ingestion of eggs on dirty fingers and possibly by inhalation of eggs on dust particles. *Enterobius* worms (Fig. 28–12) live in the large intestine, migrating to the anus where they discharge their eggs (color plates 67, 68, and 69). The resulting inflammation and itching often produce sleeplessness and behavioral disorders. Diagnosis is often made from the transparent tape test (Fig. 28–13). There are effective medications for the treatment of pinworms.

Larva Migrans

The developing forms of some roundworms migrate through body tissues and may cause serious disease (Fig. 28–14). In a recent case, a five-year-old boy became blind in one eye after inadvertently eating soil that had been contaminated by dog feces. His disease, **larva migrans,** was due to *Toxocara canis*, an intestinal roundworm of dogs that resembles *Ascaris lumbricoides* of humans. In some parts of the world it is a frequent cause of blindness acquired in childhood. Infestation is the result of ingesting the worm eggs on food or other contaminated material. Preschool children are especially

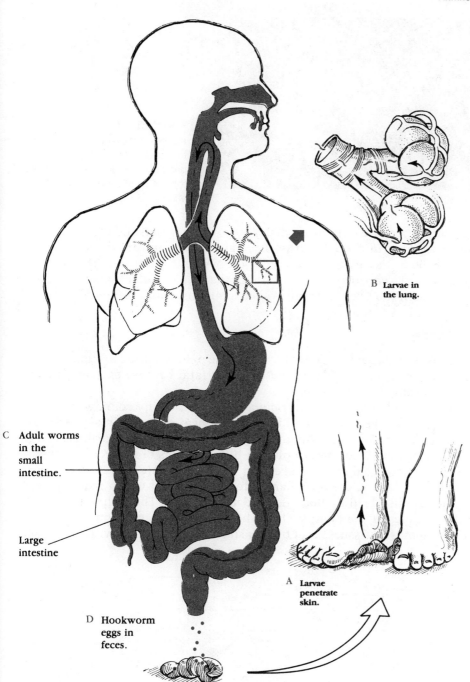

B Larvae in
   the lung.

C Adult worms
   in the
   small
   intestine.

Large
intestine

A Larvae
   penetrate
   skin.

D Hookworm
   eggs in
   feces.

Figure 28–10
Life cycle of the hookworm.

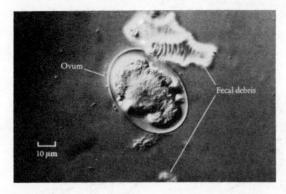

Figure 28–11
Hookworm ovum. The egg is in the four-celled stage character-istically found in the feces. (Courtesy of F. Schoenknecht.)

susceptible because of their tendency to put rocks, dirt, and other "interesting" materials in their mouths. The eggs hatch, and the larvae migrate through body tissues and eventually die without maturing. The resulting inflammatory reaction produces damage that is particularly alarming in the eye, although the liver, lung, and other tissues are more commonly affected.

Trichinosis

Trichinosis is characterized by fever, muscle pain, swelling around the eyes, and sometimes a rash. Occasional cases are fatal because of damage to the heart or brain. The disease occurs worldwide.

The cause of trichinosis is *Trichinella spiralis*, a tiny roundworm 1 to 4 mm long that lives in the small intestine of meat-eating animals, especially rats, pigs, bears, dogs, and humans (Fig. 28–15). The female worm discharges her living young into the lymph and blood vessels of the host's intestine without an intervening egg stage, and these larvae are carried to all parts of the body. They cause an inflammatory reaction in which many of them die and are dissolved by body defense mechanisms. Those that arrive alive in the muscles of the host, such as the heart and diaphragm, often lodge there. The muscle tissues react by encasing these organisms with scar tissue.

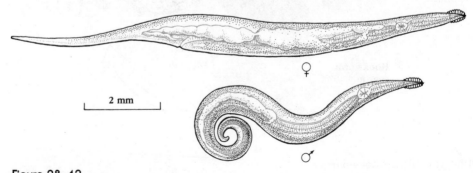

Figure 28–12
Male and female pinworms.

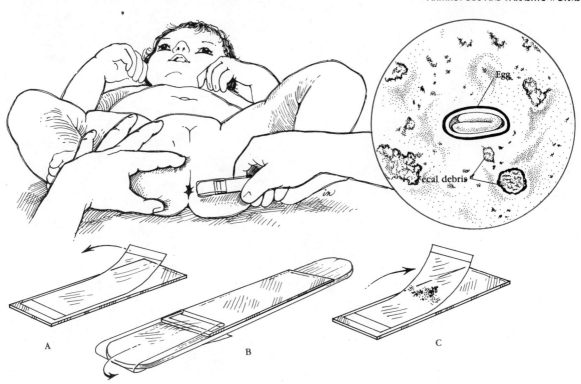

Egg

Fecal debris

A

B

C

Figure 28–13
Diagnosis of pinworm infestation by the use of transparent tape.

The worms then stay alive for months or years within the muscle. If the flesh of an infested pig or other animal host is eaten by humans or animals, the digestive juices of the new host release the larvae from their cases, permitting them to burrow into the new host's intestinal lining. There they mature, and the females begin producing larvae in the new host to complete the life cycle of the worm. Each female adult *Trichinella* may live four months or more and produce 1500 young. The penetration of these young worms into the tissues of the host's body is responsible for the symptoms of trichinosis. Recently, Alaskan black bear meat was responsible for 27 cases of trichinosis. Of those eating the improperly cooked meat, 90 percent had evidence of the disease. Generally, 50 to 250 cases of trichinosis per year are reported in the United States. Diagnosis is made by biopsy and by testing the patient's blood for antibodies to the worms.

There is no satisfactory treatment for severe trichinosis. Prevention of trichinosis in humans depends on adequate and thorough cooking of meat so that all parts reach at least 150°F. Pork has been the chief offender, presumably because pigs were fed uncooked garbage containing meat scraps, a practice now prohibited by law. However, beef ground in a machine previously used for pork has also resulted in cases of trichinosis, and bear meat

Figure 28–14
Cutaneous larva migrans. The larvae penetrate the skin directly and wander in the subcutaneous tissue for a time before dying.

is a notorious cause. Government inspection of meats does not detect *Trichinella*-infested meats, and adequate destruction of the worm larvae by freezing, while possible, is often impractical because of the low temperatures and long time required.

Strongyloidiasis

Infestation with *Strongyloides stercoralis* is uncommon in the United States, although the organism is endemic in some rural areas of the South. Generally, patients have either no symptoms or symptoms resembling those of

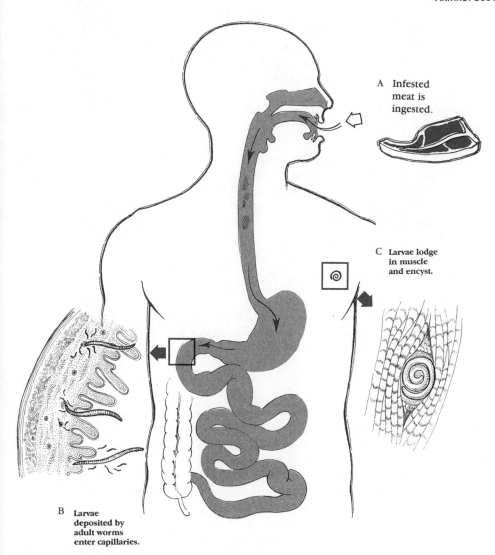

A Infested
   meat is
   ingested.

C Larvae lodge
   in muscle
   and encyst.

B Larvae
   deposited by
   adult worms
   enter capillaries.

Figure 28–15
Life cycle of *Trichinella spiralis.*

hookworm disease. Unlike hookworm, the eggs of *S. stercoralis* hatch while they are still in the intestine, and the larvae are passed with the feces (color plates 70 and 71). In soil they may mature and reproduce continuously without the human host. However, at times infectious larvae are produced that can penetrate the human skin. They then find their way to the lung and intestine, as do hookworm and *Ascaris.* Unfortunately, larvae hatching in the intestine can penetrate the wall of the intestine or perianal* skin, enter blood

*perianal—around the anus

vessels, and follow the same route to the intestine as those soil larva that penetrate the skin. Thus, *S. stercoralis* can perpetuate itself in the human host without an intervening sojourn to the soil. As a consequence, infestations can last indefinitely and pose a potential threat of fatal illness that is often totally unsuspected. These fatalities occur when the host's immunity is compromised by disease or medication such as cancer chemotherapy, whereupon great numbers of the larva enter the bloodstream, causing shock and secondary invasion by intestinal bacteria.

Diagnosis of strongyloidiasis is generally accomplished by finding the larvae in specimens of feces (the eggs are not found in feces). It is often necessary to examine several specimens of feces before finding the larvae. Even then, about one of every five cases of strongyloidiasis will fail to show fecal larvae.

Treatment with thiabendazole is usually effective in eliminating or reducing the infestation.

PLATYHELMINTHES (FLATWORMS)

Flatworm parasites of humans fall into two major groups: **flukes,** which are relatively short, flat, bilaterally symmetrical worms that generally attach by one or more sucking discs, and **tapeworms,** which are usually longer than flukes and ribbonlike in appearance.

Schistosomiasis

Schistosomiasis is one of the most common of the diseases caused by parasitic worms; its worldwide incidence is estimated to be more than 200 million cases. Besides being widespread in Africa and Asia, in the Western Hemisphere the disease is common in the Caribbean area and South America. Approximately 400,000 infected people live in the United States, having emigrated from Latin America and other areas of the world where the disease is endemic. The disease has been on the increase worldwide because of the building of new dams and irrigation projects in developing countries; these constructions provide ideal breeding places for the snail host of the parasite. For example, in the three years after the building of the Kainji dam in the northern part of Nigeria, the number of people living in the area and infected with *Schistosoma mansoni* doubled. People living in the United States are not at risk because the specific species of snail host required by the fluke does not exist in this country.

Schistosomiasis is a disease caused by flukes of the genus *Schistosoma*. It is often a chronic, slowly progressing illness resulting in damage and loss of function of the liver, with resulting malnutrition, weakness, and accumulation of excess fluid in the abdominal cavity.

Schistosoma mansoni (Fig. 28–16), a common cause of schistosomiasis, is about 10 mm long and lives in the small veins of the human intestinal wall. Its life cycle is depicted in Figure 28–17. The female worm discharges eggs (color plate 72), some of which are carried into the liver by the flow of blood from the intestine, while others rupture through the blood vessels and ad-

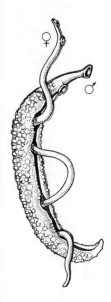

Figure 28–16
Adult *Schistosoma mansoni.*

jacent intestinal lining to enter the feces. The major symptoms of the disease are caused by the eggs, many of which lodge in the liver, producing inflammation and scarring and eventual loss of liver function.

The infectious forms (cercariae*) of *S. mansoni* (color plate 73) burrow through the skin of people who wade in infested waters. The worms then penetrate to the blood vessels beneath the skin and leave their tails at the body surface. The worms are carried by the blood flow to the intestinal veins, where they become mature and complete their life cycle. The worms can live for more than 25 years and continue to produce eggs and consequent liver damage.

Measures used in the control of schistosomiasis include drug treatment of infected individuals, sanitary disposal of feces, and chemical and biological control of snail hosts.

The disease "swimmers' itch," common in parts of North America and other areas of the world, is caused by the cercariae of schistosomes of birds and other animals. These cercariae penetrate the skin and then die, since they are unable to complete their life cycle in humans. In their short duration under the skin, however, they create an irritating itch, especially on fingers and toes.

Tapeworm Infestation

People become infested with tapeworms mainly by eating infested, inadequately cooked beef, pork, or freshwater fish. The adult worms live by ab-

---

*cercariae—plural of cercaria, the fork-tailed larval form of *Schistosoma* species; the developmental stage of schistosomes that penetrates human skin

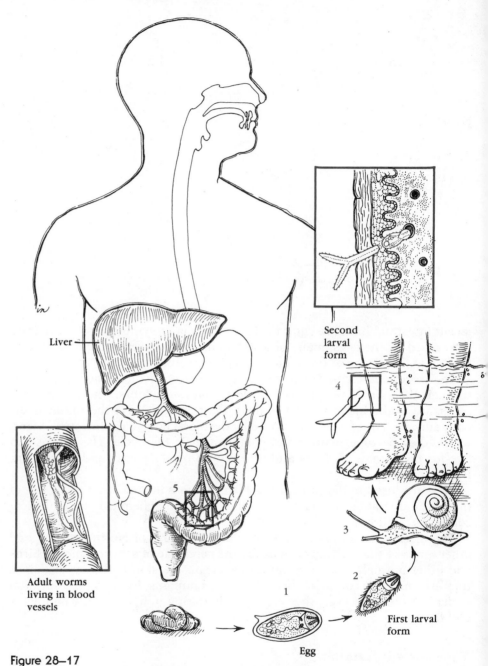

Liver

Second
larval
form

Adult worms
living in blood
vessels

First larval
form

Egg

Figure 28–17

Life cycle of *Schistosoma mansoni*. (1) Eggs deposited in the feces in fresh water. (2) Larval form living free in fresh water. (3) Larva enters body of a snail and multiplies. (4) Infectious form released from the snail penetrates the human skin. (5) Mature adults develop in the blood vessels draining the large intestine, and eggs are deposited. (6) While many eggs penetrate the intestine and enter the feces, some are carried by the bloodstream to the liver, where they cause inflammation and produce extensive scarring.

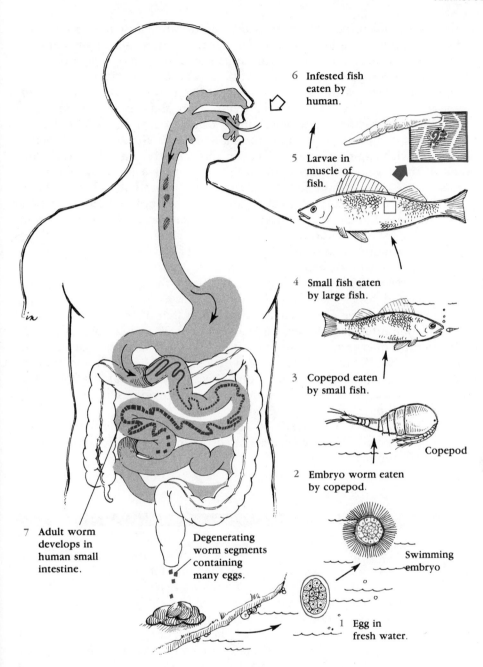

6 Infested fish eaten by human.

5 Larvae in muscle of fish.

4 Small fish eaten by large fish.

3 Copepod eaten by small fish.

Copepod

2 Embryo worm eaten by copepod.

Swimming embryo

7 Adult worm develops in human small intestine.

Degenerating worm segments containing many eggs.

1 Egg in fresh water.

Figure 28–18
Life cycle of *Diphyllobothrium latum.* It may infest people who eat raw freshwater fish. Note that three or more intermediate hosts are involved.

sorbing juices from the human intestine and usually cause few or no symptoms. However, some large tapeworm species sometimes absorb enough vitamin B_{12} to cause anemia, and in unusual circumstances, tapeworm developmental stages can invade human tissue and cause serious disease.

Figure 28–18 gives the life cycle of the fish tapeworm *Diphyllobothrium latum*, and color plate 74 shows the ovum. Note that, as with many other multicellular parasites, the earlier stages of tapeworm development occur in **intermediate hosts** (for example, the snail and fish with *D. latum*), whereas adulthood with egg production occurs in the **definitive host** (for example, a human). The developing stages of *D. latum* are invasive and grow within the tissues of the intermediate host, while the adult simply attaches to the intestinal lining of the definitive host. Humans can sometimes become infected by the intermediate developmental forms of tapeworms, producing serious disease. This usually occurs with accidental ingestion of the tapeworm eggs but may also happen as a result of the practice of applying raw meat to open wounds in the mistaken belief that the meat will aid recovery. If tapeworm larvae are present in the meat, they may penetrate the tissues via the wound.

The beef and pork tapeworms, *Taenia saginata* and *T. solium* (color plate 75), respectively, have a simpler life cycle than that of *D. latum*, since they have only one intermediate host (the cow or pig). The three species are distinguished by microscopic examination of the worm segments or eggs discharged in feces. *T. solium* is fortunately rare in the United States. It is the most dangerous of the three because on occasion the eggs hatch before being discharged in the feces. When this happens, the emerging larvae invade the various organs of the body, including the brain.

Control of tapeworms depends on adequate cooking of meat and fish and on proper disposal of human feces. Effective medicines are available for treating infested people.

SUMMARY

The importance of arthropod parasites lies mainly in their susceptibility to microorganisms and viruses that are pathogenic for humans. Mosquitoes may be infected with such agents as *Plasmodium* species (which cause malaria) and viruses responsible for yellow fever and equine encephalitis. Certain fleas may carry the enterobacterium *Yersinia pestis*, the cause of plague, and lice may carry *Rickettsia prowazekii*, the cause of typhus. Arachnids such as ticks may also be hosts for microbial pathogens including *Rickettsia rickettsii* (the cause of Rocky Mountain spotted fever), *Coxiella burnetii* (the cause of Q fever), and *Francisella tularensis* (the cause of tularemia). The mite *Sarcoptes scabiei* is responsible for scabies. Other mites are vectors of rickettsial diseases such as rickettsial pox.

In some instances, maintenance of an arthropod-borne microorganism requires a vertebrate host, whereas in other cases the microorganism can be passed transovarially from generation to generation of the arthropod host. Some arthropods tolerate their infection without apparent harm.

The wormlike multicellular parasites can cause serious disease. Some examples of such pathogenic organisms are *Necator, Ancylostoma, Enterobius, Toxocara, Trichinella* and *Strongyloides* (roundworms), and *Schistosoma, Diphyllobothrium,* and *Taenia* (flatworms).

SELF-QUIZ

1. Control of malaria is more difficult because of
   a. inability to detect *P. vivax* in stained blood smears
   b. mosquitoes resistant to insecticides
   c. malaria protozoa resistant to antimalarial medicines
   d. b and c
   e. all of the above
2. Human malarial parasites are transmitted directly
   a. from mosquito to human
   b. from human to mosquito
   c. from mosquito to mosquito
   d. all of the above
   e. a and b
3. Which one of the following statements about plague is *false?*
   a. It does not occur in the United States
   b. It can spread by the respiratory route.
   c. The rat flea *Xenopsylla cheopsis* is a vector.
   d. The causative organism is a Gram-negative facultatively anaerobic rod.
   e. The epidemic disease was once called the "Black Death."
4. In an area where ticks are common it is important to check your body frequently for the presence of ticks because
   a. prompt removal minimizes the chance of contracting Rocky Mountain spotted fever
   b. tick bites are painful
   c. ticks transmit strongyloidiasis
   d. ticks must feed on humans in order to survive
   e. paralysis is a common consequence of tick bites
5. The causative agent of scabies is a
   a. rickettsia
   b. fluke
   c. mite
   d. buffalo gnat
   e. flea
6. Hookworm disease is acquired through
   a. infection of wounds
   b. inhalation

      c. penetration of larvae through skin
      d. eating undercooked pork
      e. the bite of an arachnid
7. To prevent trichinosis, you should
      a. eat only USDA-inspected meats
      b. avoid eating undercooked meat
      c. wear shoes
      d. boil your drinking water
      e. eat lots of bear meat
8. The organ of the body most damaged by *Schistosoma mansoni* is the
      a. brain
      b. intestine
      c. lung
      d. pancreas
      e. liver

QUESTIONS FOR DISCUSSION

1. Why can chloroquine alone by used to treat congenital malaria without using primaquine?
2. Why are the parasitic diseases called the "cancers of the developing nations?"

FURTHER READING

Benenson, Abram S.: *Control of Communicable Diseases in Man,* 12th ed. An Official Report of the American Public Health Association. Washington, D.C., 1975.

Faust, E. C., Beaver, P. C., and Jung, R. C.: *Animal Agents and Vectors of Human Disease,* 4th ed. Philadelphia: Lea & Febiger, 1975.

Harrison, G.: *Mosquitoes, Malaria and Man: A History of the Hostilities Since 1880.* New York: E. P. Dutton, 1978. Fascinating historical account of the steps in unraveling the mysteries of malaria.

Havenden, H. G.: "Rocky Mountain Spotted Fever." *American Journal of Nursing,* 76:419–421 (1976).

Mahmoud, A.: "Schistosomiasis." *New England Journal of Medicine,* 297:1329–1331 (1977).

Orkin, M., and Maiback, H.: "This Scabies Pandemic." *New England Journal of Medicine,* 298:496–498 (1978).

Report of the Committee on Infectious Diseases. Evanston, Illinois: American Academy of Pediatrics, 1977.

Roberts, A.: "A History of the Plague in England." *Nursing Times,* 75 (May 19, 1979).

Schultz, M. G.: "Parasitic Diseases." *New England Journal of Medicine,* 297:1259–1261 (1977).

Warren, K.S.: "Geographic Medicine in Practice." *Hospital Practice, 15* (January 1980). Impressive electron micrographs of *Schistosoma mansoni.*

Zaman, V.: *Atlas of Medical Parasitology.* Philadelphia: Lea & Febiger, 1979. A color atlas displaying the morphology, life cycles, and clinical aspects of important protozoa, helminths, and arthropods.

Chapter 29

EPIDEMIOLOGY

I have met just as strong a stream of sewer air coming up the back staircase of a grand London house from the sink, as I have ever met at Scutari.[1]

Florence Nightingale

This quotation from Florence Nightingale describes conditions in many London homes in the 1860's. The sanitary disposal of sewage in the town was just being considered, and the typical sewer was simply an elongated cesspool[2] with an overflow at one end. Conditions such as these were common and illustrate the lack of sanitary practices that contributed to the many epidemics of the day.

OBJECTIVES

To know
1. The importance of epidemiology.
2. The factors influencing epidemics.
3. The significance of reservoirs and vectors.
4. How to control epidemic diseases.
5. The importance and control of nosocomial infections.
6. The purpose of the public health system and how it works.

---

[1]Scutari is a city in northwestern Albania. Florence Nightingale was referring to the primitive sanitary conditions that existed during the Crimean War. Quotation from *Notes on Nursing* by F. Nightingale, re-published in 1969 by Dover Publications.
[2]A deep hole in the ground to receive sewage.

595

EPIDEMIOLOGY OF INFECTIOUS DISEASES

Epidemiology is the study of factors that influence the frequency and distribution of disease. Since epidemics are influenced by the nature of the population, one aspect of epidemiology is to determine who is at risk and why. As the population of the world increases, the importance of the epidemiology of microbial infections becomes heightened, not only because people are crowded close together, which facilitates the spread of infectious diseases, but also because of mass production and distribution practices. A single food supplier, for example, may send products not only to a community but also to several states or nations. If the food is contaminated with a pathogenic microorganism, the resulting illness may be widespread. Furthermore, efforts employed in infectious disease control, such as vaccination or chlorination of a water supply, often involve millions of people in a single program. People tend to rely increasingly on these large-scale measures, as well as on a highly organized public health system, to control infectious diseases. This reliance is based on an advanced technological and social organization, and war and natural disasters are likely to disrupt it. Even a single irrational or ignorant act may affect thousands of people.

EPIDEMICS: DETECTION AND GENESIS

An epidemic is a rapidly progressing outbreak of an infectious disease through a population of susceptible individuals. The following sections describe some of the factors involved in the origin, spread, and control of epidemic diseases of microbial origin.

Determining the Cause of an Epidemic

One of the most famous cholera epidemics occurred in London in 1854 because a water source known as the Broad Street pump was contaminated. Dr. John Snow had the pump handle removed after observing that almost everyone with the disease obtained their drinking water from this pump. The epidemic ceased shortly thereafter, and Snow took this as proof of his idea that the Broad Street pump was the source of this epidemic. Other doctors were not convinced, since disease-causing "germs" had not yet been discovered.

In 1892 the people in Hamburg, Germany, used water from the Elbe River without any purification treatment. Right on the border of Hamburg was the town of Altona, a part of Prussia where the government had set up a water filtration plant. Cholera broke out in Hamburg, and cases were found on the Hamburg side of the street, but no cases were found in Altona (Fig. 29–1). A better demonstration of how the *Vibrio cholerae* bacterium was carried in the water supply could not have been planned. Those who had

Figure 29–1
On the left are depicted cholera victims in Hamburg, Germany, in 1892, who used water straight from the Elbe River. On the right is the town of Altona, whose citizens drank the same water after it had been purified.

raised questions about the importance of clean water were silenced, and the water system of Hamburg was redesigned.

One may suspect an epidemic if an unusually large number of cases of a given disease appear within a short time. However, many different agents* may sometimes produce the same disease. In addition, a single agent may sometimes produce a wide variety of diseases. For example, more than 80 different infectious agents can produce the common cold, whereas a single agent such as poliovirus can produce illness ranging from the symptoms of a cold to fatal paralysis. To determine the cause or origin of a particular epidemic, one must first define the characteristics of the people who seem to be part of the epidemic, and then look for activities or other factors they have in common, such as eating the same food, seeing the same doctor, or visiting the same hotel. It is also usually helpful to determine how frequently or infrequently these associations are found in the portion of the population that was not affected by the epidemic.

---

*agents—this convenient term includes both microorganisms and viruses

Identification of Infectious Agents

Verifying the existence of an epidemic sometimes requires exact species identification of the causative infectious agent. In one large hospital, for example, it had been the practice to name a group of lactose-nonfermenting enterobacteria "paracolons." Several enterobacterial species fall into this group and occur in cultures from sick people, and it did not seem particularly remarkable that patients occasionally died with such organisms in their bloodstream. With the introduction of more precise identification techniques in the hospital's microbiology laboratory, however, it became clear that all the paracolons from blood cultures belonged to a single species, *Serratia marcescens*. This finding prompted special investigations and led to the discovery that hospital equipment contaminated with *Serratia* was infecting the patients.

S. pyogenes antigens, page 300

The value to epidemiology of identification of types *within* a species of organisms is shown by another episode. An unusually large number of *Streptococcus pyogenes* infections had appeared over the course of several months, but because they had been reported from different hospitals, among patients attended by many different medical and other personnel, it was not clear whether an epidemic existed. When the surface antigens of the streptococci were identified by using antisera, it was found that about three quarters of the patients had been infected with the same type of *S. pyogenes*, whereas the remainder had been infected by miscellaneous types of the organism. Furthermore, all the patients infected with the first type of streptococcus had been attended by the same physician, who proved to be the disease carrier.

Epidemics due to *Staphylococcus aureus* or *Pseudomonas aeruginosa* are even more difficult to trace than streptococcal epidemics because these two species are so common among the normal body flora and in the environment. Individual strains of staphylococci can be identified by bacteriophage typing in which a set of 20 or more antistaphylococcal bacteriophages is tested against cultures of *S. aureus*. Since different strains of this bacterium are killed by different bacteriophages of the set, it is often possible to distinguish one *S. aureus* strain from another by this technique (Fig. 29–2). Pseudomonas strains can be distinguished by bacteriocin* typing, a technique in which unknown strains are tested for production of bacteriocins active against a known set of *Pseudomonas aeruginosa* strains (Fig. 29–3). Alternatively, the unknown strain can be tested for susceptibility to bacteriocins from known strains. Even greater precision can be obtained by using two or more techniques together, for example, using both antibiotic susceptibility patterns (antibiograms) and bacteriophage typing for *S. aureus* (Table 29–1). The first eight patients listed in Table 29–1 were very likely to have had the same epidemic strain of staphylococcus because all the isolates of the organism had the same antibiogram and the same phage type. Patients 9 and 10 and

---

*bacteriocin—a substance—usually a large protein—produced by one strain of bacterium, which kills certain other strains of that or closely related species

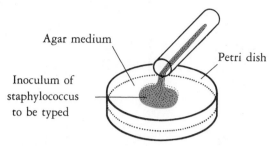

(a) Spreading an inoculum of *S. aureus* over the surface of agar medium.

(b) Dropping bacteriophage suspensions on the inoculated surface. Twenty-three different suspensions are thus deposited in a fixed pattern. After incubation, different patterns of lysis are seen with different strains of *S. aureus*.

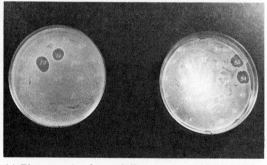

(c) Photograph of two different strains of *S. aureus*, 53/54 and 80/81.

Figure 29–2
Bacteriophage typing of *Staphylococcus aureus.*

patients 11 and 12 were probably not part of the epidemic, since their isolates differed in either phage type or antibiogram.

Strains of bacteria may also differ in their ability to metabolize various chemicals (see Chapter 4). Identification of strains of an organism by patterns of their biochemical activity against a variety of substrates is called **biotyping.** For example, one strain of *Escherichia coli* may ferment the sugar alcohol sorbitol, whereas another strain of *E. coli* will not.

Biochemical tests, page 82

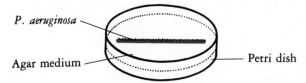

(a) Strain of *P. aeruginosa* inoculated in a straight line and allowed to grow.

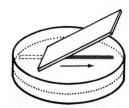

(b) The growth of *P. aeruginosa* is being scraped off the medium. Any remaining bacteria are subsequently killed with chloroform vapor. Bacteriocin that has diffused into the agar from the bacteria remains behind. Known strains of *P. aeruginosa* are then streaked at right angles to the area from which growth has been removed. The known strains are allowed to grow.

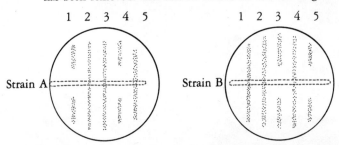

(c) Two different strains of *P. aeruginosa* distinguished by their differing pattern of killing against the same set of five known strains. *P. aeruginosa* strain A kills known cultures 1 and 4; strain B, 3 and 5.

Figure 29–3
Bacteriocin typing of *Pseudomonas aeruginosa.*

Reservoirs of Infectious Microorganisms and Viruses

The reservoir of an infectious agent consists of all the places where the agent is found. For example, the reservoirs for diphtheria, syphilis, and typhoid fever are infected humans; for botulism and coccidioidomycosis, certain soils are the reservoir; and for more than 100 other diseases, infected animals are the reservoirs. The spread of a disease from its reservoirs may involve **vectors** (insects or other animals) or **vehicles** (food or other nonliving materials). Humans may also spread infectious agents through direct person-to-person contact.

TABLE 29–1   USE OF ANTIBIOGRAM AND BACTERIOPHAGE TYPE IN DISTINGUISHING
EPIDEMIC STRAINS OF STAPHYLOCOCCI

| Patient | Staphylococcus | |
	Antibiogram[a]	Bacteriophage type
1	RRSRR	80/81
2	RRSRR	80/81
3	RRSRR	80/81
4	RRSRR	80/81
5	RRSRR	80/81
6	RRSRR	80/81
7	RRSRR	80/81
8	RRSRR	80/81
9	RRSRR	54/83
10	RRSRR	54/83
11	RSSSR	80/81
12	RSSSS	80/81

[a]R = resistant; S = sensitive to the antibiotics penicillin, tetracycline, chloramphenicol, erythromycin, and streptomycin, respectively.

When infectious agents spread from sources relatively far from inhabited areas, it is sometimes possible to detect their spread before they involve humans significantly. Encephalitis viruses of wild birds and mosquitoes, for example, may be detected by placing a "sentinel" (such as a chicken) in a cage and exposing it to the mosquito vectors. The sentinel animals can then be tested periodically to see whether the encephalitis arbovirus has infected them. The use of sentinels is especially helpful in such epidemic diseases as encephalitis, because this disease can spread extensively within the human community without being detected, owing to the high ratio of asymptomatic infected individuals to those with overt disease.

The time it takes to detect the spread of an infectious agent is also influenced by the **incubation period** of the disease. Diseases with relatively long incubation periods can spread extensively before the first cases of the disease appear. The **dose** (amount) of the infecting agent received by susceptible individuals is also important in epidemiology, because smaller doses generally result in longer incubation periods and higher percentages of asymptomatic infection.

The extent and speed of the spread of an epidemic are influenced by the **mode of transmission** of the disease. A dramatic example of the importance of this factor was the spread of typhoid fever from a ski resort in Switzerland. As many as 10,000 people had been exposed to drinking water containing small numbers of typhoid bacilli. The long incubation period of the disease allowed widespread dissemination of the causative *Salmonella typhi* organisms by skiers flying home to various parts of the world before they became ill. As a result, more than 430 cases of typhoid fever developed in at least six countries.

Air as a Vehicle of Infectious Agents

People discharge microorganisms into the air by sneezing, coughing, talking, and singing. Many organisms are enclosed within large droplets of saliva or mucus, most of which fall quickly to the ground within a distance of three or four feet from the mouth or nose. These large-sized droplets are particularly important as sources of contamination in crowded locations such as schools and military barracks. Thus, desks or beds in such locations should ideally be spaced more than four feet and preferably eight to ten feet apart to minimize the transfer of infectious agents in this manner.[3] Smaller droplets of mucus or saliva dry quickly, leaving one or two organisms attached to a thin coat of the dried material. These "droplet nuclei" can remain suspended in the air by small air currents for indefinite periods, and because of their small size they may escape the trapping mechanisms of the upper airways and enter the lungs. However, under usual conditions of good ventilation, such droplet nuclei are rapidly diluted by the movements of the air, and only highly infectious agents, such as the viruses of measles and smallpox, are readily disseminated in this manner.

On the other hand, the unreliability of the dilution factor of air under conditions of *inadequate* ventilation is demonstrated by a hospital epidemic of tuberculosis. In this epidemic, a patient with an unsuspected case of tuberculosis had been cared for in a large ward* for two and a half days; during this time, 13 of 44 people present in the ward became infected with the tubercle bacillus. As is often the case, the air conditioning unit in the ward had merely recirculated most of the room's air without filtering it. None of the 13 individuals developed active tuberculosis, since they were identified by skin testing and given antituberculosis treatment soon after tuberculosis was diagnosed in the patient.

Large buildings present special problems in the airborne transmission of disease agents. Respiratory pathogens are disseminated not only as a result of sneezing, coughing, and talking, but people with skin infections such as boils may discharge the pathogens from these lesions as body movement shakes clothing and rubs off skin cells. Passage of flatus* disseminates intestinal species. The number of organisms in the air can be estimated by using machines that propel measured volumes of this air against the surface of a medium contained in Petri dishes. This technique has shown that the number of colonies that develop per cubic foot of air sampled rises in proportion to the number of people in a room. Partly for this reason, most modern public buildings have elaborate ventilation systems that change the air constantly. In many hospitals the air flow can be regulated so that it is supplied to the operating room under a slight pressure, preventing contam-

---

[3]This principle was stated by Florence Nightingale in her book *Notes on Hospitals*, published in the 1800's.

*ward—a section in a hospital such as a wing or, as in this case, a large room with many hospital beds

*flatus—intestinal gas passed from the rectum.

inated air in the corridors from flowing into the room. Microbiology laboratories may be kept under a slight vacuum so that air flows in from the corridors, and microorganisms and viruses cannot escape to other parts of the building. Special contamination problems may arise from the pumping action of elevators and the chimneylike effect of laundry chutes extending from the warm air (and dirty laundry) areas in the basement of a building. Air conditioning systems themselves may be sources of infectious agents, since the locations of their air intakes may be such that they draw in contaminated dust. In addition, molds, actinomycetes, and organisms such as *Legionella pneumophila* might grow in humidifying devices* within the ductwork of an air conditioning system and be distributed throughout a building.

L. pneumophila, page 330

The survival of organisms in air varies greatly with the species of organism and with air conditions such as relative humidity, temperature, and exposure to light. Subdued light promotes survival of microorganisms, but the effects of temperature and humidity vary with the species. Organisms in still air and those in large particles eventually fall to the floors of buildings but are readily resuspended in the air on dust particles. Thus, sweeping is not permitted in most hospitals; floors are mopped with a dampened mop instead. Although the use of ultraviolet light or filtration can markedly reduce the number of viable organisms in air, these methods are generally expensive and unnecessary. Under usual conditions, satisfactory control of airborne infection is achieved by good ventilation and dust control. Filtration and ultraviolet irradiation do, however, have important uses in laboratory work and in some special types of hospital units where the risk of microbial contamination is great.

Water and Food as Vehicles

"The first possibility of rural cleanliness lies in water supply." So said Florence Nightingale to the Medical Officer of Health in England in 1891.[4]

Water has been responsible for numerous epidemics, frequently involving thousands of people. Although agents often infect very large numbers of people, the percentage of those exposed in whom diseases actually do develop is generally low. Moreover, the incubation period of waterborne epidemic diseases is apt to be long. These features of waterborne epidemics generally result from the enormous dilution of the infecting agent by oceans, lakes, and streams. However, infectious agents, like certain other water pollutants, can be concentrated biologically, as with the trapping of waterborne hepatitis virus by oysters and other shellfish.

Attack rates (the number of cases developing per hundred people exposed) of foodborne infections tend to be higher than those of infections borne by water because of the lack of dilution; food may also be a nutrient for the

---

[4]Strauss, M.: *Familiar Medical Quotations.* Boston: Little, Brown, 1968.

*humidifying devices—machines used to add water vapor to the incoming air

infectious agent, allowing its multiplication. A sizeable increase in numbers of the agent may therefore occur before the food is eaten. Feces, directly or indirectly, are the most common origin of water contamination. Foods can be infected at their source (pork with *Trichinella*, eggs with *Salmonella*, milk with *Brucella*) or during their preparation, as by the ameba *Entamoeba histolytica* or by bacteria such as *Salmonella* species and *Staphylococcus aureus*.

Person-to-Person Transmission

Puerperal fever, page 495

In 1848 Ignaz Philipp Semmelweis (see Chapter 24) observed that women who had their babies at the hospital with help from doctors were four times as likely to contract puerperal fever as those who delivered at home and were attended by midwives. This discovery, along with many more years of case documentation, led him to try handwashing with a calcium chloride solution between autopsy and delivery rooms and before a patient's examination. This practice led to a dramatic decrease in the number of infections. Unfortunately, the work of Semmelweis was not taken seriously, and it was not until the discoveries of Pasteur and Koch and the development of the germ theory that infection control practices such as handwashing became routine in hospitals.

Hands transfer disease-producing microorganisms and viruses from one person to another. Handwashing, which is a fairly simple routine involving the physical removal of microorganisms, is the first and most important step in preventing the spread of infection (Fig. 29–4). Reducing the numbers of potentially pathogenic microbes on the hands reduces the possibility of transferring an infectious dose* of an organism. Hands can easily become contaminated from contact with the nose, with areas of skin infection, or with feces. *Shigella dysenteriae* has been shown to contaminate the hands through toilet tissue. **Fomites** are inanimate objects such as furniture, doorknobs, and towels that can transmit infection when they become contaminated by unwashed hands or other items.

S. dysenteriae, page 315

Handwashing involves varying techniques, depending on the task to be carried out. The hand scrub performed by medical personnel before participating in an operation or when working in the newborn nursery, intensive care, or isolation unit takes ten minutes and employs an antimicrobial agent such as chlorhexidine.[5] Even when sterile rubber gloves are going to be worn, a thorough hand scrub with an antimicrobial agent is necessary because of the possibility that these thin gloves may become punctured inadvertently during surgical procedures. A lesser amount of time combined with more frequent handwashing with plain soap is required for those involved in lab-

---

[5]Effective against both Gram-negative and Gram-positive organisms.

*infectious dose—a number of microorganisms or viruses sufficient to establish an infection; varies from a few to several million, depending on the virulence of the infectious agent and on host defenses

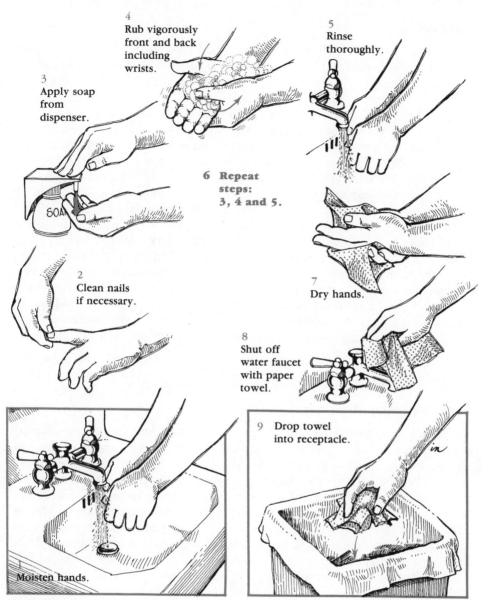

3
Apply soap
from
dispenser.

4
Rub vigorously
front and back
including
wrists.

5
Rinse
thoroughly.

6 **Repeat
steps:
3, 4 and 5.**

2
Clean nails
if necessary.

7
Dry hands.

8
Shut off
water faucet
with paper
towel.

9 Drop towel
into receptacle.

Moisten hands.

Figure 29–4
Handwashing technique used by medical personnel to minimize the spread of infection.

oratory work, dental work, or routine patient care. Using a manicure stick
to remove dirt from around the fingernails and employing a soft brush are
often necessary. A 30-second handwashing markedly decreases the number
of pathogens such as *S. aureus*, pseudomonads, *E. coli*, and the common cold
viruses, so that transmission of an infectious dose to others becomes unlikely.

Carriers

A carrier is an apparently healthy individual who harbors a potential pathogen and can spread it to the environment or to other susceptible individuals. In contrast to those who are ill, these people, the unsuspected disseminators of epidemic agents, move freely about their daily activities. Among the population of carriers are people still in the incubation period of a disease who have not yet become symptomatic, those who have recovered from an illness but continue to excrete a pathogen, and transient and chronic carriers who are colonized by a potential pathogen without ever having been ill. With many diseases (poliomyelitis is an example), the infectious agent continues to be present in body secretions during convalescence and gradually decreases in concentration over periods ranging from days to months. **Transient carriers** of infectious diseases are those who are contaminated with an agent for only a brief time, whereas **chronic carriers** continue to excrete an infectious agent constantly or intermittently for months, years, or even a lifetime. After a disease such as typhoid fever has been controlled in a population, for example, the presence of chronic carriers poses a threat to its recurrence that continues for many years. Under conditions of good sanitation, where spread of disease is difficult, deaths from arteriosclerosis* and other diseases of advancing age may gradually eliminate chronic carriers.

FACTORS INFLUENCING THE CHARACTER OF EPIDEMICS

Acquired Immunity

Influenza, page 450

The attack rate (percentage of the population that becomes ill or infected) of an epidemic disease agent is influenced by previous exposure of the population to that agent or to agents closely related to it. Influenza, for example, is unlikely to spread very widely in a population in which 90 percent of the people have previously been infected by the same agent and have acquired antibody to it. A strictly human disease such as influenza requires a continuous source of susceptible hosts, or else it disappears from a population. Resistance to epidemic disease may also be acquired by exposure to related infectious agents, as exposure to cowpox virus protected the milkmaids of Jenner's time against smallpox. Partly because of acquired immunity, a higher percentage of old people in a population gives a better chance of low disease rates with many infectious agents, because older people have acquired active immunity through exposure to a disease agent in years past and thus act as barriers to the spread of the agent. A high proportion of young children, on the other hand, may provide a better chance of epidemic spread, although some maternal antibodies cross the placenta, giving protection to an infant for the first three to six months of its life. This protection of an infant depends on the immune status of its mother and the nature of

---

*arteriosclerosis—hardening of the arteries

the infectious agent; antibodies protective against *Bordetella pertussis* (the cause of whooping cough), for example, do not cross the placenta. The persistence of acquired immunity to an infectious disease also varies greatly with the nature of its causative agent. The protection may be lifelong for natural infections with measles virus or may last only about two years with a cold virus (rhinovirus).

B. pertussis, pages 313 and 446

Genetic Background

Natural immunity varies with genetic background. Differences in the incidence of epidemic disease among different racial groups are often pronounced, but it is frequently difficult to determine the relative importance of genetic, cultural, and environmental factors to these differences. In a few instances the genetic basis for resistance to infectious diseases is known. For example, 70 percent of black people lack the erythrocytic antigen required for attachment by *Plasmodium vivax* and thus are immune to malaria caused by these protozoa. One type of schistosomiasis also depends on the presence of an inherited tissue antigen, meaning that only those born with the antigen are likely to suffer this form of schistosomiasis.

Malaria, page 565

Environmental and Cultural Factors

Malnutrition, overcrowding, and fatigue foster susceptibility to infectious diseases and their spread. Infectious diseases have generally been more of a problem in the poorer areas of a country where individuals are crowded together without proper food or sanitation. In the 1890's, people were fascinated with a new idea, "the germ theory of disease." The middle and upper classes of society, observing that the slums were filled with sickness, avoided these areas as much as possible. In fact, all public places were viewed with suspicion, and fears of contracting diseases were very real. Popular magazine articles from 1900 to 1904 included "Menace of the Barber Shop," "Disease from Public Laundries," "Infection and Postage Stamps," "Contagion by Telephone," and "Books Spread Contagion." But it was not long before people realized that one could not be guaranteed a healthy life by living outside the poorer areas and ignoring them, because pathogenic microorganisms and viruses could readily spread from places like slums where sanitary conditions were neglected. With this incentive, public health crusaders were able to bring about changes resulting in many of the rules and regulations of our present public health system.

The distribution of diseases is also influenced by religious and cultural practices. For example, in the ski resort epidemic of typhoid fever mentioned earlier, cases occurred only in tourists because the local people rarely drink water. Groups of Scandinavian, Japanese, and Jewish people who eat traditional dishes made from raw freshwater fish are prone to acquire tapeworm disease. The incidence of breast-feeding correlates with a lowered attack rate of *E. coli* diarrhea in infants.

E. coli diarrhea, page 476

Virulence of the Epidemic Strain

Meningitis, page 515

Marked variations in the incidence of infectious disease often appear with epidemic spread of different strains of the same microbial species. For instance, the spread of *N. meningitidis* in a population will produce a low incidence of disease at one time, whereas at another time a different strain will produce many cases of meningitis. Similarly, one bacteriophage type of *S. aureus* may spread among hospital personnel and patients, causing little disease, whereas another type of the organism may cause boils, pneumonia, and many infections of surgical wounds. Such differences in the rate and intensity of an infection are probably often due to differences in the virulence of strains within the pathogenic species.

Intensity of Exposure

The development of infection and disease may be low or absent if an individual is exposed only to small numbers of the microbial pathogen. On the other hand, there is probably no infection for which immunity is absolute. An unusually large infectious dose of a pathogen may produce serious or even fatal disease in a person who has immunity to ordinary doses of the pathogen. *Even immunized persons should therefore take precautions to minimize exposure to infectious agents.* This principle is especially important to laboratory workers, who may be exposed to extremely large numbers of pathogenic organisms or viruses as a result of accidents or faulty technique, and to medical workers who attend patients with infectious diseases.

APPROACHES TO CONTROL OF INFECTIOUS DISEASES

Enhancement of Host Resistance

Tuberculosis, page 446

Methods that promote good general health appear to result in increased resistance to many infections (such as tuberculosis), and when infection occurs it is more likely to be asymptomatic or to result in mild disease. Host resistance can also be increased by inducing specific immunity against pathogenic agents by using vaccines and toxoids. In the United States, among children entering kindergarten in 1974, 73 percent had been immunized against poliomyelitis, 85 percent against diphtheria, pertussis, and tetanus, and 72 percent against measles. Thus, large numbers of children were inadequately immunized, mostly because people tend to neglect to get booster injections to maintain their immunity once the possibility of recurrent natural exposure to a pathogen becomes low. For this reason, many states now exclude children from school until they present up-to-date immunization records. This requirement has resulted in present-day immunization rates well above 90 percent for schoolchildren in many areas of the country. Another method of increasing host resistance is passive immunization (administration of gamma globulin to prevent infectious hepatitis, for example). Medications can also temporarily help control epidemic disease, such as in a nursery or hospital. As an example of the latter, individuals exposed to influenza A can be protected by giving amantadine, a medicine that interferes with influenza A virus penetration into host cells.

Amantadine, page 200

Reduction of Reservoirs and Vectors

Control of infectious agents can sometimes be achieved most readily by attacking their reservoir. In the case of *Plasmodium vivax* (a cause of malaria) or the encephalitis viruses, reduction of the mosquito host population reduces both the reservoir and the vector. Searching out and treating patients with malaria also decreases the reservoir. In the case of *Schistosoma* species (parasitic worms), treating patients and poisoning the intermediate host snail with copper sulfate help to decrease the reservoir. Carriers of *Salmonella typhi* can be detected by screening populations to see who might have asymptomatic infection revealed by the presence of antibodies to the Vi capsular antigen of this organism. Instituting treatment for those proved by cultures to be carriers of the disease further limits the reservoir. Control of tuberculosis can similarly be achieved through identification of cases by the combined use of x-rays and skin tests, followed by medical treatment. Modern processing of sewage eliminates infectious agents, thereby rendering it unlikely as a source of epidemic disease. Purification of water sources by filtration and chlorination and destruction of foods or other materials containing pathogenic agents are also highly effective.

Isolation Procedures in Hospitals

Hospitals in the late 1800's placed all patients with communicable diseases in an area away from others in the hospital. Cross-infection became common, however, since these patients were not separated from one another, nor were aseptic procedures practiced, such as washing one's hands after caring for one patient and before going on to the next. In 1889 conditions were slightly improved by grouping together people with the same diseases in separate wards or hospitals and by practicing the aseptic procedures that were by then being published in nursing textbooks. Another advance occurred in 1910, when individual patients in multiple-bed rooms were isolated from one another in cubicles.* Patients were treated as if they each had separate rooms, with hospital personnel changing gowns between patients, washing their hands between patient contacts, and disinfecting contaminated articles.

In the past 30 years, the contagious disease areas of hospitals have been converted to accept all types of patients. Now there is no longer a need for separate hospitals or a contagious disease area of a general hospital, and patients with communicable illnesses are cared for along with others in the general hospital with appropriate isolation procedures (Appendix IX). Some

---

*cubicles—small compartments

hospitals have enough private rooms to enable isolation patients to have their own room, while some older hospitals still use the cubicle system in multiple-bed rooms.

It is the responsibility of the infection control nurse and committee in each hospital to develop and recommend isolation policies and procedures and to see that these are reviewed and modified as often as necessary. All medical personnel are instructed in the importance of following these procedures when in contact with patients. The patients are also told why special precautions are being taken and how long they may last.

Isolation procedures generally include putting on a mask, gloves, and hospital gown, time-consuming measures which require planning ahead when added to other nursing and medical tasks. As a result, doctors and nurses usually spend less time with the patients that are isolated than with others. Thus such patients often have a real sense of being "isolated," both physically and psychologically. For this reason, patients are placed in isolation only when necessary and are removed from isolation as soon as possible.

Nosocomial Infections

A **nosocomial infection** is an infection acquired during hospitalization. It can be acquired as a result of diagnostic or therapeutic procedures such as catheterization of the bladder or a blood vessel, or surgery. The pathogens may come from the hospital environment, from contact with medical personnel, or from the patient's own normal flora. The illness following the infection may range from being mild enough to go unnoticed to being fatal and, because of the incubation period, may appear only after the patient is discharged from the hospital. Many factors determine which microorganism or virus will be responsible for the infection, including the virulence and number of organisms, the length of time the person is exposed to the microorganism, and the state of the patient's host defenses. It is estimated that 2 to 10 percent of all hospitalized patients in this country acquire a nosocomial infection. This amounts to roughly one million patients each year in the United States and results in six billion dollars in additional hospital costs! In fact, these infections make up at least half of all the cases of infectious diseases seen in the hospitals of technologically advanced countries.

Nosocomial infections have been a problem since hospitals began, but they have only recently been recognized and dealt with scientifically. A hospital can be regarded as a high-population-density community made up of unusually susceptible people, into which the most virulent microbial pathogens of a region are continually introduced. One of the differences in hospitals today compared with the first hospitals is the extensive use of antimicrobial medications. This use results in the selection of those microbial strains that resist antibiotic therapy. Frequently, members of the hospital staff become carriers of these strains. Furthermore, interference with the normal host defenses of patients is a common occurrence in hospitals. For example, surgery causes a breach of the skin barrier, allowing opportunistic pathogens to reach underlying tissues. Antibiotics suppress the normal microbial flora, allowing colonization by resistant pathogens, while certain medications impair the inflammatory and immune responses.

The increasing problem of nosocomial infections led to the requirement, starting in 1976, that all hospitals desiring accreditation by the Joint Commission on Accreditation of Hospitals hire an individual to be responsible for identifying and controlling nosocomial infections. This individual is usually a registered nurse known as a nurse epidemiologist or infection control nurse. The nurse epidemiologist generally reports to an infection control committee composed of representatives of the various professionals in the hospital who are concerned with infection control. These include nurses, doctors, dietitians, engineers, and housekeeping and microbiology laboratory personnel.

Some of the usual responsibilities of the nurse epidemiologist are

1. Reviewing results of cultures from hospitalized patients.
2. Checking admissions records for infections acquired in the community that might be transmitted to other patients.
3. Seeing that proper precautions and isolation methods are being used with suspected and known infections in hospital patients.
4. Teaching patients, their families, and hospital staff the proper measures for control of the spread of infectious diseases.
5. Conducting an infection control orientation program for new employees.
6. Checking any infections that could have originated from hospital staff.
7. Reporting infections to the regional health department.
8. Being in contact with supervisors in departments such as housekeeping, dietary, and x-ray.
9. Investigating outbreaks of infections in the hospital to determine their source, mode of spread, and effective control measures.
10. Preparing and presenting information for the infection control committee, including (a) community-acquired infections that might threaten to spread through the hospital, (b) nosocomial infections—numbers, causes, locations in the hospital, types of infection (wound, respiratory, septicemia, and so on), (c) potential problems such as faulty ventilation or disposal of contaminated materials, and (d) new information on infection control from other hospitals and state and national agencies.

The Public Health Network

Infectious disease control nationwide depends heavily on a network of people and agencies across the country that watch for developments that might influence the spread of infectious diseases.

Each state has a public health laboratory that is involved in infection surveillance and control as well as other health-related activities. The activities of the communicable disease section of a state laboratory often include specialized laboratory services for the examination of specimens or cultures submitted by physicians, local health departments, hospital laboratories, epidemiologists, sanitarians, and others. Examples of these laboratory services are

1. Enteric pathogen laboratory: isolation and identification of *Salmonella*, *Shigella*, *Yersinia*, *Campylobacter*, *Vibrio*, and other species.
2. Reference laboratory: identification or confirmation of identification of pure cultures submitted by personnel from other laboratories who are unable to be certain about the identity of a microorganism.
3. Serology laboratory: testing for antibody against syphilis, brucellosis, leptospirosis, or tularemia in specimens of serum from patients.
4. Virology-rickettsiology laboratory: testing for poliomyelitis, influenza, rickettsiae, and chlamydiae.
5. Fluorescence microscopy laboratory: identification of certain species, including *Legionella pneumophila*, by fluorescent antibody tests.
6. Parasitology laboratory: identification of parasitic protozoa, worms, and others.
7. Mycobacteria and fungus laboratory: identification of mycobacteria and testing of their antibiotic sensitivity; identification of pathogenic fungi.
8. Food microbiology laboratory: examination of foods for food-poisoning bacteria, botulism toxin, and algal (*Gonyaulax*) toxins of shellfish.

Gonyaulax species, page 367

Results from these laboratories that are of public health significance are shared with other public health agencies in the state and also with the National Center for Disease Control and the World Health Organization.

Other sections of a state public health laboratory deal with environmental health matters, providing such services as the testing of water supplies for the presence of pathogens, advice on laboratory safety and design, and assistance in handling outbreaks of infectious diseases. Other programs enhance the reliability of laboratories in the state, ensuring that they meet standards for certification and sending them cultures of known pathogens to see whether they can identify the cultures properly.

The National Center for Disease Control is part of the U.S. Department of Health, Education and Welfare and is located in Atlanta, Georgia. It provides support for many national laboratories as well as several in other countries and collects data on diseases of public health importance. The number of new cases of tuberculosis, hepatitis, venereal disease, rabies, chickenpox, measles, mumps, malaria, and a number of others is collected each week and published in a pamphlet called *Morbidity and Mortality Weekly Report*. The Center also conducts research relating to infectious diseases and can dispatch teams to various states or to other parts of the world to assist with controlling epidemics. The Center also works with state laboratories to provide refresher courses which update the knowledge of laboratory and infection control personnel.

SUMMARY

Epidemiology is the study of factors that influence the frequency and distribution of disease. As the population of the world grows and people depend more and more on mass production and distribution of foods and other substances, large-scale public health measures will become even more important to protect our health.

Epidemic spread of microorganisms and viruses that cause infectious diseases is not always easy to detect, and precise identification of the causative agent is usually very important. Knowledge of the reservoirs that are the sources of infectious agents is often vital to effective control. Vectors, air, water, food, and people are the principal ways in which infectious diseases are transmitted. Carriers of infectious disease agents are very important and often must be identified in order to control epidemics. Control of infectious disease involves reducing the reservoir, interfering with transmission by practicing good aseptic technique, protecting susceptible hosts by increasing their resistance through vaccines, and isolating those people found to be infectious. Nosocomial infections are recognized as problems that can be prevented or solved when approached scientifically. The public health system, with its many agencies, is an effective and important contributor to the good health of a community.

SELF-QUIZ

1. Which of these statements regarding epidemiology is *false?*
   a. Epidemiology is the study of factors that influence the frequency and distribution of disease.
   b. In trying to identify the causes of an epidemic, it is generally important to identify the agent as to species and even strains within the species.
   c. During an epidemic, one can assume all cases of a disease are caused by the same pathogen.
   d. We are more vulnerable to epidemics during times of social disorganization.
   e. Epidemics can originate from pathogens in water or food.
2. A "sentinel" is used
   a. to define the reservoir of a pathogenic microorganism
   b. in the laboratory to determine different types within species
   c. for the early detection of encephalitis viruses in a community
   d. as an alarm to detect pathogens in a defective air conditioning humidifier
   e. to prevent the escape of nosocomial infections into the community
3. The number of cases of symptomatic foodborne infections is generally greater than for waterborne infections because

     a. people have more exposure to food than water

     b. more virulent organisms live in food

     c. the infecting dose is higher with food because pathogens grow in it

     d. a small infecting dose causes more symptomatic infections than a large infecting dose

     e. the toxins of waterborne pathogens are inactivated by water

4. Which of these statements regarding handwashing is *true?*

     a. A disinfectant is essential for handwashing soap in hospitals.

     b. Since antibiotics are so plentiful, you do not have to be careful in washing your hands.

     c. Surgeons wear sterile rubber gloves and therefore do not have to wash their hands before surgery.

     d. Handwashing is an important means of preventing spread of infectious agents in hospitals.

     e. Handwashing for 10 minutes reliably removes all pathogens.

5. A nosocomial infection is

     a. one that can reliably be prevented when proper precautions are taken

     b. one that cannot arise from the patient's own normal flora

     c. reliably considered to be bacterial

     d. one that is acquired during a stay in the hospital

     e. an infection caused by a fungus of the genus *Nosocom*

6. The scope of an epidemic can be influenced by

     a. the nature of the vector or vehicle

     b. the incubation period of the infectious agent

     c. the infecting dose

     d. the age and personal habits of the population involved

     e. all of the above

7. In characterizing an epidemic strain of microorganisms, the following can be useful:

     a. antibiotic sensitivity testing

     b. bacteriophage testing

     c. testing with antibodies (serotyping)

     d. biochemical tests (biotyping)

     e. all of the above

8. Factors promoting epidemics in hospitals include

     a. presence of virulent pathogens

     b. patients with weakened host defenses

     c. equipment and procedures that bypass host defenses

     d. all of the above

     e. none of the above

QUESTIONS FOR DISCUSSION

1. Is epidemic disease occurring in a Laotian village of any concern to us?

2. Smallpox, one of the great epidemic diseases of all time, has now been eliminated from the populations of the world. Do you think we should now destroy all the samples of the smallpox virus that are kept frozen in various laboratories?

3. How would you go about choosing the next epidemic disease to be eliminated from the planet?

FURTHER READING

Barrett-Conner, E., Brandt, S. L., Simon, H. L., and Dechairo, D. C.: *Epidemiology for the Infection Control Nurse.* St. Louis: C. V. Mosby Co., 1978.

Bennett, J. V., and Brachman, P. S.: *Hospital Infections.* Boston: Little, Brown & Co., 1979.

Ehrenreich, B., and English, D.: *Complaints and Disorders: The Sexual Politics of Sickness.* Glass Mountain Pamphlet No. 2. Old Westbury, New York: The Feminist Press, 1973.

Infection Control in the Hospital, 4th ed., Chicago: American Hospital Association, 1979.

McNeill, W. H.: *Plagues and Peoples.* Garden City, New York: Anchor Press/Doubleday, 1976.

Nadolny, M. D.: "Infection Control in Hospitals." *American Journal of Nursing* 80:430–434 (1980).

Nightingale, F.: *Notes on Nursing.* New York: Dover Publications, 1969 (unabridged republication of first American edition as published by D. Appleton & Co., 1860).

Reinary, J. A.: "Nosocomial Infections." *Clinical Symposia—Ciba, 30* (1978).

Part V

APPLIED MICROBIOLOGY

Chapter 30

MICROORGANISMS IN THE ENVIRONMENT

It is Sunday morning. The sun is up, and the tide will be at one of its lowest points for the year. You can practically taste the steamed clams and clam chowder that you will make from the fresh clams you hope to get at the beach. As you dress, you switch on the radio to listen to the news. The announcer says: "All beaches in the Puget Sound area will be closed because of 'red tide.'" There will be no fresh clams today.

Red tide is caused by certain red-colored species of algae within the Dinoflagellate group. They can be so numerous as to actually cause the water to appear red. These Dinoflagellates, such as *Gonyaulax catenella,* produce a toxin causing paralysis and sometimes death in humans who eat shellfish that have been feeding on these organisms. The toxin is not destroyed by cooking. The shellfish that ingest and concentrate these organisms from the plankton are not affected by the poison.

OBJECTIVES

To know
1. The functions of some important beneficial microorganisms in the soil.
2. Some plant pathogens in the soil and how they are transmitted.
3. Some human pathogens that are soil microorganisms.
4. The meaning and importance of biomagnification.
5. The roles of various microorganisms that inhabit water.
6. Some important human pathogens transmitted via water.
7. The methods generally used to detect fecal contamination of water.
8. Examples of human poisoning resulting from toxins produced by aquatic microorganisms or by their metabolic activity.
9. The principles of metropolitan waste treatment.

Soil and water provide ideal conditions for the growth of many species of microorganisms, including a wide variety that are pathogenic for plants, animals, and humans. Many of these pathogens can be disseminated through the air. The control of pathogens and other microorganisms in the environment is a full-time occupation for many hundreds of thousands of people, such as those whose jobs involve waste disposal and food preparation. Indeed, this activity is a necessary part of everyone's daily life. In light of the tremendous expenditure of time and money needed to control certain environmental microorganisms, would it be better to destroy all microorganisms in the environment? Would the world be a better place without so many bacteria, algae, fungi, and protozoa? Consider these questions as you read this chapter.

MICROORGANISMS IN THE SOIL

Beneficial Microorganisms in the Soil

Microorganisms in the soil exist as complex populations made up largely of fungi and bacteria, but algae and protozoa are also plentiful. The nature of the particular soil governs the populations present; the pH, temperature, and amounts of water and nutrients are critical factors. Organic materials within the soil are broken down by microorganisms, allowing their constituent elements, such as nitrogen, carbon, oxygen, phosphorus, and sulfur, to be recycled.

Almost any compound generated by biological processes is biodegradable, that is, capable of being broken down by living microorganisms. The more complex molecules, such as certain ring-shaped (aromatic) chemical compounds, are degraded only by a few species of organisms. Simpler structures can generally be broken down by a large number of different species. Generally speaking, animal flesh is degraded mainly by bacteria, and plant materials by fungi.

Many soil microorganisms play an indispensable role in the cycling of both organic and inorganic compounds. Some of them can fix nitrogen from the air, producing nitrogen-containing nutrients required by plants and animals (Fig. 30–1). Nitrogen is constantly being lost from the soil and must be replaced. The gaseous form, N_2, so abundant in air, cannot be used directly by plants or animals; however, it can be changed into usable forms by either free-living bacteria or bacteria that have a symbiotic relationship with certain plants. The free-living nitrogen-fixers of soil come from many different genera, as indicated in Table 30–1. Members of the genus *Rhizobium* are particularly important symbionts, growing as nodules on the roots of clover, peas, and other leguminous plants (Fig. 30–1 and Table 30–2). The symbiotic nitrogen-fixing bacteria are much more efficient than free-living organisms in adding nitrogen to the soil. Some examples of their efficiency are given in Table 30–3. As indicated in the table, some nonleguminous plants, such as alder trees, also form symbiotic relationships with nitrogen-fixing bacteria.

Certain microorganisms originally isolated from soil are the main sources of important antibiotics. Others are used in industry to produce compounds such as organic acids and vitamins.

Figure 30–1

Root nodules resulting from infection of roots of a leguminous plant by *Rhizobium*. The host plant must be a legume. (The scale represents 1 cm.) (Courtesy of F. J. Bergersey.)

TABLE 30–1   GENERA OF FREE-LIVING BACTERIA CONTAINING SPECIES
CAPABLE OF FIXING NITROGEN

Genus	Special features
Azotobacter	Abundant in grasslands with soils of near-neutral pH
Beijerinckia	Found mainly in tropical climates, in acid soils
Chlorobium	Green anaerobic photosynthesizers that deposit sulfur
Clostridium	Anaerobic nitrogen-fixers
Cyanobacteria (many genera)	Aerobic photosynthesizers
Enterobacter	Facultative nitrogen-fixers
Rhodospirillum	Helical anaerobic photosynthetic bacteria

Microbial Pathogens in the Soil

In addition to innumerable soil microorganisms that are beneficial and even essential, pathogenic microorganisms that are harmful for plants and animals may be present in the soil. Relatively few species of soil microorganisms can attack living plants or animals, but these few can be of major importance.

Plant Pathogens

Fungi are by far the most numerous of the plant pathogens, although plants are also susceptible to infections by bacteria and viruses. These different microorganisms routinely claim about 10 percent of the total cash yield of crops. It is no wonder that so much effort is spent in developing strains of plants that resist infection naturally (color plate 76).

Some of the fungi commonly pathogenic for plants are listed in Table 30–4. Representatives of all the major groups of fungi are included. Particularly important to the human economy are the rusts and smuts. These Basidiomycetes include more than 20,000 species of rusts and more than 1000 species of smuts. Various species of rusts and smuts are parasites of most of the important crops of grains and forest and ornamental plants. They produce huge numbers of spores on the surfaces of their host plants,

TABLE 30–2   EXAMPLES OF SYMBIOSIS BETWEEN SPECIES
OF RHIZOBIUM AND LEGUMINOUS PLANTS

Species of Rhizobium	Group of leguminous plants
R. trifolii	Clover
R. leguminosarum	Peas
R. meliloti	Alfalfa
R. japonicum	Soybeans
R. lupini	Lupines
R. phaseoli	Beans

TABLE 30–3   QUANTITIES OF NITROGEN FIXED BY MICROORGANISMS

Group	Species or habitat	N₂ fixed per acre per year (lb)
Nodulated legumes	Alfalfa	113–297
	Soybean	57–105
	Red clover	75–171
Nodulated nonlegumes	Alder tree	200
Cyanobacteria	Arid soil in Australia	3
	Paddy field in India	30
Free-living heterotrophs	Soil under wheat	14
	Soil under grass	22
	Rain forest in Nigeria	65

From M. Alexander, *Microbial Ecology.* New York: John Wiley & Sons, 1971, p. 428.

giving the appearance of rust or smut on the plant (hence their names). These spores are often airborne over exceedingly long distances and in incredible numbers. The spread of rusts, in particular, may be limited to some extent by the need for two unrelated hosts in order to complete the life cycle of these fungi. For example, the stem rust of wheat must first grow on the barberry plant before it can attack wheat. This is fortunate, because the stem rust fungus produces as many as 10 billion spores per acre of wheat, and the spores can be carried in air currents for hundreds or even thousands of miles, as far as from Texas to north central Canada. At one time, about a million of these spores per square foot were detected near Fargo, North Dakota, when there was no wheat rust disease for hundreds of miles around, indicating that this large number of spores had arrived by air.

TABLE 30–4   SOME FUNGAL DISEASES OF PLANTS

Disease	Host	Group of fungus responsible
Brown spot	Corn	Phycomycete
Soft rot	Sweet potato	Phycomycete
Powdery scab	Potato	Phycomycete
Chestnut blight	American chestnut	Ascomycete
Ergot	Rye	Ascomycete
Scab	Apple	Ascomycete
Powdery mildew	Rose	Ascomycete
Peach leaf curl	Peach	Ascomycete
Dutch elm disease	Dutch elm	Ascomycete
Scab	Peach	Deuteromycete
Late blight	Celery	Deuteromycete
White pine blister rust	White pine	Basidiomycete
Leaf rust	Wheat	Basidiomycete

TABLE 30–5   SOME BACTERIAL DISEASES OF PLANTS

Disease	*Host*	*Species responsible*
Crown gall tumor	Most deciduous plants	*Agrobacterium tumefaciens*
Bacterial wilt	Tobacco Tomato	*Pseudomonas solanacearum*
Bacterial spot	Tomato	*Xanthomonas vesicatoria*
Bacterial canker	Tomato	*Corynebacterium michiganense*
Fire blight	Apple Pear	*Erwinia amylovora*

Some bacterial diseases of plants are listed in Table 30–5. The bacterially induced crown gall tumor of plants presents an interesting and much-studied disease and model of tumor growth (Fig. 30–2). This plant tumor is caused by *Agrobacterium tumefaciens*, a Gram-negative, flagellated, rod-shaped organism, closely related to some members of the beneficial nitrogen-fixing symbiotic genus *Rhizobium*. In crown gall tumors, killing the causative bacteria does not stop the development of the tumor once it has begun to form, showing that living bacteria are not needed for continuing tumor development. The explanation is that a small fragment of a plasmid, present in all

Figure 30–2
Crown gall tumor induced by *Agrobacterium tumefaciens* on the stem of a sunflower plant.

TABLE 30-6   SOME VIRAL DISEASES OF PLANTS

Disease	Natural hosts
Tobacco mosaic disease	Tobacco
	Tomato
Cucumber mosaic disease	Cucumber
	Delphinium
	Lupin
	Dahlia
Wound tumors	Clover

virulent strains of *A. tumefaciens*, is transferred from the bacterial cell to the plant, where the prokaryotic DNA becomes integrated into the eukaryotic DNA of the plant nucleus. A few genes in the plasmid fragment are responsible for the plant tumor. This points to a fundamental similarity between these plant tumors and certain animal tumors that are caused by the insertion of viral genes into the chromosomes of animals.

In addition to fungi and bacteria, viruses are also common pathogens of plants, as indicated in Table 30-6.

Pathogens for Animals and People

Microorganisms that are normally harmless inhabitants of soil may cause disease when they enter mammalian hosts. This is the case for a number of fungi and bacteria, especially some of the endospore-forming bacteria. Some of the soil fungi that infect people, often with the subsequent production of disease, are listed in Table 30-7, along with some characteristics of the infections. A few important disease-producing bacteria found in the soil are also included in Table 30-7. Another example of a soil organism that can cause human disease is the polymorphic protozoan *Naegleria*.

BIOMAGNIFICATION

A large number of complex hydrocarbons, which often contain chlorine atoms and are chemically distinct from naturally occurring compounds, are not biodegradable (Fig. 30-3). Many of these compounds, including certain herbicides and pesticides, are fat-soluble and accumulate in the fat of organisms in the food chain,* producing harmful effects in some cases. Table 30-8 shows how a tiny amount (0.00005 parts per million) of the nonbiodegradable pesticide DDT* in water is concentrated in floating organisms called plankton, further concentrated in minnows that feed on the plankton, and even more concentrated in various birds that eat the minnows or other small animals in the food chain. The final result in the merganser* is a highly dangerous concentration of DDT—22.8 parts per million. This progressive

---

*food chain—a series of organisms, each of which provides the food supply for the next

*DDT—an insecticide, dichlorodiphenyl trichloroethane

*merganser—type of fish-eating duck

TABLE 30–7   CHARACTERISTICS OF SOME HUMAN DISEASES CAUSED BY SOIL
MICROORGANISMS

Causative organism	Disease	Characteristics
Fungi		
Coccidioides immitis	Coccidioidomycosis	Lung or generalized infection
Histoplasma capsulatum	Histoplasmosis	Lung or generalized infection
Sporothrix schenckii	Sporotrichosis	Lesions in skin and underlying tissues, occurring along a lymphatic vessel
Bacteria		
Clostridium botulinum	Botulism	Paralysis caused by bacterial exotoxin
Nocardia asteroides	Nocardiosis	Lung and generalized infections

increase in concentration along the food chain is called **biological magnification,** or simply **biomagnification.**

MICROORGANISMS IN WATER

Beneficial Microorganisms in Water

As in soil, microorganisms in water play vital roles in various ecosystems and in the recycling of essential elements. For example, photosynthetic microorganisms present in water transform energy from the sun into chemical energy, and most of them also produce oxygen, essential for animal life. Others participate in the degradation and recycling of compounds.

Many aquatic microorganisms have unique capacities for exploiting their environments. The concentration of nutrients in water is usually quite low compared with that in solid media; thus, organisms that have special abilities have been selected for aquatic survival. Many algae and aquatic bac-

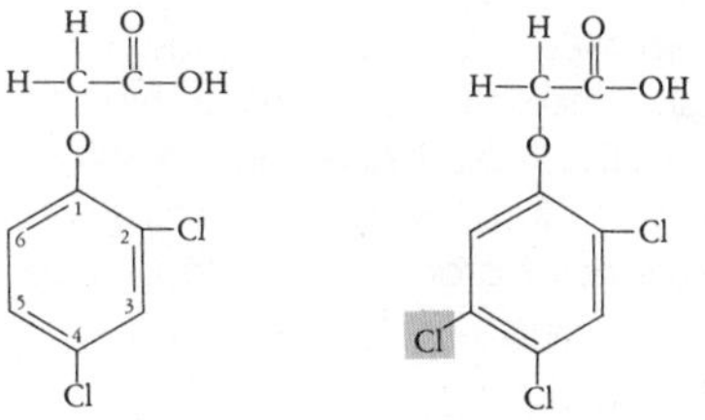

Figure 30–3
Chlorine atoms on hydrocarbon compounds tend to make the compounds nonbiodegradable. The ring structure on the left represents the herbicide 2,4-D, which is usually biodegraded within a few weeks after being applied to the soil. On the right, the ring structure of 2,4,5-T has an additional chlorine atom (shaded area) on the molecule. This compound is relatively nonbiodegradable and is often present in the soil for more than a year after application.

TABLE 30–8   FOOD CHAIN CONCENTRATION OF DDT

	Parts per million DDT residues
Water	0.00005
Plankton	0.04
Silverside minnow	0.23
Heron (feeds on small animals)	3.57
Herring gull (scavenger)	6.00
Fish hawk (Osprey) egg	13.8
Merganser (fish-eating duck)	22.8

From E. P. Odum, *Fundamentals of Ecology.* Philadelphia: W. B. Saunders, 1971, p. 74.

teria, for example, attach to stones, plants, or other solid surfaces in shallow water and continually withdraw nutrients from the water flowing past them (Fig. 30–4). Others, found in deeper waters, contain gases that allow them to float; thus, they can use more energy from sunlight than if they were at lower depths.

The flora of fresh and salt waters varies from aerobic and often photosynthetic microorganisms in the upper layers of the water to many anaerobes in the lower depths and sediments at the bottom of bodies of water (Fig. 30–5). Algae and protozoa constitute a large portion of aquatic flora, although various bacteria—aerobic and anaerobic, photosynthetic and non-photosynthetic—are also represented, depending on the type of habitat.

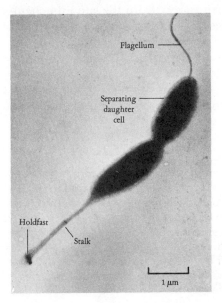

Flagellum

Separating daughter cell

Holdfast

Stalk

1 μm

Figure 30–4
Caulobacter species. This organism has a stalk and a holdfast, which attaches to solid materials in water, permitting the bacterium to extract adequate nutrients from water containing very low concentrations of organic materials. (Courtesy of J. T. Staley and J. P. Dalmasso.)

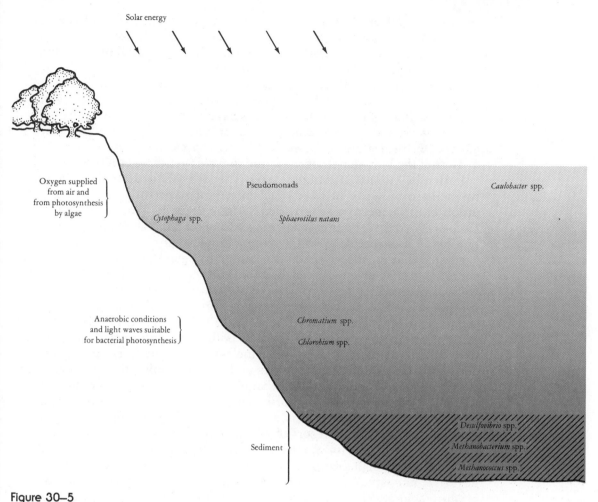

Solar energy

Oxygen supplied
from air and
from photosynthesis
by algae

Cytophaga spp.

Pseudomonads

Sphaerotilus natans

Caulobacter spp.

Anaerobic conditions
and light waves suitable
for bacterial photosynthesis

Chromatium spp.

Chlorobium spp.

Sediment

Desulfovibrio spp.

Methanobacterium spp.

Methanococcus spp.

Figure 30–5
Representative bacteria in a freshwater lake. Aerobic, often photosynthetic organisms are found in the upper layers of the lake. Anaerobic, photosynthetic bacteria of the genera *Chromatium* and *Chlorobium* inhabit lower levels of the lake. The sediment at the bottom of the lake contains anaerobic species of genera such as *Desulfovibrio*, *Methanobacterium*, *Methanococcus*, and *Clostridium*, which help to recycle sulfur, carbon, and nitrogen from compounds in the sediment.

Microbial Pathogens in Water

A few bacterial species found in the sea are pathogenic or nonpathogenic inhabitants of marine animals or plants. For example, one group of vibrios is of major importance as fish pathogens. These bacteria have been found in about 50 species of fish found in many different parts of the world. Mycobacteria are pathogenic for some marine and freshwater fish as well as for people; however, the species of mycobacteria that infect fish are in most cases distinct from those that cause similar human diseases, and vice versa. *Mycobacterium marinum* causes fish tuberculosis, but although *M. marinum* can infect humans, the disease it produces is quite different from human tuberculosis, which is caused by *M. tuberculosis.*

The salt-requiring *Vibrio parahemolyticus* (also known as *Beneckea parahemolyticus*) is abundant in seawater in warm seasons and contaminates various seafoods. Fish, crabs, and clams have been the source of many human infections around the world, and in Japan about two thirds of all cases of gastroenteritis are said to be caused by this species.

Most of the human pathogens found in water get there as a result of water pollution by human or animal wastes. The idea that certain diseases can be transmitted by water has been recognized for some time. One of the early proofs of this concept was provided by John Snow in London in 1854 in the famous Broad Street pump incident (see Chapter 29). Snow traced the source of a cholera outbreak to water from a single pump on Broad Street and suggested that the disease had been caused by agents able to multiply in the water. This conclusion was reached some 30 years before the cholera vibrios were discovered! As soon as methods for culturing bacteria became widely available, it was easy to prove that water can transmit many different pathogens, some of which are listed in Table 30–9. Thus, great care is now taken in most countries to ensure that water pollution is minimized by adequate treatment of disposed wastes. Nevertheless, large waterborne epidemics occur from time to time, even in cities with water supplies that are well monitored for fecal contamination. Examples include the 1965 epidemic of salmonellosis in Riverside, California, involving more than 16,000 cases of gastroenteritis and three deaths, and epidemics of typhoid fever in Aberdeen, Scotland, and in the ski resort of Zermatt, Switzerland.

Broad Street pump incident, page 596

The usual methods for monitoring water supplies depend upon the detection of the coliform group of bacteria. The coliforms are aerobic, Gram-negative, non–spore-forming rod-shaped bacteria that ferment lactose to produce acid and gas within 48 hours at 35°C. This group includes *Escherichia coli*, a common organism found in fecal material. It is assumed that if these bacteria are found in appreciable numbers in a water supply then the water has been contaminated with fecal material. Since pathogenic organisms such as *Shigella* and *Salmonella* are more difficult to isolate from

TABLE 30–9   SOME HUMAN PATHOGENS COMMONLY TRANSMITTED BY CONTAMINATED WATER

Pathogen	*Disease produced*
Vibrio cholerae	Cholera
Salmonella typhi	Typhoid fever
Salmonella enteritidis	Gastroenteritis
Shigella species	Dysentery
Polioviruses	Poliomyelitis
Hepatitis A virus	Liver disease (hepatitis)

water samples, testing for the coliforms is carried out instead. During purification procedures, these pathogens behave in the same way that *E. coli* does, so it is generally safe to assume that if *E. coli* is not detected, the pathogens will not be present either. Figures 30–6 and 30–7 show the procedure that is used most often in the testing of water supplies for contamination.

Water Microorganisms that Cause Poisoning

Several kinds of poisoning occur as a result of eating seafood contaminated with certain microorganisms. Scombroid fish poisoning is one such disease.

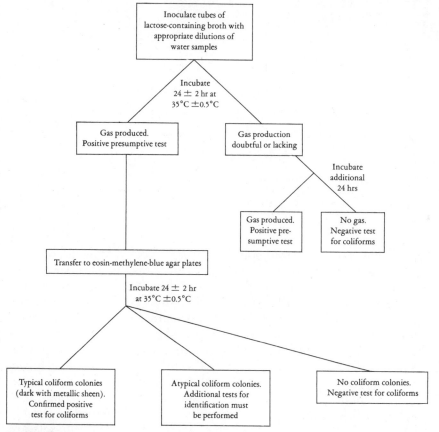

Figure 30–6
Analysis of water for fecal contamination. Method recommended by the United States Public Health Service for testing for coliforms in drinking water. Coliform organisms are all aerobic and facultative, Gram-negative, non–spore-forming, rod-shaped bacteria that ferment lactose with gas formation within 48 hours at 35°C. (Adapted from *Standard Methods for Examination of Water and Waste Water*, 13th ed. American Public Health Association, 1971.)

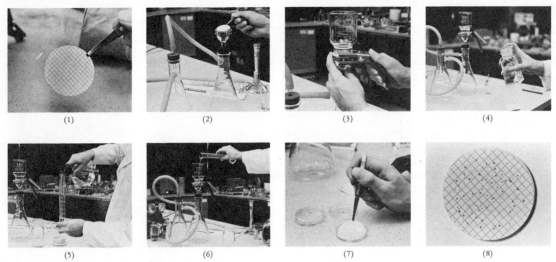

(1) (2) (3) (4)

(5) (6) (7) (8)

1. Sterile filter with grids for counting is handled with sterile forceps.
2. The membrane is placed on the filtering apparatus.
3. The apparatus is assembled.
4. A water sample is mixed well.
5. A portion of the sample is measured, and if necessary it is diluted with sterile diluent.
6. The portion is filtered by vacuum. Any bacteria that are present are retained on the filter.
7. The filter is placed on a dish of medium designed to select for coliforms. Nutrients can diffuse through the filter and colonies are formed on the gridded membrane where they can be counted readily.
8. The number of colonies on the filter reflects the number of coliform bacteria present in the original sample.

Figure 30–7
Analysis of water for fecal contamination. Cellulose acetate membrane (Millipore filter) method. (Courtesy of the Millipore Corporation.)

More than 200 cases of scombroid poisoning were reported to health officials in the United States over a recent four-year period. This disease results from ingesting heat-stable toxins produced by bacterial action on tuna and related species of the family Scombridae (hence the name scombroid poisoning). Bacteria grow in tuna that are not properly refrigerated, causing high levels of histamine in the meat. Histamine causes the symptoms, which last for a few hours. Ciguatera fish poisoning also results from eating fish contaminated with a heat-stable toxin, but this toxin is thought to be made by algae. Ciguatera poisoning occurs most often in Florida, Hawaii, and other warm areas and it affects people who have eaten contaminated red snapper, barracuda, and other fish common to warm waters. Paralytic shellfish poisoning is more likely to occur in cooler climates, where clams and mussels are found. In recent years it has been reported in Washington State, where it necessitated the closing of many beaches to clam diggers. This poisoning is caused by a neurotoxin produced by algae associated with "red tides." Shellfish concentrate the toxin, so that large amounts may be ingested. Symptoms usually subside within 24 hours after ingesting the very stable toxin; however, the disease can be fatal.

THE MICROBIOLOGY OF WASTE TREATMENT

Human and animal wastes in sewage are the source of most water contamination and of the pathogens in water listed in Table 30–9. Thus, it is obvious that sewage must be treated adequately to prevent the dispersal of pathogenic organisms. Less obvious, but equally important, is the need to treat wastes of all kinds in a manner that will permit them to be broken down and their essential materials to be recycled. Sewage contains not only human wastes but also food wastes, industrial wastes, cleaning compounds, toxic metals and other poisons, and an almost infinite variety of other matter. Most of this material is degraded either aerobically or anaerobically by microorganisms. Figure 30–8 diagrams the steps in waste treatment employed by most large cities.

During the aerobic treatment of sewage, microbial oxidation of organic compounds produces carbon dioxide and inorganic nitrogen-containing nutrients that could be used by plants. The usefulness of reclaimed solids from wastes, however, is limited at times by the presence in the solids of toxic materials and of potentially pathogenic viruses, such as polioviruses, that can survive during sewage treatment. Viruses are sometimes enclosed within small particles of material during the treatment process and thus are able to survive even treatment with chlorine.

During the anaerobic treatment of sewage, anaerobic microorganisms ferment organic compounds. The products of this fermentation are subsequently used through anaerobic respiration. The methane bacteria have an important function, converting the small breakdown products formed by other bacteria from organic carbon compounds into carbon dioxide and methane, the major component of natural gas.

SUMMARY

A wide variety of microorganisms, including many either beneficial or deleterious for humans, thrive in the soil, water, and air of the environment. Among the beneficial soil microorganisms are those that function in nitrogen fixation and many that are essential in the recycling of both organic and inorganic compounds. Soil microorganisms also include species pathogenic for plants, animals, and people. Some of the human pathogens found in soil are the fungi that cause histoplasmosis, coccidioidomycosis, and other mycoses, and the bacteria responsible for botulism, anthrax, and other diseases.

Microorganisms in water also play beneficial roles as primary producers in the food chain, and in the degradation of materials. But water can also serve as a source of pathogens. Among the human pathogens that can be disseminated by water are many of the bacteria that cause gastroenteritis and the bacteria that cause typhoid fever and cholera. Other pathogens spread by water include protozoa that cause gastrointestinal disease. Vi-

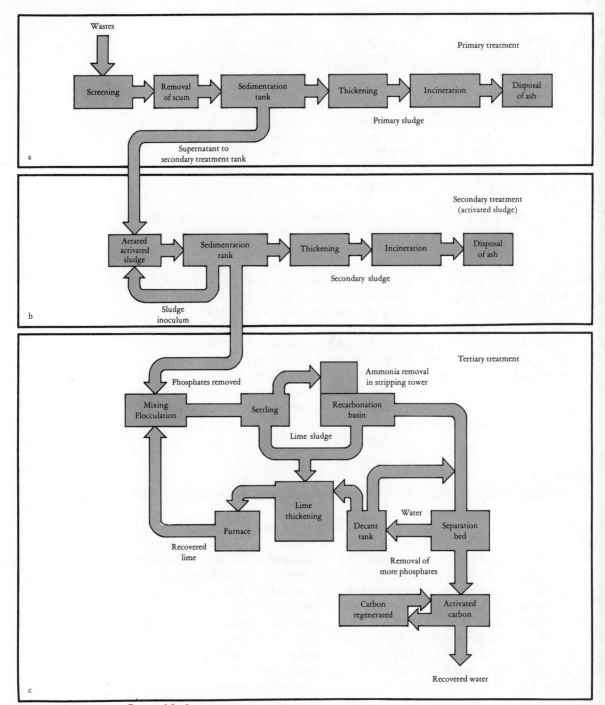

Figure 30—8
Metropolitan waste treatment. The most advanced waste treatment schemes utilize tertiary as well as primary and secondary treatments. (a) Primary treatment. (b) Secondary treatment. (c) Tertiary treatment.

ruses, for example those causing polio and hepatitis, also survive well in water.

Microbial degradative capacities are put to good use in various aerobic and anaerobic methods of waste disposal.

SELF-QUIZ

1. Bacteria of the genus *Rhizobium* have all the following characteristics *except*
   a. they are symbiotic with certain plants
   b. they form root nodules
   c. they are free-living nitrogen-fixers
   d. they add nitrogen to the soil with high efficiency
   e. they grow in association with legumes
2. Rusts and smuts are soil fungi belonging to the class
   a. Phycomycetes
   b. Basidiomycetes
   c. Ascomycetes
   d. Deuteromycetes
   e. Fungi Imperfecti
3. The Gram-negative rod *Agrobacterium tumefaciens*
   a. is nonmotile
   b. causes tumors in plants
   c. may be pathogenic for humans
   d. is not known to contain any plasmids
   e. is a free-living nitrogen-fixer
4. Human pathogens that normally are inhabitants of the soil include all of the following *except*
   a. *Agrobacterium tumefaciens*
   b. *Nocardia asteroides*
   c. *Clostridium botulinum*
   d. *Coccidioides immitis*
   e. *Histoplasma capsulatum*
5. Contaminated seafood is a source of all the following microbially induced diseases *except*
   a. *Vibrio parahemolyticus* gastroenteritis
   b. *Salmonella* gastroenteritis
   c. scombroid poisoning
   d. crown gall tumor
   e. ciguatera poisoning

QUESTIONS FOR DISCUSSION

1. How might genetic engineering, involving plasmids, be used to decrease the use of chemical fertilizers while increasing food production?

2. What difference in optimal growth conditions would be expected between water microorganisms pathogenic for marine fish and those pathogenic for people? Would the same media be used, and the same cultural conditions, for isolating both kinds of pathogens?
3. What microorganisms other than *E. coli* could serve as indicators of fecal contamination? Why are these not used?

FURTHER READING

Alexander, M.: *Microbial Ecology.* New York: John Wiley & Sons, 1971. A broad view of microbial ecology, including the oceans and fresh water as natural ecosystems.

Bascom, W.: "The Disposal of Waste in the Ocean." *Scientific American* (August 1974). This article presents rational approaches to the safe disposal of wastes into the ocean.

Brill, W.: "Biological Nitrogen Fixation." *Scientific American* (March 1977).

Brock, T. D.: *Principles of Microbial Ecology.* Englewood Cliffs, N.J.: Prentice-Hall, 1966. An excellent text on microbial ecology.

Mitchell, R.: *Introduction to Environmental Microbiology.* Englewood Cliffs, N.J.: Prentice-Hall, 1974. Written for students of environmental engineering and the basic sciences.

Odum, E. P.: *Fundamentals of Ecology*, 3d ed. Philadelphia: W. B. Saunders Co., 1971. Includes chapters on fresh water ecology and marine ecology.

Scientific American (September 1976). The entire issue is devoted to food and agriculture. An article of special interest is "The Cycles of Plant and Animal Nutrition" by J. Janik, C. Noller, and C. Rhykerd.

Standard Methods for Examination of Water and Wastewater, 13th ed. Washington, D.C.: American Public Health Association.

Taber, W. A.: "Wastewater Microbiology." *Annual Review of Microbiology* 30:263–277 (1976). A review of wastewater microbiology.

Tsutomu, H.: *Microbial Life in the Soil: An Introduction.* New York: Marcel Dekker, 1973. Introductory text covering the full range of topics associated with microbial life in the soil. It is a rather advanced, concisely written book that refers to numerous original research papers.

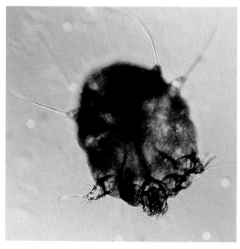

COLOR PLATE 63 *Sarcoptes scabiei* (scabies mite). (×125) (Courtesy of S. Eng and F. Schoenknecht.)

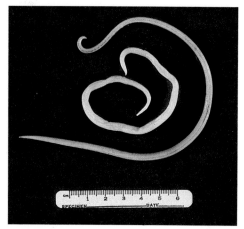

COLOR PLATE 64 *Ascaris lumbricoides*. Adult worms; the female is on the outside and the male is on the inside. (Courtesy of S. Eng and F. Schoenknecht.)

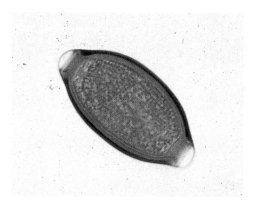

COLOR PLATE 65 *Trichuris trichiura* ovum in feces. (×500) (Courtesy of S. Eng and F. Schoenknecht.)

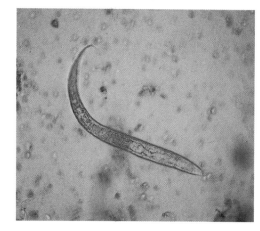

COLOR PLATE 66 Hookworm larva. (×125) (Courtesy of S. Eng and F. Schoenknecht.)

COLOR PLATE 67 *Enterobius vermicularis* (pinworm). (×10) Stained preparation showing the thousands of eggs that are released from a single worm. (Courtesy of S. Eng and F. Schoenknecht.)

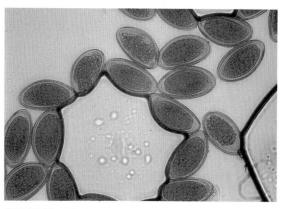

COLOR PLATE 68 *Enterobius vermicularis*. Stained preparation. (×125) (Courtesy of S. Eng and F. Schoenknecht.)

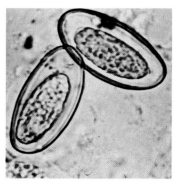

COLOR PLATE 69 *Enterobius vermicularis* ova. Transparent tape preparation. (×250) (Courtesy of S. Eng and F. Schoenknecht.)

COLOR PLATE 70 *Strongyloides stercoralis* in feces. Stained preparation. (×50) (Courtesy of S. Eng and F. Schoenknecht.)

COLOR PLATE 71 *Strongyloides stercoralis* larva (wet mount). (×125) (Courtesy of S. Eng and F. Schoenknecht.)

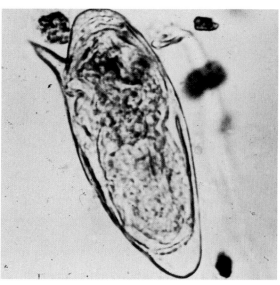

COLOR PLATE 72 *Schistosoma mansoni* ovum (wet mount). (×150) (Courtesy of S. Eng and F. Schoenknecht.)

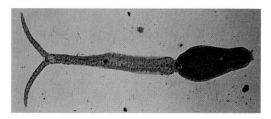

COLOR PLATE 73 *Schistosoma mansoni* (cercaria). (×75) This is the stage that penetrates the skin. (Courtesy of S. Eng and F. Schoenknecht.)

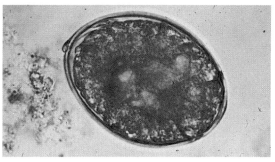

COLOR PLATE 74 *Diphyllobothrium latum* ovum. (×300) (Courtesy of S. Eng and F. Schoenknecht.)

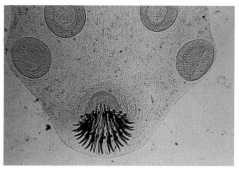

COLOR PLATE 75 *Taenia solium* scolex (head). Pork tapeworm. (×25) (Courtesy of S. Eng and F. Schoenknecht.)

COLOR PLATE 76 **A.** The black spot fungus on rose leaves. This fungus, which occurs in wet climates, causes the leaves to fall prematurely. **B.** The fungus *Venturia inaequalis* causes apple scab. Note that the areas of scab formation develop poorly and are corklike in appearance. These fungi can cause serious disease problems in orchards if control measures are not used. (Courtesy of M. Nester.)

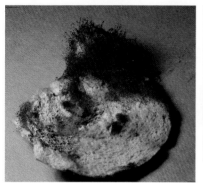

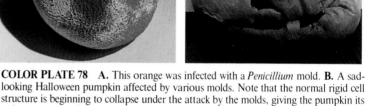

COLOR PLATE 77 The mold *Rhizopus stolonifer,* commonly found on bread.

COLOR PLATE 78 **A.** This orange was infected with a *Penicillium* mold. **B.** A sad-looking Halloween pumpkin affected by various molds. Note that the normal rigid cell structure is beginning to collapse under the attack by the molds, giving the pumpkin its sad appearance. (Courtesy of M. Nester.)

COLOR PLATE 79 **A.** A tomato found molding in the refrigerator. Some molds grow well at cool temperatures. **B.** A number of different molds were found growing inside this potato. Fungi can damage food crops extensively. (**A** and **B** courtesy of M. Nester.)

COLOR PLATE 80 These two large cheddar cheeses were produced at the University of Wisconsin in Wausau, Wisconsin. One weighed 500 pounds and the other weighed 3000 pounds. (Courtesy of A. Braceman.)

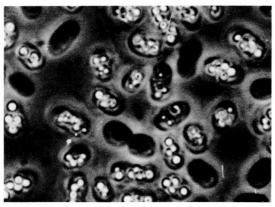

COLOR PLATE 81 Phase contrast microscopy. The round, bright areas are sulfur granules of a photosynthetic bacterium. (×2500) (Courtesy of J. Staley.)

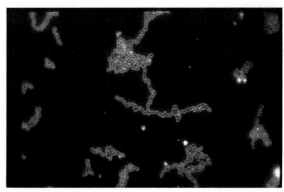

COLOR PLATE 82 Group A *Streptococcus* as viewed by dark-field microscopy. (Courtesy of C. E. Roberts, Jr.)

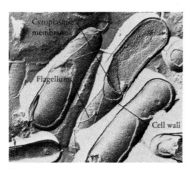

COLOR PLATE 83 Freeze-etched preparation of *Bacillus subtilis.* (Courtesy of S. Holt.)

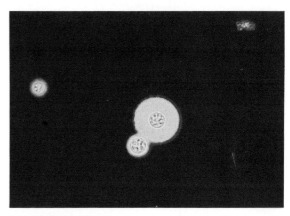

COLOR PLATE 84 *Cryptococcus neoformans.* Negative stain with India ink. The ink is unable to penetrate the surface of the cell, thereby outlining the cell.

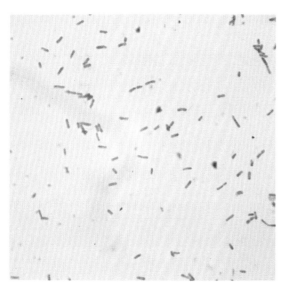

COLOR PLATE 85 *Escherichia coli* stained with methylene blue.

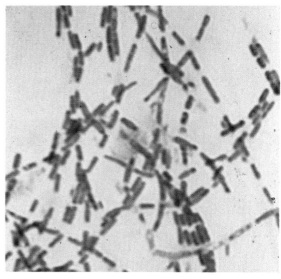

COLOR PLATE 86 Endospores (red bodies) in *Bacillus anthracis.*

Chapter 31

MICROORGANISMS AND FOOD

Many years ago, near Roquefort in France, a shepherd boy carrying bread and farmer's cheese for his lunch set out for his day's work. During the day, a storm forced him to take refuge in a moist, moldy cave among the rocks, and when he left, he was so concerned with the safety of his sheep that he forgot his bread and cheese. Weeks later, he returned by chance to the same cave and to his surprise found his somewhat tasteless farmer's cheese transformed by molds into a blue-veined delicacy. He spread the word to others in his village, who began to use these molds to make more of this delicious sheeps'-milk cheese, which they called Roquefort cheese. Or so the story goes. Whether it actually happened this way is not known, but it is a certainty that molds of the genus *Penicillium* are responsible for the almost miraculous transformations that occur in Roquefort cheese. This is just one of the ways that microorganisms interact with foods.

OBJECTIVES

To know
1. The factors that influence food spoilage by microorganisms.
2. What is meant by the moisture content of food and how this affects microbial spoilage of food.
3. The basic principles of methods for preserving food from microbial spoilage and some of the ways these principles are used.
4. Some of the microorganisms and microbial activities responsible for the production of bread, cheese, dairy products, and pickles.
5. How microorganisms contribute to the production of alcoholic beverages.

Microorganisms act upon food and beverages in a variety of ways. Among the more important of these ways are the spoilage of food and the transformation of food and beverages by fermentation or other microbial processes. The spread of diseases by foodborne pathogens or their products is discussed in other chapters. The large, easily visible algae and fungi may themselves serve as food. For example, some of the edible algae, such as the Japanese *nori* and other seaweeds, are harvested, and mushrooms are fungi that are enjoyed by many people. However, several of their microscopic relatives are also edible or are important in food production. Even the tiny bacteria are important, as in the bacterial culture of yogurt. Microorganisms as sources of single-cell protein represent an even more important potential source of food, now and in the future, as discussed in Chapter 32.

This chapter will also consider some methods of preserving food from microbial spoilage.

FOOD SPOILAGE BY MICROORGANISMS

Everyone has experienced the annoyance of opening a loaf of bread and finding it covered with black "whiskers" of bread mold or picking up a tempting peach from the fruit bowl only to find the underside brown, soft, and spoiled. These kinds of spoilage result from fungal invasion of the food, by species of *Rhizopus* on the bread (color plate 77) and species of *Candida, Penicillium,* or other fungi on the fruit (color plate 78). It is not surprising that foods for humans make excellent media for the growth of many different fungi and bacteria from the environment.

Food spoilage is influenced by the number of spoilage organisms present, the nature of the food, the acidity (pH), the amount of water and oxygen available, and the temperature. Because most bacteria grow well at room temperature or higher and poorly at lower temperatures, storage at refrigerator temperatures helps to retard food spoilage. However, **psychrophiles** ("cold-loving" bacteria) such as some species of *Pseudomonas, Achromobacter, Alcaligenes,* and *Flavobacterium,* and some yeasts and molds thrive at these lower temperatures (color plate 79). These organisms can spoil refrigerated food and even some frozen food.

Most microorganisms need water in order to live. Different microorganisms need different amounts of moisture. Although there are some fungi and yeasts known as **xerophiles** that can survive at very low moisture levels, food spoilage can generally be prevented by drying. Fruits and vegetables have been dried for centuries to preserve them. Today we dry such foods as milk and eggs so that they will keep for a long time. Interestingly, preserving foods by means of either high salt or high sugar solutions has very much the same effect as drying them. The salt or sugar causes the water to pass out of the microbial cell and inhibits its growth.

A few common examples of microbial species that contribute to food spoilage are listed in Table 31–1. Notice the difference between the species that

TABLE 31–1   COMMON EXAMPLES OF MICROBIAL SPECIES THAT CONTRIBUTE
TO FOOD SPOILAGE

Food	Organism	Type of spoilage
Bread	Aspergillus niger Neurospora sitophila Mucor sp. Rhizopus nigricans	Bread mold
Milk (raw)	Streptococcus lactis S. cremoris Lactobacillus bulgaricus	Sour milk
Milk (pasteurized)	L. thermophilus Microbacterium lacticum S. thermophilus	Sour milk
Chicken	Achromobacter sp. Pseudomonas sp. Alcaligenes sp. Flavobacterium sp.	Sliminess
Fish	Achromobacter sp. Flavobacterium sp. Micrococcus sp. Pseudomonas sp. Serratia sp.	Characteristic "fishy" odor

Adapted from Miller, B. M., and Litsky, W.: *Industrial Microbiology.* New York: McGraw-Hill, 1976, pp. 268, 271.

sour raw milk and those that sour pasteurized milk (Fig. 31–1). As would be expected, heat-resistant, acid-forming bacteria are the culprits in spoiling pasteurized milk, since others are eliminated by pasteurization. Notice also in Table 31–1 that species of *Pseudomonas, Achromobacter, Flavobacterium,* and others can spoil a wide variety of foods. In fish, these bacteria convert trimethylamine oxide in the muscle tissue to trimethylamine, the compound that gives spoiled fish their characteristic fishy odor.

FOOD PRESERVATION

Methods for preserving food from microbial spoilage depend upon either killing microorganisms and preventing their entrance into food or supplying conditions inhibitory to the common spoilage organism.

From ancient times, people have used many ingenious ways to preserve food, including drying, salting, and pickling (Fig. 31–2). Dried foods are usually well preserved because their very low moisture content does not permit growth of most microorganisms. Hard cheeses also have a low moisture content and have long been used to preserve milk nutrients. Salting discourages all but the halophilic ("salt-loving") microrganisms. Pickling combines a high salt content with a pH so low (very acidic) that it discourages the growth of most bacteria. Preserves and jellies have such a high sugar

Microbial Growth In Raw Milk At Room Temperature

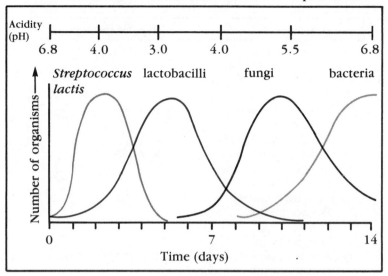

Figure 31–1

Growth of microbial populations in raw milk at room temperature. Unpasteurized milk naturally contains *Streptococcus lactis* and other bacteria that multiply efficiently at room temperature, producing enough lactic acid and other acids to decrease the pH from near neutral (6.8) to acid (4.0 to 4.5) within a few days. As the acidity becomes greater, growth of lactobacilli, which are also normally present in the milk, is favored. The lactobacilli produce even more acid, further decreasing the pH to a very acidic 3.0 to 3.5. This acid pH discourages the growth of most bacteria but permits many fungi to grow. The fungi actually oxidize the acids, releasing energy and producing carbon dioxide and water. As the acidity is lost, species of *Pseudomonas*, some spore-forming bacteria, and other organisms digest the proteins and fats and putrify the milk.

content that most bacteria cannot multiply, although many fungi can grow in these foods (Fig. 31–3).

Despite these time-honored methods for preserving food, at the beginning of the nineteenth century the need for better techniques of preservation became apparent. In 1810, Napoleon awarded a prize of 12,000 francs to Nicholas Appert for developing the method we now know as canning, i.e., heating foods in sealed containers. Later, in the 1890's and into the twentieth century, research continued on heat-resistant, spore-forming organisms and on ways to eliminate them from canned foods. Also during the nineteenth century, Pasteur introduced the technique of mild heating of wine or food (pasteurization) to destroy harmful or unwanted microorganisms without seriously affecting the taste of the food or beverage.

Chemicals other than salt, sugar, and acids are also useful for preserving

Figure 31-2
Producing dried food in a homemade food drier. Using a source of heat (the four lightbulbs) and a means of circulating the air (hair drier), these plums were dried sufficiently within two days to permit them to be stored at room temperature without spoilage. The water content of the plums was reduced to the point where most microorganisms could not find a suitable habitat. (Courtesy of M. Nester.)

food. A look at the labels on soft drinks, candies, and margarine containers will reveal that benzoic acid or sodium benzoate is often added as a preservative. Sodium diacetate added to bakery products discourages mold growth in bread and other baked goods, as does calcium propionate or sodium propionate. Sodium nitrite, used in curing meats and fish, effectively inhibits certain anaerobic, spore-forming bacteria, but it has been disparaged recently as a possible carcinogen. The Food and Drug Administration constantly reviews these and other food additives to determine whether they are safe and prohibits the use in foods of any found to be harmful.

In addition to the processes of food preservation by heating, drying, salting, pickling, and adding antimicrobial chemicals, technical advances in recent decades have made it possible to use a number of different forms of radiation for preserving food. Types of radiation often used for this purpose are ultraviolet light and gamma radiation.

Figure 31–3
Other means of preserving food include pickling with high concentrations of salt and vinegar (acid pH) and canning fruits and jellies with high sugar content. Both acid pH and high sugar content discourage most microorganisms. In addition, these preserves were heated before being sealed in the jars. (Courtesy of M. Nester.)

TRANSFORMATION OF FOOD BY FERMENTATION OR OTHER MICROBIAL PROCESSES

Microbial transformations of food have been used by humans as far back as history has recorded and no doubt even earlier. A current film shows giraffes, wart hogs, and a host of other African animals feeding on trees full of over-ripe, naturally fermented fruits; the feast is so tasty that the animals gorge themselves until they literally reel away, providing comic sequences of drunken giraffes and almost comatose wart hogs. Undoubtedly, early people had similar experiences, although it is to be hoped that they were intelligent enough not to overindulge to quite the same degree. Instead, they learned about this means of using naturally occurring yeasts to ferment the sugar in fruit juices into alcohol, thereby converting the juice into wine, which can be kept for long periods of time without spoiling.

Yeasts similar to those involved in wine-making are also responsible for raising (or leavening) bread; besides being able to produce ethyl alcohol, they also generate bubbles of carbon dioxide. They were being used long before the existence of microorganisms was even suspected.

Breads, cheeses, yogurt, sour cream, butter, sausages, pickles, and sauerkraut are some of the food products of microbial fermentation or other metabolic processes. As would be expected, the microorganisms involved are widespread in nature, especially in the fruits, vegetables, and grains acted upon. Although in recent years humans may have selected improved strains, the microorganisms occurring under natural conditions can perform the appropriate food transformations. The majority of these processes involve species of either a *Saccharomyces* yeast, an *Aspergillus* mold, or some of the bacteria that produce lactic acid.

Bread

Bread can be leavened chemically by adding baking powder to the bread dough or microbiologically by adding cultures of the yeast *Saccharomyces cerevisiae* (Fig. 31–4). Both methods result in the evolution of carbon dioxide

A

B

C

D

Figure 31–4

Bread made with yeast. (a) The ingredients of a loaf of bread include yeast, milk, oil, flour, salt, and sugar. (b) After the dough is thoroughly mixed, it is put in a bowl and allowed to rise. During this time the yeast grows and produces alcohol and CO_2. The dough rises or increases in volume until it nearly fills the bowl (c). The gases generated create air pockets in the bread, producing its texture. Following this period of rising, the dough is shaped into loaves and then baked. Note the holes caused by the production of CO_2 in the finished loaf of bread (d). (Courtesy of M. Nester.)

(CO_2), which makes the bread rise. However, a taste of the products of each of these methods gives evidence that the activities of the yeast improve the texture of bread and yield aromatic compounds that greatly enhance the flavor of bread. The alcohol produced along with CO_2 vaporizes and escapes during baking.

Cheese

Cheese-making represents a combination of applied microbiology and pure art. The varieties of cheese seem almost endless, with more than 200 different kinds being made in France alone. These are all produced by the same basic mechanisms of coagulating milk by rennet or similar enzymes in the presence of microbially produced lactic acid, and then molding and ripening the curds of coagulated milk solids. The final product is categorized as soft, medium, or hard and as ripened by either molds or bacteria. Some 1200 quarts of milk are required to make a 200-pound wheel of cheese (color plate 80), so in this manner the nutrients of large amounts of milk are compressed and preserved for long periods of time and greatly improved in flavor in the process. Table 31–2 gives a few examples of some microorganisms that transform milk into cheese.

Yogurt and Other Dairy Products

Another popular form of soured and preserved milk is yogurt, prepared by heating milk, cooling it, and then inoculating it with *Streptococcus thermophilus* and *Lactobacillus bulgaricus*. Other forms are acidophilus milk, sour cream, and buttermilk. Bacteria are also important in butter production and help to produce the characteristic butter flavor.

TABLE 31–2 SOME MICROORGANISMS THAT PARTICIPATE IN THE PRODUCTION OF CHEESES

Stage of production	Microorganisms involved
Initial souring of milk, lactose fermented to lactic acid	*Streptococcus cremoris, S. lactis, Lactobacillus bulgaricus,* other lactic-acid bacteria
Rennet enzyme added to coagulated solids of the sour milk	
Molding and ripening, which varies with the type of cheese:	
Emmentaler (Swiss)	*L. helveticus* *Propionibacterium* sp.
Cheddar	*L. plantarum* *L. casei*
Camembert	*Penicillium camemberti*
Roquefort	*P. roqueforti*

Products of Acid-Producing Fermentations

Lactic acid bacteria help to preserve and flavor certain kinds of sausage by carrying out lactic acid fermentation. The Italian pepperoni and German Thüringer sausages are examples. The production of pickles and sauerkraut also depends on lactic acid fermentation by species of *Lactobacillus*, aided and augmented by other microbial activities. Even the vinegar used for pickling is the product of oxidation of the ethyl alcohol in apple cider, wine, or other substances by species of *Acetobacter*, producing acetic acid.

MICROBIAL CONTRIBUTIONS TO THE PRODUCTION OF ALCOHOLIC BEVERAGES

Different varieties of the yeast *Saccharomyces cerevisiae* or related yeasts used for making bread, or similar species of yeasts, are used to make wine, beer, and cider, in which alcohol is an important constituent of the final product. Wines are the result of yeast fermentation of grape juice or other fruit juices by *S. cerevisiae*, var. *ellipsoideus*. Such natural fermentation gives wine an alcohol content usually less than 14 percent. Most dry table wines contain about 12½ percent alcohol, and some of the sweeter wines have less. Sherry, port, and other wines higher in alcohol content are fortified with extra alcohol by adding brandy.

For the preparation of red wines, blue grapes are crushed, about 100 ppm of sulfite is added to kill most potential spoilage organisms, cultures of the appropriate yeast are added, and the mixture is allowed to ferment under controlled conditions for five to eight days before the liquid is pressed from the solids. White wines are prepared by pressing the grape juice from the pigment-containing solids before fermentation. After fermentation is complete, the wine is clarified and stabilized; the exact procedures depend on the kind of wine being made. Some wines with a naturally high acidity are improved by an additional microbial fermentation of malic acid to lactic acid, a weaker acid. This reaction is carried out by the Gram-positive micrococci *Micrococcus malolacticus* or *M. acidovorax*.

Beers differ from wines in that the substrate for fermentation is derived from grains, and malt is used to convert the starch in grains to fermentable sugars. Amylases, which degrade starches, and other enzymes produced by germinating barley are the important constituents of malt. The action of these enzymes on starch produces a carbohydrate mixture called wort, which is boiled with hops, cooled, and fermented by *S. cerevisiae* or the similar *S. carlsbergiensis*. Whereas *S. cerevisiae* is used to produce ales, *S. carlsbergiensis* is the yeast of choice for making lager beers. Hops contribute to the flavor and bitterness of beer, and their resins are preservatives. American lager beers usually contain an average of 3.8 percent alcohol and 4.3 percent dextrins. The latter carbohydrates can be removed by enzyme degradation using fungal enzymes, resulting in the widely advertised low-calorie or "light" beers. Figures for 1968 indicate that Americans consumed about 110 million barrels (12.87 × 10^9 liters) or about 16 gallons per person, making these microbial transformations an important factor in our economy.

Of course, distilled alcoholic beverages, or spirits, also have their origin in yeast transformations of fermentable substrates into alcohol. In the case of gins and whiskies, the starting material is grain; for brandies, fruit juices are fermented; and for rum, molasses is fermented. Scotch whisky, for example, is made from barley, following the same initial steps as for preparing beer to convert the starches to sugars; bourbon whiskey uses corn as the basic starting material. Other microbial processes may contribute to the final product. For example, sour mash whiskey attains its highly prized flavor from a bacterial lactic acid fermentation before the alcoholic fermentation by yeast.

SUMMARY

Factors that influence the growth of spoilage microorganisms in food include the numbers of organisms initially present, the nature and pH of the food, the moisture content of the food, amounts of oxygen available, and the temperature. Among the bacteria that commonly spoil food are species of *Pseudomonas*, *Achromobacter*, *Flavobacterium*, and others. Many fungi can also spoil food, notably species of *Aspergillus*, *Rhizopus*, *Penicillium*, *Candida*, and others.

Methods for preserving food from microbial spoilage include various techniques for heating, drying, salting, pickling, preserving with a high sugar content, adding antimicrobial chemicals, and using irradiation.

Microorganisms function importantly in the transformation of food by fermentation or other metabolic processes. Bread, cheese, yogurt, sour cream, butter, sausages, pickles, vinegar, and sauerkraut are all products of microbial transformations. Beverages such as wine, beer, and spirits all result from alcoholic fermentation by yeasts.

SELF-QUIZ

1. Psychrophilic bacteria important in food spoilage include species of all the following genera *except*
   a. *Pseudomonas*
   b. *Actinomyces*
   c. *Alcaligenes*
   d. *Achromobacter*
   e. *Flavobacterium*
2. Of the following, which condition is most likely to encourage food preservation?
   a. alkaline pH
   b. neutral pH
   c. storage at refrigerator temperatures

   d. low moisture content
   e. moderate salt content
3. Wine results from the fermentation of fruit juices by
   a. *Micrococcus malolacticus*
   b. *M. acidovorax*
   c. *Saccharomyces carlsbergiensis*
   d. *S. cerevisiae*, var. *ellipsoideus*
   e. *Candida albicans*
4. All of the following are used in producing beer *except*
   a. grain
   b. *S. carlsbergiensis*
   c. amylase
   d. hops
   e. lactic acid bacteria
5. Microorganisms of primary importance in transforming foods by fermentation or other microbial processes include species of
   a. *Streptococcus, Staphylococcus, Clostridium*
   b. *Saccharomyces, Aspergillus*, lactic-acid bacteria
   c. *Saccharomyces, Staphylococcus, Clostridium*
   d. *Clostridium, Streptococcus, Bacillus*
   e. *Saccharomyces, Candida, Aspergillus*

QUESTIONS FOR DISCUSSION

1. What disadvantages are there in preserving foods by heat treatment, and how can these disadvantages be overcome?
2. What conditions of food preparation might lead to microbial food poisoning? How does this relate to the metabolism of the organisms involved?
3. What kinds of development could lead to improvements in food microbiology? For example, could the production of wine be made better, cheaper, or more efficient by developments in microbiology?

FURTHER READING

Dixon, B.: *Magnificent Microbes*. New York: Atheneum Press, 1979.

Hobbs, B. C., and Gilbert, R. J.: *Food Poisoning and Food Hygiene*, 4th ed. London: Edward Arnold Publishers Ltd., 1978.

Kharatyan, S.: "Microbes as Food for Humans." *Annual Review of Microbiology* 32:301–327 (1978).

Lawrence, R. C., and Thomas, T. D.: "The Fermentation of Milk by Lactic Acid Bacteria." In Bull, A. T., et al. (eds.): *Microbial Technology: Current State, Future Prospects*. New York: Cambridge University Press, 1979, pp. 187–219.

Miller, B. M., and Litsky, W.: *Industrial Microbiology*. New York: McGraw-Hill, 1976.

Rossmore, H. W.: *The Microbes, Our Unseen Friends*. Detroit: Wayne State University Press, 1976.

Chapter 32

COMMERCIAL APPLICATIONS OF MICROBIOLOGY AND IMMUNOLOGY

In the basement of Louis Pasteur's boyhood home in Dôle, France, sat smelly vats of tanning leather. Pasteur's father, Jean Joseph Pasteur, was a tanner by trade. Leather tanning, the commercial process by which leather is preserved and made usable for a variety of products, depends in part on the action of bacteria to produce enzymes that help preserve and soften the skins. This process is called "bating." In earlier times, the source of these enzymes was dog feces, which were rubbed onto the hides. Because there are a variety of bacteria in dog feces, the results were a bit uneven. Today enzymes prepared commercially from bacteria are purified and therefore give more predictable results. Here again we can see the usefulness of microorganisms in everyday life.

OBJECTIVES

To know
1. Some general principles of industrial and commercial microbiology.
2. Some microbial processes used in industry.
3. A few commercial applications of immunology.
4. The meaning of single-cell protein and its place in industry.
5. The principles of microbial mining.

Although it is beyond the scope of this book to detail the technology of industrial applications of microbiology and immunology, it is appropriate to point out some of the many ways these sciences have been applied commercially. Some examples of these applications have become evident from previous chapters concerning microorganisms and foods, beverages, wastes, and the environment. This chapter will briefly consider some general principles of industrial microbiology and some important examples of commercial applications of microbiology and immunology.

GENERAL PRINCIPLES OF COMMERCIAL MICROBIOLOGY

As is true for any commercial venture, microbiological processes must be economically feasible if they are to be used by industry. Not only must the desired product be produced by microbial means but it also must be produced more cheaply or more advantageously than by other means. For example, acetone and butanol used to be produced quite successfully by microbial fermentation; however, this method became obsolete when more economical methods for chemical production were developed.

Political, economic, and microbiological factors all interact to determine the feasibility of production, as exemplified by the development of the penicillin industry. The initial discovery by Fleming in 1928 that some cultures of *Penicillium* mold killed staphylococci was not put to use commercially until more than a decade later. The *Penicillium* strains used by Fleming produced only small quantities of the active agent, and the technology and money to develop his discovery were not available during the Depression. With the advent of World War II, the demand for antimicrobial agents became intense, and money and improved technology became available. Thus by 1943, penicillin was being produced commercially.

In applying microbiology commercially, an important early step is to select a strain of organism that produces the desired compound efficiently. Then efforts are directed toward developing or selecting even higher-producing strains. In the case of penicillin, original production was begun by using *P. notatum,* but in the early years of development, a strain of *P. chrysogenum* isolated from a moldy melon in Peoria, Illinois, was found to be much more efficient at producing penicillin than was the original *P. notatum.* At present, mutants of this strain that produced increased yields are still used in industry.

Other commercially important cultures have been isolated from natural sources, but often appropriate cultures are obtained by selecting mutant strains that lack normal mechanisms to control production of the desired compound. Exposing *P. chrysogenum* to mutagens and selecting strains that gave increased yields of penicillin have made it possible to improve the production of penicillin tremendously. For example, in France alone the yield was increased from 20 units/ml in 1943 to between 12,000 and 15,000 units/ml in 1972, largely as a result of mutant selection. Of course, through genetic engineering, it is now possible to alter the genetic information of microorganisms to produce desired strains.

In addition to selecting high-yield strains of microorganisms and maintaining these strains, another essential factor in industry is determining the most economically advantageous medium for the job and the optimum conditions for production of the desired compound. The disposal or possible use of waste products that are generated is another critical factor.

BIOSYNTHESIS OF ORGANIC ACIDS AND VITAMINS

Microorganisms can synthesize a variety of organic compounds including (among many others) simple organic acids, such as citric acid; amino acids, such as glutamic acid and lysine; and vitamins, such as vitamin B$_{12}$ and vitamin C (ascorbic acid).

Citric Acid

Citric acid is widely used; more than 100,000 tons are produced annually. It is added to many foods and cosmetics and can act as a chelating agent,* taking up metal ions in such uses as the cleaning of oil pipes and the tanning of hides. To produce this acid, one grows specially selected strains of the mold *Aspergillus niger* on sugar-containing materials, such as molasses or sucrose. The citric acid that is produced is precipitated from the growth medium by adding lime, and the citric acid is purified and crystallized. Other methods are sometimes used to remove the citric acid from the medium, and other microorganisms and substrates* can also be used. Some yeast strains of *Candida lipolytica* are used to produce large quantities of citric acid from paraffins.

Glutamic Acid

Glutamic acid, an amino acid, is used in food preparation as monosodium glutamate, or MSG. Mutants of a strain of *Corynebacterium* from soil, called *C. glutamicum*, use sugars such as glucose or molasses to produce the glutamic acid.

Vitamins

Vitamin B$_{12}$ is an essential nutrient and must be ingested by humans; however, many bacteria can synthesize it from simple nutrients. Commercially, certain species of *Bacillus*, *Pseudomonas*, *Streptomyces*, and *Propionibacterium* are used for B$_{12}$ production; strains of *Propionibacterium*, a Gram-positive bacterium, give the best yields.

---

*chelating agent—a compound that removes metallic ions from a solution by combining them with certain organic compounds

*substrates—the substances on which enzymes act to form a product

Riboflavin and precursors of vitamins A and C are commonly produced in industry by either bacteria or yeasts.

BIOSYNTHESIS OF ENZYMES

It is clear from earlier discussions of microbial metabolism that microorganisms produce enzymes of almost endless variety, some of them involved in synthesizing compounds and others essential for degradation. Many of these enzymes play important roles in industrial processes. Some examples of microbial enzymes produced by industry, and their uses, are described here.

Glucose Isomerase

Glucose isomerase converts glucose to the much sweeter sugar, fructose. Strains of *Bacillus* and *Arthrobacter* are used commercially to obtain the enzyme. With the withdrawal from the market of cyclamates and other artificial sweeteners, the demand for fructose, and consequently for glucose isomerase, has greatly increased.

Amylases

Amylases, produced by strains of *Aspergillus* mold, are used to degrade starch to sugars that can be fermented. These enzymes are used to make chocolate syrups from cocoa, to prepare corn syrup, to break down certain starches (dextrins) in beer (producing the "low-calorie" beers), as medicines to aid digestion, and for a number of other industrial processes. Amylases made by strains of *Bacillus* are more temperature-resistant than the fungal enzymes and are used in the textile industry.

Proteases

Proteases—enzymes that degrade proteins—have many applications in industry. Some of them are also used in enzyme detergents; others are used to tan leather. In the textile industry, proteases are used to remove coatings from certain fibers, and in the photographic industry, they remove the gelatin base from used x-ray films so that the silver in these films can be recovered. Proteases can also be used to tenderize meat and for many other purposes.

Beta-galactosidase

Beta-galactosidase, an enzyme produced commercially by some species on the yeasts *Saccharomyces* and *Candida*, has been used medically to degrade lactose in milk for patients who do not have an essential lactose-degrading enzyme. Lactose is a sugar made up of glucose and galactose. Normally, ingested lactose is split in the gastrointestinal tract into glucose and galactose, and the galactose is absorbed into the body and further broken down by enzymes. When the necessary enzymes are lacking as the result of a genetic defect, galactose levels in the blood rise, and serious or even fatal physical and mental defects in affected infants eventually result. The commercially produced enzymes can be used to break down lactose in milk, a life-saving measure for babies who have this genetic defect. Beta-galactosidase is also used in the dairy industry to prevent undesirable lactose crystal formation in ice cream.

Other Enzymes

Rennins, used to coagulate milk for making cheese, may be produced commercially by certain molds, such as species of *Mucor*. *Aspergillus* and *Penicillium* synthesize pectinases, enzymes that break down pectin. These enzymes are especially useful in the coffee industry for removing the red berry from the coffee bean in a manner superior to the natural fermentation that has been employed traditionally to prepare coffee beans.

These examples represent only a few of the many microbial enzymes, but they suggest the multiple applications of the enzymes in industry. Techniques for immobilizing enzymes by attaching them to inert substances such as glass beads or synthetic polymers have made even more uses of microbial enzymes possible in recent years.

BIOCONVERSION OF STEROIDS

Steroids are lipid compounds with a characteristic chemical structure, shown in Figure 32–1. Important steroids include cholesterol, corticosteroids, and progesterone and other hormones. The basic steroid molecules, readily obtained by chemical synthesis or from plants, are converted by microorganisms into compounds useful in humans. For example, microbes are used to change or bioconvert a chemical group at position II in Figure

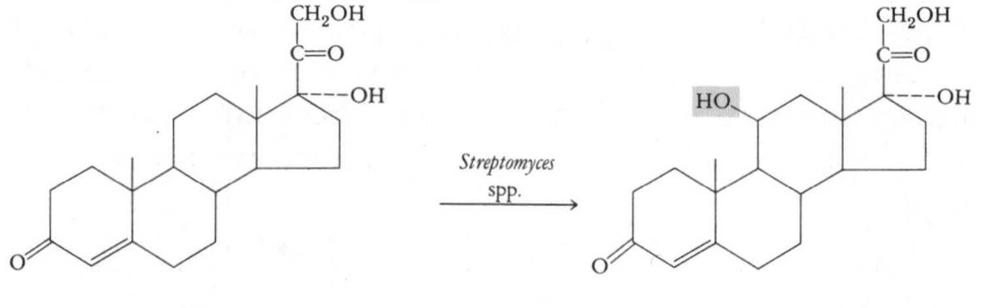

Compound S

Steroid

Figure 32–1
Steroid transformation. Special strains of bacteria or molds can enzymatically convert compound S, an inactive substance widely distributed in nature, to valuable steroids such as hydrocortisone.

32–1, as a step in the production of the powerful antiinflammatory agent hydrocortisone from the chemically produced precursor molecule.

ANTIBIOTICS

About 40 years ago, penicillin was the first of the antibiotics to be produced commercially on a large scale. Since that time, a number of developments have permitted increasingly efficient production, as mentioned earlier. Mutant strains of *P. chrysogenum* are used to produce benzylpenicillin, which is then altered chemically to yield various semisynthetic penicillins. An interesting intermediate step is the conversion of benzylpenicillin to an active state using an enzyme from *E. coli* before the chemical conversion (Fig. 32–2).

Streptomyces strains of various species have yielded antibiotics such as streptomycin, the tetracyclines, chloramphenicol, and the antifungal agent amphotericin B. The same principles of mutant selection and media development used to maximize the yields of penicillins have been applied to commercial production of these and other antibiotics.

VACCINES, TOXOIDS, AND ANTITOXINS

One of the first benefits of immunology was the ability to immunize effectively against a number of infectious diseases. Over the years, vaccine development has improved as more has been learned about immune responses. The uses of vaccines, toxoids, and antitoxins are discussed in Chapter 12. Commercial production of these substances is closely regulated by the government, and the development of new products requires extensive testing for safety before the product is approved. Progress is being made in isolating the important antigens that cause protective responses. These are isolated from whole organisms or mixtures of substances. For example, polysaccha-

---

Penicillium chrysogenum mutant grown in medium containing glucose, growth factors from corn steep liquor, and mineral salts

↓ Incubation period

Mycelium is removed and benzylpenicillin is extracted from the broth

↓

Penicillin acylase enzyme from *Escherichia coli* is used to produce the intermediate 6-aminopenicillamic acid or 6-APA

↓

Various side groups are introduced into the 6-APA molecule by chemical means to produce the semisynthetic penicillins

---

Figure 32–2
Steps in the commercial production of semisynthetic penicillins.

Note

rides of meningococci and pneumococci that induce protective antibodies have been isolated and used to prepare effective vaccines.

Progress is also being made in developing safe and efficient adjuvants* for human use. It has been found, for example, that the portion of mycobacteria most active as an adjuvant is a small fragment of the cell wall called MDP (a derivative of muramic acid). This material is much safer to use than the whole mycobacterium, which cannot be used in humans because this organism causes severe inflammatory reactions.

Advances are also being made in the production of viral vaccines. Specific antigens are being identified and isolated for immunization. For example, a vaccine against influenza is prepared by using only two components of the virus (hemagglutinin* and neuraminidase*) known to be the most important antigens for inducing immunity against the whole virus (see Chapter 12).

ANTIBODY PRODUCTION BY CELL CULTURES

A remarkable method called the **hybridoma technique** for in-vitro antibody production has recently been developed. In this method, a normal cell producing the desired antibody is fused with a myeloma* cell capable of surviving in culture and producing huge quantities of immunoglobulin. This clone of cells is propagated to produce large numbers of hybrid cells that synthesize and secrete the desired antibodies, all identical to each other. This technique has been enormously useful in research, and it shows great promise in industry. At present, it is being used commercially to produce highly specific antibodies for laboratory use.

PRODUCTION OF SINGLE-CELL PROTEIN

American industries are developing new ways to use microorganisms commercially to recycle some of the three to four billion tons of waste materials generated in the United States each year. One way is the production of single-cell protein. Wastes or cheap materials are converted to valuable proteins by the metabolic activities of microorganisms. For example, yeasts are grown on wastes from corn processing and used as animal feed. Fungi of two genera work together to produce feeds from potato starch water; yeasts of one genus produce enzymes that degrade the starch to sugars, and other yeasts, of the genus *Candida*, use the sugars to produce proteins. The photosynthetic organism *Spirulina maxima* produces protein by using bicarbonate, carbon dioxide, and sunlight. This process is used commercially in Mexico.

---

*adjuvants—substances used to increase the efficiency of the immune response when injected simultaneously with an antigen

*hemagglutinin—substance that causes erythrocytes to clump together

*neuraminidase—enzyme that degrades derivatives of neuraminic acid; neuraminic acid derivatives occur in erythrocyte membranes and other body substances

*myeloma—tumor composed of immunoglobulin-producing cells; typically involves the bone marrow

EXTRACTION OF MINERALS

In this age of declining supplies of necessary materials, including some of the minerals, it is of interest that microorganisms can be used commercially to extract minerals from the earth. Microorganisms are used to recover copper, uranium, zinc, and other substances from insoluble compounds by a combination of chemical and metabolic processes.

This process of "microbial mining" depends on the activities of the acidophilic* thiobacilli, which, as their name implies, prefer acidic conditions and utilize sulfur. Of primary importance is *Thiobacillus ferrooxidans;* however, other organisms also participate, some of them thermophiles. *Thiobacillus ferrooxidans* oxidizes sulfur, sulfides, and ferrous iron to obtain energy; it can use atmospheric CO_2 as a carbon source and dissolved ammonia or nitrates as a nitrogen source. It can even fix nitrogen from the air if necessary. It grows well on iron pyrites (FeS_2), which are usually found together with valuable minerals such as uranium and copper. The organisms attach firmly to the pyrite and produce, from the FeS_2, ferric sulfate and sulfuric acid, a mixture that dissolves many other minerals, allowing them to be extracted, or leached,* from ores. This mining process is called sulfide leaching. In copper deposits, low-grade ores or the wastes of richer ores can be leached in this manner, releasing copper sulfate that is readily recovered. Water containing the copper sulfate is passed over scrap iron, on which the copper is deposited in a layer and from which it can be scraped. The aqueous solution left after the copper is removed contains ferrous iron; this fluid is collected, and microorganisms are used to oxidize the ferrous iron to ferric iron, which is recycled to leach more copper from the ore. About 10 percent of the copper produced in the United States is obtained by microbial mining with this kind of leaching process.

Uranium associated with iron pyrites is readily leached from ore by ferric sulfate produced by bacteria. The insoluble oxide of uranium is converted to soluble uranyl sulfate, which is then recovered from solution by chemical means. Microbial mining takes longer than conventional mining methods—more than twice as long for uranium—but it is cheaper and requires less energy. As this process is being used more and more, the future of mining has been projected as follows:

The ore extraction plant of the future could have the appearance of a present-day water-treatment plant: situated in an environment of comparative quiet and tranquility and free from the dirt and spoil heaps normally associated with mining operations, while far below ground millions of microbes are carrying out the tasks which today are characterized by the roar of machinery and the ring of pick and shovel on rock.[1]

---

[1]Manchee, R.: "Microbial Mining." *Trends in Biochemical Sciences* (April 1979).

*acidophilic—preferring or requiring an acidic environment

*leached—extracted by dissolving in a liquid

SUMMARY

The sciences of microbiology and immunology have many applications in industry. An important principle of commercial microbiology is that the microbial procedure used must be economically advantageous if it is to be successful. Necessary steps in developing economically feasible applications are the selection of strains efficient in the desired microbial reaction, the further development of even more efficient strains, and the choice of media that give maximal production of the desired product.

Among the microbial processes currently important to industry are the following: biosynthesis of organic acids and vitamins, including citric acid, glutamic acid, lysine, and vitamins A, B_{12}, and C; biosynthesis of a multitude of enzymes, including (among others) glucose isomerase, amylases, proteases, beta-galactosidase, and rennins; bioconversion of plant or chemical steroids to compounds useful in humans; and the production of antibiotics.

Practical applications in the area of immunology include the production of vaccines and other immunizing agents and a new method for the synthesis of specific antibodies by cultures of hybrid cells—the hybridoma technique.

The production of single-cell protein by microorganisms is now economically feasible. Microbial processes also contribute substantially to commercial extraction of a number of minerals, including copper, uranium, and zinc.

SELF-QUIZ

1. At present, the commercial production of penicillin uses a high-producing mutant strain of
   a. *Penicillium notatum*
   b. *P. chrysogenum*
   c. *P. griseofulvum*
   d. *Aspergillus flavus*
   e. *Streptomyces* species
2. Soil microorganisms can be used to produce all the following substances *except*
   a. glutamic acid
   b. vitamin B_{12}
   c. glucose isomerase
   d. enzyme detergents
   e. agar
3. Fungi are commercially important in producing all the following substances *except*
   a. rennins
   b. pectinases
   c. streptomycin

d. single-cell protein
e. vitamins
4. An organism of primary importance in microbial mining is
   a. *Thiobacillus ferrooxidans*
   b. *Thiothrix ferrooxidans*
   c. *Thiothrix schenckii*
   d. *Thermoactinomyces* species
   e. *Actinomyces israelii*
5. Glucose is converted to fructose by the enzyme glucose isomerase, produced commercially by strains of
   a. *Aerobacter* and *Brevibacterium*
   b. *Bacillus* and *Arthrobacter*
   c. *Saccharomyces* species
   d. *Candida* species
   e. *Streptomyces* species

QUESTIONS FOR DISCUSSION

1. How might industry profit from genetic engineering, or how has it already profited?
2. The Supreme Court has ruled that a scientist can patent a strain of
   microorganisms capable of degrading oil spills. Why is this decision important?
3. What place is there for microbiologists in the mining industry? What
   kind of microbial preparation would be most valuable for mining?

FURTHER READING

Bull, A. T., Fellwood, A. C., and Rutledge, C., (eds.): *Microbial Technology: Current State, Future Prospect.* New York: Cambridge University Press, 1979.

Dixon, B.: *Magnificent Microbes.* New York: Atheneum Press, 1976.

Gilbert, W., and Villa-Kanoroff, L.: "Useful Proteins from Recombinant Bacteria." *Scientific American* (April 1980).

Manchee, R.: "Microbial Mining." *Trends in Biochemical Sciences* (April 1979). Describes the use of bacteria to obtain various metals, especially in low-grade
ores.

Miller, B. M., and Litsky, W.: *Industrial Microbiology.* New York: McGraw-Hill Book
Co., 1976.

Milstein, C.: "Monoclonal Antibodies." *Scientific American* (October 1980).

Riviere, J.: *Industrial Application of Microbiology.* (Translated and edited by M. O.
Moss and J. E. Smith.) New York: John Wiley & Sons, 1977.

Ryther, J. H., and Goldman, J. C.: Microbes as Food in Mariculture." *Annual Review
of Microbiology* 29:429–443 (1975). A review restricted to the role of microorganisms as food, directly or indirectly, from commercially produced marine
organisms.

Appendix I

BIOCHEMISTRY OF THE MOLECULES OF LIFE

FORMATION OF MOLECULES: CHEMICAL BONDS

Most atoms can associate with other atoms to form **molecules.** There are several kinds of bonds, differing in strength and in the kinds of atoms they can unite. The most important strong bonds are the **covalent bonds.** Covalent bonds are formed by the sharing of electrons. These bonds are indicated by a line between the atoms: H—H, C—C, C—H, C—O. The most important weak bonds are **hydrogen bonds,** indicated by $\cdots$ between atoms. These bonds are formed only between hydrogen atoms covalently bonded to either O (oxygen) or N (nitrogen), and N or O atoms covalently bonded to H or C (carbon). A single covalent bond is strong enough to hold a molecule together, although a single hydrogen bond is not. However, many hydrogen bonds acting as a group are able to hold molecules together firmly. The bonds that hold the two strands of DNA together are hydrogen bonds. Hydrogen bonds can form only when the atoms being joined are very close to each other. This closeness is possible only when the atoms joined together are **complementary** to one another, such that a positive charge on one atom is matched by a negative charge on the atom to which it bonds. Complementarity plays a very important role in the interaction between the components of nucleic acids.

IMPORTANT MOLECULES OF LIFE

Small Molecules

All cells contain a variety of small molecules and atoms, both organic* and inorganic.* About 1 percent of the total weight of a bacterial cell is composed

---

*organic—any molecule containing a carbon atom except CO_2 and CO

*inorganic—any molecule not containing a carbon atom; also CO_2 and CO

of inorganic ions,* principally Na^+, K^+, Mg^{2+}, Ca^{2+}, Fe^{2+}, Cl^-, PO_4^{3-}, and SO_4^{2-}. Small organic molecules consist mainly of compounds that are being metabolized or have accumulated as a result of metabolism and those that serve as the building blocks of various macromolecules. Alcohols, aldehydes, and organic acids are the most frequent products of metabolism. Small molecules can be classified according to the important group of atoms (called a **functional group**) they contain (Table I–1).

Macromolecules and Their Subunits

Three major classes of biologically important macromolecules are proteins, nucleic acids, and polysaccharides. The lipids, another class of molecules not nearly as large as the macromolecules, will be discussed briefly.

All macromolecules are composed of **subunits,** with each subunit having features similar to those of the other subunits of the macromolecule. For example, amino acids are the subunits of proteins. Although there are more than 20 different amino acids, each one is characterized by a carboxyl group and an amino group (Table I–1) located in the same position in each molecule.

PROTEINS

Twenty different major **amino acids** are commonly found in **proteins.** Their names and chemical formulas are given in Figure I–1. All amino acids have several features in common: a terminal carboxyl group (—COOH) and an amino group (—NH_2) are bonded to the same carbon atom to which a side chain (shaded) that is characteristic for each amino acid is also bonded. The side chain, which gives individuality to each amino acid, may be a very simple group such as —H or —CH_3, a longer chain of carbon atoms, or one of various kinds of ring structures, such as the benzene ring found in tyrosine and phenylalanine. The amino acids are bonded together by **peptide bonds,** a unique type of covalent linkage formed when the carboxyl group reacts with the amino group of the adjacent amino acid (Fig. I–2).

NUCLEIC ACIDS

Nucleic acids—both ribonucleic acid (RNA) and deoxyribonucleic acid (DNA)—are long linear molecules in which the subunits are called **nucleotides.** Each nucleotide is composed of a nitrogen-containing ring compound, known as a base, covalently bonded to a five-carbon sugar molecule, which in turn is bonded to a phosphate molecule. There are five different nitrogen-containing bases. These can be divided into two groups, purines and pyrimidines, depending on their structure (Fig. I–3). The two types of nucleic acids,

---

*ions—atoms or groups of atoms that are positively or negatively charged as a result of having lost or gained one or more electrons

TABLE I–1 CLASSIFICATION OF SMALL MOLECULES BY FUNCTIONAL GROUPS

Group	Name of group	Examples of group in biologically important molecule		Class of molecule found in
(methyl structure)	Methyl	(methane structure)	Methane	
—O—H	Hydroxyl	(ethanol structure)	Ethanol (ethyl alcohol)	Alcohols
(carboxyl structure)	Carboxyl	(acetic acid structure)	Acetic acid	Acids
(amino structure)	Amino	(glycine structure)	Glycine	Amines and amino acids
(ketone structure)	Ketone	(acetone structure)	Acetone	Ketones
(aldehyde structure)	Aldehyde	(acetaldehyde structure)	Acetaldehyde	Aldehydes
—S—H	Sulfhydryl	(cysteine structure)	Cysteine	A few amino acids
(ketone structure) and (carboxyl structure)	Ketone and Carboxyl	(pyruvic acid structure)	Pyruvic acid	Keto acids
—O—H and (carboxyl structure)	Hydroxyl and Carboxyl	(lactic acid structure)	Lactic acid	Hydroxy acids

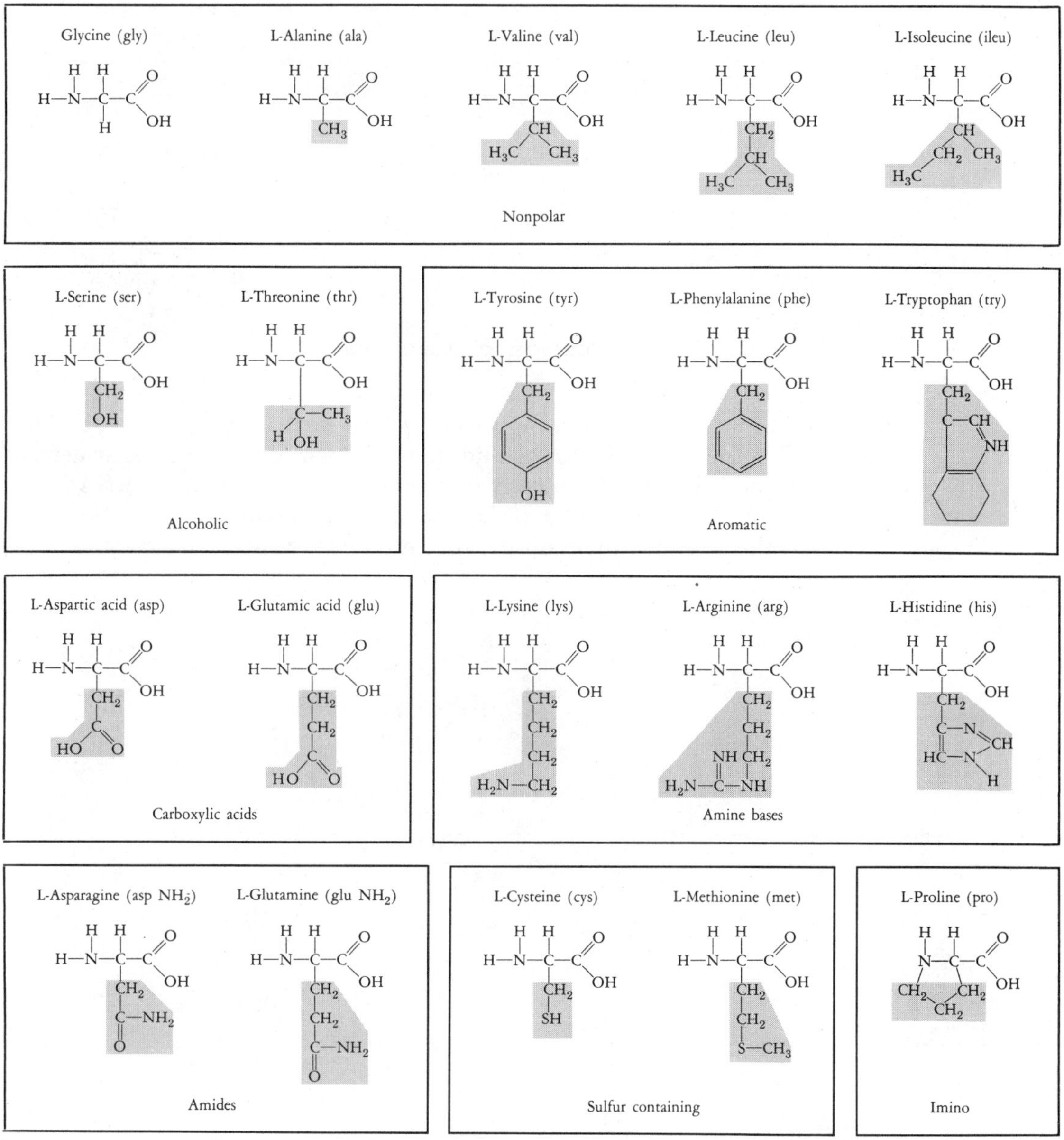

Figure I–1

Amino acids. All amino acids have one feature in common: a carboxyl group and an amino group on the carbon atom next to the carboxyl group. Although the remainder of the molecule differs for each amino acid, they can be grouped because of certain features they have in common.

Glycine Serine Glycylserine (*dipeptide*)

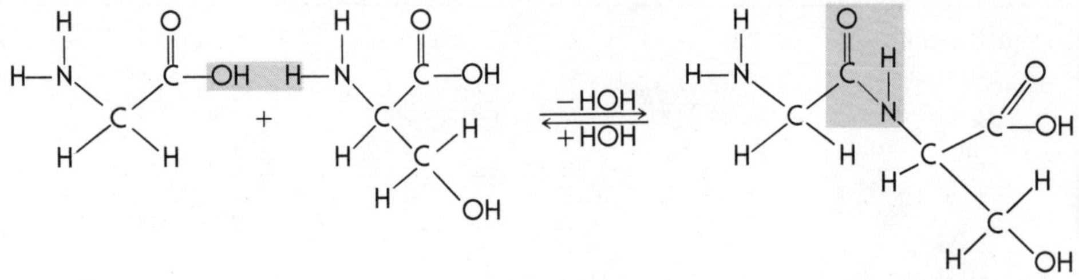

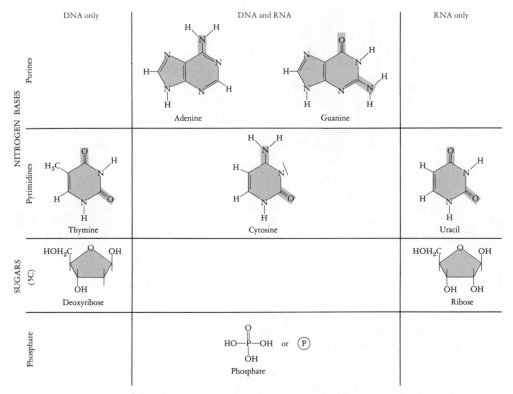

Figure I–2
Formation of a peptide bond between glycine and serine.

RNA and DNA, differ in the composition of their nucleotides. The nucleotides of RNA contain the sugar **ribose,** and those of DNA contain the sugar **deoxyribose.** An additional difference is that the pyrimidine **uracil** in RNA is replaced by **thymine** in DNA. Both DNA and RNA contain the same two purines (**adenine** and **guanine**) and the pyrimidine **cytosine.** Thus, each nucleic

Figure I–3
Components of RNA and DNA. Carbon atoms are not represented in ring structures.

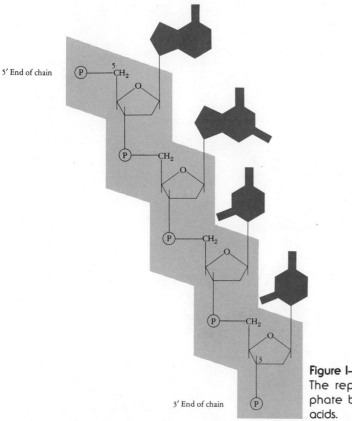

5′ End of chain

3′ End of chain

Figure I—4
The repeating sugar-phosphate backbone of nucleic acids.

acid is composed of four particular kinds of nucleotide subunits (Fig. I–3). The subunits are joined together by a covalent bond between the **phosphate** of one nucleotide and the **sugar** of the adjacent nucleotide. This phosphate bridge joins the number-three carbon atom of one sugar to the number-five carbon atom of the other, so that a backbone composed of alternating sugar and phosphate molecules is formed (Fig. I–4).

The length of the nucleic acid molecule varies, depending on whether the macromolecule is RNA or DNA. RNA is considerably shorter than DNA. Furthermore, DNA is a double-stranded molecule, in which the strands are wound around each other to form a double helix (Fig. I–5). In general, RNA is composed of only a single strand of nucleotides. There are three kinds of RNA: **ribosomal RNA, messenger RNA** (mRNA) and **transfer RNA** (tRNA).

POLYSACCHARIDES

Carbohydrates are compounds containing principally carbon, hydrogen, and oxygen atoms in a ratio of approximately 1:2:1. The carbohydrate macromolecules, the **polysaccharides,** are long linear molecules whose subunits

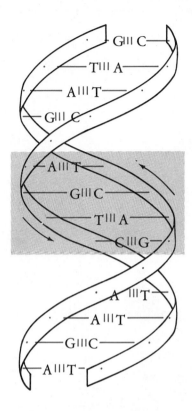

A

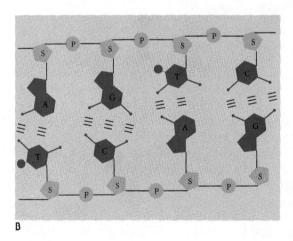

B

Figure I–5

(a) Helical structure of DNA and (b) its appearance when not twisted. The helical DNA molecule can be looked on as a spiral staircase, in which the sugar-phosphate repeating sequences form the two railings and the purine and pyrimidine bases represent the stairs. The railings move in opposite directions, one going up, the other down. If the helix is untwisted, the molecule assumes the shape of a ladder. The fact that the two strands run in opposite directions is most readily seen in this untwisted form. The two strands are held together by hydrogen bonding.

are **monosaccharides.** Monosaccharides are classified by the number of carbon atoms they contain. The most common in nature are the five-carbon pentoses and the six-carbon hexoses (Fig. I–6). The five-carbon pentose sugars are represented by the sugars in nucleic acids, **ribose** and **deoxyribose**

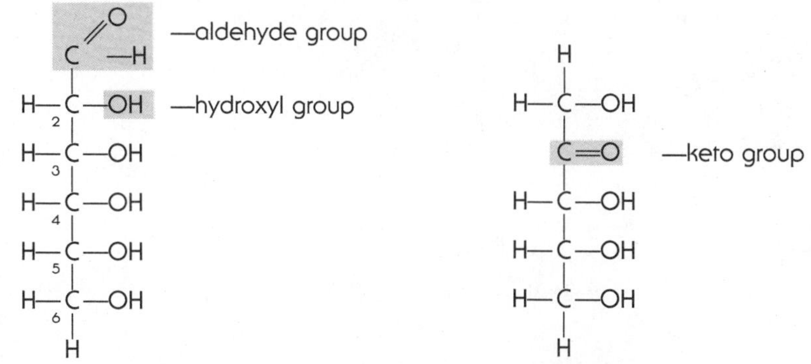

6-Carbon Sugar 5-Carbon Sugar

Figure I–6
General formulas for a 6-carbon and a 5-carbon sugar.

(Fig. I–3). Common hexoses are **glucose, galactose,** and **mannose** (Fig. I–7). The chief distinguishing feature of carbohydrates is that they contain a large number of alcohol groups (—O—H), which accounts for their ratio of carbon, hydrogen, and oxygen atoms. Carbohydrates usually contain an aldehyde (—$\overset{O}{\overset{\|}{C}}$—H) and, less commonly, a ketone group (—$\overset{O}{\overset{\|}{C}}$—). The amino sugars contain an amino group (—NH_2).

The general structural feature of a polysaccharide is the repeating monosaccharide subunit, the most frequent being glucose. However, unlike pro-

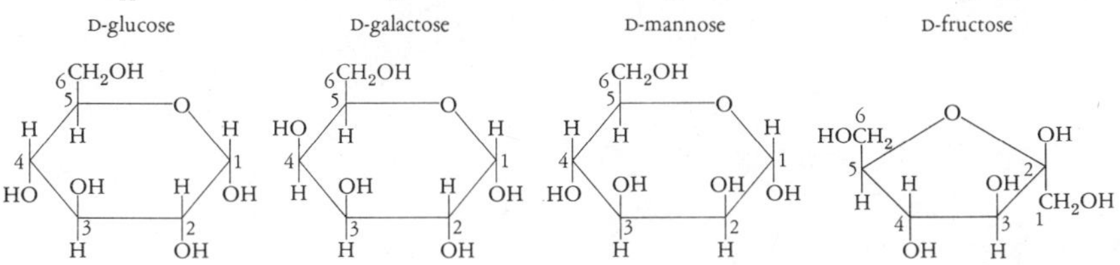

Figure I–7
Formulas of some common sugars. The ring structures arise as a result of the reaction of the —OH group on carbon atom 5 with the aldehyde group on carbon atom 1, resulting in an O bridge between carbon atoms 1 and 5. Note that fructose is unusual in that it has a keto group on carbon atom 2.

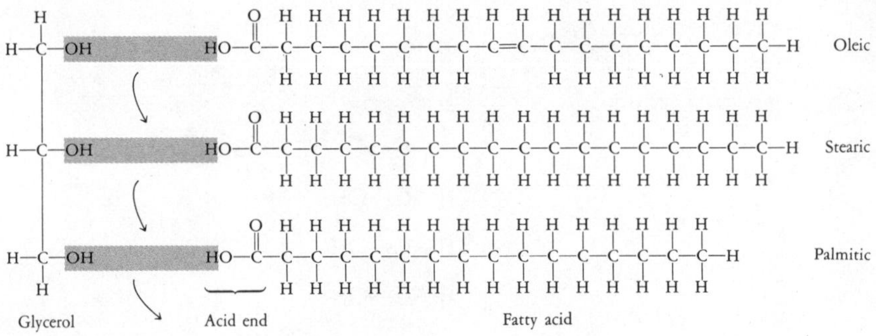

Figure I—8
Formation and general formula of a fat. An ester bond is formed between the alcohol group of the glycerol and the acid group of the fatty acid.

teins and nucleic acids, polysaccharides can have side chains branching from the main linear chain.

LIPIDS

The **lipids** represent a very heterogeneous group of biologically important molecules, all of which have the property of being only slightly soluble in water but readily soluble in most organic solvents such as ether, benzene, and chloroform. The two general classes are the **simple** and the **compound lipids.** The most common of the simple lipids are the **fats,** a combination of fatty acids and glycerol in which the bond forms between the carboxyl group of the acid and the hydroxyl group of the glycerol. The general formula for a fat is given in Figure I–8. Compound lipids contain other elements, such as sulfur, nitrogen, or phosphorus, in addition to the carbon, hydrogen, and oxygen found in simple lipids. **Phospholipids** contain the fatty acids and glycerol and, in addition, a phosphate molecule bonded to a nitrogen-containing compound. Another group of compound lipids, the **lipoproteins,** are loose associations of proteins and lipids held together by weak bonding forces. **Lipopolysaccharides** are molecules of lipids associated with polysaccharides through covalent bonds.

Appendix II

TECHNIQUES OF MICROBIOLOGY

MICROSCOPIC TECHNIQUES: INSTRUMENTS

Compound Microscope

The most commonly used instrument for observing cells is the **light microscope,** so named because it illuminates its object by visible light. There are many variations of this instrument. Van Leeuwenhoek's light microscope, consisting of only a single magnifying lens, was called a **simple microscope.** The light microscopes most commonly used today have two sets of lenses, an objective and an ocular, and are called **compound microscopes.** The magnification achieved with such a microscope is the product of the magnification of each of the individual lenses. Most compound microscopes have a number of objective lenses, thereby making possible a number of different magnifications with the same instrument. A modern compound microscope, with its major components labeled, is shown in Figure II–1. Even a very good compound microscope can magnify objects only up to 2000 times.

The usefulness of a microscope depends not so much on the degree of magnification but rather on the ability of the microscope to separate clearly or **resolve** two objects that are very close together. The **resolving power** of a microscope is defined as the minimum distance that can exist between two points such that the points are observed as being separate. The resolving power determines how much detail can actually be seen. It depends on the quality of the lens, the magnification, and the preparation of the specimens under observation. The maximum resolving power of the best light microscope functioning under optimal conditions is 0.2 μm (about 0.000008 inch) because the resolving power is limited by the wavelength* of the light em-

---

*wavelength—light may be considered to travel in waves; the distance between adjacent peaks of two waves is the wavelength of the light

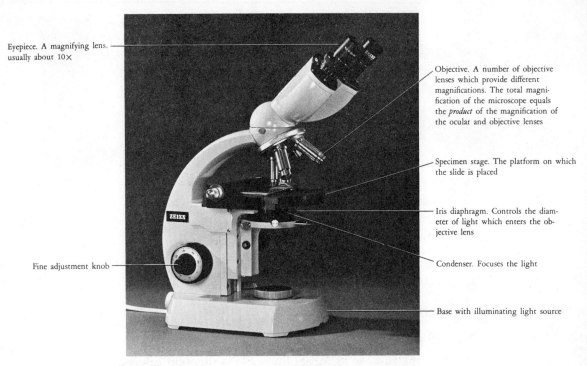

Eyepiece. A magnifying lens, usually about 10×

Objective. A number of objective lenses which provide different magnifications. The total magnification of the microscope equals the *product* of the magnification of the ocular and objective lenses

Specimen stage. The platform on which the slide is placed

Iris diaphragm. Controls the diameter of light which enters the objective lens

Fine adjustment knob

Condenser. Focuses the light

Base with illuminating light source

Figure II–1
Compound microscope. (Courtesy of Carl Zeiss, Inc.)

ployed for illumination. The shorter the wavelength, the greater the resolution that can be attained. The unaided human eye has a resolving power of about 200 μm (about 0.008 inch).

Phase Contrast Microscopy

Although bacteria are often large enough to be seen with a light microscope when suspended in a drop of liquid, they are very difficult to observe because they are transparent, usually colorless, and sometimes so tiny that they may be difficult to find. To overcome the problem of transparency in all types of cells, a special type of light microscope called the **phase contrast microscope** has been devised. The microscope, probably the one most commonly used in research laboratories for observing living microbes, is a variation of the compound microscope. It has special optical* devices that increase the contrast between the microbes and the surrounding medium. Since cells are denser than the surrounding medium, the illuminating light is slowed down more as it passes through the specimen than it is in the medium surrounding the specimen. Thus, even though the cells are not magnified to any greater

*optical—made to give help in seeing

extent, they nevertheless stand out from the background and are clearly visible. By this means it is possible to observe living organisms clearly and to study their movements in the medium in which they are growing. Even internal structures are discernible by phase contrast microscopy (color plate 81).

Dark-field Microscopy

Another commonly employed method for achieving a marked contrast between living organisms and the background is **dark-field microscopy.** In this technique, light is directed toward the specimen at an angle, so that only light that is scattered by the specimen enters the objective lens and is visualized. The background is completely dark except for the objects being viewed, which are brilliantly illuminated (color plate 82). This technique makes visible objects and cells that are invisible by ordinary light microscopy. When dark-field microscopy is used, *Escherichia coli* is clearly visible at magnifications of only 100-fold. This technique is especially useful for observing very thin cells, in particular the organism that causes syphilis, *Treponema pallidum*, which is barely visible using ordinary light microscopy. The advantage of using dark-field microscopy for observing living bacteria is that it permits the viewer to estimate their true size, shape, and mobility.

Fluorescence Microscopy

Another type of microscope important in laboratories concerned with identifying microorganisms is the **fluorescence microscope.** This microscope is used to visualize objects that fluoresce, that is, emit light when a light of a different wavelength impinges on them. The fluorescence may be a natural property of the specimen being viewed, or it may result from a fluorescent compound being bound to a normally nonfluorescing material. A special lamp supplies the ultraviolet light, which reaches the object stained with the fluorescent compound. When stimulated by ultraviolet light, the fluorescent material emits a visible yellow-green light (color plate 35).

Electron Microscopy

To achieve resolving powers significantly greater than those of the light microscope, a new type of microscope, the **transmission electron microscope,** was constructed. In this microscope, a beam of electrons, which is analogous to the light source, is focused by magnetic fields,* which function as lenses. Some of the electrons pass through the specimen, others are scattered, and still others cannot pass. The electrons impinge on an electron-sensitive screen, creating an image that is determined by the ability of the electrons

*magnetic fields—attractive forces near a magnet

to pass through various parts of the object being viewed. A resolving power of 0.003 μm (about 600 times better than the light microscope) can be achieved with this instrument. To clarify the details of the cell structure, investigators often slice the specimen to be viewed into very thin sections, commonly with a diamond knife, and observe these "thin sections" (each approximately 0.02 μm thick). In the past, specimens had to be fixed* so that they were "bone dry" and extremely thin before they could be observed with the electron microscope. Since the fixation process may introduce a large number of artifacts* into the preparation, one of the main jobs of electron microscopists is to interpret whether what they see is actually present in the living cell or merely a product that results from vigorous treatment in the fixation process.

A technique called freeze-etching has been devised that circumvents the need for chemical fixation and thereby presumably overcomes the problem of artifacts. The material is frozen and thin sections are chipped off. The surface of a section is then coated with a layer of carbon thin enough to be viewed with the electron microscope. The pictures from such freeze-etched preparations can be quite dramatic (color plate 83).

A significant advance in microscopic techniques has been made in recent years with the development of a modified electron microscope, the **scanning electron microscope.** The specimen to be observed is coated with a thin film of metal. The electron beams scan back and forth over the surface (none pass through the surface, in contrast to transmission EM) and are reflected back into the viewing chamber. The most significant aspect of this technique is that relatively large specimens can be viewed, and a dramatic three-dimensional figure can be observed.

MICROSCOPIC TECHNIQUES: STAINING

Since living, transparent, and often motile single-celled organisms are difficult to see, they are frequently killed and then treated with one or more dyes that have a special affinity* for one or more cellular components. Such treatment results in the entire organism, or part of it, appearing in marked contrast to the unstained background. In the staining procedure, a drop of liquid containing the organisms is placed on a glass slide and allowed to dry. The dried film is then "fixed" onto the slide, either with a chemical fixative* or more commonly by passing it over a flame to coagulate the cell protein. The dyes, or stains, are then applied to the "fixed" organisms. A wide variety of stains and staining procedures are currently employed, each one having its own particular use. Some dyes will stain only a particular

*fixed—prepared and immobilized firmly

*artifacts—properties not present originally

*affinity—attraction that results in binding

*fixative—any preparation that results in the attachment of bacteria to the slide

cell component. Other staining procedures will stain one group but not another group of organisms, thereby serving to divide organisms into groups by their staining characteristics. Stains can be divided into two major types by their affinity for cell components. **Positive stains** have a strong affinity for one or more cell components and color these components when added to fixed cell preparations. **Negative stains** cannot penetrate the cell envelope and thus make the cell highly visible by providing a contrasting dark background (color plate 84). Negative stains are generally used on living cells to demonstrate surface structures that are not stained well by positive stains.

Simple Staining Procedure

In simple staining, only a single dye is applied to the cells. This technique may be used to stain the entire cell or specific structures in the cell. The dye methylene blue is frequently applied to a fixed suspension of bacteria to stain entire cells a blue color without staining the background material (color plate 85). This basic dye binds primarily to the nucleic acids of the cell, and little evidence of other internal structures is revealed. Acidic dyes, which include safranin, acid fuchsin, and Congo red, stain basic compounds in the cell, primarily certain proteins carrying a basic charge. Sudan black is a dye that is very soluble in fat and is frequently used to identify the presence and location of fat droplets in bacteria (color plate 12).

Differential Staining Techniques

It is possible to separate bacteria into groups, depending on their ability to take up and retain certain dyes. The most widely used staining procedure, capable of dividing virtually all bacteria into one of two groups, is the **Gram stain** (Fig. II–2). The procedure involves the application of a basic purple dye, usually gentian or crystal violet, which stains all bacteria able to absorb this dye. Next, a dilute solution of iodine is added, which serves to decrease the solubility of the purple dye within the cell by combining with the dye to form a dye-iodine complex. An organic solvent such as ethanol, which readily removes the purple dye-iodine complex from some but not other species of bacteria, is added next. A red stain is then applied. Those bacteria that are decolorized by the ethanol appear red, or **Gram-negative** (color plate 10). Those that retain the purple dye appear purple, or **Gram-positive** (color plate 9).

Another differential staining procedure, the **acid-fast stain,** is used to characterize a small group of organisms that resist decolorization with an acidic solution of alcohol after being stained with a basic dye. Only a few groups of bacteria retain the basic dye—that is, are acid-fast—but because one acid-fast group causes tuberculosis, this stain has proved to be extremely valuable in diagnostic laboratories concerned with detecting these organisms (color plate 1).

Differential stains are also employed to stain specific cell structures such

STEPS IN STAINING   APPEARANCE OF CELLS

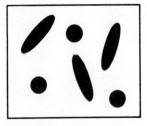

Cells fixed
by heating

Shape of cells
becomes distorted
and cells shrink
in size

Basic dye, crystal
violet applied

All cells
stain purple

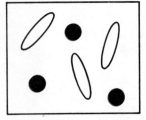

Addition of
iodine

All cells
remain purple

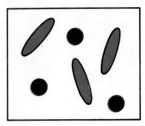

Addition of
alcohol

Gram-positive cells
remain purple;
Gram-negative cells
become colorless

Addition of
counterstain,
often safranin
(red dye)

Gram-positive cells
remain purple;
Gram-negative cells
appear red

Figure II–2
The steps in the Gram stain.

as flagella, endospores (color plate 86), and nuclear bodies. The staining procedure is different for each structure and takes advantage of the chemical composition and properties of the structure.

Because staining generally requires that the cells be heat-dried and killed, their morphology can easily be distorted, producing artifacts. The fixation process tends to reduce the size of the cells, whereas the addition of dyes tends to increase their size. Indeed, flagella (organelles of locomotion) are so thin that they can be observed with a light microscope only if their size is increased by a specific staining procedure.

METHODS USED TO STUDY VIRUSES

Virus Purification

It is necessary to have quantities of purified viruses for many purposes. These include chemical and physical studies as well as the production of vaccines. With bacterial or plant viruses, such purification usually is relatively easy because it is possible to obtain large quantities of infected cells. The purification process consists of separating the virus particles from the constituents of the host cell. One method that is often used involves breakage of the host cells and sedimentation of their contents in an ultracentrifuge.* Centrifugation at low speed removes particles of cell debris, whereas higher centrifugation speeds are necessary to sediment the smaller virus particles.

With animal viruses, the primary difficulty is not so much in purifying the virions* but in obtaining enough infected cells. Some viruses still defy efforts to cultivate them in surroundings other than animal hosts, while others are more conveniently grown in cell cultures or embryonated chicken eggs.*

Culture of Human and Other Host Cells

Human and animal cells have been grown under laboratory conditions since the early part of this century. However, the methods used for culturing them have become routine only during the last 25 years, since the advent of antibiotics (used to suppress growth of bacteria and fungi in cell cultures) and more sophisticated aseptic* techniques. Currently, almost any type of animal cell that normally divides in the body can be grown or maintained in culture. Such cell cultures provide the most convenient means of studying the growth of animal viruses or of producing them in large quantities.

Before the development of methods for cell culture, the most useful host for the cultivation of animal viruses was the embryonated chicken egg. Embryonated eggs are infected through an opening made in the eggshell about a week or two after fertilization. Depending upon the virus in question, growth occurs in one or more of the embryonic membranes, in the yolk sac, or in the embryo itself.

---

*ultracentrifuge—a machine that rotates the samples at an extremely high speed

*virions—virus particles consisting of nucleic acid surrounded by a protein coat

*embryonated chicken eggs—chicken eggs containing a living chick (the embryo)

*aseptic—not leading to contamination

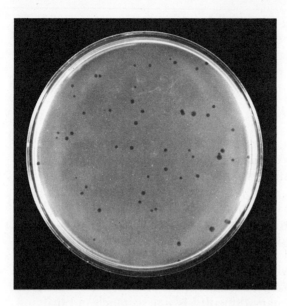

Figure II—3
Phage plaques. The opaque background represents a lawn of bacteria. Each clear area is a plaque formed as a result of lysis of the bacteria by the progeny of a single phage particle.

Quantification of Viruses

For any detailed study of viruses, it is necessary to develop methods for determining their number accurately. The methods used vary greatly, ranging from the counting of virions to studying the consequences of their interaction with living cells. A summary of the most commonly used methods follows.

Plaque Method. The detection of plaques is the most flexible, the most general, and the most informative method for quantification of virions because it is simple and reproducible. The method was first perfected for bacterial viruses and consists of the following operations: A sample of bacteriophage is mixed with a concentrated suspension of host bacteria and a few milliliters of melted agar at about 44°C. The agar is then poured into a Petri dish containing a hardened layer of nutrient agar on which it solidifies into a thin sheet containing a random distribution of viruses and bacteria. Each virus particle infects a bacterium, multiplies, and releases several hundred new virions. These infect other bacteria in the immediate vicinity, which again release virions. After a few multiplication cycles, all of the bacteria in an area surrounding the initial virus particle are destroyed. During the same period of time, uninfected bacteria in regions of the Petri dish without viruses multiply rapidly, giving a dense, opaque background. Thus the net result of the plaque assay consists of clear plaque areas contrasting with areas of bacterial growth (Fig. II–3). Because each of the dilute virus plaques corresponds to a single virus particle in the initial sample suspension, the number of plaques is a direct assay for virus concentration. The plaque count is analogous to the bacterial colony count.

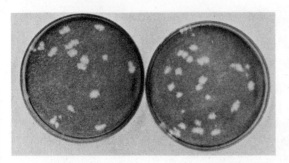

Figure II–4
Plaques caused by infection of human embryonic tonsil cells with herpesvirus, type I. (Courtesy of B. B. Wentworth. B. B. Wentworth and L. French, *Proc. Soc. Exp. Biol. Med.,* *131*:590, 1969.)

Plaque assays are equally useful for quantifying animal viruses that are lytic (Fig. II–4). In these assays, the viruses added to animal tissue cell cultures produce clear plaques in the monolayer* of cells.

Electron Microscopic Counting. If reasonably pure preparations of virions are obtainable, their concentration may be readily determined by counting the virus particles in a specimen prepared for the electron microscope. A great disadvantage inherent in this method is that it cannot distinguish between infective and noninfective virus particles.

Quantal Assays. Quantal assays can yield an approximation of virus concentration. Several dilutions of the virus-containing preparation are made and administered to a number of animals, cell cultures, or chick embryos, depending on the host specificity of the virus. The titer, or end-point, of virus is taken to be that dilution at which 50 percent of the inoculated hosts are infected. This is referred to as either the ID_{50} (infective dose) or the LD_{50} (lethal dose).

Hemagglutination. A large variety of viruses are able to agglutinate erythrocytes by chemical reactions between virions and surface components of the cells (hemagglutination). In this process, a virion attaches to two erythrocytes simultaneously and causes them to clump together. Sufficiently high concentrations of virus cause large clumps of erythrocytes to form (Fig. 10–16). Hemagglutination can be measured by mixing serial dilutions* of the viral suspension with a standard erythrocyte suspension. The highest dilution showing complete agglutination is taken as the end-point, or **titer.** One group of animal viruses able to agglutinate erythrocytes is the myxoviruses, exemplified by the influenza virus.

Figure 10–16, page 243

---

*monolayer—single layer; animal cells grow as a monolayer in tissue culture

*serial dilutions—dilutions in which the amount of material decreases by a constant fraction

Appendix III

PATHWAY OF GLUCOSE DEGRADATION THROUGH THE GLYCOLYTIC AND TCA CYCLES

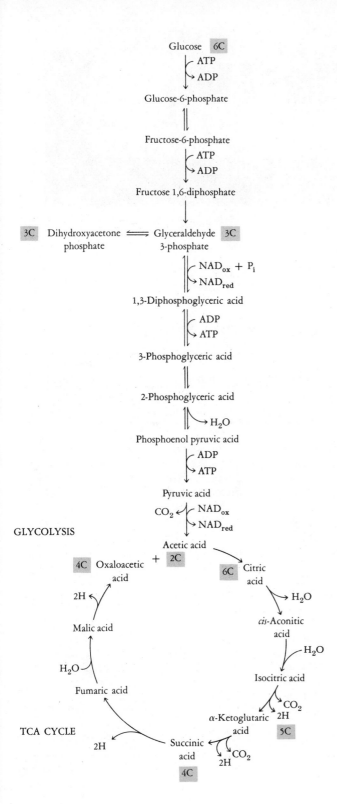

Appendix IV

RECOMMENDATIONS FOR CHEMICAL DISINFECTION AND STERILIZATION OF INSTRUMENTS

| Equipment | Disinfecting treatment | | Sterilizing treatment |
	Category A microorganisms	Category B microorganisms	Category C microorganisms
Smooth, hard	1A for 10 min	1B for 15 min	2 for 18 hr
	2 for 5 min	2 for 10 min	7 for 3–12 hr[a]
	3 for 10 min	4B for 20 min	8 for 12 hr
	4A for 10 min	5B for 20 min	9 for 10 hr
	5A for 5 min[b]	6B for 20 min[c]	
	6A for 5 min	8 for 15 min	
	8 for 5 min	9 for 15 min	
	9 for 5 min		
Rubber tubing and catheters, completely filled	3 for 10 min	4B for 20 min	7 for 3–12 hr[a]
	4A for 10 min	5B for 20 min	
	5A for 5 min	9 for 15 min	
Polyethylene tubing and catheters, completely filled	1A for 10 min	1B for 15 min	2 for 12 hr
	3 for 10 min	4B for 20 min	7 for 3–12 hr[a]
	4A for 10 min	5B for 20 min	8 for 12 hr
	5A for 10 min	9 for 15 min	9 for 10 hr
Lensed instruments	3 for 10 min	8 for 15 min	7 for 3–12 hr[a]
	4A for 10 min	9 for 15 min	8 for 12 hr
	5A for 10 min		9 for 10 hr
Thermometers, oral and rectal[c]	1C for 10 min	1C for 15 min	2 for 12 hr
			7 for 3–12 hr[a]
			8 for 12 hr
			9 for 10 hr

Table continued on the next page

676

RECOMMENDATIONS FOR CHEMICAL DISINFECTION AND STERILIZATION OF INSTRUMENTS (*Cont.*)

	Disinfecting treatment		*Sterilizing treatment*
Equipment	*Category A microorganisms*	*Category B microorganisms*	*Category C microorganisms*
Hinged instruments[d]	1A for 15 min 2 for 10 min 3 for 20 min 4A for 20 min 5A for 15 min 8 for 10 min 9 for 10 min	1B for 20 min 2 for 15 min 4B for 30 min 5B for 30 min 8 for 20 min 9 for 20 min	7 for 3–12 hr[a] 8 for 12 hr 9 for 10 hr
Inhalation and anesthesia equipment	1A for 15 min 3 for 20 min 9 for 5 min	1B for 20 min 9 for 20 min	7 for 3–12 hr[a] 9 for 10 hr

[a]Preferably washed with soap and water and thoroughly wiped before disinfection or sterilization. Alcohol-iodine solutions remove markings on poor-grade thermometers.

[b]To prevent corrosion, 0.2% sodium nitrite should be present in alcohols, formalin, formaldehyde-alcohol, quaternary ammonium, and iodophor solutions, and 0.5% sodium bicarbonate should be present in phenolic solutions.

[c]Must first be cleaned to be free of organic salt.

[d]Very little direct observation has been possible; use heat whenever possible.

Courtesy of Earle H. Spaulding and George F. Mallison. From *Infection Control in the Hospital*, 4th ed., published by the American Hospital Association, copyright 1979.

Key to Chemical Compounds:

1A	70–90% ethyl alcohol or isopropyl alcohol*
1B	70–90% ethyl alcohol
1C	Compound 1B plus 0.2% iodine
2	20% formalin plus 70% alcohol solution*
3	1:500 aqueous quaternary ammonium solution*
4A	Iodophor (100 ppm available iodine)*
4B	Iodophor (500 ppm available iodine)*
5A	1% aqueous phenolic solution*
5B	2% aqueous phenolic solution*
6A	Sodium hypochlorite (100 ppm available chlorine)
6B	Sodium hypochlorite (1000 ppm available chlorine)
7	Ethylene oxide gas
8	20% aqueous formalin
9	2% aqueous activated glutaraldehyde*

Key to Categories of Microorganisms:

A	Vegetative bacteria and fungi, influenza viruses
B	A plus tubercle bacilli and enteroviruses
C	A plus B hepatitis viruses,† bacterial and some fungal spores

*Time required depends on procedure used.

†Not recommended for metal instruments.

Appendix V

THE PHENOL COEFFICIENT

Most of the disinfectants that are sold in interstate commerce are required by law to be rated according to their antimicrobial activity in comparison with pure phenol. The laboratory methods used for comparing these disinfectants with phenol are rigorously defined. For example, the tests must be carried out against specific known strains of bacteria such as *Salmonella typhi* ATCC* 6539, *Staphylococcus aureus* ATCC 6538, or *Pseudomonas aeruginosa* ATCC 15442. The composition of the media, type of glassware, temperature, and time of incubation, and other factors are also carefully defined. A series of test tubes containing progressively greater dilutions of the disinfectant is prepared, and a prescribed amount of suspension of the test organism is added to each tube. After 5, 10, and 15 minutes of incubation at 20°C, samples of the mixture are removed and tested for living organisms. This is done by inoculating the samples into tubes containing a broth medium that neutralizes the disinfectant. The samples are then incubated for two days to see if any growth of the test bacterium occurs. From these results, the maximum dilution of disinfectant that kills the organisms *in 10 minutes but not in 5 minutes* is determined. Under the conditions of this test, pure phenol kills *Salmonella typhi* ATCC 6539 in a dilution of about 1:90 (equivalent to 1 gram of phenol, in 90 grams of water). Most disinfectants are far more active than phenol and therefore are able to kill the bacterium in much higher dilutions than 1:90. For example, under the same conditions, another disinfectant might kill *S. typhi* organisms in a dilution of 1:900.

A measure of the activity of a disinfectant relative to that of phenol is obtained by dividing the maximum dilution that kills the organisms in *10*

---

*ATCC—American Type Culture Collection, a non-profit organization devoted to the preservation of reference cultures

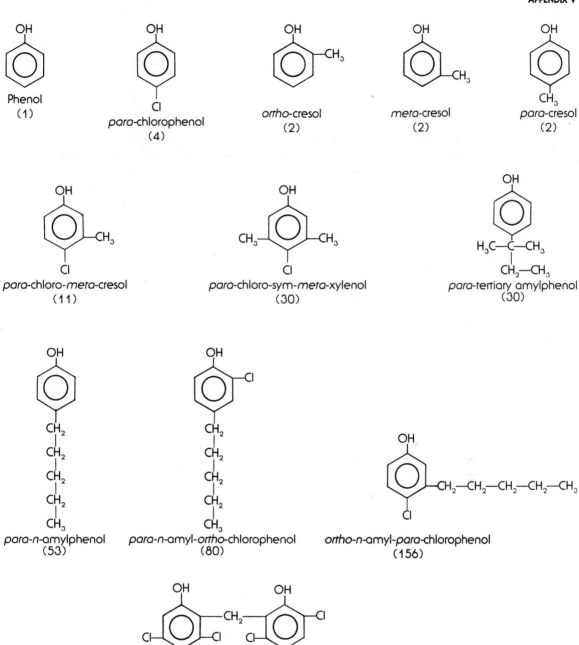

Figure V–1
Formulas of some phenolic compounds. Their approximate phenol coefficients as determined with *Salmonella typhi* are given in parentheses.

minutes by the corresponding killing dilution of phenol (900 ÷ 90 in the previous example). The resulting number (10) is called the **phenol coefficient** of the disinfectant, defined as the ratio of the dilution of disinfectant that kills in *10* minutes to the dilution of phenol that does the same.

Examples of some phenolics and their phenol coefficients are shown in Figure V–1. Phenol coefficients can differ widely depending on the test organism. Hexachlorophene, for example, has a phenol coefficient of more than 50 against *Staphylococcus aureus* but less than 15 against *Salmonella typhi*. The phenol coefficient provides a guide to the antimicrobial activity of a disinfectant, but the conditions of the tests described above may differ greatly from the conditions under which the disinfectant is actually used; an example is the operating room of a hospital. Therefore, additional tests are usually necessary to ensure that a given disinfectant will perform satisfactorily under conditions of use.

Appendix VI

MEDICINES THAT ACT AGAINST EUKARYOTIC CELLS

ANTIFUNGAL MEDICINES

Only a few medicines are available for treatment of serious fungal infections, and these have a low therapeutic ratio. The most important ones are shown in Table VI–1, and some examples are discussed below. Many more medicines are available for treating superficial infections of body surfaces. Most kinds of fungal infections must be treated for weeks or months to obtain a cure.

*Medicines Used for Deep Infections.** Amphotericin B is an antibiotic produced by *Streptomyces nodosus*. It is one of a group of polyene antibiotics, so called because of their chemical structure (Fig. VI–1). The site of action of amphotericin is the cytoplasmic membrane of the fungal cell, where it binds to chemical components, called sterols, of the membranes. This reaction causes potassium to leak from the cell, and the cell dies. Amphotericin is the most effective medicine available for use against invading fungi of the genera* *Aspergillus, Blastomyces, Candida, Coccidioides, Cryptococcus, Histoplasma, Mucor,* and *Paracoccidioides*. It is, however, quite toxic; most patients have one or more side effects such as fever, headache, nausea, vomiting, painful joints, inflammation of the veins receiving injections of the medicine, and kidney damage. Amphotericin is not absorbed from the gastrointestinal tract and must be given by injection for treatment of deep infections.

---

*genera—plural of genus

*deep infections—those that involve other than superficial tissues of the body

681

TABLE VI–1   ANTIFUNGAL MEDICINES FOR DEEP INFECTIONS

Medicine (how given)	Origin	Principal genera used against	Side effects
Amphotericin B (intravenous or injected into joints or spinal fluid)	Antibiotic	Aspergillus, Blastomyces, Candida, Coccidioides, Cryptococcus, Histoplasma, Mucor, Paracoccidioides, Sporothrix	Fever, nausea, vomiting, phlebitis,* kidney damage and anemia* are common
Flucytosine (oral)	Synthetic	Candida, Phialophora, Cladosporium, Cryptococcus	Occasionally diarrhea, skin rash, liver injury, decreased leukocytes
Miconazole (intravenous)	Synthetic	Coccidioides, Paracoccidioides	Phlebitis, gastrointestinal symptoms, and itching are common
Potassium Iodide (oral)	—	Sporothrix	Swollen saliva glands, acne-like rash are frequent
Sulfonamide (oral)	Synthetic	Paracoccidioides	Rare allergic reactions, depression of blood cell formation or gastrointestinal disturbances
Hydroxystilbamidine (intravenous)	Synthetic	Blastomyces	Loss of appetite and nausea are common; occasional liver injury, skin rash, numbness, and tingling

*phlebitis—inflamed veins; in this case the veins into which the medicine is injected
*anemia—insufficient erythrocytes

Flucytosine (also called 5-fluorocytosine) is a synthetic chemical derivative of cytosine, a natural component of nucleic acids. Enzymes within fungal cells convert flucytosine to a highly toxic substance, 5-fluorouracil, which prevents normal nucleic acid synthesis from taking place. Flucytosine is much less toxic than amphotericin, and it can be taken by mouth. Unfortunately, resistant fungal mutants are commonplace, and flucytosine is used mostly in combination with amphotericin or as a backup medicine in patients who are unable to tolerate amphotericin. Side effects include injury to the liver and a decrease in the number of leukocytes.

Miconazole is another synthetic antifungal medicine that can be used as an alternative to amphotericin in a few kinds of deep fungal infections. It too must be given intravenously, and toxic side effects are common, includ-

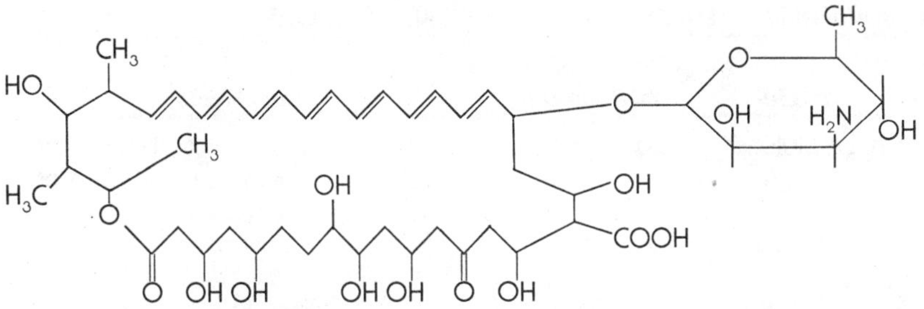

Figure VI–1
Amphotericin B, a polyene antibiotic.

ing inflamed* veins in about 30 percent of the patients and gastrointestinal symptoms in about 20 percent. Itching is also a common side effect.

Iodine compounds are effective in treating sporotrichosis and a rare phycomycete infection that occurs in tropical countries. The medicine usually used is potassium iodide, and it is taken orally. Side effects include an acne-like rash and swollen salivary glands.*

Medicines Used for Superficial Infections. Except for potassium iodine, all the substances used for deep infections are also used for superficial fungal infections such as those that may involve the alimentary tract,* skin, vagina, and bronchial tubes. In addition, a number of other substances are employed (Table VI–2), and some of the most important of these are mentioned below.

Griseofulvin (Fig. VI–2) is an antibiotic produced by a species of *Penicillium* and is rare among microbially produced substances in that its molecule contains chlorine. Griseofulvin is also notable among medicines used for superficial fungal infections because it is given by mouth, is absorbed from the intestinal tract, enters the bloodstream, and is carried to the deep skin layers, where it attaches avidly to cells producing keratin, the horny substance that predominates in hair, nails, and superficial skin layers. Griseofulvin makes the keratin resistant to attack by dermatophytes, and it is useful only against this group of fungi. It is the only medicine useful against infections of the fingernails and toenails. Gastrointestinal upsets sometimes occur as a side effect, and occasionally allergic reactions or suppression of leukocyte production by the bone marrow occurs.

Nystatin is another antifungal antibiotic and is related chemically to amphotericin. Like amphotericin, it is produced by a species of *Streptomyces*. It has a similar mode of action to that of amphotericin. However, because of its toxicity and poor solubility, it can be used only on body surfaces,

---

*inflamed—red, swollen, tender, hot

*salivary glands—the glands that produce saliva

*alimentary tract—mouth, throat, esophagus, stomach, and intestine

TABLE VI–2   ANTIFUNGAL MEDICINES FOR SUPERFICIAL INFECTIONS

Medicine (how given)	Origin	Principal genera used against	Side effects
Amphotericin B (topical)	Antibiotic	*Candida*	Occasionally, itching and burning in the area of application
Nystatin (topical)	Antibiotic	*Candida*	Usually none
Miconazole (topical)	Synthetic	*Candida, Epidermophyton, Malassezia, Trichophyton*	Occasionally, itching and burning in the area of application
Clotrimazole (topical)	Synthethic	*Candida, Epidermophyton, Malassezia, Microsporum, Trichophyton*	Rarely, redness, blistering, itching in area of application
Griseofulvin (oral)	Antibiotic	*Epidermophyton, Microsporum, Trichophyton*	Occasionally, gastrointestinal upsets and allergic reactions; rarely, decreased leukocytes
Tolnaftate (topical)	Synthetic	*Epidermophyton, Malassezia, Microsporum, Trichophyton*	Usually none
Haloprogin (topical)	Synthetic	*Epidermophyton, Malassezia, Microsporum, Trichophyton*	Occasionally, itching and burning in area of application
Natamycin (topical)	Antibiotic	*Fusarium*	Usually none

including the mouth, intestine, skin, and vagina. Its usefulness is restricted to treatment of dermatomycoses* and infections due to strains of *Candida*. There are usually no side effects.

Many other medicines, such as undecylenic, benzoic, and salicylic acids, sodium thiosulfate, ammoniated mercury, and potassium permanganate, are frequently used to treat superficial fungal infections of the skin. Such substances have little or no selective toxicity and are essentially mild antiseptics. Their value is secondary to general measures to improve the health of the skin, such as cleanliness and access to the air.

---

*dermatomycoses—superficial skin infections caused by species of *Epidermophyton*, *Microsporum*, and *Trichophyton*

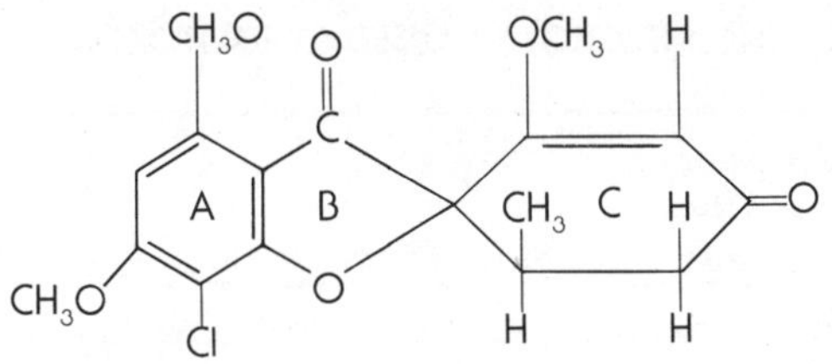

Figure VI–2
Griseofulvin, an antifungal medicine.

MEDICINES ACTIVE AGAINST PROTOZOA

Table VI–3 and Figure VI–3 present some of the medicines used in diseases caused by protozoa. A few are already familiar because of their use against fungi or bacteria, but the majority are useful only against protozoa. Most have a low therapeutic ratio. For most, the mode of action is not completely known.

TABLE VI–3   SOME MEDICINES USED IN TREATING INFECTIONS CAUSED BY PROTOZOA

Medicine	Disease	Genus of causative protozoa	Side effects
Amodiaquine	Malaria	*Plasmodium*	Occasionally vomiting, diarrhea, dizziness
Amphotericin	Amebic meningitis	*Naegleria*	Frequently, kidney damage, fever, nausea, phlebitis; occasionally, anemia
	American leishmaniasis	*Leishmania*	
Chloroquine	Amebic abscess	*Entamoeba*	Occasionally itching, vomiting, hair loss, vision problems, anemia
	Malaria	*Plasmodium*	
Dehydroemetine and emetine	Amebic abscess, amebic dysentery	*Entamoeba*	Frequently cardiac irritability, chest pain, muscle weakness, severe irritation at injection sites
Diiodohydroxyquin	Amebic dysentery	*Entamoeba*	Occasionally, rash, nausea, diarrhea; rarely, loss of vision in children under prolonged treatment
Melarsoprol	African sleeping sickness	*Trypanosoma*	Frequently heart, kidney, nerve, and brain damage; abdominal pain; vomiting
Metronidazole	Vaginitis	*Trichomonas*	Frequently nausea, headache, metallic taste; occasionally vomiting, diarrhea, dizziness
	Giardiasis	*Giardia*	

Table continued on the following page

TABLE VI–3   SOME MEDICINES USED IN TREATING INFECTIONS CAUSED BY PROTOZOA
              (*Cont.*)

Medicine	Disease	Genus of causative protozoa	Side effects
Nifurtimox	Chagas' disease	*Trypanosoma*	Frequently vomiting, weight loss, weakness, nervous system abnormalities
Pentamidine	Pneumonia	*Pneumocystis*	Frequently, drop in blood pressure; vomiting; kidney damage; blood disorders
Primaquine	Malaria (liver stage)	*Plasmodium*	Occasionally, blood disorders and gastrointestinal upset
Pyrimethamine	Malaria (prevention and treatment of chloroquine-resistant forms)	*Plasmodium*	Occasionally, blood abnormalities
	Toxoplasmosis	*Toxoplasma*	
Quinacrine	Giardiasis	*Giardia*	Frequently yellow skin, vomiting, nervous system disorders; occasionally blood disorders, skin disorders, allergic reactions
Quinine	Malaria (chloroquine-resistant forms)	*Plasmodium*	Frequently headache, ringing in ears, and vision abnormalities; occasionally heart irritability, blood disorders
Stibogluconate	Leishmaniasis	*Leishmania*	Frequently painful joints and muscle aches; occasionally heart damage, cramps and diarrhea, itching, rash
Sulfonamides (sulfadiazine, sulfamethoxazole, trisulfapyrimidines)	Malaria (chloroquine-resistant)	*Plasmodium*	Usually result of use along with another medicine such as quinine, pyrimethamine, or trimethoprim
	Pneumonia	*Pneumocystis*	Occasionally, gastrointestinal upset; rarely, serious allergic reactions or blood disorders
Suramin	Toxoplasmosis African trypanosomiasis	*Toxoplasma* *Trypanosoma*	Frequently, vomiting, abnormal skin sensation; occasionally, kidney damage, serious blood abnormalities, shock

Figure VI–3
Some medicines used in treatment of infections caused by protozoa.

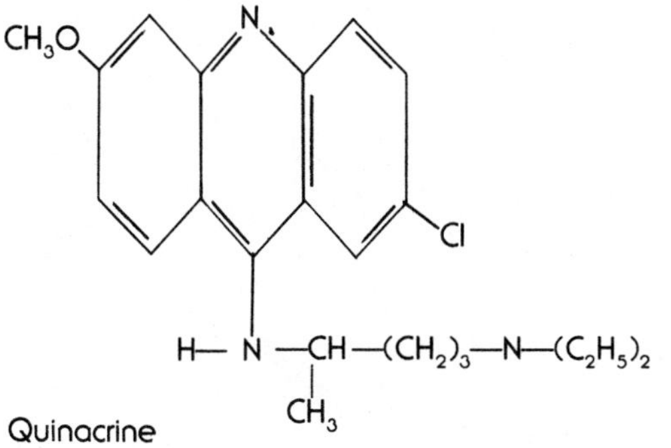

Quinacrine

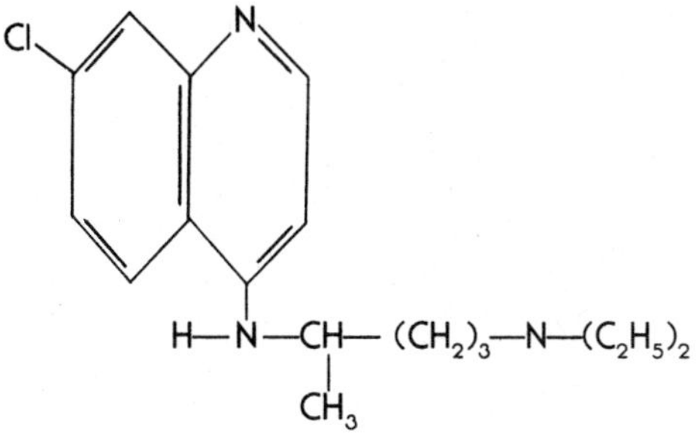

Chloroquine

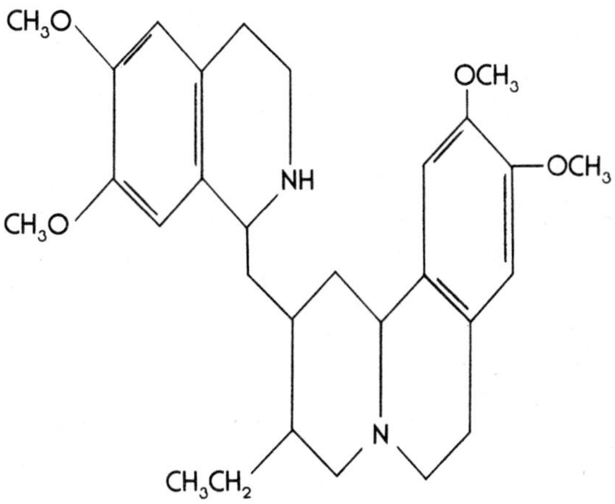

Emetine

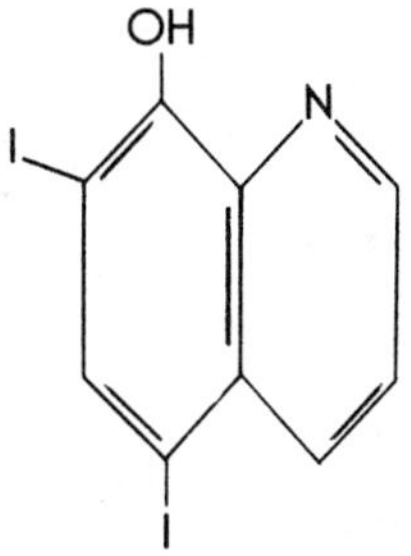

Diiodohydroxyquine

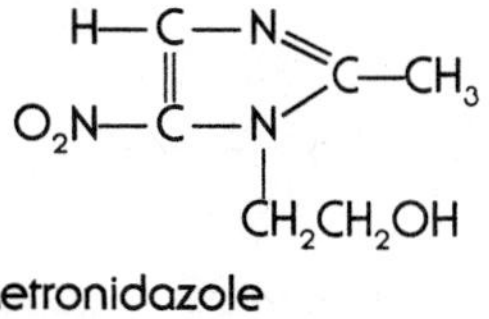

Metronidazole

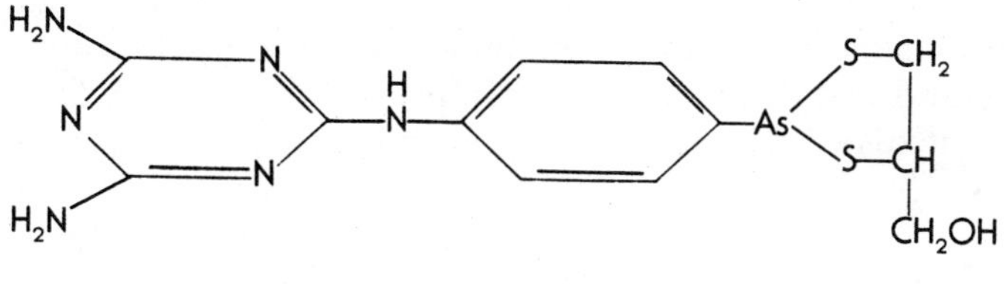

Melarsoprol

Figure continued on the following page

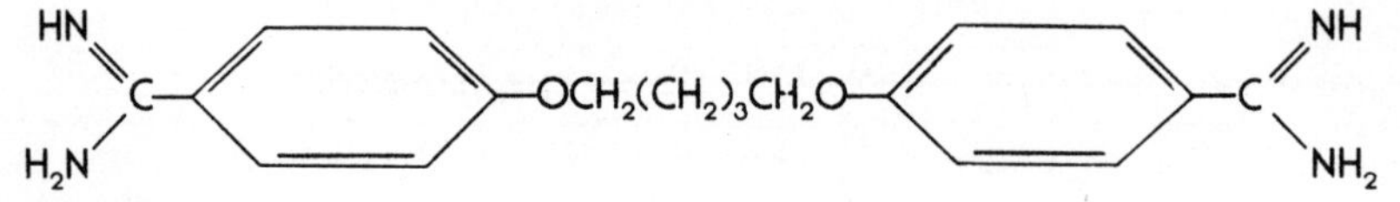

Pentamidine

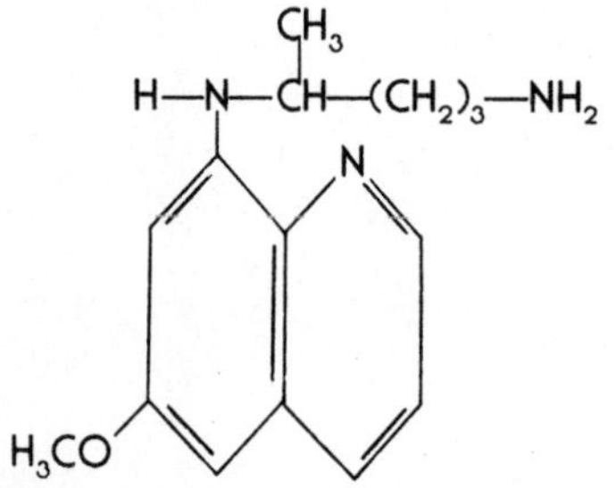

Primaquine

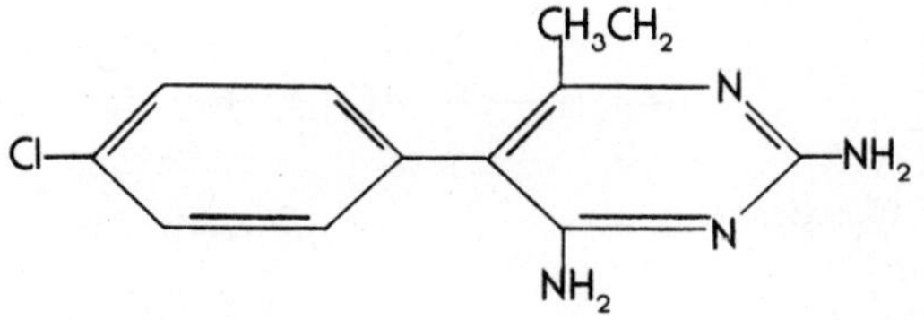

Pyrimethamine

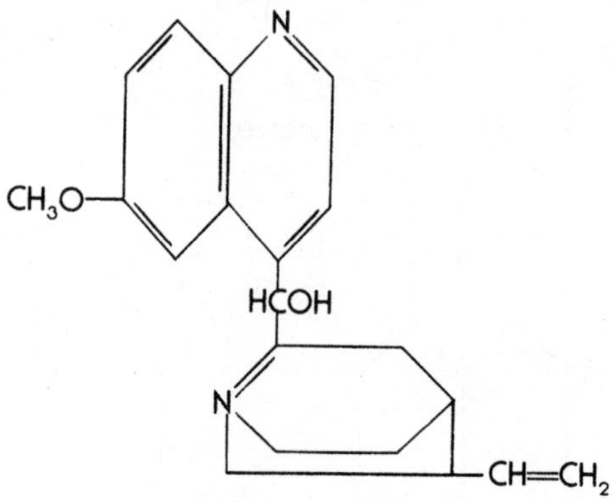

Figure VI–3 (Cont.)　　　　Quinine

ANTITUMOR MEDICINES

Perhaps the ultimate therapeutic challenge is to find a medicine that has a highly selective activity against cancer cells and spares normal cells of the body. All the medicines discovered so far fall short of this goal, being highly toxic to normal as well as malignant cells. They have, however, along with surgery and radiation, markedly improved the outlook for several types of malignant diseases.* Some of the antibiotics used in cancer chemotherapy

---

*malignant diseases—cancers, leukemias, and similar conditions involving unrestrained growth of abnormal body cells

TABLE VI–4 ANTIBIOTICS USED AGAINST MALIGNANCIES

Antibiotic	Source	Uses	Mode of action	Side effects
Dactinomycin	*Streptomyces parvullus*	Certain tumors of kidney, muscle, testis, uterus, and other organs	Binds with DNA, causing inhibition of replication and transcription	Fatigue, painful muscles, sore mouth and throat, nausea, vomiting, abdominal pain, diarrhea, severe anemia, loss of hair
Daunorubicin	*Streptomyces peuceticus*	One type of acute leukemia	Similar to dactinomycin	Nausea, vomiting, sore mouth, loss of hair, severe impairment of blood cell formation, heart damage
Doxyrubicin	*Streptomyces peuceticus*	Some kinds of acute leukemia; kidney, nerve, bone, breast, ovary, bladder, and other tumors	Similar to dactinomycin	Similar to daunorubicin
Mitomycin	*Streptomyces carspitosus*	Cancer of stomach and pancreas	Reacts irreversibly with DNA, thus interfering with normal transcription	Severe impairment of blood cell formation; sore mouth, nausea, vomiting; occasionally lung and kidney damage

are presented in Table VI–4, and daunorubicin, used in treating acute leukemias, is shown in Figure VI–4. They are frequently used together with synthetic antimetabolites.*

---

*antimetabolites—substances that interfere with specific chemical reactions inside cells

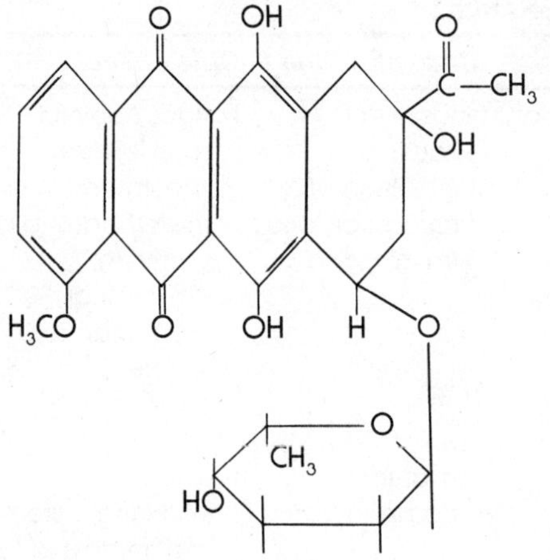

Figure VI—4
Structure of daunorubicin, an antibiotic useful in treating acute leukemias.

Appendix VII

CLASSIFICATION OF BACTERIA

With microorganisms, as with all living organisms, some scheme of classification is necessary to be able to identify them. Bacteria are difficult to classify into groups because their evolution cannot be traced accurately, they have many methods for genetic transfer that enable them to change rapidly, and they have simple structures that are similar among many different kinds of microorganisms. A recent study suggests that the evolutionary relationships among bacteria can be determined by comparing the sequence of nucleotides in the small subunits of their DNA. Nevertheless, at the present time bacteria are classified into genera which are made up of species on the basis of as many characteristics as can be observed.

In classifying bacteria, a clone* of the organism must first be isolated, and a number of properties, or characters, of the isolate must be determined. Important characters include, among others, the morphology (whether cocci or rods, with endospores or flagella), physiological properties (aerobic or anaerobic growth), biochemical capabilities (fermentation of sugars, production of certain enzymes), cultural and nutritional characteristics, drug sensitivities, presence of specific antigens, and genetic characteristics (ability to transfer genes, composition of the DNA). After many bits of information have been collected about a clone of bacteria, each bit can be coded by giving it an appropriate number. Even the presence or absence of a character can be assigned a number. For example, if endospores are absent, a value of 0 is assigned. If endospores are present, a value of 1 is assigned. Having given numbers to the characters, it is simple to analyze the data mathematically, usually by computer analysis, to determine how closely various bacteria

---

*clone—a family of cells derived from a single parental cell by repeated divisions

resemble each other. This allows the classification of organisms into genera of closely related species, families of genera, and orders of related families. Although there is great variability, often species within a genus will have about 80 percent similar characteristics, while genera within a family will share about 60 percent characteristics.

Once bacteria are classified by many different characters, it is often possible to identify organisms by testing only a few key properties. One important clue in identifying organisms is their source. For example, it is known from extensive past experience that *Staphylococcus aureus* lives in the anterior part of the nose of many humans but not normally in milk. Thus, Gram-positive cocci isolated from milk are not likely to be *S. aureus*, whereas this species is much more likely to be isolated from the nose. If bacteria from a nose culture grow as golden-colored, creamy colonies of Gram-positive cocci in clusters on blood agar, forming zones of beta-hemolysis (clearing) around each colony, *S. aureus* is strongly suspected because of the colonial characteristics and morphology of the organisms. If, in addition, the cocci grow in broth with a high salt content, produce acid by fermenting mannitol, and give a positive coagulase test, the organism is generally identified as *S. aureus*. It should be remembered, however, that the original classification of organisms demands the determination of many different characters; identification of already-classified bacteria may require determination of only a few key properties.

The grouping of bacteria given in *Bergey's Manual of Determinative Bacteriology* represents an interim classification continually being updated as new information reveals previously unrecognized relationships. The kingdom Prokaryotae, defined as comprising all prokaryotic cells, is subdivided into cyanobacteria (Division I) and bacteria (Division II). The *Manual* is concerned with further classification of Division II, which is composed of 19 parts. These parts are given below; the genera mentioned in this text are listed beneath each part.

Part 1:   Phototrophic bacteria (18 genera)

        Rhodospirillum
        Chlorobium

Part 2:   Gliding bacteria (27 genera)

Part 3:   Sheathed bacteria (7 genera)

Part 4:   Budding and/or appendaged (17 genera)

Part 5:   Spirochetes (5 genera)

        *Treponema**

---

*Medically important genera.

Borrelia*
Leptospira*

Part 6:   Spiral and curved bacteria (6 genera)

Spirillum*
Campylobacter*
Bdellovibrio

Part 7:   Gram-negative aerobic rods and cocci (20 genera)

Pseudomonas*	*Beijerinckia*
Azotobacter	*Alcaligenes*
Rhizobium	*Brucella**
Agrobacterium	*Bordetella**
Halobacterium	*Francisella**
Acetobacter	*Xanthomonas*
Legionella*	

Part 8:   Gram-negative facultatively anaerobic rods (26 genera)

Escherichia*	*Erwinia*
Salmonella*	*Klebsiella**
Shigella*	*Streptobacillus**
Enterobacter*	*Flavobacterium**
Serratia*	*Haemophilus**
Proteus*	*Pasteurella**
Yersinia*	*Moraxella**
Vibrio*	

Part 9:   Gram-negative anaerobic bacteria (9 genera)

Bacteroides*
Fusobacterium*
Leptotrichia

Part 10:   Gram-negative cocci and coccobacilli (6 genera)

Neisseria*
Acinetobacter*
Branhamella*

Part 11:   Gram-negative anaerobic cocci (3 genera)

Veillonella*

Part 12:   Gram-negative chemolithotrophic bacteria (17 genera)

Thiobacillus
Acetobacter

Part 13: Methane-producing bacteria (3 genera)

Part 14: Gram-positive cocci (12 genera)

> *Micrococcus**
> *Staphylococcus**
> *Streptococcus**
> *Peptostreptococcus**

Part 15: Endospore-forming rods and cocci (6 genera)

> *Bacillus**
> *Clostridium**

Part 16: Gram-positive asporogenous rod-shaped bacteria (4 genera)

> *Lactobacillus*
> *Listeria**

Part 17: Actinomycetes and related organisms (39 genera)

> *Corynebacterium**   *Nocardia**
> *Arthrobacter*   *Streptomyces**
> *Propionibacterium*   *Rothia*
> *Actinomyces**   *Bacterionema*
> *Mycobacterium**

Part 18: Rickettsiae (18 genera)

> *Rickettsia**
> *Coxiella**
> *Chlamydia**

Part 19: Mycoplasmas (4 genera)

> *Mycoplasma**
> *Spiroplasma*

Appendix VIII

GROUPS OF ANIMAL VIRUSES

No universal clasification of animal viruses has been generally agreed upon, but most of the various groups of viruses infecting humans and other animals can be divided into 13 categories, 5 consisting of DNA viruses and 8 of RNA viruses. The viruses containing DNA are shown in Table VIII–1.

TABLE VIII–1   DNA VIRUSES

Group	Important characteristics	Examples
1. Pox viruses	Very large ovoid viruses of complex structure; unlike other DNA viruses they replicate in the cytoplasm	Cause smallpox, cowpox, rabbit myxomatosis, and economically important diseases of domestic fowl
2. Herpesviruses	Medium to large-sized, enveloped viruses that frequently cause latent infections; some cause tumors	Human oral and genital herpes simplex, cytomegalovirus infections, varicella zoster, and EB virus infections of human beings; infections of numerous warm- and cold-blooded vertebrates
3. Adenoviruses	Medium-sized viruses; some cause persistent infections of tonsil tissue; base ratios vary widely, although the appearance is uniform within the group	About 40 types known to infect the respiratory and intestinal tracts of humans; frequent cause of conjunctivitis or sore throat; many additional types infect other animals
4. Papovaviruses	Small viruses, often cause cancer in animals other than humans. Name derives from *papilloma, polyoma, vacuolating* viruses	Viruses that cause human warts and certain degenerative brain diseases
5. Parvoviruses	Very small viruses; contain single-stranded DNA in some cases; some require a helper virus in order to propagate	Infect rodents, swine, and arthropods; found in humans in association with adenovirus helpers, but their pathological significance is uncertain

Characteristics of RNA virus groups are presented in Table VIII–2

TABLE VIII–2   RNA VIRUSES

Group	Important characteristics	Examples
1. Picornaviruses	Very diverse group ot small, viruses; their name derives from "pico," meaning small, and RNA; one subgroup, the enteroviruses, is acid-resistant; the second subgroup, the rhinoviruses, is sensitive to acid	Enteroviruses infect the intestine and a variety of body tissues; about 70 types infect humans, including polio viruses, Coxsackie viruses and ECHO viruses; rhinoviruses principally infect the respiratory tract and are the major cause of human colds
2. Togaviruses	Large, diverse group of medium-sized, enveloped viruses, many of which are transmitted by arthropods which they also often infect	Cause human infections such as rubella, yellow fever, equine encephalitis, dengue, and others; numerous other vertebrates and invertebrates are susceptible to togavirus infections
3. Myxoviruses	Medium-sized viruses, usually showing projecting spikes	The influenza viruses of humans and other animals
4. Paramyxo-viruses	Similar to myxoviruses but somewhat larger	Human rubeola and mumps; Newcastle disease of chickens; distemper of dogs
5. Reoviruses	Unlike other animal viruses, reoviruses contain double-stranded RNA; strongly resemble certain plant viruses	One group, the rotaviruses, is the principal cause of vomiting and diarrhea in children; Colorado tick fever virus falls into another reovirus subgroup; numerous other animals are susceptible to reoviruses
6. Rhabdoviruses	Bullet-shaped viruses	Includes rabies virus and other viruses of mammals, fish, and insects
7. Retroviruses	Generally resemble myxoviruses in size and shape but possess the enzyme reverse transcriptase; one subgroup, the oncorna (*oncogenic* RNA) viruses, transforms cells and causes tumors; a second subgroup causes certain slow virus infections of animals	Rous sarcoma virus of chickens; mouse mammary tumor virus; cat leukemia virus; visna virus of sheep
8. Arenaviruses	Medium-sized viruses with a granular appearance	Lassa fever virus of humans; lymphocytic choriomeningitis virus of mice

Appendix IX

ISOLATION TECHNIQUES FOR HOSPITALS

ENTERIC PRECAUTIONS

1. Visitors should report to Nurses' Station before entering the room.
2. Private room—necessary for children only.
3. Gowns—must be worn by all individuals having direct contact with the patient.
4. Masks—not necessary.
5. Hands—must be washed on entering and leaving the room.
6. Gloves—must be worn by all individuals having direct contact with the patient or with articles contaminated with fecal material.
7. Articles—special precautions necessary for articles contaminated with urine and feces. Articles must be disinfected or discarded.

Diseases Requiring Enteric Precautions

Cholera
Diarrhea (acute illness with suspected infectious etiology)
Enterocolitis (staphylococcal)
Gastroenteritis caused by (a) enteropathogenic or enterotoxic *Escherichia coli*, (b) *Salmonella* species, or (c) *Yersinia enterocolitica*
Hepatitis (viral), type A, B, or unspecified
Typhoid fever (*Salmonella typhi*)

RESPIRATORY ISOLATION

1. Visitors should report to Nurses' Station before entering the room.
2. Private room—necessary. Door must be kept closed.

3. Gowns—not necessary.
4. Masks—must be worn by any person entering the room unless that person is not susceptible to the disease.
5. Hands—must be washed on entering and leaving the room.
6. Gloves—not necessary.
7. Articles—those contaminated with secretions must be disinfected.

Diseases Requiring Respiratory Isolation

Measles (rubeola)
Meningococcal meningitis
Meningococcemia
Mumps
Pertussis (whooping cough)
Rubella (German measles)
Tuberculosis, pulmonary (including tuberculosis of the respiratory tract—suspected or sputum-positive smear)

WOUND AND SKIN PRECAUTIONS

1. Visitors should report to Nurses' Station before entering the room.
2. Private room—desirable.
3. Gowns—must be worn by all individuals having direct contact with the patient.
4. Masks—not necessary except during dressing changes.
5. Hands—must be washed on entering and leaving the room.
6. Gloves—must be worn by all individuals having direct contact with the infected area.
7. Articles—special precautions necessary for instruments, dressings, and linen.

Note: Special dressing techniques should be used when changing dressings.

Diseases Requiring Wound and Skin Precautions

Burns that are infected, except those infected with *Staphylococcus aureus* or group A streptococcus that are not covered or not adequately contained by dressings
Gas gangrene (due to *Clostridium perfringens*)
Herpes zoster, localized
Plague (bubonic)
Puerperal sepsis—group A streptococcus, vaginal discharge
Wound and skin infections that are covered by dressings and the discharge is adequately contained, including those infected with *Staphylococcus aureus* or group A streptococcus. Minor wound infections, such as stitch abscesses, need only secretion precautions.

GLOSSARY

abscess (ab'-sess)   A localized collection of pus surrounded by inflamed tissue.

acellular (ay-sell'-you-lar)   Containing no cells.

Acetabularia (a-seh-tab-you-lair'-ee-a)

Acetobacter (a-see'-toe-back'-ter)

Achromobacter (ā-krōm'-oh-back-ter)

acid-fast bacteria   Bacteria that retain certain dyes even after washing with acid. Mycobacteria are acid-fast bacteria.

acidophilic (a-see-do-fill'-ik)   Preferring or requiring an acidic environment.

Acinetobacter calcoaceticus (a-sin-et'-oh-back-ter kal-koh-a-seat'-i-kus)

acquired response   A highly specific antibody response often referred to as "the immune response."

Actinomyces israelii (ak-tin-oh-my'-seez is-rail'-ee-eye)

activation energy   The energy required to elevate molecules from one energy level where they are nonreactive to a higher energy level where they react spontaneously.

active site   The sequence of amino acids on an enzyme molecule that is concerned with binding and changing the substrate. Also called the *catalytic site.*

active transport   The energy-requiring process by which molecules are carried across cell boundaries. Often the molecules are moved from a lower to a higher concentration.

acute   Short lived as opposed to chronic.

adenoviruses (ad'-eh-no viruses)

adjuvants (aj'-e-vent)   Substances used to increase the efficiency of the immune response when injected simultaneously with an antigen.

adsorption (viral)   The interaction of a virus with specific receptors on host cells, resulting in attachment of the virus to the cell surface.

aerobic organisms (abbreviated **aerobes**)   Organisms that can use oxygen as the final hydrogen acceptor during metabolism.

aerosol   A fine suspension of bacteria in air.

aerotolerant microorganisms   Microorganisms that can survive and grow in 20 percent oxygen, although they do not use it for metabolism.

affinity   Chemical attraction between two substances.

aflatoxin   Any of several toxins produced by molds such as *Aspergillus flavus*; some are suspected of being carcinogenic.

agar (ah'-grr)   A polysaccharide obtained from various species of seaweeds used to gel materials (especially microbiological media). It is not broken down by most bacteria.

agents   Convenient term used to indicate both microorganisms and viruses.

Agrobacterium tumefaciens (ag-roh-back-teer'-ee-um too-me-faysh'-ee-enz)

Alcaligenes (al-kah-lij'-en-eez)

alimentary tract   The mouth, throat, esophagus, stomach, and intestines.

allergen (al'-er-jen)   A substance that induces allergy.

allergy (al'-er-jee)   Also known as hypersensitivity. An immunological response to an antigen that results in damage to the individual. It usually results in an impaired function of a tissue.

allosteric site (al-oh-steer'-ik site)   The sequence of amino acids on the first enzyme of a biosynthetic pathway to which the end-product of the pathway attaches. This attachment prevents the enzyme from functioning.

alveoli (al-vee'-oh-lye)   Air cells of the lung.

amino acids   The subunits of a protein molecule.

anaerobic organisms   Organisms that do not use oxygen as the final hydrogen acceptor during metabolism; organisms that can grow only in the absence of oxygen (air).

anaphylactic shock   Shock produced by a hypersensitive reaction to an allergen.

anemia   A condition characterized by having less than the normal amount of hemoglobin.

antibiotic   A chemical substance produced by certain molds and bacteria that inhibits the growth of or kills other microorganisms.

antibody   A protein produced by the body in response to the presence of a foreign substance.

anticodon (an-tee-koh'-don)   The three nucleotides in transfer RNA that are complementary to a sequence of three nucleotides in messenger RNA (a codon).

antigen (an'-ti-jen)   A substance that can incite the production of specific antibodies and can combine with those antibodies.

antigenic drift   The slight changes occurring in the antigens of a virus; specific antibodies made to the antigen before the change occurred are now only partially effective.

antimetabolites (an-tee-meh-tab'-oh-lites) Substances that interfere with specific chemical reactions inside cells.

antiphagocytic (an-tee-fag-oh-sit'-ik)   Prevents phagocytosis or ingestion by a cell specialized to "eat" foreign matter.

antiseptic (an-ti-sep'-tik)   A substance that will inhibit or kill microbes. This term often implies that the substance is sufficiently nontoxic to be applied to the skin or other tissues.

arteriosclerosis (ar-teer'-ee-oh-skler-oh'-siss) Hardening of the arteries.

arthrospores (ar'-throw-spores)   Fungal spores formed by breaking apart of the cells composing septated hyphae.

artifacts   Properties not present originally. An apparent structure in a cell produced by staining or other technical procedure.

Ascomycete (ass'-koh-my-seat)

aseptic   Not leading to contamination.

aseptically   By sterile techniques; avoiding introduction of any organisms.

asexually   Produced without sexual action or differentiation.

Aspergillus niger (ass-per-jill'-us nye'-jer)

assay   To analyze for one or more components.

asymptomatic   Without apparent symptoms.

ATCC   Abbreviation for American Type Culture Collection, a nonprofit organization devoted to the preservation of reference cultures of all types of microorganisms and tissues.

atmospheric pressure   The force (approximately 15 pounds per square inch at sea level) exerted on the earth by the mass of air that surrounds it.

ATP   Abbreviation for adenosine triphosphate (a-den'-oh-sin tri-fos'-fate).   The major chemical storage form of energy in cells. The energy is stored in two high-energy phosphate bonds that release energy when hydrolyzed.

attenuated   Modified to be incapable of causing disease, although often still able to stimulate antibody production.

autoclave   A device employing steam under pressure for sterilizing materials stable to heat and moisture.

autoimmune disease   An illness caused by hypersensitivity to some constituents of one's own body.

autolyze   To break up or destroy, usually by a cell's own enzymes; self-destruction.

autotroph (aw'-tow-trofe)   An organism that can use CO_2 as its main source of carbon.

auxotroph (awx'-oh-trofe)   Term applied to a laboratory-derived mutant of a microorganism that requires an organic supplement other than a carbon and energy source for growth. These supplements include amino acids, purines, pyrimidines, or vitamins for growth.

avirulent (ay-veer'-you-lent)   Lacking virulence. When used to describe microbes, it means an agent is lacking those properties that normally promote the ability of an agent to cause disease.

Azotobacter (ay-zoh'-tow-back-ter)

Bacillus anthracis (bah-sill'-us an-thra'-siss)

B. cereus (B. seer'-ee-us)

B. stearothermophilus (B. steer-oh-ther-mo'-fill-us)

bactericidal   Capable of killing bacteria.

bacteriophage (back-tear'-ee-oh-fayj)   A virus that infects bacteria; often abbreviated "phage."

Bacterionema (back-tear-ee-oh-nee'-mah)

bacteriostatic   An agent that inhibits the growth of bacteria without killing them.

Bacteroides fragilis (back-ter-oid'-eez frah'-jill-us)

B. melaninogenicus (B. mel-an-in-oh-jen'-i-kus)

Balantidium coli (bal-an-tid'-ee-um coh'-lee)

Basidiomycete (baa-sid'-ee-oh-my-seat)

Beijerinckia (bye-er-ink'-ee-a)

binary fission   An asexual reproductive process in which one cell splits into two independent daughter cells.

biodegradable (bye-oh-dee-grade'-a-bull)   Can be broken down by the action of living organisms.

biomagnification   The phenomenon by which the concentration of a chemical substance such as a pesticide increases in organisms the higher the organism is on the food chain.

biopsy   The surgical removal of a small

amount of tissue for microscopic examination.

Blastomyces dermatitidis (blast-oh-my'-seez dermah'-ti-ti-dis)

blocked Prevented from functioning.

blood agar Agar medium containing blood.

Bordetella pertussis (bor-deh-tell'-ah per-tus'-siss)

Borrelia recurrentis (bor-rell'-ee-ah ree-kur-ren'-tiss)

botulism The disease caused by the toxin of *Clostridium botulinum.*

bovine Having to do with cows.

Branhamella catarrhalis (bran-hem-el'-ah cat-a-rah'-lis)

bronchioles Branches of the bronchi (of the lungs).

Brucella abortus (brew-sell'-ah ah-bore'-tus)

B. melitensis (B. mel-i-ten'-siss)

buffer A substance in a solution capable of neutralizing both acids and bases and thereby maintaining the original acidity or basicity of the solution.

calculus (kal'-kew-lus) Calcified dental plaque.

Campylobacter fetus (kam-pie'-lo-back-ter fee'-tus)

cancer Groups of abnormally growing cells that can spread from their site of origin, also termed *malignant tumors.*

Candida albicans (kan'-did-ah al'-bi-kanz)

C. lipolytica (C. lih-poh-lih'-tih-ka)

capsid (kap'-sid) The protein coat that surrounds and protects the nucleic acid of a virus.

capsule A loose-fitting gelatinous structure that surrounds some bacteria. Colonies of these bacteria often appear moist and slimy.

carbohydrate Compounds containing principally carbon, hydrogen, and oxygen atoms in a ratio of 1:2:1.

carbon skeleton The carbon atoms bonded to one another; does not include the other atoms in the molecule.

carbuncle A painful skin infection that penetrates into the deeper layers.

carcinogens (kar-sin'-oh-jens) Cancer-causing agents.

cariogenic (kar-ee-oh-jen'-ik) Causing dental caries.

catalyst (kat'-a-list) A substance that accelerates the rate of a chemical reaction without being altered or depleted in the process.

Cephalosporium (sef-all-oh-spore'-ee-um)

cercaria (ser-care'-ee-ah) (pl. **cercariae**) The forked-tailed larval form of the *Schistosoma* species; the developmental stage of schistosomes that penetrate human skin.

cerebrospinal fluid Fluid that surrounds the brain and spinal cord, formed within the ventricles of the brain.

chancre (shan'-ker) A primary sore or ulcer at the site of entry of a pathogen, the initial lesion of syphilis.

chancroid (shan'-kroid) A venereal disease caused by the bacterium *Haemophilus ducreyi.*

chelating agent A compound that removes metallic ions from a solution by combining them with certain organic compounds.

chemical bond energy The energy present in the bonds joining atoms together.

chemotaxis Movement of an organism in response to chemicals in the environment.

chemotherapy Treatment of disease by chemicals.

Chlamydia trachomatis (klah-mid'-ee-ah trahko'-ma-tiss)

Chlamydomonas (klah-mid-oh-moan'-ass)

Chlorobium (klor-oh'-bee-um)

chlorophylls Green pigments necessary for photosynthesis.

chloroplast A plastid that contains chlorophyll. It is the site of photosynthesis and starch formation.

chromosome A DNA-containing linear body of the cell nuclei of plants and animals responsible for determining and transmitting hereditary characteristics.

chronic Of long duration, prolonged, lingering.

Ciliata (sill-ee-ah'-tah)

cilium (sill'-ee-um) A short, projecting hairlike organelle of locomotion of eukaryotic organisms.

citric acid A 6-carbon compound with three carboxyl (—COOH) groups.

clones Families of cells derived from a single parental cell by repeated divisions.

Clostridium botulinum (kloss-trid'-ee-um botyou-line'-um)

Cl. difficile (cl. dif-fi-seal')

Cl. perfringens (Cl. per-fringe'-enz)

Cl. tetani (Cl. tet'-an-eye)

Coccidioides immitis (kock-sid-ee-oid'-eez im'-mi-tiss)

codon (koh'-don)   Three nucleotides that specify a particular amino acid in a protein or that start or stop protein synthesis.

coenzyme   A small organic molecule that transfers small molecules from one enzyme to another.

coitus (koh-ee'-tus)   Sexual intercourse.

colon   Large intestine.

colonial   Characteristics of a colony of bacteria, such as its appearance.

colonization   Establishment of a site of reproduction of microorganisms on a material, animal, or person without necessarily resulting in tissue invasion or damage.

competitive inhibition   The inhibition of enzyme activity caused by the competition of the inhibitor with the substrate for the active (catalytic) site on the enzyme.

complement (C)   A system of at least 11 serum proteins that act in sequence, producing certain biological effects concerned with inflammation and the immune response.

congenital   Existing at birth.

conidium (kon-id'-ee-um) (pl. **conidia**)   An asexual fungal spore produced on a structure called a conidiophore.

conjugation   A mechanism of gene transfer that involves cell-to-cell contact in bacteria.

conjunctiva (kon-junk-tye'-vah)   The mucous membrane that lines the inner surface of the eyelids and exposed surfaces of the eyeball.

contagious (kun-tay'-jus)   Transmissible by contact; catching (used to describe the infection).

Corynebacterium diphtheriae (koh-ryne'-nee-back-teer-ee-um dif-theer'-ee-eye)

C. glutamicum (C. glue-tam'-i-kum)

C. pseudodiphtheriticum (C. sood-oh-dif-ther-it'-ik-um)

covalent bond   A strong chemical bond formed by the sharing of electrons.

Coxiella burnetii (kox-see-ell'-ah bur-net'-ee-ee)

cross-wall   The cell wall that forms in the middle of a cell to produce two cells. It appears during cell division.

Cryptococcus (krip-toe-kock'-kus)

cubicle   Small compartment.

cultural   Refers to conditions necessary for bacterial growth in the laboratory.

cutaneous   Of or pertaining to the skin.

cyanobacterium (sye-an'-oh-back-teer-ee-um) (pl. **cyanobacteria**)   The blue-green algae.

cyst   Dormant resting cells characterized by a thickened cell wall.

cytopathic effects (CPE)   Observable changes in cells in vitro produced by viral action. For example, lysis or fusion of cells.

cytoplasmic membrane   The flexible structure immediately surrounding the cytoplasm in all cells.

cytoplasmic streaming   The means by which cytoplasm moves about in a eukaryotic cell to allow the distribution of materials within the cell.

DDT   An insecticide, dichlorodiphenyl trichloroethane.

death phase   The stage in which the number of viable bacteria in a population decreases at an exponential rate.

decontamination   Removing or inactivating pathogenic microorganisms or their toxic products.

degrade (also **degradation, degradative**)   To break up or destroy.

demineralize   To remove minerals, such as those that form part of the tooth enamel.

denature   To destroy the three-dimensional structure of a protein molecule.

dental caries   Tooth decay.

dental plaque   Spongelike collections of bacteria on the teeth.

derivative   A molecule made from another molecule by chemical alteration.

dermatomycoses (der-ma-toe-my-koh'-seez)   Superficial skin infections caused by species of *Epidermophyton* and *Trichophyton*.

dermatophyte   A fungus parasitic to the skin or skin appendages such as hair, nails, and so on.

dermis   The portion of the skin lying just underneath the surface layers and containing nerves, blood vessels, and connective tissue.

differentiation   Modification of different parts of the body for performance of particular functions; specialization of parts or organs.

dimorphic (dye-more'-fick)   Capable of existing as a mold under some conditions and as a yeast under other conditions.

diphtheroids (dif'-ther-oidz)   Bacteria that somewhat resemble diphtheria bacteria except that they do not produce diphtheria toxin.

diploid   Condition in which each chromosome is represented twice.

disinfectant   An agent that kills or inhibits the growth of pathogenic microbes but does not necessarily sterilize the material being treated.

disinfection   Killing or inhibiting pathogenic microbes on or in a material by using a disinfectant.

DNA   The abbreviation for deoxyribonucleic acid.

ectoparasites   Parasites that attack the skin of the host.

electrolytes   Acids, bases, and salts dissolved in body fluids.

electron   A negatively charged particle that is part of all atoms.

EM   Abbreviation for electron microscope.

embryonated chicken eggs   Chicken eggs that contain a developing chick.

emulsify   To convert into an emulsion such as fat distributed in water in very small droplets so that the fat stays evenly distributed and does not separate.

encapsulated   Enclosed by a coating such as the protein coat of the virus.

encephalitis   (en-sef-ah-lye'-tis)   Inflammation of the brain.

endemic   Constantly present in a population.

endocarditis   Inflammation of the lining of the heart or its valves.

endospores   A very resistant form of a resting bacterial cell that develops from a vegetative cell by a series of biochemical reactions called *sporulation*.

endotoxin   A poisonous substance present in Gram-negative bacteria (such as those that cause typhoid fever) as an integral part of the cell wall.

enriched medium   Having supplementary nutrients such as those found in nutrient broth.

enrichment culture   A technique used for isolating an organism from a mixed culture by manipulating conditions to favor the growth of the organism sought while minimizing the growth of the other organisms present.

Entamoeba histolytica   (en-tah-mee'-bah his-toh-lit'-ik-ah)

enterobacteria   Any of a family of Gram-negative rod-shaped bacteria that ferment glucose. They include such bacteria as the salmonellae and other bacteria that live in the colon.

enzyme   A protein that speeds up a biochemical reaction.

epidemic   Affecting many individuals within a region or population.

epidermis   The outermost skin layers.

Epidermophyton   (eh-pee-der'-moh-fy-ton)

erythrocytes   (ee-rith'-row-sites)   Red blood cells.

Escherichia coli   (esh-er-ee'-she-ah koh'-lie)

etiological   Causative.

eubacteria   "True bacteria" as opposed to archaebacteria.

Euglena   (you-gleen'-ah)

eukaryotic   (you'-carry-o-tik)   Complex cells characterized by a nuclear membrane, mitochondria, and chromosomes. All living cells except bacteria and the cyanobacteria are eukaryotic.

excise   To become free of; the opposite of *integrate*.

exotoxin   A soluble poisonous substance excreted by a microorganism.

extruded   Forced through a small opening, as material ejected from a cell without lysing the cell.

exudate   Pus.

facultative anaerobe   An organism that can use oxygen if it is present but can also grow in its absence.

FAD   Abbreviation for flavin adenine dinucleotide, a derivative of the vitamin riboflavin.

fastidious   Exacting; used to refer to organisms that require many growth factors.

feedback inhibition   Another term for allosteric or end-product inhibition. The inhibition of the first enzyme of a biosynthetic pathway by the end-product of the pathway.

fermentation   The metabolic process in which the final hydrogen acceptor is an organic compound.

F factor   Extrachromosomal DNA that codes for pilus synthesis.
  F⁺ cell   Donor cell; transfers plasmid.
  F⁻ cell   Recipient cell; receives plasmid.
  F′ cell   A cell containing a plasmid that is a combination of the chromosome and the F factor.

fibrous tissue   Scar tissue.

fixative   A technique that results in the attachment of bacteria to a microscope slide or that hardens cells so that they are not distorted in the process.

flagellum (fla-jell'-um) (pl. **flagella**)   A long whiplike organelle of locomotion made up of intertwined molecules of protein.

flatus   Intestinal gas passed from the rectum.

Flavobacterium meningosepticum (flay'-voh-back-teer-ee-um men-in-joh-sep'-ti-kum)

flocculation test   A kind of precipitation reaction used between antibody and antigen such as in testing for syphilis.

flora   The living organisms in a particular environment.

fluoresce   To emit light when a light of a different wavelength impinges on the object.

fomite   An inanimate object such as a book, tool, or towel that can act as a transmitter of pathogenic microorganisms even though it does not support their growth.

food chain   A series of organisms, each providing the food supply for the next.

formalin   Approximately 37 percent solution of formaldehyde in water.

Francisella tularensis (fran-siss-sell'-ah too-lah-ren'-siss)

free living   In the absence of living cells.

furuncle   A boil, an infection that penetrates deeper than the dermis layer of the skin.

Fusobacterium nucleatum (fu'-zoh-back-teer-ee-um new-klee-ah'-tum)

galactosides   Molecules that contain the sugar galactose.

gametes (gam'-eats)   Haploid cells that fuse with other gametes to form the diploid zygote in sexual reproduction.

ganglia   Groups of nerve cells.

gastroenteritis (gas'-trow-en-ter-ite'-is)   Inflammation of the gastrointestinal tract, usually characterized by diarrhea, nausea, and vomiting.

gel   Material having the consistency of jelly.

generation time   The average time required for the population of cells in actively growing culture to double in number.

genes   The subunits of chromosomes, each of which codes for one protein.

genital   Of or pertaining to the reproductive system.

genome (jeen'-ome)   All the genetic material in the cell.

genotype   The sum total of the genetic constitution of an organism. Much of the genotype is not expressed at any one time.

genus (pl. **genera**)   A category of related organisms, usually containing several species. The first name of an organism in the Binomial System of Classification.

germicide   An agent that kills microorganisms or viruses.

germinate   Begin to grow.

germination   The sum total of the biochemical and morphological changes that an endospore or other resting cell undergoes before becoming a vegetative cell.

Giardia lamblia (jee-are'-dee-ah lamb'-lee-ah)

giardiasis (jee-are-die'-a-siss)   The disease caused by *Giardia lamblia*.

gingival crevice (jin'-ji-val)   The space between the tooth and the gum.

gingivitis (jin-ji-vye'-tiss)   Inflamed gums.

globulins   Proteins found in the blood, some of which are involved in the immune response.

glomerulonephritis (glom-er-you-low-neh-fry'-tiss)   A kidney disease characterized by inflamed capillary loops through which portions of blood plasma are normally filtered.

glycocalyx (gly-koh-kay'-licks)   A mass of tangled fibers of polysaccharide that originate from the bacterial cell surface and help the bacteria attach to several surfaces.

glycogen (gly'-koh-jen)   A polysaccharide composed of glucose molecules linked together in a characteristic manner.

glycolysis (gly-kol'-i-siss)   Breakdown of carbohydrates. The initial sequence by which this occurs is often called the Meyerhof-Embden-Parnas pathway, named for the scientists who did the most to determine the order of its reactions.

glycoproteins   Proteins with sugar molecules attached.

Gonyaulax (gon-ee-ow'-lax)

Gram-negative bacteria   Bacteria that lose the purple dye (gentian or crystal violet) first applied in the staining of bacteria by the Gram method.

Gram-positive bacteria   Bacteria that retain the purple dye (gentian or crystal violet) that is first applied in staining bacteria.

granules   Cytoplasmic particles such as reserve food supplies that bacteria store as high molecular weight compounds.

granuloma (gran-you-lome'-ah)   A localized

collection of macrophages and other cells found around a foreign body or an infectious agent in the tissues.

Haemophilus ducreyi (hee-moff'-ill-us due-kray'-ee)

H. influenzae (H. in-flew-en'-zee)

halophile   An organism that requires a high salt (NaCl) medium.

haploid   Condition in which each chromosome is represented in a given cell only once.

hapten (hap'-ten or hap'-teen)   Substance that can combine with specific antibodies but cannot incite the production of those antibodies unless it is attached to a large carrier molecule.

heat labile   Destroyed by heating, for example, 100°C for 20 minutes.

heavy metals   Metals such as lead, mercury, and zinc (does not refer to rock-and-roll music).

hemagglutinin (heem-ah-glue'-tin-nin)   A substance that causes erythrocytes to clump together.

hemin (hee'-min)   A crystalline salt derived from blood by heating.

hemolysis (hee-mol'-i-siss)   Lysis of erythrocytes with liberation of hemoglobin.
  alpha hemolysis   A greenish clear area is seen around these colonies growing on blood agar.
  beta hemolysis   A clear area around these colonies.

hemophiliac (hee-moh-fill'-ee-ack)   A person with a blood coagulation disorder, usually inherited.

hemorrhage   To bleed.

hepatic   Referring to the liver.

herbicide   A chemical that kills plants.

Herellea vaginicola (her-ell'-ee-ah vaj-ji-na-ko'-lah)

herpes viruses (herp'-eez viruses)

heterophile antibodies   Antibodies that react with cells of another species.

heterotroph   An organism that obtains energy from organic compounds and that uses an organic compound as its main source of carbon.

Histoplasma capsulatum (hiss-toe-plaz'-mah cap-sue-lah'-tum)

holdfast   Rootlike structures on some algae that allow them to cling to rocks or other fixed structures in the water.

homologous   Matching in characteristics; in genetics, one of the pair of genes or chromosomes in diploid eukaryotic cells that carry information for the same traits.

host   An organism on or in which smaller organisms or viruses live, feed, or reproduce.

humidifying devices   Apparatus used to add water vapor to the incoming air.

hyaluronic acid (hye'-al-you-ron-ik acid)   A mucopolysaccharide gel that fills intercellular spaces in tissues and also occurs as a capsular substance of some bacteria.

hydrogen bond   A weak attraction (bond) between an atom with a strong attraction for electrons and a hydrogen atom covalently bonded to another atom that attracts the electron of the hydrogen atom.

hydrogen ion   (H^+)   hydrogen atom that has lost its electron (negative charge) and thus has a positive charge.

hyperbaric oxygen treatment   The patient is placed in a special chamber and breathes pure oxygen under three times normal pressure.

hypersensitive   Allergic.

hypha (high'-fah) (pl. **hyphae**)   Filament of fungal cells. A single filament of a fungus. A large number of hyphal filaments constitute a mycelium.

Immunoglobulin   (Ig)   Antibodies or antibody-like protein molecules, usually part of the gamma globulin portion of blood.

impetigo (im-pa-tie'-go)   A bacterial disease of the skin, often highly contagious, characterized by small blisters, weeping fluid, and formation of crusts.

incubate   Place under suitable conditions for growth, such as in a temperature-controlled cabinet called an *incubator*.

indicator   A substance that changes from one color to another in response to a specific chemical reaction.

indurated   Thickened.

inert   Chemically inactive.

infectious dose$_{50}$ (ID_{50})   The number of microorganisms or viruses sufficient to establish an infection in 50 percent of susceptible hosts. This varies from a few to several million, depending on the virulence of the infectious agent and on the host defenses.

infestation   A word often used in place of "infection" when talking about multicellular parasites living on or in the body of a human being.

infiltrate   Pass into.

inflammation   A nonspecific response of tissue to injury, characterized by redness, heat, swelling, and pain in the infected area.

inflammatory response   A nonspecific response to injury, characterized by redness, heat, swelling, and pain in the affected area.

inhibited   Prevented.

innate response   A nonspecific response to an inciting agent that does not require the organism to have had previous exposure to this agent.

inorganic   Refers to all chemicals not containing carbon, plus CO and CO_2.

inserted   Put into; incorporated into the chromosome.

inspiration   The act of drawing air into the lungs.

insulin   A hormone released into the blood from certain pancreatic cells. Insulin deficiency results in diabetes.

integrate   To join by strong chemical bonds. In genetics, refers to genetic material joined to the chromosome.

interface   The surface forming the shared boundary between the air and liquid.

interferons   Proteins produced by virus-infected cells (or by cells in response to certain other stimuli) that prevent the replication of viruses in other cells.

intermediates   In metabolism, any compound representing one of the biochemical steps in synthesis or breakdown of a substance, e.g., during the glycolytic or TCA cycle.

interstitial fluid   Fluid found between the cells of the body.

intravenous catheters   Plastic tubes inserted into a vein to provide medications, fluids, or nutrients.

ion   An atom or group of atoms that carry a positive or negative charge.

Isospora belli   (eye-soss-pore'-ah bell'-ee)

jaundice   (jawn'-diss)   Yellowing of the skin caused by bile pigments.

keratin   A durable protein found in the hair, nails, and skin.

Klebsiella   (kleb-see-ell'-ah)

Lactobacillus bulgaricus   (lack'-toe-ba-sill-lus bul-gair'-i-kus)

lactose fermenting   Can degrade the sugar lactose with the resulting production of acid.

lag phase   The stage in the growth of a bacterial culture characterized by extensive synthesis of nucleic acid and protein but no increase in the number of viable cells.

lance   To cut open.

larva (pl. **larvae**)   Immature stage in the development of worms, insects, and certain other creatures.

latent   Present and capable of becoming active but not currently active.

leached   The process in which minerals are extracted by dissolving them in a liquid.

Legionella pneumophila   (lee-jon-ell'-ah new-moh-fill'ah)

Leptospira interrogans   (lep-toe-spire'-ah in-ter-roh'-ganz)

Leptotrichia   (lep-toe-tree'-kee-ah)

lethal dose$_{50}$ (LD$_{50}$)   The dilution of virus or other test agent that kills 50 percent of the inoculated hosts.

leukocytes   (lou'-koh-sites)   White blood cells.

L forms   Bacteria that can no longer synthesize normally the peptidoglycan portion of their cell wall. They are resistant to penicillin.

linear   Having two ends.

lipid   Any of a diverse group of organic substances that are relatively insoluble in water but are soluble in alcohol, ether, chloroform, or other fat solvents.

lipopolysaccharide   (lip'-o-poly-sak'-a-ride)   The macromolecule formed by bonding of a lipid molecule to a polysaccharide.

lipoprotein   (lip-o-pro'-teen)   The macromolecule formed by complexing lipid and protein molecules.

liter   Metric unit of volume, slightly larger than a quart.

log phase   The stage of growth of a bacterial culture when the cells are multiplying exponentially.

low oxidation reduction potential   Condition in which there is little or no oxygen or other electron recipient available.

lymph   Fluid found in lymphatic vessels, originates from plasma.

lymphangitis   Inflammation of the lymphatic vessels.

lymph nodes   Bean-shaped bodies into which the lymphatic vessels drain. They trap foreign material and are one site of antibody production.

lymphocyte (limf'-o-site)   Small, round or oval lymphoid cell with a large nucleus and scanty cytoplasm. Lymphocytes are inactive cells, but when stimulated they can transform into active, dividing cells. They are found in blood and lymph as well as in many body tissues.

lymphoid (lim'-foid)   Pertaining to the lymphocytes or the lymphatic tissue.

lymphokines   Substances produced by T lymphocytes that participate in cell-mediated immunity in a nonspecific manner.

lymphoma (lim-fo'-ma)   Cancer that arises from the lymphoid tissue.

lyse   Dissolve, break up (the cell wall).

lysis   Disruption or disintegration of cells.

lysogenic conversion   The change in the properties of bacteria as a result of their carrying a prophage; for example, the conversion of a non–toxin-producing bacterium to a toxin-producing bacterium by the action of a prophage.

lysozyme   An enzyme that degrades the peptidoglycan layer of a bacterial cell.

lytic (lit'-ick)   Degradative, leading to destruction (i.e., of the cell walls).

macrophage (mack'-row-fayj or mack'-row-fahge)   A large mononuclear phagocyte.

malaise (may'-layze)   A vague feeling of being ill.

malignant tissues   Cancers, lymphomas, and other abnormal collections of cells showing unrestrained growth.

Mastigophora (mas-ti-gaw'-for-ah)

mastoids   An air-filled bulge of the temporal bone behind the ear.

MBC   Abbreviation for minimal bactericidal concentration.

medium (pl. **media**)   Any material used for growing organisms.

meiosis (my-o'-siss)   The process by which the chromosome number is reduced from diploid (2N) to haploid (1N). Segregation and reassortment of the genes occur in this process, and gametes or spores may be produced as the end-product of meiosis.

meninges   Certain membranes covering the brain and spinal cord.

meningitis   Inflammation of the meninges.

mesophiles   Bacteria that grow between 20°C and 50°C.

metabolic processes   Biochemical reactions catalyzed by the enzymes of the cell.

metabolite   Any product of metabolism.

metachromatic staining   Situation in which binding of a dye to substances within a cell changes the color of the dye, for example, from blue to red.

metazoan   Multicellular animal that shows tissue differentiation.

MIC   Abbreviation for minimal inhibitory concentration.

microaerophilic organisms   Organisms that require low concentrations of oxygen for growth (lower than the concentration in air).

microbial burden   The number of bacteria stuck to the cells of the mucous membranes or other body surfaces.

Micrococcus acidovorax (my-kroh-kock'-us a-sid-oh-vor'-ax)

M. malolacticus (M. mah-lo-lak'-ti-kus)

M. radiodurans (M. ray-dee-oh'-dur-anz)

microgram   One thousandth of a milligram; one millionth of a gram.

Micromonospora (my-kroh-mono-spore'-ah)

microorganism   Living organisms too small to be seen without a magnifying instrument.

Microsporum (my-kroh-spore'-um)

mitochondria (my-tow-kon'-dree-ah)   Highly complex structures approximately the size of bacteria found in eukaryotic cells and used to generate energy.

mitosis (my-toe'-sis)   A process of chromosome duplication. The chromosomes divide longitudinally, and the daughter chromosomes then separate to form two identical daughter nuclei.

molecular weight   The relative weight of an atom or molecule based on a scale in which the hydrogen (H) atom is assigned a weight of 1.0.

monolayer   Single layer; animal cells grow as a monolayer in tissue culture.

mononucleosis   A viral disease that markedly increases the number of mononuclear leukocytes in the blood stream. Symptoms include fever, sore throat, and enlarged lymph nodes and spleen.

Moraxella lacunata (more-ax-ell'-ah lak-u-nah'-tah)

morphology   The form and structure of an organism.

motor nerve cells   Nerve cells responsible for controlling muscles.

M protein   A protein in the outer layer of *Strep-*

tococcus pyogenes; the protein differs from one strain to another.

mucins Slimy sugar proteins like those found in saliva.

Mucor (mu'-kor)

multivalent vaccine Immunizing agent composed of several strains or species of pathogens.

mutagen, mutagenic agent Any agent that increases the frequency at which DNA is altered, thereby increasing the frequency of mutations.

mutation A modification in the base sequence of DNA in a gene altering the protein coded by the gene.

mycelium (my-seal'-ee-um) (pl. **mycelia**) A tangled matlike mass of fungal hyphae exemplified by the cottony growth of molds; the collection of hyphae composing a fungus.

Mycobacterium bovis (my-koh-back-teer'-ee-um boh'-viss)

M. leprae (M. lep'-ree)

M. tuberculosis (M. too-ber-kew-loh'-siss)

Mycoplasma hominis (my-koh-plaz'-mah om-i-niss)

M. pneumoniae (M. new-moan'-ee-eye)

mycorrhiza (my-koh-rise'-ah) Certain symbiotic fungi that live on the roots of plants.

mycoses (my-koh'-seas) Diseases caused by fungi.

myeloma Tumor composed of immunoglobulin-producing cells; typically involves the bone marrow.

myocarditis Inflammatory disease of the heart muscle.

NAD Abbreviation for nicotinamide adenine dinucleotide.

nanometer One billionth (10^{-9}) of a meter.

narcotics Drugs that in moderate doses dull the senses, relieve pain, and induce sleep but in excessive doses cause stupor, coma, or convulsions.

nasopharynx (nay-zo-fair'-inks) The area behind the nasal chamber and above the throat.

natural Occurring in nature.

necrotic (neh-krot'-ik) Dead; used when referring to dead cells or tissues that are in contact with living cells, as necrotic (dead) tissue in wounds.

Neisseria gonorrhoeae (nye-seer'-ee-ah gahn-oh-ree'-eye)

N. meningitidis (N. men-in-jit'-id-iss)

neuraminidase An enzyme that degrades the hemagglutinin receptors on erythrocytes to allow the elution of the attached virus.

Neurospora crassa (new-rah'-spor-ah kras'-sah)

neurotoxin A toxin produced by certain bacteria or other organisms that acts on the nervous system.

neutral Neither acid nor alkaline.

neutralizing antibodies Immunoglobulins produced in response to infection. They react specifically with a virus or toxin and make it noninfectious or inactive.

Nocardia asteroides (noh-kar'-dee-ah ass-ter-oid'-eez)

nomenclature In biology, refers to the system by which organisms are named, i.e., the binomial system. Each organism is given a genus and a species name.

normal flora That group of microorganisms that colonize the body's surfaces including the intestine. These organisms under normal circumstances do not cause disease of the host, but they can cause severe disease in the presence of wounds of other abnormal conditions.

nucleoprotein A macromolecule formed by complexing nucleic acid and protein.

nucleotides The basic subunits of ribonucleic acid consisting of a purine or pyrimidine covalently bonded to ribose, which is covalently bound to a phosphate molecule.

obligate (strict) aerobes Those organisms that have an absolute requirement for oxygen gas. They grow best in flasks that are continuously shaken to mix oxygen into the medium.

obligate (strict) anaerobes Those organisms that cannot use oxygen. Traces of oxygen actually kill some members of this group.

obligate intracellular parasites Parasites that grow only inside living cells.

oncogenic (on-ko-jen'-ic) Tumor causing.

opportunist An organism that can cause disease only in hosts with impaired defense mechanisms that might result, for example, from wounds, alcoholism, and so on.

opsonins (op'-soh-nins) Antibodies and other materials such as complement that aid in phagocytosis; substances that cause increased phagocytosis.

optical Made to aid sight.

organ A structure composed of different tissues coordinated to perform a special function.

organelles (or-gan-ells′) Characteristic structures that perform a specific function in a cell.

organic compounds Any carbon-containing compounds except CO and CO_2.

oropharynx Part of the throat just behind the mouth.

osmosis (os-moh′-sis) The passage of a solvent through a membrane from a dilute solution into a more concentrated one.

osmotic lysis The disruption or disintegration of a cell caused by osmosis.

osmotic pressure The pressure exerted by water on the cytoplasmic membrane because of the difference in the concentration of molecules on each side of the membrane.

oxidation The removal of an electron.

oxidative phosphorylation The generation of energy in the form of ATP, which results from the passage of (hydrogen) electrons through the electron transport chain to a final (hydrogen) electron acceptor.

PABA Abbreviation for *para*-aminobenzoic acid, a growth factor for certain bacteria; a precursor of the vitamin folic acid.

pandemic A worldwide epidemic.

papillomas Benign tumors consisting of nipplelike protrusions of tissue covered with skin cells or mucous membrane.

papovaviruses (pap-oh′-vah-viruses)

Paracoccidioides (par-ah-kok-sid-i-oid′-eez)

paramecium A protozoa of the Ciliata class.

paraplegics Individuals paralyzed in the lower half of the body.

parasite An organism living on another organism and gaining benefit at the expense of the host.

parvoviruses (par′-voh-viruses)

passive diffusion The free flowing of molecules into and out of the cell so that the concentration of any particular molecule is the same on the inside as it is on the outside of the cell.

Pasteurella multocida (pass-ture-ell′-ah mull-toss′-id-ah)

P. pestis (P. pes′-tiss)

P. tularensis (P. too-lar-en′-siss)

pasteurization The heating of food or other

substances under controlled conditions of time and temperature (for example, 63°C for 30 minutes) to kill pathogens and to reduce the total number of microorganisms without damaging the substance pasteurized.

pathogenesis (path-oh-jen′-is-iss) The process by which disease develops.

pathogenic Able to cause disease.

pellicle (pell′-e-kull) Thickened, elastic cell covering, not rigid like a cell wall.

penicillinase An enzyme that destroys penicillin.

Penicillium chrysogenum (pen-eh-sill′-ee-um kry-sah′-je-num)

P. notatum (P. noh-tah-tum)

pepsin Protein-splitting enzyme found in the stomach.

peptide bond Bond formed between the carboxyl (—COOH) group of one amino acid and the NH_2 group of the adjacent amino acid during protein synthesis.

peptides Short chains of amino acids.

peptidoglycan layer The rigid backbone of the bacterial cell wall consisting of two major subunits, N-acetyl muramic acid and N-acetyl glucosamine, and a number of amino acids.

Peptostreptococcus (pep′-toh-strep-to-kock′-us)

pericarditis Inflammation of the pericardium.

pericardium The sac enclosing the heart.

periodontal disease Abnormality of the tissues surrounding the base of the tooth.

periplasmic space The very narrow space that lies between the cell wall and the cytoplasmic membrane.

peristalsis The progressive wavelike contraction of the intestinal walls that moves the contents in the proper direction.

permeases A system of enzymes concerned with the transport of nutrients into the cell.

pesticides A general class of agents that kills either plant or animal pests.

Petri dish A two-part dish of glass or plastic used to incubate bacterial cultures on media solidified with agar. It is named for its inventor, Julius Petri.

pH A scale of 0–14 that expresses the acidity or alkalinity of a solution.

phage (faj or fayj) The abbreviation used for bacteriophage.

phagocytic cells Cells that protect the host by ingesting foreign particles such as bacteria, viruses, and so on.

phagocytized   To be ingested by a cell.

phenol coefficient   A number indicating the antimicrobial activity of a disinfectant compared with pure phenol. The larger the value, the more active the agent under the conditions of the test.

phenotype (feen'-o-type)   The sum total of the observable properties of an organism resulting from the interaction of the environment with the genotype.

phlebitis   Inflamed veins.

photooxidation   A chemical reaction occurring because of absorption of light energy in the presence of oxygen. It is sometimes responsible for the death of microbes.

photosynthesis   The sum total of the metabolic processes by which light energy is used to convert CO_2 and a reduced inorganic compound to cytoplasm: $6 HX + 6 CO_2 \rightarrow C_6H_{12}O_6 + 6 X$.

phycomycete (fye-koh-my'-seat)   Kind of fungus.

phylum (pl. **phyla**)   A large division of related families in the scheme used to classify living organisms. The classification scheme is subdivided as follows: kingdom → subkingdom → phylum → class → order → family → genus → species.

phytoplankton   The floating and swimming algae and prokaryotic organisms of lakes and oceans.

picornaviruses (pick'-orn-ah-viruses)

pili   Short hairlike appendages on many Gram-negative bacteria.

pilosebaceous gland (pile-oh-seh-bay'-see-us)   The tiny gland that secretes an oily substance into the hair follicle.

pinocytosis (pine-o-cy-toe'-is)   Cellular ingestion of liquids; a process similar to phagocytosis.

Pityrosporum (pit-ee-roh-spore'-um)

plaque (plack)   A clear area in a monolayer culture of cells. Viral plaques are created by viral lysis of infected cells within the clear area. In dentistry, a plaque is a collection of bacteria tightly adhering to a tooth surface and responsible for dental caries.

plasma cell   A lymphoid cell specialized to produce immunoglobulins in large quantities.

plasmid   An extrachromosomal circular DNA molecule that carries genetic information.

plasmodium (plaz-mode'-ee-um) (pl. **plasmodia**)   A mass of cytoplasm, enclosed by only a cytoplasmic membrane and therefore free to ooze over a substrate. An example is the acellular slime molds.

plastid   Cytoplastic organelle found in eukaryotic cells, e.g., chloroplasts and mitochondria.

plate   A common laboratory name for Petri dish.

pleomorphic (plee-oh-more'-fick)   Occurring in various shapes.

pleurisy   Inflammation of the membrane that lines the lungs and thorax.

pneumonitis   Inflammation of the lungs.

Pneumocystis carinii (new-moh-siss'-tiss ka-ree'-nee-ee)

polymerized   Joined together into a large molecule by chemical bonds.

polymorphic (polly-more'-fick)   Having different distinct forms at various stages of the life cycle.

polypeptide   Another term for a protein; refers to the peptide bonds joining the amino acids.

polysaccharide (polly-sak'-kar-ride)   Long chains of monosaccharide subunits bonded together.

polyvalent vaccine   Vaccine containing a number of different strains of organism.

precursor   One that comes before.

primates   The group of mammals that includes monkeys and human beings.

progeny   Descendants.

prokaryotic (pro-karry-o'-tick)   The simple cell type characterized by the lack of a nuclear membrane, the absence of mitochondria, and a haploid state.

prophylaxis (pro-fi-lak'-sis)   Something done to reduce the possibility of disease.

Propionibacterium acnes (proh-pee-on-ih-backteer'-ee-um ak'-neez)

protective isolation   Set of procedures used to prevent pathogens that might be in the environment from infecting a patient.

Proteus mirabilis (proh'-tee-us mih-rab'-ill-us)

protoplast (pro'-tow-plast)   A cell that has its rigid cell wall removed.

prototroph (pro'-tow-trofe)   An organism without organic growth requirements other than a source of carbon and energy.

protozoa   A group of single-celled eukaryotic organisms with often complex life cycles; includes some serious parasites of humans.

pseudohyphae (sue-dough-high'-fee)   Cellular filaments resembling hyphae but formed by adhering elongated yeast cells formed by budding.

pseudomonads   A general term given to organisms closely related to *Pseudomonas aeruginosa*.

Pseudomonas aeruginosa   (sue-dough-moan'-ass ā-rue-jin-oh'-sa)

psychrophile   Organisms that grow best between −5°C and 20°C.

pure culture   A culture containing only a single strain of an organism.

purines and pyrimidines   Organic nitrogen-containing bases that serve as the building blocks of DNA and RNA molecules.

purulent   Containing pus.

pus   Thick, opaque, often yellowish fluid that forms at the site of a wound. It contains leukocytes (mainly PMNs), tissue debris, and microorganisms.

pyogenic   (pie'-oh-jen-ik)   Producing pus.

quarantine   Any isolation or restriction on travel imposed to keep contagious diseases from spreading.

radioisotopes   Radioactive compounds.

reduction   The addition of an electron or hydrogen atom.

regurgitate   To vomit, throw up.

repair enzymes   Enzymes possessed by some cells that can correct damage done to DNA by radiation.

repression   The inhibition of gene function.

reservoir   Source of a disease-producing organism.

resolving power   In looking through a microscope, the minimum distance that can exist between two points such that the points are observed as separate.

respiration   (res-spir-ray'-shun)   The sum total of metabolic steps in the degradation of foodstuffs when the electron (hydrogen) acceptor is an inorganic compound.

reverse transcriptase   An enzyme coded by RNA-containing tumor viruses that synthesizes DNA complementary to an RNA template.

Rhizobium   (rye-zoh'-bee-um)

Rhodospirillum   (row-dough-spur-ill'-um)

ribosomes   Cellular "workbenches" on which proteins are synthesized.

Rickettsia   (rik-ett'-see-ah)

RNA   Abbreviation for ribonucleic acid.
  mRNA   Messenger RNA.
  tRNA   Transfer RNA.

Rothia   (row'-thee-ah)

Saccharomyces cerevisiae   (sack-a-row-my'-seas sair-a-vis'-e-eye)

salinity   The amount of salt.

salivary glands   Glands that produce saliva, found in the region of the mouth.

Salmonella cholerae-suis   (sall-moh-nell'-ah kol'-er-ay-sues)

S. enteritidis   (S. en-ter-it'-id-iss)

S. typhi   (S. tye'-fee)

salmonellosis   Disease caused by Gram-negative rods of the genus *Salmonella*.

sanitization   Reducing the microbial populations on objects to acceptably safe public health levels.

saprophyte   (sap'-roh-fyte)   A microorganism that lives on dead organic material.

Sarcodina   (sar-koh-dee'-nah)

Sargassum   (sar-gas'-um)

satellitism   The phenomenon in which a bacterial colony supplies a required substance for other bacteria.

secondary infection   Invasion by a second microorganism after the host has been weakened by a previous infection.

self-limiting   Does not continue to spread even without intervention.

semiconservative replication   A type of DNA replication in which new daughter strands complementary to each of the original strands are synthesized in each round of replication.

semipermeable   The passage of some but not other material through a membrane.

sensitivity testing   Use of laboratory procedures to find out whether an antimicrobial medicine is active against a given microorganism at achievable concentrations.

septicemia   Illness caused by microorganisms circulating in the blood stream; "blood poisoning."

serial dilutions   Several or more dilutions of a substance in which the amount of material decreases progressively, usually by a constant fraction.

serology   Having to do with serum, especially serum-containing antibody.

serotype, of the same serotype   Having the same antigens as detected by using known antisera.

Serratia marcescens   (ser-ah'-tee-ah mar-sess'-enz)

serum   The fluid portion of the blood that remains after the blood clots.

sexual Involving the exchange of genetic material.

Shigella boydii (shig-ell'-ah boy'-dee-eye)

S. dysenteriae (S. diss-en-tear'-ee-eye)

S. flexneri (S. flex'-ner-eye)

S. sonnei (S. sonn'-ee-eye)

shigellosis Infection with pathogenic organisms of the genus *Shigella*.

shock A state of insufficient blood pressure to supply adequate blood flow to meet the oxygen needs of vital body tissues.

sickle cell anemia A condition in which an individual produces abnormal hemoglobin molecules resulting in an altered erythrocyte shape.

single cell protein (abbreviated **SCP**) The protein contained in members of the microbial world that may serve as a source of food.

sinuses Bony cavities found in the area of the nose.

somatostatin A growth hormone.

species The second name given to an organism in the Binomial System of Classification.

spectrum of activity Range of different kinds of organisms against which a medicine is active.

spherule (sphere'-youl) Large, round, thick-walled structure containing many fungal spores, characteristic of the tissue phase of *Coccidioides immitis*.

Spirillum minus (spy-rill'-um my'-nus)

spirochetes (spy'-roh-keets) Any of the group of bacteria belonging to the *Spirochaeta*, slender, spirally undulating bacteria that include those that cause syphilis.

Spirulena maxima (spy-roo-lean'-ah max'-i-ma)

spontaneous abortion Commonly known as a miscarriage of a fetus.

sporogenesis The process by which a vegetative cell produces a spore.

Sporothrix schenckii (spore'-oh-thrix shenk'-e-eye)

Sporozoa (spore-oh-zoh'-ah)

Staphylococcus aureus (staff-il-oh-kock'-us or'-ee-us)

starch A large molecule consisting of glucose subunits polymerized into long chains.

stasis (stay'-sis) Stagnation due to fluid staying in one place instead of flowing naturally.

static Not moving.

stationary phase The stage of a culture's growth in which the number of viable cells remains constant.

sterilization Rendering an object or substance free of all viable microorganisms and viruses.

sterol Special components of some lipids; cholesterol is a common sterol.

strains Varieties within a species of bacteria.

Streptobacillus moniliformis (strep-toe-ba-sill'-us mon-ill-ih-for'-miss)

Streptococcus agalactiae (strep-toe-kock'-us a-gal-ac'-tee-eye)

S. carlsbergiensis (S. karls-berg-ee-en'-siss)

S. epidermidis (S. eh-pi-derm'-mih-diss)

S. mutans (S. mew'-tans)

S. pneumoniae (S. new-moan'-ee-eye)

S. pyogenes (S. pie-aw'-gen-ease)

S. thermophilus (S. ther-moh'-fill-us)

Streptomyces antibioticus (strep-toe-my'-seez anti-bi-au'-ti-kuss)

S. griseus (S. gree'-see-us)

S. mediterranei (S. med-it-ter-an'-ee-eye)

S. venezuelae (S. ven-ez-u-ayl'-eye)

structural proteins Proteins that form part of a virus, such as those of the capsid.

subclinical Without symptoms.

subcutaneous Below the skin.

substrate The substance on which an enzyme acts to form the product; surface on which an organism will grow.

substrate phosphorylation The transfer of a high energy bond in a substrate to ADP to form ATP.

sucrose A sugar consisting of a molecule of glucose bonded to fructose, i.e., common table sugar.

sulfa drugs Medicinal drugs that include and have a close chemical relationship to sulfanilamide.

superinfection An infection added to one already present; secondary infection.

suppressor cells A subset of lymphocytes that suppresses certain immune responses.

surface tension A property of liquids in which their surfaces resemble a thin elastic membrane under tension.

sutures (su'-chores) Special thread that surgeons use for stitching parts of the body together.

symbionts Organisms of different species that live together in an intimate relationship.

symbiotic relationship   Intimate relationship between members of different species.

symptomatic treatment   Treating symptoms brought on by the disease when the disease itself cannot be treated.

synergistic action   The cooperative action of two distinct organisms such that the growth of one depends on the growth of the other.

synthetic   Made outside the cells.

synthetic medium   A medium in which every component's chemical composition and quantity have been identified.

systemic   Pertaining to the whole body.

TCA cycle   Abbreviation for *tricarboxylic acid cycle* (Krebs or citric acid cycle).

thermophiles   Organisms that prefer or require a higher temperature for growth.

Thiobacillus ferrooxidans   (thigh-oh-bah-sill'-us fair-oh-ox'-ih-danz)

thymectomy   Surgical removal of the thymus gland.

tissue   A group of cells that associate into a structural subunit and perform a specific function.

tissue culture   A culture of plant or animal cells that grow in an enriched medium.

titer   (tight'-er)   A measure of the concentration of a substance in solution; for example, the amount of a specific antibody in serum, usually measured as the highest dilution of serum that will give a positive test for that antibody. The titer is often expressed as the reciprocal of its dilution; thus, a serum that gives a positive test when diluted 1:256, but not 1:512, is said to have a titer of 256.

toxic   Poisonous.

toxin   Poisonous chemical substance originating from microbes, plants, or animals.

Toxoplasma gondii   (tox-oh-plaz'-mah gahn'-dee-ee)

tracheobronchial system   The windpipe, bronchial tubes, and their branches.

transcription   The transference of genetic information coded in DNA into informational messenger RNA.

transductants   Recipient cells that have received and expressed donor DNA gained by transduction.

transduction   Transfer of genetic information by temperate phages.

transformation   Modification of the properties of mammalian cells; commonly caused by certain viruses. Not to be confused with bacterial transformation, which is the process of "naked" DNA transfer from one bacterial cell to another.

translation   The process by which genetic information in the messenger RNA directs the order of the amino acids in the protein during protein synthesis.

transovarially   Passing from one generation to another through infection of the egg by the parasite.

transposons   Genes that move from one DNA molecule to another within a cell. So-called "jumping" genes.

trauma   An injury (as a wound) to living tissue caused by an external agent.

tremor   Involuntary shaking of parts of the body.

Treponema pallidum   (trep-oh-nee'-mah pal'-ee-dum)

treponemes   Refers to spirochete bacteria like *Treponema pallidum*, which causes syphilis.

Trichomonas vaginalis   (trick-oh-moan'-us vaj'-in-alice)

Trichophyton   (trick-oh-phye'-ton)

trophozoite   (trowf'-o-zoe'-ite)   The vegetative form of some protozoa.

tuberculoid   Resembling tuberculosis, as "tuberculoid leprosy."

tumor   Tissue that grows independently of surrounding tissue. *Neoplasm* is a synonym.

tyndallization   (tyn-dal-i-za'-shun)   Repeated controlled heating at relatively low temperatures that will kill spore-forming bacteria suspended in nutrient fluid. For example, 100°C for 30 minutes on each of three days or 60°C for one hour on each of five days.

ultracentrifuge   A machine that rotates samples at extremely high speeds.

ultraviolet light   (abbreviated **UV**)   The electromagnetic radiation with wavelengths between 175 and 350 nm (shorter than visible light).

ureter   The tube that drains the kidney and connects the kidney to the bladder.

urethra   The tube that empties the bladder to the outside of the body.

vaccine   A preparation of microorganisms or viruses used to immunize a person or animal against a particular disease.

vacuole   A membrane-enclosed cavity in the cytoplasm of a cell.

vascular   Referring to vessels of the blood system.

vector   An organism such as a mosquito that transmits disease.

vegetative cell   The growing or feeding form of a cell.

vehicle   An inanimate carrier of an infectious agent from one host to another.

Veillonella   (vah-yon-ell'-ah)

vena cava   Largest vein in the body. It conducts blood into the right side of the heart.

venereal disease   Contagious disease typically acquired during sexual intercourse.

vesicle   A small membrane-enclosed sac or small blister.

viable   Living.

Vibrio cholerae   (vib'-ree-oh kahl'-er-eye)

V. parahemolyticus   (V. pair-ah-hee-moh-lit'-ih-kus)

viremia   Presence of viruses in the blood.

virion   A viral particle consisting of nucleic acid surrounded by a protein coat.

virulence   The relative capacity of a pathogen to overcome body defenses; the sum of properties of an organism that enhances pathogenicity.

virus   A nonliving, submicroscopic infectious agent.

viscosity   State of being sticky or gummy.

vulvovaginitis   Inflammation of both the vulva and the vagina.

ward   A section of a hospital such as a wing or a large room with many hospital beds.

wavelength   The distance between adjacent peaks of two waves of light.

wetting agent   A substance that allows water to adhere to the surface of other materials.

wheal   Red, fluid-filled bumps that itch, as in the allergic reaction known as hives.

wild type   Occurring in nature.

xerophiles   (zer'-oh-files)   Fungi and yeasts that can survive at very low moisture levels.

Yersinia pestis   (yer-sin'-ee-ah pess'-tiss)

zygote   (zye'goat)   A cell formed by the sexual union of a male and a female gamete.

ANSWERS TO SELF-QUIZZES

Chapter One
1.b 2.b 3.d 4.d

Chapter Two
1.c 2.b 3.e 4.b 5.a

Chapter Three
1.b 2.b 3.d 4.a 5.b

Chapter Four
1.d 2.e 3.a 4.b 5.a

Chapter Five
1.a and d
2.b 3.c 4.a 5.d

Chapter Six
1.a 2.c 3.c 4.a 5.d

Chapter Seven
1.d 2.a 3.a 4.c 5.b 6.b

Chapter Eight
1.d 2.d 3.d 4.b 5.d 6.b
7.b

Chapter Nine
1.d 2.c 3.b 4.c 5.d

Chapter Ten
1.a 2.d 3.d 4.e 5.a

Chapter Eleven
1.e 2.c 3.d 4.c 5.b

Chapter Twelve
1.d 2.e 3.e 4.d 5.d

Chapter Thirteen
1.b 2.d 3.b 4.c 5.e

Chapter Fourteen
1.a 2.c 3.a 4.d 5.b

Chapter Fifteen
1.e 2.b 3.c 4.e 5.a

Chapter Sixteen
1.e 2.e 3.a 4.a 5.d

Chapter Seventeen
1.a 2.d 3.c 4.d 5.a

Chapter Eighteen
1.d 2.b 3.a 4.d 5.e

Chapter Nineteen
1.b 2.c 3.a 4.c 5.a

Chapter Twenty
1.b 2.b 3.c 4.a 5.d 6.a
7.d

Chapter Twenty-one
1.b 2.e 3.e 4.d 5.d 6.e
7.b 8.c

Chapter Twenty-two
1.c 2.a 3.c 4.c 5.c 6.b
7.b 8.e

Chapter Twenty-three
1.e 2.a 3.a 4.d 5.e 6.d
7.c 8.a

Chapter Twenty-four
1.c 2.e 3.c 4.a 5.c 6.e
7.d 8.a

Chapter Twenty-five
1.a 2.b 3.c 4.e 5.d 6.a
7.d 8.e

Chapter Twenty-six
1.d 2.a 3.e 4.b 5.e

Chapter Twenty-seven
1.a 2.c 3.b 4.d 5.a 6.c
7.e 8.d

Chapter Twenty-eight
1.d 2.d 3.a 4.a 5.c 6.c
7.b 8.e

Chapter Twenty-nine
1.c 2.c 3.c 4.d 5.d 6.e
7.e 8.d

Chapter Thirty
1.c 2.b 3.b 4.a 5.d

Chapter Thirty-one
1.b 2.d 3.d 4.e 5.b

Chapter Thirty-two
1.b 2.e 3.c 4.a 5.b

INDEX

Major references are indicated by boldface page numbers.